AF689868

Nonlinear Guided Wave Optics

A testbed for extreme waves

Nonlinear Guided Wave Optics

A testbed for extreme waves

Stefan Wabnitz, Editor

Università di Brescia, Dipartimento di Ingegneria dell'Informazione, Brescia, Italy, and Istituto Nazionale di Ottica del CNR, Brescia, Italy, and Novosibirsk State University, Novosibirsk, Russia

IOP Publishing, Bristol, UK

© IOP Publishing Ltd 2017

All rights reserved. No part of this publication may be reproduced, stored in a retrieval system or transmitted in any form or by any means, electronic, mechanical, photocopying, recording or otherwise, without the prior permission of the publisher, or as expressly permitted by law or under terms agreed with the appropriate rights organization. Multiple copying is permitted in accordance with the terms of licences issued by the Copyright Licensing Agency, the Copyright Clearance Centre and other reproduction rights organisations.

Permission to make use of IOP Publishing content other than as set out above may be sought at permissions@iop.org.

Stefan Wabnitz, Editor has asserted his right to be identified as the author of this work in accordance with sections 77 and 78 of the Copyright, Designs and Patents Act 1988.

ISBN 978-0-7503-1460-2 (ebook)
ISBN 978-0-7503-1458-9 (print)
ISBN 978-0-7503-1459-6 (mobi)

DOI 10.1088/978-0-7503-1460-2

Version: 20180101

IOP Expanding Physics
ISSN 2053-2563 (online)
ISSN 2054-7315 (print)

British Library Cataloguing-in-Publication Data: A catalogue record for this book is available from the British Library.

Published by IOP Publishing, wholly owned by The Institute of Physics, London

IOP Publishing, Temple Circus, Temple Way, Bristol, BS1 6HG, UK

US Office: IOP Publishing, Inc., 190 North Independence Mall West, Suite 601, Philadelphia, PA 19106, USA

To my wife Olga,
and my children Mario and Sasha.

Contents

Preface

In spite of their presumed rarity, the occurrence of *extreme events* is today perceived as an overhang that may suddenly affect each and every human life. Among extreme events, we may classify wars, revolutions, epidemics, stock market crashes, tsunamis, earthquakes, hurricanes, giant rogue ocean waves, and even fatal events such as the heart attack. Despite their disparate origins, perhaps all extreme events share one common feature: nobody could ever forecast them in advance. In hindsight, people quickly adjust to the occurrence of an extreme event, to the point of considering it inescapable. Another common characteristic of extreme events, is that their sudden and brief occurrence may lead to long term consequences, and often to paradigm shifts in human society.

If we want to develop any capability to forecast extreme events, it is necessary to build a theory that can be confronted with past observations. To this end, it would be very interesting if one could exploit analogue physical systems where the fast acquisition of large amounts of data under well controlled conditions can be carried out. The development of such a testbed can be based on the observation that many different phenomena in physical and social sciences can be described in terms of the same universal statistical distributions. These statistics are characterized by the presence of extreme events, which are associated with the properties of the tail ends of the corresponding distributions. Consider for example hydrodynamics (turbulence, hurricanes, tsunamis), geosciences (earthquakes, floods, landslides), economics (financial markets), social sciences (distributions of populations), medical sciences (neuronal avalanches, epileptic seizure), material failures, power grid and computer networks (black-outs) environment and climate sciences (forest fires, evolution and competition of animal and plant species). There is an ongoing philosophical debate: should we consider extreme events as essentially unpredictable, as out springs of the popular *black swan theory, a manifestation of self-organized criticality?*

Such a viewpoint is inherently pessimistic: the lack of predictability ultimately entails a lack of accountability for scientists, who were unable to make any prior warning about the occurrence of the extreme event. Fortunately, a new theory of extreme events has been emerging over the past few years, that associates their presence with the generation of coherent wave structures, such as for example nonlinear waves or *solitons*, which do not belong to the same population of random linear waves. The appearance of extreme events in the form of solitons is then subject to their own '*dragon-king*' kind of statistical distribution. The important consequence of this new point of view is that predictability of the occurrence of the extreme event, based on the observation of a suitable precursor, should indeed be possible. This may entail the possibility of controlling, or even suppressing, the emergence of the extreme event, if the observation of precursors was able to trigger a proper feedback signal.

To give an example, of great interest today is the study of the mechanism of extreme or rogue waves in oceanography. In the regime of deep waters, the encounter with a rogue wave may destroy a ship. Whereas in the regime of shallow waters, catastrophic damage may be produced by the on-shore arrival of a tsunami. The advance forecast of such an extreme wave could save many human lives, and have great impact on our economies.

The scope of this book is to highlight the use of optical fiber and waveguide experiments, where optical pulse propagation can be modelled by simple, and yet universal nonlinear evolution equations (e.g., the nonlinear Schrödinger equation (NLSE), the nonlinear shallow water equation (NLSWE), the Ginzburg–Landau equation (GLE), etc), as a simply accessible test bed for exploring, in a well-controlled manner, the dynamics of complex systems phenomena that exhibit extreme events. As we shall see throughout the sixteen chapters of this book, nonlinear guided wave optics provides a practical testbed for the accessible statistical study of extreme event generation in the many diverse domains of science and technology.

A key advantage of the guided wave optics platform is the possibility of controlling not only the boundary conditions for the excitation of extreme waves, but also the number of spatial dimensions involved in the temporal evolution of the waves. Chapter 1 analyses precisely the role of spatial dimensions in the emergence, and statistics, of rogue wave phenomena in semiconductor lasers or optical parametric oscillators in the presence of an external forcing. Some extreme wave phenomena, such as turbulent vortices, may only occur in the presence of two transverse spatial dimensions.

The number of spatial dimensions is not the only means to control the degrees of freedom for extreme waves in nonlinear guided waves optics. Another important possibility is provided by the number of waves, or components of a vector wave system. Chapter 2 discusses the explicit analytic rogue wave solutions, and their stability in the presence of noise, for the system of three coupled resonant waves, which describes stimulated Brillouin or transient Raman scattering in optical fibers, or parametric mixing in quadratic nonlinear waveguides.

Besides rogue waves, another important category of extreme waves where guided wave optics has the capability of providing a controllable testbed are shock waves. Since their discovery by Ernst Mach 230 years ago, who observed a shock front generated by the motion of a body traveling at supersonic speed, shock waves have been observed in many physical systems ranging from hydrodynamics to astrophysics, dispersive gas dynamics, plasma physics, granular systems, and Bose–Einstein condensates. Chapter 3 introduces a new formalism, based on the so-called Gamow vectors, to describe shock wave dynamics in optical nonlocal nonlinear defocusing media. This approach permits establishing that optical shock waves are an intrinsically time-irreversible phenomenon. Moreover, when considering the quantum nature of light, the new formalism is used to describe the spontaneous squeezing of a photon fluid in a nonlocal nonlinear medium, which may have important applications to optical metrology, and lead to analogies with quantum gravity.

In the limiting case where the highly nonlinear and instantaneous response of the medium dominates over the dispersive behavior, the hydrodynamic approximation can be made, which reduces the description of shock wave generation in optical fibers to the nonlinear shallow water regime of oceanography. Chapter 4 illustrates experiments where the water dynamics of dam-breaking are emulated by studying the temporal dynamics of an equivalent photon fluid, represented by a suitably prepared optical waveform propagating in the normal dispersion regime of a short piece of optical fiber.

A particularly useful testbed for exploring complex rogue wave and extreme event dynamics and their statistics is provided by fiber lasers. In these devices, dissipative temporal light structures are spontaneously formed from quantum noise, and their continuous recirculation through the cavity permits the rapid collection of statistical data. Chapters 5 and 6 present an overview of the complex dynamics of extreme events and rogue waves occurring in mode-locked fiber lasers. For fiber lasers operating in the normal dispersion regime, chapter 5 shows that stimulated Raman scattering may disrupt stable laser emission, and lead to noise-like pulses with long-tailed statistics. In these lasers, transient 'soliton explosions' and rogue dark solitons may appear in the cavity field build-up phase. Moreover, chapter 6 reveals the emergence, in the anomalous dispersion regime of the cavity, of incoherent dissipative solitons. These consist of chaotic soliton bunches mixed with noise-like pulses.

A crucial question arises: can rogue waves and extreme events be predicted before their occurrence, and if yes, to what degree? Although such a formidable challenge remains to date largely unsolved, it is fundamental to unveil the conditions that may lead to rogue and extreme waves formation. The necessary, but not sufficient, condition for rogue wave formation is the instability, with respect to spatiotemporal modulations, of the steady state wave pattern (e.g., a flat sea). Chapter 7 analyzes and compares the emergence of modulation instabilities in hydrodynamics and in fiber optics, where the instabilities may be initiated by noise or by small coherent perturbations, and connects the presence of these instabilities with the formation of giant, isolated rogue waves.

Extreme events are often the result of wave dynamics in physical systems characterized by the simultaneous presence of disorder and nonlinearity. As an example of such a testbed in the context of well controllable optical experiments, chapter 8 presents a statistical study of the extreme waves arising in a strongly turbulent scenario of waves propagating in photorefractive ferroelectric crystals. In this context, experiments permit one to analyze the role of system dimensionality and incoherence of the optical field in the appearance of extreme waves.

An important degree of freedom of light waves in optical fibers is provided by their state of polarization. Chapter 9 shows that the coherent coupling between polarization components of a light pulse within a fiber laser may result in complex polarization instabilities, which originate the formation of both bright and dark rogue pulses with the characteristic statistical properties of extreme waves.

Pulse propagation in the optical fiber testbed may be described with very good accuracy by means of a completely integrable (or exactly solvable) nonlinear wave evolution equation, the nonlinear Schrödinger equation. Chapter 10 takes advantage of the integrability of the underlying wave propagation model, and presents a whole hierarchy of analytical solutions that can be associated with the emergence of high intensity extreme peaks of light, which correspond to the class of rogue wave solutions.

Coming back to the main question, concerning the predictability of rogue or extreme waves, it is very difficult to answer this question in many scenarios that exhibit rogue waves, for example in oceanography, due to the scarcity of available data and measurements. But our fast data collection capabilities are strongly enhanced when using the testbed of nonlinear and fiber optics. Chapter 11 addresses the connection between rogue wave predictability and the presence of heavy tail statistical distributions, in the frame of both unidimensional propagation in single mode fibers, and multidimensional beam filamentation in water and gas cells. A careful statistical analysis of these experiments permits one to conclude that predictability and the presence of heavy tails statistics are largely unrelated. In fact, predictability is associated with some degree of deterministic, although complex, behavior in the wave evolution. A small-scale degree of predictability can be observed for ocean waves and for multiple filamentation events, whereas it is largely absent for fiber optics rogue pulses emerging in supercontinuum generation as a result of quantum noise fluctuations that initiate the nonlinear pulse dynamics.

The evolution of noisy waves in a physical system described by an integrable nonlinear wave equation leads to a phenomenology known as 'integrable turbulence'. Chapter 12 presents theory and experiments that characterize the emergence of rogue wave events in the evolution of light pulses propagating in optical fibers in a regime which is analogous to the shallow water regime of waves near a beach. In this context, different fast measurement techniques have been used to resolve the temporal dynamics of these waves.

The polarization degree of freedom can be exploited in the optical fiber testbed for the generation of rogue waves also in the absence of an optical cavity. Chapter 13 presents a numerical study of rogue wave events generated from initially noisy pulses that propagate in birefringent optical fibers either in the normal or the anomalous dispersion regime, and discusses the influence of the various fiber parameters to control the properties of the statistical distribution of extreme events.

An emerging platform for the controlled study of complex spatiotemporal nonlinear wave dynamics is that of multimode optical fibers. By varying the number of modes, which can be achieved simply by tuning the wavelength of the laser, one may explore the whole range of dynamical behaviors ranging from the one-dimensional case of a single mode fiber, up to the limit case of nearly free space propagation with two transverse dimensions leading to filamentation phenomena. Chapter 14 describes recent experimental studies of the nonlinear wave dynamics in highly multimode, passive and active optical fibers. This platform permits the study of the emergence of extreme waves, in the form of stable nonlinear coherent attractors characterized by low-order transverse mode patterns.

The effect of initial random noise in driving the dynamics of rogue and extreme wave processes is extensively investigated in chapter 15. Here, the quantum noise seeding of modulation instability in optical fibers is shown to lead to a variety of rogue waveforms, which can be very well approximated in terms of the analytical rogue wave solutions that have been presented in chapter 10. Various ultrafast measurement techniques, similar to those employed in the studies of chapter 12, have permitted capture in real-time of the dynamics of rogue waves emerging from noise-seeded modulation instability.

Another guided wave platform that has emerged over the past few years for the study of complex spatiotemporal optical wave dynamics is that of externally driven cavities and microcavities. The nonlinear response of such driven resonators leads to the generation of optical frequency combs and cavity solitons, which may have widespread use in the near future for optical communications and metrology applications. Chapter 16, the final chapter, analyzes dissipative rogue wave formation phenomena of different dimensionalities in both passive Kerr and active (semiconductor laser) cavities. The computation of the probability distribution of pulse height shows that a rogue wave can be generated for moderate injected field intensities, which facilitates their experimental observation.

In summary, the content of this book provides an up-to-date overview of the capabilities of the nonlinear guided optical wave platform for the study of extreme wave phenomena and their statistics. The strength of this testbed mainly rests on the possibility of precisely controlling the input conditions and the parameters of the physical system, and on the availability of advanced data acquisition techniques that permit extensive and fast characterization of the statistical properties of the extreme events distribution. It is our hope that these advances will foster progress in other fields of science, technology and human activities at large, where the analysis and forecast of extreme events could have a tremendous role for the progress of the wealth and health of mankind.

The value of this book is testified by its impressive team of contributors, renowned scientists in the field of rogue and extreme nonlinear optical waves. Therefore, I would like to conclude by thanking all members of this team for sharing their time and effort in producing such a comprehensive manuscript in a timely manner.

Stefan Wabnitz
Novosibirsk, Russian Federation, November 15, 2017

Acknowledgments

The Editor acknowledges support from the Ministry of Education and Science of the Russian Federation (Grant No. 14.Y26.31.0017).

Editor biography

Stefan Wabnitz

Stefan Wabnitz obtained the Laurea Degree in Electronics Engineering from the University of Rome 'La Sapienza' in 1982, the MS in Electrical Engineering from Caltech in 1983, and the PhD in Applied Electromagnetism from the Italian Ministry of Education in 1988. He was with the Ugo Bordoni Foundation between 1985 and 1996. In 1996 he became full professor in Physics at the University of Burgundy in Dijon, France. Between 1999 and 2003 he was with Alcatel Research and Innovation Labs in France and with Xtera Communications in Texas. Since 2007 he is full professor the Department of Information Engineering of the University of Brescia, Italy. His research activities involve nonlinear propagation effects in optical communications and information processing devices. He is the author and co-author of over 700 international refereed papers, conference presentations, and book chapters. He is the Deputy Editor of *Elsevier Optical Fiber Technology*, a Fellow member of the Optical Society of America, and senior member of IEEE-Photonics Society.

List of contributors

N Akhmediev The Australian National University, Canberra, Australia

Adrian Ankiewicz The Australian National University, Canberra, Australia

Stéphane Barland Université de Nice, Sophia Antipolis, France

Fabio Baronio Istituto Nazionale di Ottica del CNR and Università di Brescia, Brescia, Italy

Alain Barthélemy Université de Limoges, Limoges, France

Maria Chiara Braidotti University of L'Aquila, L'Aquila, Italy

Massimo Brambilla CNR-IFN and Dipartimento di Fisica Interateneo Politecnico e Università di Bari, Bari, Italy

Neil G R Broderick The University of Auckland, Auckland, New Zealand

A Chabchoub The University of Sydney, Sydney, Australia

Shihua Chen Southeast University, Nanjing, China

Marcel G Clerc Universidad de Chile, Santiago, Chile

Aurélien Coillet Université Bourgogne Franche-Comté, Dijon, France

Lorenzo Columbo CNR-IFN and Dipartimento di Fisica Interateneo Politecnico e Università di Bari, Bari, Italy

Matteo Conforti Université de Lille, Lille, France

Claudio Conti University Sapienza, Rome, Italy

Vincent Couderc Université de Limoges, Limoges, France

S Coulibaly Université de Lille, Lille, France

Eugenio DelRe Università di Roma 'La Sapienza', Rome, Italy

Ayhan Demircan Leibniz Universität, Hannover Germany

L Drouzi Université de Lille, Lille, France

John M. Dudley Université Bourgogne Franche-Comté, Dijon, France

Pierre Ealczak Université de Nice, Sophia Antipolis, France

Gennady El Loughborough University, Loughborough, UK

Miro Erkintalo The University of Auckland, Auckland, New Zealand

Bruno Garbin Université de Nice, Sophia Antipolis, France

A Gelash Novosibirsk State University, Novosibirsk, Russia

Silvia Gentilini University Sapienza, Rome, Italy

Goëry Genty Tampere University of Technology, Tampere, Finland

Christopher J Gibson University of Strathclyde, Glasgow, UK

Massimo Giudici Université de Nice, Sophia Antipolis, France

Philippe Grelu Université Bourgogne Franche-Comté, Dijon, France

François Gustave Université de Nice, Sophia Antipolis, France

Saïd Hamdi Université Bourgogne Franche-Comté, Dijon, France

V Kalashnikov Vienna University of Technology, Vienna, Austria

H Kbashi Aston University, Birmingham, UK

Edmund J R Kelleher University of British Columbia, Vancouver, Canada, and Department of Physics, Imperial College London, London, UK

B Kibler Université Bourgogne Franche-Comté, Dijon, France

SV Kolpakov Aston University, Birmingham, UK

Katarzyna Krupa Università di Brescia, Brescia, Italy

Alexandre Kudlinski Université de Lille, Lille, France

K Laabidi Université Mohamed Premier, Oujda, Morocco

Giulia Marcucci University Sapienza, Rome, Italy

A Martinez Aston University, Birmingham, UK

Cristina Masoller Universitat Politècnica de Catalunya, Terrassa, Spain

Guy Millot Université Bourgogne Franche-Comté, Dijon, France

C Mou Aston University, Birmingham, UK

Arnaud Mussot Université de Lille, Lille, France

Mikko Närhi Tampere University of Technology, Tampere, Finland

K Nithyanandan Université Bourgogne Franche-Comte, Dijon, France

Miguel Onorato Università degli Studi di Torino, Torino, Italy

Gian-Luca Opo University of Strathclyde, Glasgow, UK

Krassimir Panajotov Vrije Universiteit Brussel, Brussels, Belgium and Institute of Solid State Physics, Sofia, Bulgaria

Antonio Picozzi Université Bourgogne Franche-Comté, Dijon, France

Davide Pierangeli Università di Roma 'La Sapienza', Rome, Italy

Franco Prati Unversità dell'Insubria, Como, Italy

Stéphane Randoux Université de Lille, Villeneuve d'Ascq, France

Cristina Rimoldi Université de Nice, Sophia Antipolis, France

Roberto José Rios-Leite Universidade Federal de Pernambuco, Recife, Brazil

Antoine F J Runge University of Southampton, Southampton, UK

SV Sergeyev Aston University, Birmingham, UK

José Maria Soto-Crespo Consejo Superior de Investigaciones Cientcas (CSIC), Madrid, Spain

Günter Steinmeyer Max-Born-Institut, Berlin, Germany

Pierre Suret Université de Lille, Villeneuve d'Ascq, France

C G L Tiofack Université de Lille, Lille, France

Giovanna Tissoni Université de Nice, Sophia Antipolis, France

Mustapha Tlidi Université libre de Bruxelles, Brussels, Belgium

Shanti Toenger Tampere University of Technology, Tampere, Finland

Alessandro Tonello Université de Limoges, Limoges, France

Jorge R Tredicce Université de la Nouvelle Calédonie, Nouvelle Calédonie, France

Stefan Wabnitz Università di Brescia, Brescia, Italy and Istituto Nazionale di Ottica del CNR, Italy and Novosibirsk State University, Novosibirsk, Russia

Robert I Woodward Macquarie University, New South Wales, Australia

Gang Xu Université de Lille, Lille, France

Allison Yao University of Strathclyde, Glasgow, UK

VE Zakharov Novosibirsk State University, Novosibirsk, Russia and Lebedev Physical Institute, Russian Academy of Sciences, Moscow, Russia, and University of Arizona, Tucson, Arizona, USA

Jordi Zamora-Munt Universitat Politècnica de Catalunya, Terrassa, Spain

IOP Publishing

Nonlinear Guided Wave Optics
A testbed for extreme waves
Stefan Wabnitz

Chapter 1

Extreme events in forced oscillatory media in zero, one and two dimensions

S Barland, M Brambilla, L Columbo, B Garbin, C J Gibson, M Giudici, F Gustave, C Masoller, G L Oppo, F Prati, C Rimoldi, J R Rios, J R Tredicce, G Tissoni, P Walczak, A M Yao and J Zamora-Munt

1.1 Introduction

A decade after the publication of the seminal paper on optical rogue waves [1], a whole new field of study has emerged, focusing on extreme waves in optical systems. While this first study obviously relied heavily on the use of the nonlinear Schrödinger equation (NLSE) to describe the envelope of water waves and that of optical waves, the field has now considerably broadened (see e.g. [2] for a recent overview). At this point, optical systems have become an excellent playground for the analysis not only of the optical equivalent of oceanic rogue waves as studied in the NLSE but more generally for the study of large and statistically (thus in some sense loosely) defined events. Thanks to this broader vision, the relevance of the field has increased a lot, in particular for the emphasis put on individuating the general physical or dynamical concepts which can lead to the emergence of heavy-tailed statistics. Examples of such mechanisms (many of which are reviewed in this volume) include solitons in integrable systems [3–6] and their dissipative counterparts [7, 8], spatiotemporal chaos [9, 30], joint role of disorder and nonlinearity [10, 11], or modulational instability [12, 13]. Of course, one key question since the very early days of the field is the role of spatial dimensionality of the system under study in the emergence of extreme waves. This issue is presently explicitly addressed in the context of optical fibers (see chapter 14 by Krupa *et al* in this volume) and photorefractive media (see chapter 8 by Pierangeli *et al* in this volume). Here we address this issue by reviewing conceptually similar systems consisting of forced oscillatory media (namely a laser with optical external forcing or an optical parametric oscillator) with different numbers of spatial dimensions. In the first section, we review experimental and numerical observations performed on a physical device whose spatial dimensions are

doi:10.1088/978-0-7503-1460-2ch1

© IOP Publishing Ltd 2017

very small, strongly constraining the spatial degrees of freedom of the slowly varying envelope of the field to the point at which the experiment can be modelled with a set of ordinary differential equations [20, 21]. In this case, the only source of complexity (besides possible stochastic effects) is that of low-dimensional chaos. In the second section, the experimental arrangement is such that the forced laser emits on many longitudinal modes, i.e. it is a 1-dimensional system, spatially extended along the propagation direction [22, 23, 25]. In the third section, we present a theoretical analysis of a 2-dimensional device, spatially extended in the dimensions transverse to propagation, where spatial coupling occurs via diffraction [29, 34].

1.2 Zero dimensions

In this section, we discuss the generation of rogue waves in a simple system composed of a semiconductor laser with continuous-wave optical injection.

A single-transverse mode Vertical-Cavity Surface-Emitting semiconductor Laser (VCSEL) is injected by a wavelength-tunable beam. VCSEL operates at around 980 nm and it features threshold current at about 0.2 mA. The laser is pumped by a current source stabilized to better than 0.1 μA, and its package is temperature stabilized to better than 0.01 K. It is important to note that the VCSEL emits a single polarization mode in the parameter range explored in our experiment. The optical injection beam can be tuned in the range 960–980 nm by steps of 51 GHz and, within each step, it can be tuned continuously in a range of 6 GHz. Its power can be adjusted up to 2 mW. The slave laser output passes through an optical isolator preventing back-reflections from the detection part of the setup. Intensity output is monitored by a detection apparatus consisting of a 6 GHz oscilloscope (LeCroy Wavemaster) and an amplified photodetector covering the 0.01–4.2 GHz frequency range. The optical spectra are detected with a scanning Fabry–Perot interferometer (free spectral range of 78 GHz and Finesse of 125).

Critical parameters of our experiment are the frequency detuning between the injected beam (ν_o) and the slave lasers (ν_s), $\Delta\nu = \nu_o - \nu_s$, the injection intensity, P_{inj}, and the slave laser pumping current, J. It is worthwhile noting that these parameters are not independent. A variation of J, for example, affects ν_s via the change of semiconductor refraction index, thus inducing a redshift of the slave laser emission for increasing J which is, for the laser used, 116 MHz μA^{-1} for current values around $J = 1$ mA. In figure 1.1 we show the optical spectrum emission of the slave laser as a function of J varied from 0.867 to 1.046 mA. This current variation implies a variation of $\Delta\nu$ between −14 GHz and +7 GHz, with $\Delta\nu = 0$ occurring at $J = 0.987$ mA (indicated with a dashed line on the graph).

We obtain extreme events in the intensity output of the slave laser in the parameter region where the slave laser gets unlocked from the master laser and develops a chaotic state through a period doubling bifurcation sequence ($J \approx 0.975$ mA).

In this region, the intensity output of the slave laser shows a series of small pulses with sporadic extremely high ones, as shown in figure 1.2(a). The corresponding power distribution of these events exhibits a long tail, thus revealing the presence of extreme events. In figure 1.2(b) we show one typical histogram of the amplitude

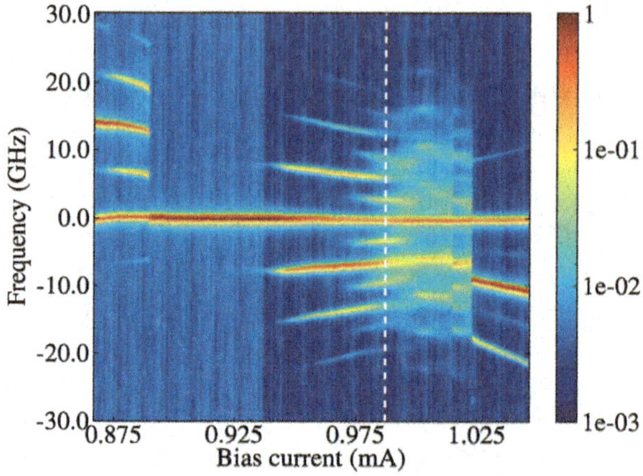

Figure 1.1. Experimental bifurcation diagram displaying the optical spectra of the slave laser versus the bias current. The dashed line denotes the zero-detuning condition between the master and the slave lasers. Reprinted with permission from [20]. Copyright 2011 by the American Physical Society.

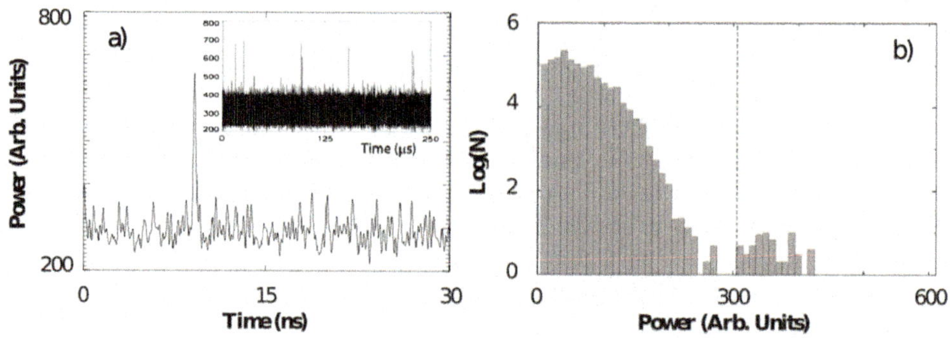

Figure 1.2. (a) Typical time series of the slave laser intensity, where rare large pulses can be observed. The experimental parameters are $J = 0.976$ mA, $\Delta\nu = -1.3$ GHz, and $P_{inj} = 1.1$ mW. (b) Power distribution of the number N of pulses emitted by the slave laser. The vertical dashed line denotes the mean value plus eight standard deviations and thus marks the limit for a pulse to be considered a rogue event. The histogram in (b) corresponds to the time series displayed in (a). Reprinted with permission from [20]. Copyright 2011 by the American Physical Society.

distribution of the pulses obtained where a number of events crosses the limit above which a pulse can be considered extreme. By increasing the slave current value, the system is pushed deeper in the chaotic region and the probability of large power pulse emission increases. Accordingly, the mean value of the peak power increases and the relative weight of the distribution decreases, to the point at which pulses no longer overcome the 'rogue wave' threshold.

This bifurcation scenario and, more specifically, the presence of these extreme events is found in a wide range of optically injected power (from 0.6 to 1.4 mW). The critical detuning for which these events occur tends to zero as P_{inj} decreases, and it can also be positive for low injected power. The range of injected power where extreme events are observed depends also on the slave laser pumping current. If J is increased to 1.321 mA, this range is shifted upwards (from 1.0 to 2.0 mW) while the critical detuning remains almost unchanged. Finally, for $J > 2$ mA, the chaotic region supporting extreme events disappears.

These experimental results are well described by a rate-equations model for the slow envelope of the complex electric field E and for the carrier density N [19].

$$\begin{aligned}\frac{dE}{dt} &= \kappa(1 + i\alpha)(N - 1)E + i\Delta\omega E + \sqrt{P_{inj}} + \sqrt{D}\,\xi(t),\\ \frac{dN}{dt} &= \gamma_N(\mu(t) - N - |E|^2),\end{aligned} \tag{1.1}$$

where κ is the field decay rate, α is the line-width enhancement factor, γ_N is the carrier decay rate, and μ is the injection current parameter, normalized such that the threshold current of the free-running laser is $\mu_{th} = 1$. $\Delta\nu = \Delta\omega/2\pi$ is the optical frequency detuning and $\Delta\omega = \omega_0 - \omega_s$, with ω_0 and ω_s being the frequencies of the master and slave lasers, respectively. P_{inj} is the constant optical injection strength. The term $D\xi(t)$ represents optical noise: ξ is a complex uncorrelated Gaussian white noise and D is the noise strength.

The model equations were numerically solved using the same parameters as in [20, 21]: $\kappa = 300$ ns^{-1}, $\alpha = 3, \gamma_N = 1$ ns^{-1}, $P_{inj} = 60$ ns^{-2}, and the other parameters were varied.

Figure 1.3 displays the time trace of the laser intensity, $I = |E|^2$, generated from random initial conditions, and the corresponding probability distribution function (PDF) of intensity values. We define a RW event when the intensity is higher than a threshold given by $\langle I\rangle + 8\sigma_I$, where $\langle I\rangle$ and σ_I are the mean and standard deviations of the intensity PDF.

As shown in [20], RWs appear in regions of the parameter space where the dynamics are chaotic. However, that is not enough to observe RWs, as there are regions of chaotic dynamics with RWs and others without RWs.

In our model, in the deterministic case ($D = 0$), RWs appear abruptly when a parameter is varied, as can be seen in figure 1.4(a). Then, two relevant questions are which special properties the chaotic regions with RWs have, and what types of bifurcations produce RWs.

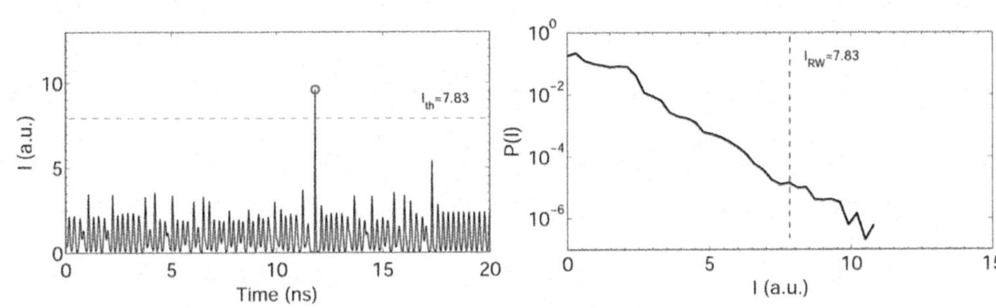

Figure 1.3. Deterministic rogue wave. (Left) Time trace of the laser intensity. The cycle indicates the detection of an RW. (Right) Probability distribution function (PDF) of intensity values. The vertical dashed line indicates the RW threshold, $\langle I\rangle + 8\sigma_I$. Parameters are as in [20], point B: $\kappa = 300$ ns^{-1}, $\alpha = 3, \gamma_N = 1$ ns^{-1}, $P_{inj} = 60$ ns^{-2}, $\Delta\nu = -1.86$GHz, $\mu = 1.96$ and $D = 0$.

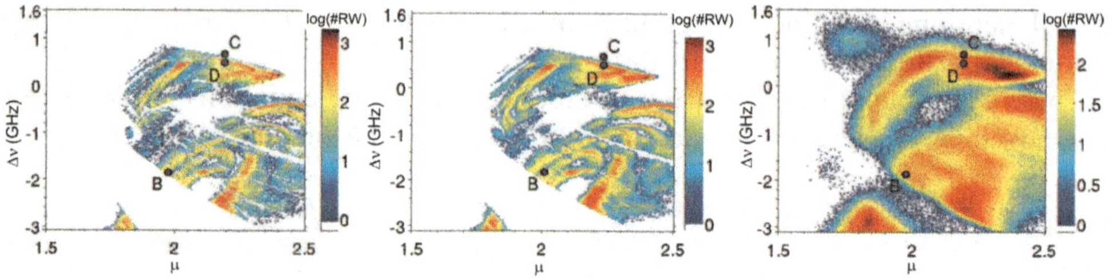

Figure 1.4. Number of RWs in the parameter space (μ, $\Delta\nu$) for different noise strengths. The color code is plotted in logarithmic scale in order to increase the contrast of the regions with a small number of RWs. In the white regions no RWs are detected. Left: $D = 0$, center: $D = 10^{-4}$ ns^{-1}, right: $D = 10^{-2}$ ns^{-1}. The set of parameters used in figure 1.3 and 1.7 are labeled as B (as in [20]: $\mu = 1.96, \Delta\nu = -1.86$ GHz); labels C and D indicate a boundary region: in point D there are deterministic RWs, while in C, RWs are induced by strong noise.

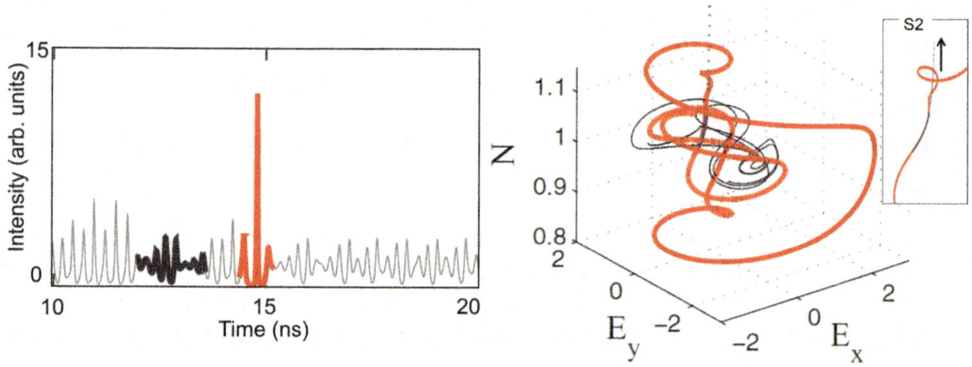

Figure 1.5. Left: intensity time-trace displaying small oscillations in the dense part of the chaotic attractor (in black) and in the sparse part of the attractor when a RW occurs. Right: plot of the two trajectory segments in the (E_x, E_y, N) phase space. The inset displays in detail the evolution along the stable manifold of S2 (dotted blue).

As shown in [21], RWs cannot occur at any point of the phase space. Only a small region of the phase space can lead to an RW when it is visited by the trajectory. In the chaotic regions with RWs, the trajectory can access a narrow region of the phase space (referred to as 'the rogue wave door'), where the stable one-dimensional manifold of a saddle point (referred to as S2) is located. Whenever the trajectory closely approaches this region, an RW is likely to occur. The mechanism is as follows: the trajectory evolves along the stable manifold towards S2, which is the low-intensity solution with high N and $I = |E^2| \sim 0$. However, S2 has a two-dimensional unstable manifold, and thus the trajectory also spirals out. Because the system accumulates carriers while approaching a low intensity during the evolution along the stable manifold of S2, a high pulse is emitted. This process is illustrated in figure 1.5, which displays the intensity time-trace and the corresponding plot in the (E_x, E_y, N) phase space during the frequently visited ('dense') part of the attractor (in black) and also during the RW (when the trajectory performs a large excursion in phase space, in red). The inset shows that the trajectory moves along the stable manifold of S2 (dotted blue) while spiraling out. The duration of the evolution along the stable manifold of S2 is irregular because it varies with the initial 'entry' conditions and is affected by numerical noise, and, in consequence, there is a wide distribution of pulse heights, as shown in figure 1.3(b).

In contrast, in the chaotic regions without RWs, the trajectory in phase space never approaches this dangerous region of the phase space where the RW door (the one-dimensional stable manifold of S2) is located.

This type of bifurcation was also analyzed in [21], and numerical evidence suggested that RWs appear as the result of a collision of the attractor (that develops from a fixed point referred to as S3) with the stable manifold of another saddle-focus point (referred to as S1). As shown in figure 1.6, after this collision the attractor expands and gains access to the phase space region where the RW door is. In other words, the attractor generated from fixed point S3 collides with the stable manifold of S1, which produces an abrupt expansion of the attractor, which in turn enables access to the region of the phase space where the stable manifold of S2 is; then, when the trajectory approaches this manifold, an RW can be triggered.

In [21] it was also shown that RWs can be predicted with a certain prediction horizon. This is shown in figure 1.7, where we plot a large number of time traces

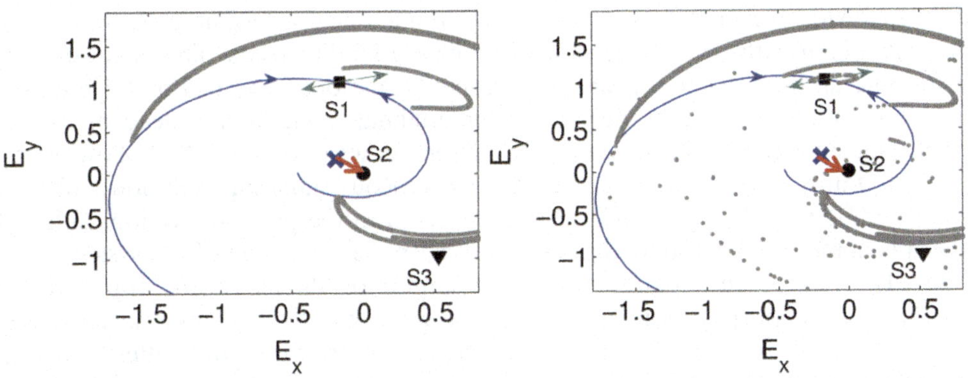

Figure 1.6. Poincarè section of the attractor with the plane $N = 1.0036$. The three fixed points (S1, S2, S3) and the stable manifold of S1 are also projected in the Poincarè plane. Left: before the collision with the stable manifold of S1, the attractor (developed from S3) is confined to a region of the phase space that does not include the stable manifold of S2. The parameters are $\mu = 2.2$, $\Delta\nu = 0.6$ GHz. Right: after the collision, the attractor expands and gains access to the region where the stable manifold of S2 is located. The parameters are $\mu = 2.2$, $\Delta\nu = 0.594$ GHz.

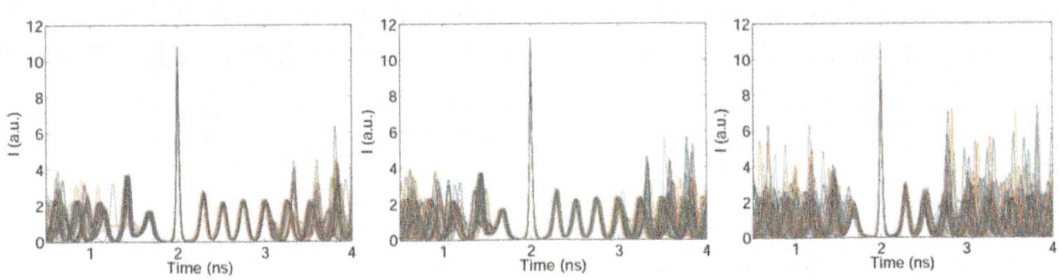

Figure 1.7. Time traces of the intensity centered at the maximum of the RW for different noise strengths. Left: $D = 0$, center: $D = 10^{-4}$ ns^{-1}, right: $D = 10^{-2}$ ns^{-1}. The parameters are $\mu = 1.96, \Delta\nu = -1.86$ GHz (point B).

centered at the maximum intensity of the RW. Note that not only before the RW, but also after, the intensity behaves in a predictable way. Here we also show the robustness of the RWs to noise. Noise reduces the horizon of predictability, as could be expected, but even with strong noise there is a considerable interval of time when the RW can be predicted. Moreover, the height of the RW is mainly unaffected by the noise strength.

To summarize, in our system, RWs can be deterministic but they can also be induced by noise in certain parameter regions. Both deterministic and stochastic RWs can be predicted within a certain time horizon. We have also found that, in the deterministic case, RWs occur when an abrupt expansion of the attractor enables access to a particular region of phase space. The three fixed points of the system are relevant for understanding the RW phenomenon: an attractor develops from one fixed point, the stable manifold of a second fixed point acts a barrier, while the stable manifold of the third fixed point acts as an 'RW door': if the trajectory closely approaches this manifold, an RW can be triggered.

1.3 One dimension

In the following, we analyze experimentally and numerically the emergence of extreme events in a laser with optical injection with one spatial dimension. This is achieved by performing an experiment conceptually similar to the one described in 1.2 but, as a slave laser, we will use a strongly multi-longitudinal mode laser instead of using a strictly single longitudinal/transverse mode laser. As we shall see in the following, the role of 0D attractors (i.e. the associated homogeneous stationary solutions) will still provide some resemblance to RW events observed in the previous section, although other phenomena will strongly differ due to the spatial extension of the system.

In 1D systems, the spatiotemporal dynamics of the laser are appropriately described by the propagation equation of the coherent field although, heuristically, the spatial profile of the field can be interpreted in terms of nonlinear interaction and competition of several empty cavity modes, as ruled by salient features of the device such as gain width and free spectral range. The system, now infinite dimensional, may exhibit a wealth of dynamical behaviors including spatiotemporal chaos, which is an ideal irregular substrate over which extreme events can be recorded as high-intensity peaks. In this instance, we adopted a model for spatially (1D) extended lasers, based on a set of rate equations which, apart from the different notations, is identical to equations (1.1), with the addition in the equation for the electric field of a propagation term containing the first derivative along the longitudinal coordinate z and a spectral filter (diffusion) term containing the second derivative [22, 23]

$$\frac{\partial E}{\partial z} + \frac{\partial E}{\partial t} - d\frac{\partial^2 E}{\partial z^2} = T[-(1 + i\theta)E + y + (1 - i\alpha)ED], \tag{1.2}$$

$$\frac{\partial D}{\partial t} = b[\mu - D(1 + |E|^2)]. \tag{1.3}$$

E and D are, respectively, the slowly varying envelope of the electric field and the excess of carrier density with respect to transparency. The cavity losses are represented by the mirror transmissivity T, y is the injected field amplitude, θ is the frequency detuning between the injected field and the closest cavity mode, b is the ratio of the roundtrip time to the carrier lifetime, and α and μ have the same meaning as in equations (1.1)

The diffusion term introduces an effective damping of short-scale spatial modulations and its weight is determined by the d coefficient, which is proportional to the squared ratio of the cavity free spectral range to the gain linewidth.

The stationary homogeneous solution $E_s = \rho_s \exp(i\phi_s)$, D_s of equations (1.2) and (1.3) is

$$y^2 = \rho_s^2[(1 - D_s)^2 + (\theta + \alpha D_s)^2], \quad \phi_s = \arctan\left(\frac{\theta + \alpha D_s}{D_s - 1}\right), \quad D_s = \frac{\mu}{1 + \rho_s^2} \tag{1.4}$$

and for parameters compatible with the experiment[1], the stationary intensity curve $I = \rho_s^2$ of the electric field as a function of the injection amplitude y is multivalued, as shown in figure 1.8(a).

Here, A, B and C indicate the three fixed points for the chosen value of y (vertical red dashed line): in particular, in the 0D model, A is a node, B is a saddle, and C is a focus [22, 23]. In the following, the phase–space trajectory of the system state will be studied in relation to these three attractors.

In figure 1.8(b), the stability scenery of the homogeneous stationary solution is plotted in the (I, n) plane, where n is the order of the longitudinal sidemode. The green domain indicates the unstable region in the presence of diffusion ($d = 10^{-6}$), while the blue domain corresponds to the choice $d = 0$, which amounts to the pure rate equation model without spectral filtering. It is clear that the diffusive term sets a limit to the number of unstable longitudinal modes but does not significantly affect the instability threshold and its local curvature.

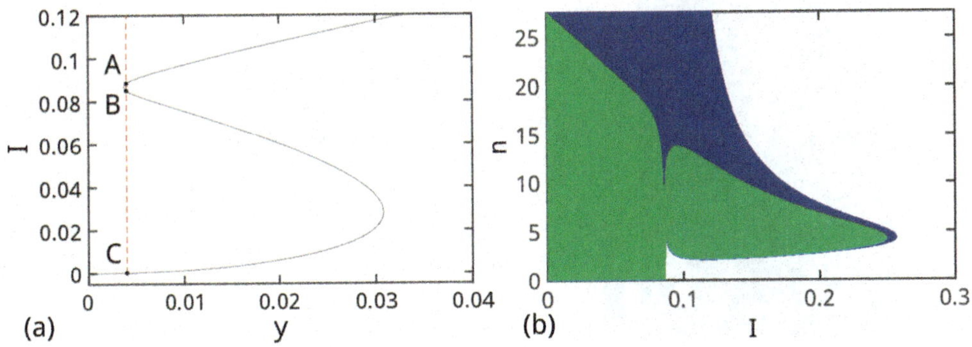

Figure 1.8. (a) Stationary homogenous solution of equations (1.2) and (1.3) for $\mu = 1.1$, $\theta = -3.04$, $\alpha = 3$, $b = 10$, $T = 0.3$. Points A, B, and C are the solutions corresponding to the value $y = 0.004$ of the injected amplitude. (b) Instability domains in the (I, n) plane according to equations (1.2) and (1.3) (in green) and for the pure rate equation model with $d = 0$ (in blue). Reprinted with permission from [50].

[1] Note that the pump threshold value for the pump parameter in the free running laser is $\mu = 1$.

The stability diagram of figure 1.8(b) is typical of the condition $\theta + \alpha < 0$, which means that the injected field is blue detuned with respect to the slave laser frequency. In this case, a multimode instability affects a part of the upper branch close to the left turning point. The homogeneous stationary solution is unstable and a roll pattern arises due to the beating of the fundamental mode with high-order sidemodes. Eventually even the rolls become unstable when the injection amplitude is brought too close to the left turning point, and more complex spatiotemporal dynamics can be observed, as in the experiments.

The experimental setup is based on a highly multimode semiconductor ring laser under coherent external forcing [22–24]. The ring laser is made of a semiconductor optical amplifier (SOA) acting as gain medium centered around 980 nm and enclosed in a 1 m long ring cavity. The anti-reflection coating on the SOA facets prevents self-lasing of the semiconductor element. An example of an experimental observation of such a dynamical regime in which an unstable roll pattern hosts high-intensity peaks appearing randomly in space and time is shown in figure 1.9.

Figure 1.9(a) shows a spatiotemporal diagram, where the horizontal axis refers to the longitudinal spatial dimension of the laser and the vertical axis represents time in units of the cavity roundtrip time $\tau_c \simeq 3.6$ ns. In this figure, we can clearly identify the unstable roll pattern in the background and high intensity pulses nucleated on top of it. In order to get the statistical signatures of this regime, we computed the PDF of the total optical power and the distribution of the peak amplitudes, respectively, represented in figures 1.9(b) and (c).

In the PDF of the power, we first distinguish a plateau (up to 10), followed by an exponential decay (up to 17), and a second exponential decay with smaller slope

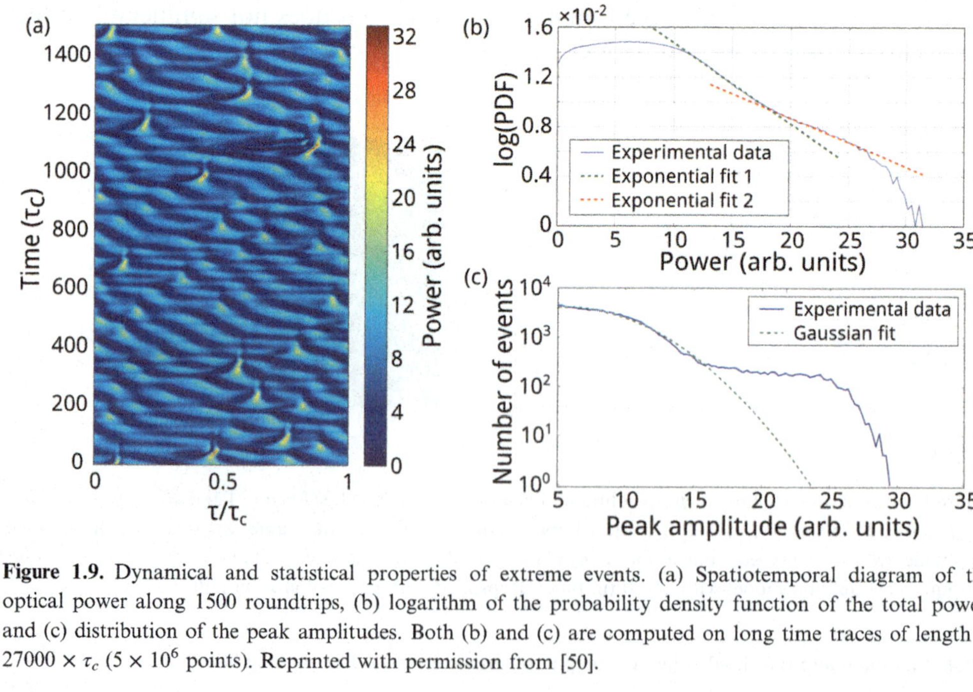

Figure 1.9. Dynamical and statistical properties of extreme events. (a) Spatiotemporal diagram of the optical power along 1500 roundtrips, (b) logarithm of the probability density function of the total power, and (c) distribution of the peak amplitudes. Both (b) and (c) are computed on long time traces of length $\simeq 27000 \times \tau_c$ (5×10^6 points). Reprinted with permission from [50].

corresponding to heavy tail statistics. The presence of this statistical deviation at high amplitudes implies that high-intensity pulses occur more often than what would be expected looking at the first exponential decay.

In the PDF of the peak heights shown in 1.9(c), the heights are the difference between each maximum and the previous minimum of the full time trace. The shape of the distribution is well fitted by a Gaussian up to 16 and presents a huge deviation beyond that point. The Gaussian part of the distribution (at low values) is related to the oscillating background, whereas the deviation appearing at high values corresponds to the presence of abnormal events [49] that we will consider as extreme events in the following.

To understand the formation of these extreme events, we focus on one particular event, statistically representative of the relevant regime, and provide a detailed analysis by looking at both phase and intensity dynamics. Figure 1.10(a) shows a

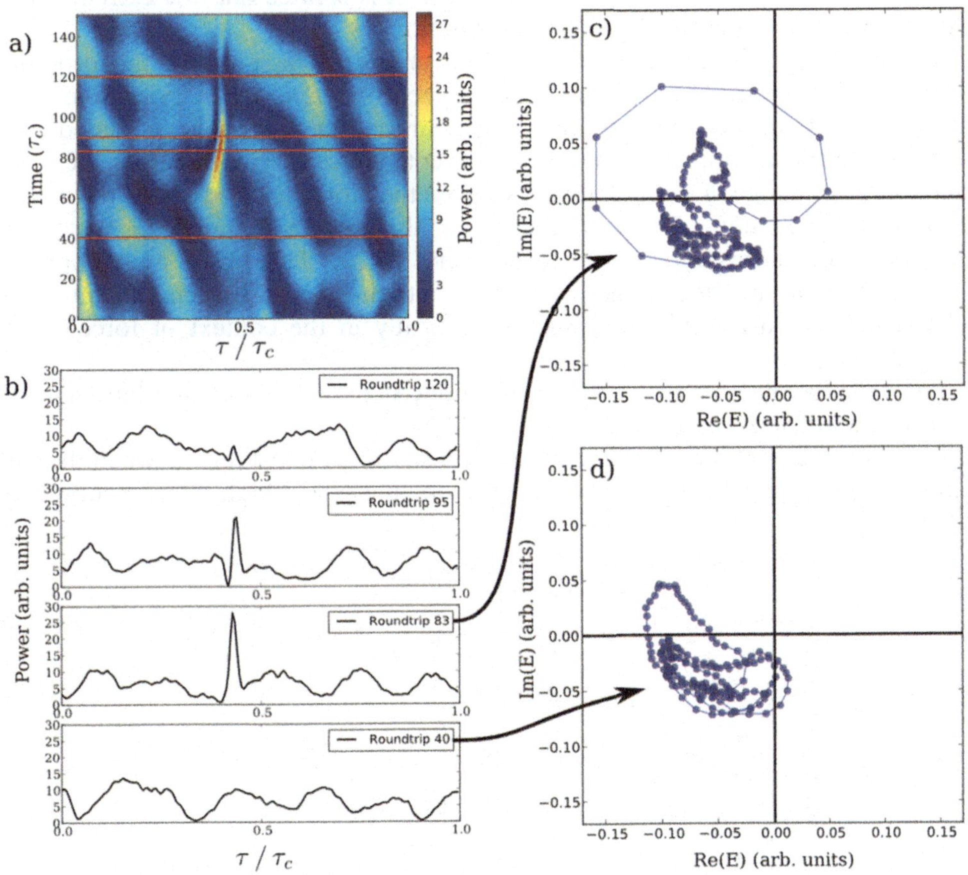

Figure 1.10. Power and phase dynamics around the formation of an extreme event. (a) Spatiotemporal diagram focused on one event, taken around roundtrip 1000 of figure 1.9(a). (b) Power profiles corresponding to roundtrips 40, 83, 95, and 120 and depicted by red horizontal lines in (a). (c) Phase trajectory in the Argand plane before the extreme event (roundtrip 40) and (d) at the maximum peak power of the event (roundtrip 83). Reprinted with permission from [50].

zoom of the spatiotemporal diagram presented in figure 1.9(a) on 150 roundtrips around an extreme event. The horizontal red lines represent the positions at which intensity profiles shown in figure 1.10(b) have been taken. The first line (roundtrip 40) shows the state of the system before the extreme event occurs, and the associated phase trajectory is presented in figure 1.10(d). At this stage of evolution, the phase remains in a bounded region of the complex plane (Argand plane).

At roundtrip 83, the extreme event has reached its maximum intensity. By comparing the height of this event with the statistics shown in figure 1.9(c), we can easily understand that this object participates in the strong statistical deviation described before. As we can see in figure 1.10(c), the high intensity peak is accompanied by a striking topological change of the phase space picture, because now the trajectory circles around the origin. The system seems to be pushed away from the bounded state and to explore a wider region of the complex plane. By crossing the origin of the complex plane, a defect is created and the extreme event acquires a counterclockwise ($+2\pi$) phase rotation.

In figure 1.11, we show the evolution of the unfolded phase around the formation of an extreme event. The time trace is centered on the extreme event maximum intensity ($t = 0$). In this representation, it is clear that the phase stays bounded up to 20 ns before the maximum of the extreme event. We can clearly identify the phase rotations as a change of slope that starts before the moment where an extreme event reaches its maximum. At each roundtrip, the system acquires 2π rotations, corresponding in this representation to steps of 2π. The phase and amplitude dynamics leading to the emergence of these extreme events are reminiscent of the defect-mediated turbulence discussed theoretically in the context of forced oscillatory media [14, 29].

The experimental findings are well reproduced by numerical simulations made with the parameters of figure 1.8 and $y = 0.004$.

In figure 1.12, we show a zoom of a typical spatiotemporal regime when the rolls pattern is unstable. The diagram is centered on an event of high intensity occurring

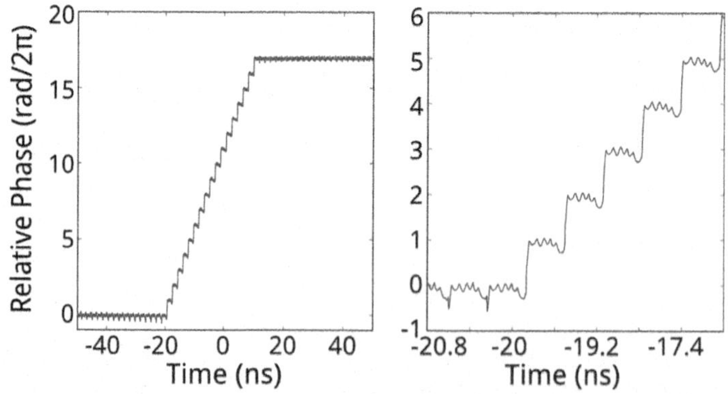

Figure 1.11. Evolution of the phase when approaching the extreme event. Left: 100 ns long time trace of the phase centered on the maximum of an extreme event showing the acquisition of positive phase rotations. Right: zoom of the same time trace at the beginning of the change of slope. The initial phase is arbitrarily set to zero. Reprinted with permission from [50].

at the point highlighted by line 3. The four temporal profiles illustrated in figure 1.12(a) correspond to the horizontal cuts highlighted in figure 1.12(b). In particular, the temporal profile from section 1 occurs 500 roundtrips before the event and corresponds to a mostly bounded regime, where we can appreciate the remnants of the roll pattern which would be stable for a larger injection amplitude. Section 3 is taken at the maximum of the high-intensity event, while sections 2 and 4 are taken, respectively, 40 roundtrips before the event and 160 roundtrips after.

For the statistical analysis, we plot in figure 1.13 the PDF of the temporal peak heights (maximum minus precedent minimum) (a) and of the overall intensity values

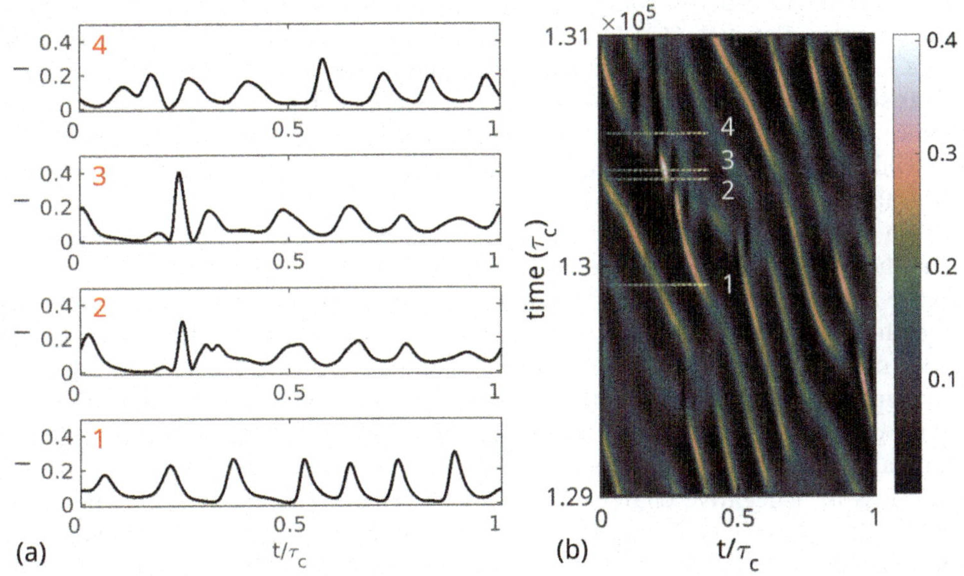

Figure 1.12. Zoom (b) of the spatiotemporal diagram centered on an event of high intensity (3) and intensity time traces (a) at fixed roundtrip corresponding to the horizontal cuts highlighted on the diagram. Reprinted with permission from [50].

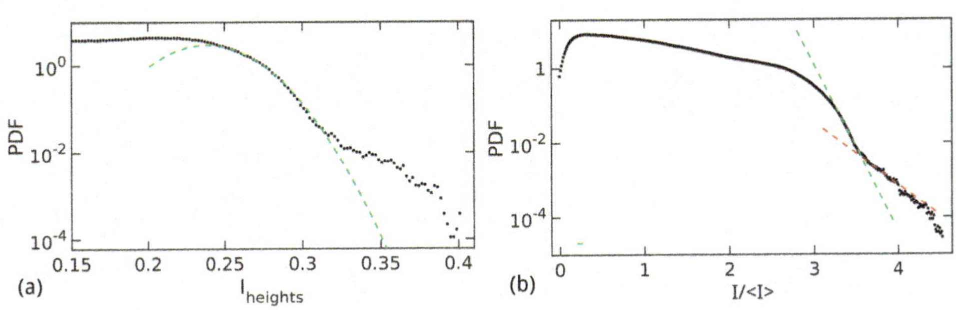

Figure 1.13. (a) PDF of the temporal peak heights, computed over 2.1×10^6 roundtrips; the green dashed line corresponds to a Gaussian fit of the initial slope. (b) PDF of all the values explored by the intensity, computed over 2.1×10^5 roundtrips. The green and red dashed lines correspond to two negative exponential fits highlighting the change of the slopes in the tail of the distribution.

(b). Similar to the corresponding experimental figure 1.9, we observe that the PDF of the heights (a) presents an initial slope that follows a Gaussian distribution (green dashed line) to be associated with the bounded state, while the events in the tail of the distribution appear more frequently than expected by Gaussian statistics. Furthermore, in the statistics of the total intensity, we notice two different negative exponential slopes in the tail of the distribution as observed in the experiments: the first slope (green dashed line fit) may be associated with the roll pattern; the second slope (red dashed line fit) highlights the heavy tail where very intense events are more frequent than expected in the bounded regime.

Let us focus now on the role of the phase dynamics in the formation of these extreme events. In order to compare with the experiment, we computed a histogram in the complex plane (Re(E), Im(E)) of all the values assumed by the electric field. Figure 1.14(a) is the experimental histogram and figure 1.14(b) is the numerical one. A clear similarity between the two histograms is noticeable, especially regarding the presence of an unstable bounded state (yellow area in (a), pink-white area in (b)) from which rarer events get detached, exploring other areas of the histogram (red area in (a), blue-green area in (b)).

From the experiments it was already clear that chirality was playing an important role in the formation of extreme events: in particular the latter seem to be associated with the appearance of a positive (counterclockwise) chiral charge (see figure 1.11). In the numerical simulations, we noticed that clockwise phase rotations also play an important role in the formation of extreme events.

In figure 1.15(a) we plotted the evolution of the phase when approaching an extreme event, in a situation similar to the experimental case: here we notice that an initial balance between the number of positive and negative chiral charges is broken by the loss of a negative charge, which gives rise to a change in the slope of the phase. A little later, an extreme event occurs (highlighted by the red vertical dashed line). In the zoom in figure 1.15(b) it is possible to notice that in the initial plateau for the phase there are both a positive and a negative chiral charge, which then get reduced to a single positive charge.

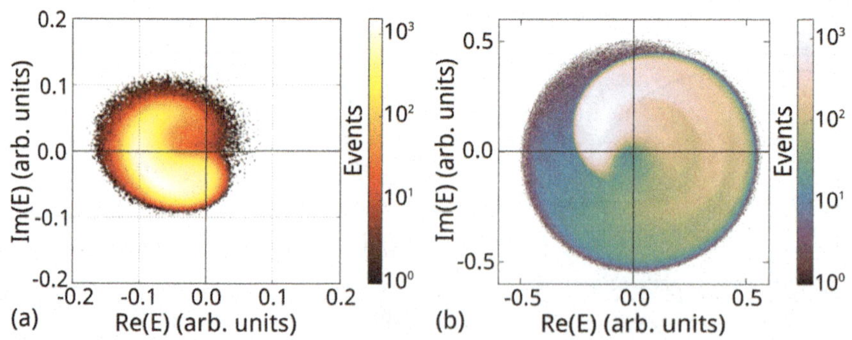

Figure 1.14. Histogram of the values assumed by the electric field in the complex plane (Re(E), Im(E)) in the experimental case (a), computed over 10^4 roundtrips, and in the numerical case (b), computed over 10^5 roundtrips, in order to maintain the same number of events associated with the bounded state. Reprinted with permission from [49].

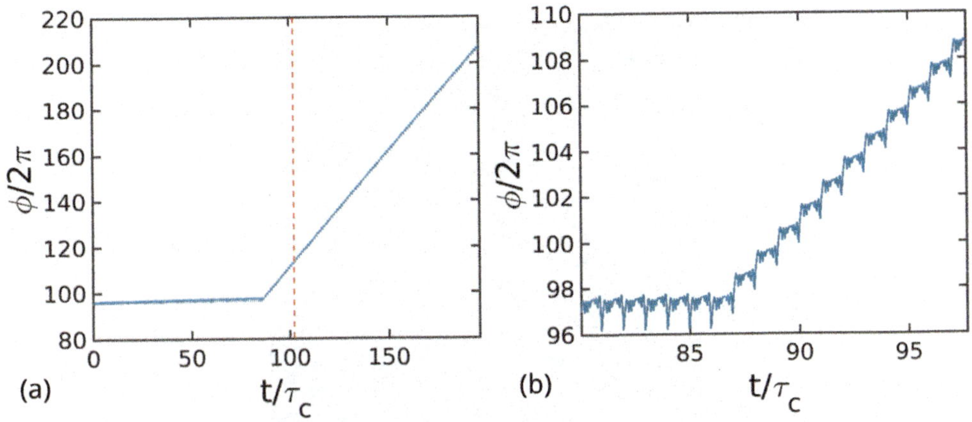

Figure 1.15. Evolution of the phase when approaching an extreme event: the extreme event (occurring at the time indicated by the red vertical dashed line) is to be associated with a change in the slope of the phase. In particular in this case, we notice an initially balanced situation between the number of positive and negative chiral charges that breaks when the negative chiral charge disappears. A little later, an extreme event occurs. Reprinted figure with permission from [49].

We can argue that the occurrence of extreme events may be associated with a global change in the slope of the phase; however, as it has been observed in the experiment, this change is not to be considered as a precursor since it can occur either before or after the extreme event itself.

We now turn our attention to a slightly different experimental setup where a Fabry–Perot resonator is used instead of a ring [25, 30]. Now the active medium is antireflection coated only on one side. A lens is inserted to focus the output field on a high reflected mirror (99%) which closes the one meter long cavity. Two 10% beam splitters are inserted inside the Fabry–Perot resonator to provide both an input for the forcing field and an output for the detection of the emitted field. The master laser is provided by a grating tunable external cavity semiconductor laser. In order to prevent all reflections from the slave laser to the master laser, an optical isolator is inserted.

In this configuration, the formation of phase soliton complexes was observed [25]: two phase solitons with one chiral charge attract and collide to form a single phase soliton with a double chiral charge, whose velocity is different from that of the single-charge phase soliton. This process can repeat, leading to complexes with high chiral charge.

As outlined in the Introduction, we hypothesized that phase soliton collisions could be linked to emergence of extreme events, so we focused on new regimes where an underlying irregular dynamics could couple with the propagation of self-organized structures. Here, we have selected a specific regime where we are able to observe the emergence of a spatiotemporally localized event with high amplitude. Figure 1.16(a) represents a spatiotemporal diagram of such a regime. We can observe mainly three different behaviors. During the first 7751 roundtrips, the system is in synchronized state. Between roundtrips 7850 and 10 060, the system exhibits a markedly different and complex dynamics. We can recognize the propagation of dissipative phase solitons, often with double charge. These coherent structures, which are spontaneously nucleated from the synchronized state, propagate and sometimes collide with

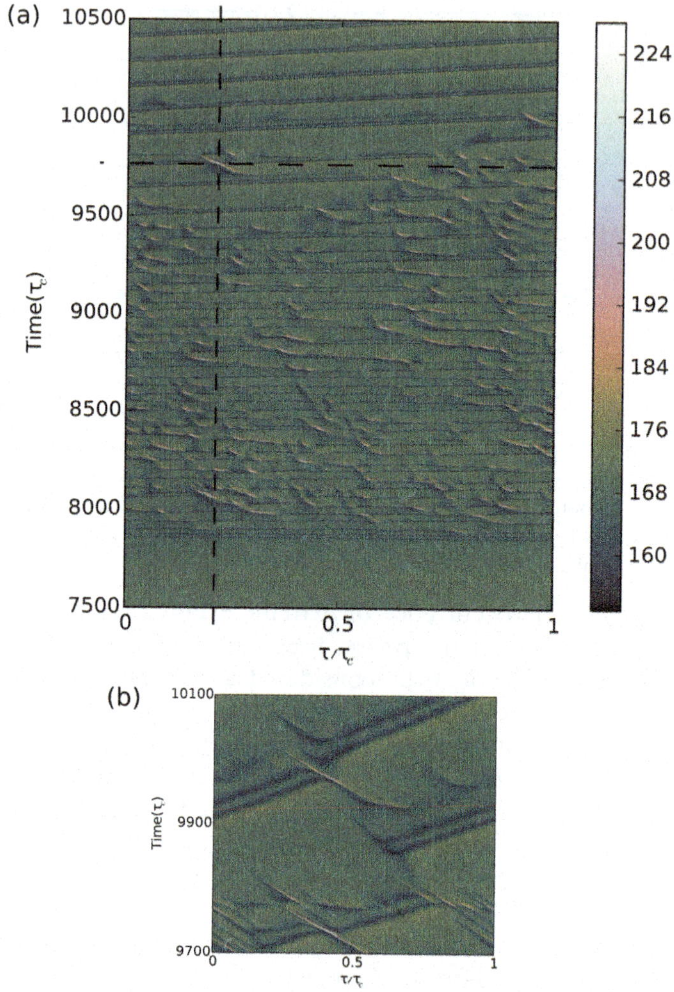

Figure 1.16. (a) Space–time diagram showing the different regimes observed in the experiment. Collisions between dissipative phase solitons and counter-propagating structures lead to the emergence of extreme events. Before 7500, the system is stably locked and after 10 500, one soliton is stable. (b) Observation of collision in a slightly modified reference frame (6450 ps roundtrip time instead of 6470). The laser threshold current is 18 mA, and here its bias current is set at 1.6 times above the threshold. Reprinted figure with permission from [50].

different structures propagating in the opposite direction in this reference frame, inducing extreme events localized in time and space.

In figure 1.16(b) we see that before the collision, the dissipative phase soliton with chiral charge of 2 collides with a different particular structure. After this collision, one hump of the soliton complex seems to be annihilated and a new hump is regenerated about ten roundtrips later. This double structure propagates and eventually collides again with a counterpropagating structure, giving rise to a new extreme event. Finally, after this complex regime (from roundtrip number 10 400 to the end of this record), a regular regime is attained, dissipative phase solitons with a chiral charge of 2 are stable and propagate without collisions.

The statistical impact of these events can be characterized by the computation of the PDF of the power fluctuations for a long time trace as is represented in figure 1.17. The black solid line represents the PDF on all the roundtrips, which is asymmetric and shows a clear heavy tail at the highest power values. The dashed red line, green line with circles, and blue line with squares correspond, respectively, to the synchronized state, turbulent regime, and soliton-propagation regime described in figure 1.16(a). In the synchronized state, where the emission is regular and continuous, the PDF results essentially from the physical and detection noise (dominant), which is stochastic. This explains why the PDF has a Gaussian form. In the opposite case, in the turbulent regime, the emergence of extreme events modifies strongly the PDF which deviates from the Gaussian statistics to exhibit heavy tail statistics. Finally, the propagation of dissipative phase solitons plays also a significant role in the modification of the PDF. Due to their particular shape, including a depression on the trailing edge [22, 23], the probability of the low power values is increased.

We performed numerical simulations in the same regime observed in the experiment, that is, where the injected field is red detuned with respect to the slave laser frequency ($\theta + \alpha > 0$). Precisely, we set $\theta = -2.7$ and assumed that the laser is well above threshold $\mu = 2$, as in the experiment, with injected amplitude $y = 0.11$. The remaining parameters are the same as in the previous simulations. The stability scenery with these parameters differs from that of figure 1.8, because the whole upper branch of the homogeneous stationary solution now is stable and the system can sustain phase solitons (PSs), provided the amplitude of the injected field is such that the fixed points A and B described above are sufficiently close, i.e. the system is

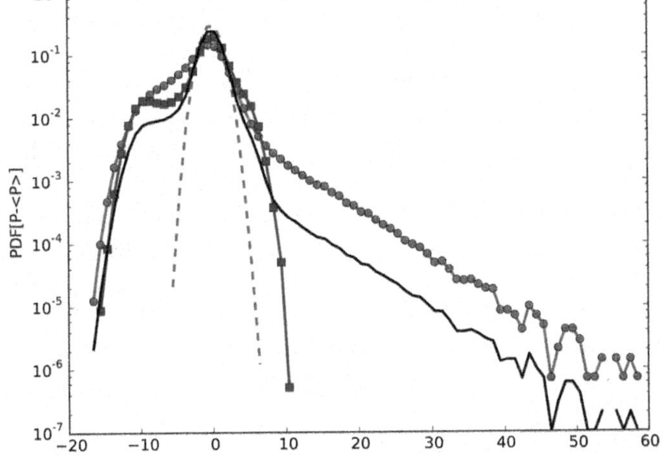

Figure 1.17. Experiment: probability density function (PDF) of the power fluctuations along the recording time trace. The PDF shows the deformation of the statistics with the emergence of localized events with high amplitude. Dashed red line: PDF from 0 to 7751 (locked state); Green line with circles: from 7850 to 10 060 (collision regime); blue line with squares: from 10 500 to 15 500 (stable soliton propagation); black line: PDF from 0 to 15 500. Reprinted figure with permission from [50].

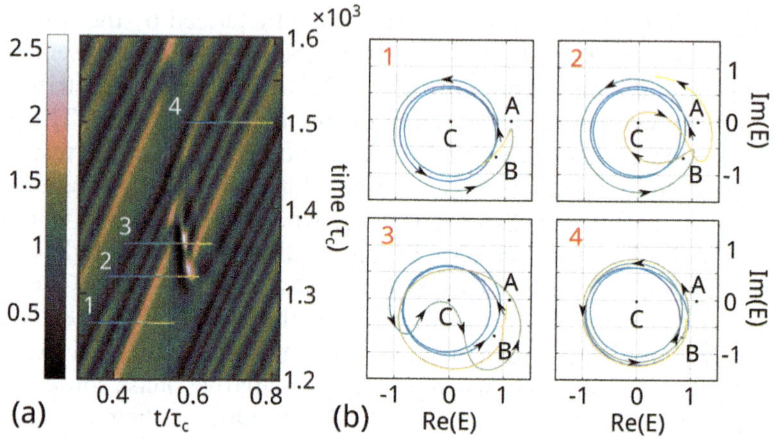

Figure 1.18. Numerical simulation: zoom of the spatiotemporal diagram of the electric field intensity around a collision (a), and trajectories in the Argand plane for the selected roundtrip sections (b). Roundtrip section 3 coincides with the occurrence of a high-intensity event, and it is preceded by a clockwise rotation at the roundtrip section 2. Reprinted figure with permission from [50].

close to the left turning point of the homogeneous stationary solution [22, 23, 25]. Phase solitons are homoclinic trajectories connecting the locked homogeneous state A with itself, through a rotation of the phase of electric field by (multiples of) 2π. Therefore, phase solitons possess a chiral charge, and in a non-instantaneous propagative (temporal) system such as ours, the chiral charge can only be positive (i.e. the phase rotates counterclockwise) [27, 23].

In figure 1.18(a), we show a zoom of the spatiotemporal diagram in a simulation where the initial condition was the homogeneous stationary solution with a superimposed phase kink of 4π (but the same kind of regime can develop starting from noise). We can notice that the spatiotemporal dynamics present some clear similarities with the experimental data. In particular, we can observe the presence of phase solitons and the emergence of some peculiar structures moving at a different velocity that eventually collide with the phase solitons, giving rise to an event of high intensity.

In figure 1.18(b), we show the phase diagrams relative to the horizontal cuts highlighted in figure 1.18(a); time grows from red to yellow. The points A, B and C are the projections in the Argand plane of the three fixed points [22, 23], respectively a node (stable for our choice of parameters), saddle, and an unstable focus. In the first frame, relative to roundtrip section 1, we can observe a phase soliton complex [25] of charge 3 that propagates inside the cavity: the trajectory of the system consists of three counterclockwise phase rotations in the Argand plane, passing only once close to point A. At roundtrip section 2, a new object emerges at the right side of the soliton complex, together with a clockwise phase rotation (charge -1) in the Argand plane. The interaction between the two structures gives rise to a high-intensity event at roundtrip section 3, where the clockwise rotation has already been lost. The interaction also has the effect of altering the phase soliton complex velocity inside the cavity, as can be noticed just after roundtrip section 3. At roundtrip section 4, the complex has regained its shape, with one additional charge, coming from the interaction.

This evidence indicates that the appearance of these short-lived pulses, with clockwise phase rotations that collide with stable phase solitons, is the basic physical mechanism responsible for the extreme events observed in this system.

1.4 Two dimensions

In the previous sections, we have seen excellent examples of the generation of extreme events due to external forcing in systems described by ordinary differential equations (section 1.2) or by partial differential equations in time and one auxiliary dimension (section 1.3). Here we describe the generation of rogue waves due to external forcing in the dynamics of 2D optical systems with gain. In particular, we investigate spatiotemporal turbulence due to interacting optical vortices in lasers and in optical parametric oscillators (OPOs) with injection [27]. Transverse 2D rogue waves have also been found in lasers with saturable absorbers [28, 8] and even in passive optical systems in the presence of pattern competition [29]. The focus of this section is, however, on the specific effect of the external forcing via an optical injection on inducing 2D RWs, i.e. high-intensity peaks with non-Gaussian statistics, in the transverse section of laser systems.

We investigate the spatiotemporal dynamics of two closely related (in terms of equations and gain threshold) optical systems: a class A laser with injected signal [30] and a singly resonant optical parametric oscillator (SROPO) with seeding [31]. The first case is closely related to systems described in the previous sections, the main difference being that in class A lasers, such as dye, quantum dots, quantum cascade, or He–Ne lasers, all material variables (for example, energy level populations and material polarization) evolve on temporal scales much shorter than that of the cavity field. The SROPO case corresponds to parametric down-conversion of a pump field into a signal and an idler field, whose frequencies sum to that of the pump, inside an optical cavity. In the singly resonant case the frequencies of the signal and idler fields are very different and only the signal field is resonated inside the cavity. In this case the material variable dynamics are instantaneous since parametric down-conversion utilizes virtual energy levels that exist for times smaller than those given by the Heisenberg time–energy uncertainty principle. In both cases we assume that an external forcing takes the form of an optical injection detuned from one of the optical cavity resonances.

For a class-A laser with injection, the model equations are

$$\frac{\partial E}{\partial t} = E_{IN} - (1 - i\omega)\, E + P\left(1 - \frac{1}{3}|E|^2\right)E + i\nabla^2 E - \Gamma(\omega + \varepsilon\nabla^2)^2 E, \tag{1.5}$$

while for the SROPO case with injection we have

$$\frac{\partial E}{\partial t} = E_{IN} - (1 - i\omega)\, E + P\ \mathrm{sinc}^2(|E|)\, E + i\nabla^2 E - \Gamma(\omega + \varepsilon\nabla^2)^2 E, \tag{1.6}$$

where E is the complex field inside the cavity, E_{IN} is the (real) amplitude of the external forcing, ω is the frequency difference between the unperturbed field and the external driver, ∇^2 is the transverse Laplacian that describes diffraction, P is the pump, ε is a small parameter (here fixed at 0.01), Γ is equal to one for the Swift–Hohenberg correction [32] and zero otherwise. Equation (1.6) reduces to equation (1.5) when the pump P is close to its threshold value of unity. We assume flat mirrors but note that in the case in which the mirrors are curved, these equations can be projected onto a set of Laguerre–Gauss modes [33], thus transforming partial differential equations into a large set of ordinary differential equations. Superposition of Laguerre–Gauss modes can, of course, generate interacting vortices in two spatial dimensions and several of the results presented subsequently may find confirmation in these mode expanded equations. We start by considering the spatially homogeneous stationary states of the two model equations. There are three parameters that can be controlled experimentally: the detuning between slave laser and injection ω, the amplitude of the injection E_{IN}, and the external pump P. For fixed values of ω and P, the stationary intensity I_S of the homogeneous states has a typical S-shaped dependence on E_{IN}^2 as displayed, for example, in figure 1.19. By increasing (decreasing) the pump value P, the S-shaped curve stretches upward (downward) while keeping its intersection with the origin fixed. The uppermost lines in the S-shaped curves of figure 1.19 correspond to the homogeneous locked states where the external driving is large enough to overcome the frequency difference with the injected device ω. When increasing E_{IN}, a saddle-node bifurcation heralds the onset of the frequency- and phase-locked homogeneous states. In the $[|E_{IN}|^2;\ I_s]$ parameter space, the coordinate of the turning point where the saddle-node bifurcation takes place is well approximated by $[\omega^2(P-1);\ P-1]$ for $\Gamma = 0$ [34]. When, instead, decreasing the parameter E_{IN}, the homogeneous solution loses stability to spatially periodic patterns with a critical wave-vector given by $k_c = \sqrt{\omega}$. In figure 1.19 the maximum and minimum intensities of the hexagonal patterns obtained numerically when reducing the external driver are displayed via a black dotted line. Although the phase of the pattern is periodically modulated in space, the stationary character of these pattern solutions demonstrates that they are locked to the frequency of the injection.

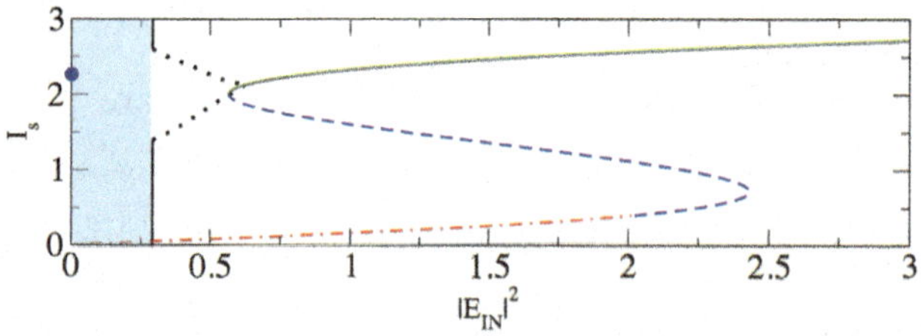

Figure 1.19. Stationary intensity I_s of the homogeneous solutions and their stability (stable = solid green, unstable with real eigenvalues = dashed blue, unstable with complex eigenvalues = dash-dotted red) for the model (1.5) with $P = 4$, $\omega = 0.53$, and $\Gamma = 1$. The black dotted lines represent the minima and maxima of stationary hexagonal patterns and the vertical lines where the optical turbulent state starts (shaded area). The blue circle is the stationary intensity of the laser with no injection.

By further decreasing the injection amplitude E_{IN}, one enters into a region of spatiotemporal chaos [27], i.e. the regime of optical turbulence of relevance here. By using the input pump P as the control parameter while keeping the detuning ω and the injection amplitude E_{IN} fixed, we demonstrate here that the transition to a turbulent state dominated by interacting vortices is a generic property of both laser and OPO models. Just above the threshold for laser or OPO signal generation, i.e. P above but close to the value 1, the system output is locked to the injection and is basically homogeneous apart from finite size diffraction rings (see figure 1.20(a)). By increasing P, an instability to a stable but periodically modulated structure with a strong central peak sets in (see figure 1.20(b)). Such spatial structure is strongly affected by the transverse boundary conditions that, in this case, are due to the finite size of both pump and injection beams. By further increasing the pump P, a symmetry breaking bifurcation takes place and the system moves into a turbulent state (see figure 1.20(c)) characterized by regions where the intensity is close to zero and others corresponding

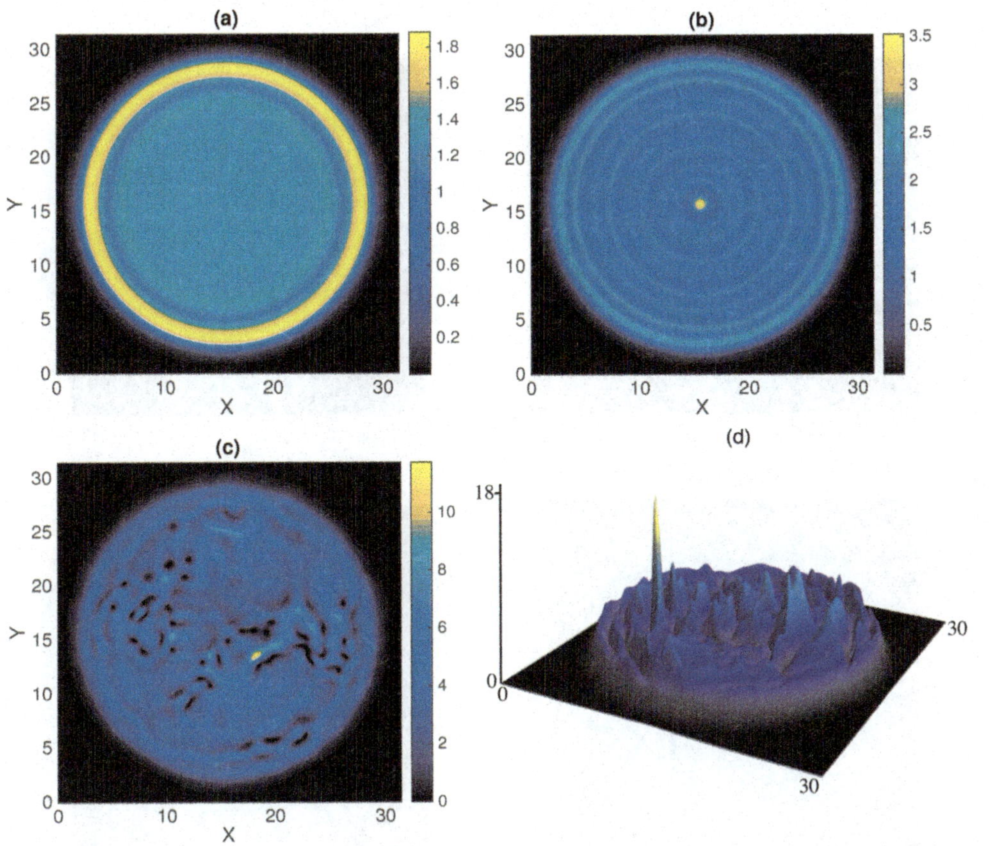

Figure 1.20. Transverse intensity profiles for the SROPO model (1.6) with $\omega = 1$, $E_{IN} = 1.2$, and $\Gamma = 0$. (a) Close-to-threshold stationary state for $P = 1.1$; (b) modulated pattern for $P = 1.96$; (c) and (d) optical vortex-mediated turbulence for $P = 4$. Panel (d) shows a 3D representation of the turbulent intensity, displaying a very large intensity peak.

to high intensity peaks (see figure 1.20(d)). Note that the intensity peaks appear at random positions in the transverse plane during the spatiotemporal evolution.

From figure 1.20(c) and (d) it is clear that in our regime of optical turbulence intensity peaks appear in the middle of regions where the intensity is very low or even zero. In order to clarify the nature of this optical turbulence, we report in figure 1.21 a turbulent regime obtained through the numerical integration of equation (1.5) and for smaller values of the detuning ω and injection E_{IN} with respect to the results displayed in figure 1.20. In this case ($\omega = 0.53$, $E_{IN} = 0.6$, and $P = 4$) there is bistability between the steady spatially modulated ring structure of figure 1.21(a) and the turbulent regime. By locally perturbing the ring structure, turbulence sets in as shown in figure 1.21(b)–(d) and remains asymptotically. In this regime, we can clearly identify points of zero intensity where lines of zero real and zero imaginary parts of the field E cross each other. These zero-intensity points are optical vortices [35] and we observe that they appear in couples of opposite topological charge (see the four vortices on the right of figure 1.21(b), the first couple being at the top and

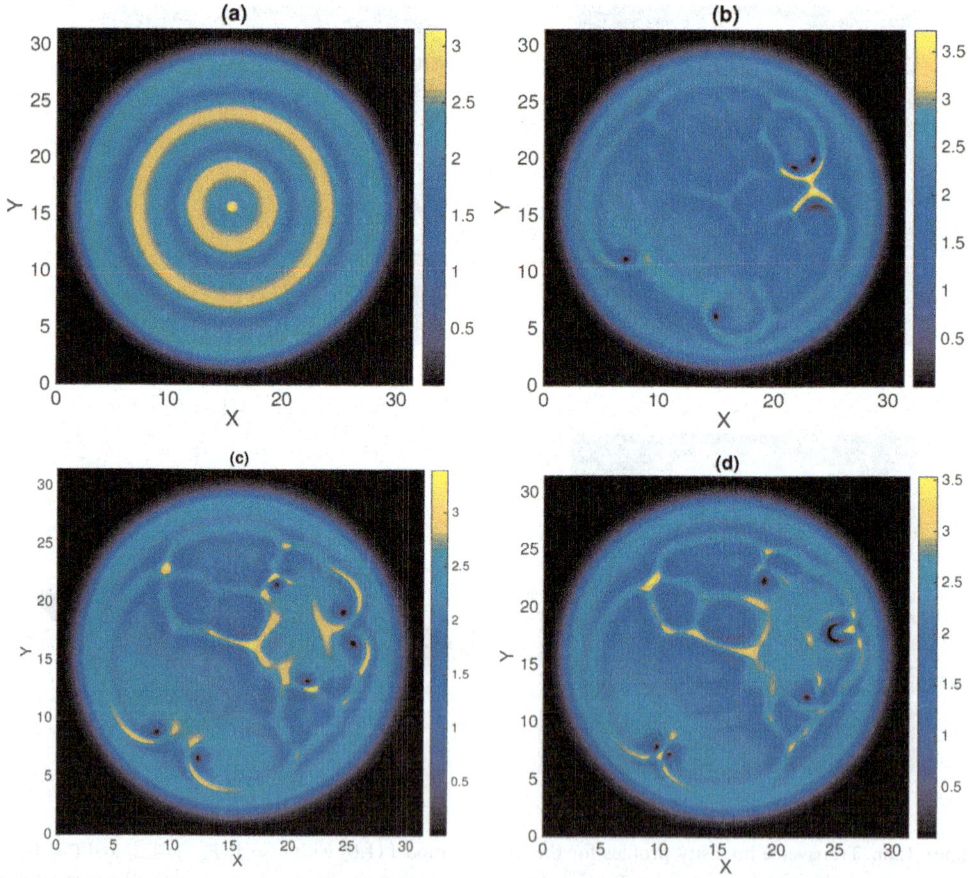

Figure 1.21. Transverse intensity profiles for equation (1.5) with $\omega = 0.53$, $E_{IN} = 0.6$, and $P = 4$. (a) Periodically modulated steady ring structure. (b)–(d) Spatiotemporal snapshots of optical vortex-mediated turbulence.

the second at the bottom), and they also disappear in collisions of oppositely charged pairs (see events in figure 1.21(d) on both right and left).

The turbulent states that we observe in figure 1.20 and 1.21 appear to be examples of defect-mediated turbulence (DMT), a configuration first described by Coullet, Gil and Lega in 1989 [36] in the complex Ginzburg–Landau (CGL) equation

$$\frac{\partial E}{\partial t} = (P - 1 + i\omega)\, E - (1 + i\alpha)|E|^2\, E + (1 + i\beta)\nabla^2 E, \tag{1.7}$$

where the control parameters are the pump P and the detuning ω as well as α and β that regulate the ratio between dispersive and absorptive nonlinearity and between diffraction and diffusion, respectively. In particular, Coullet, Gil, and Lega describe the onset of DMT when the condition $(1 + \alpha\beta) < 0$ is met and the defects are pointwise vortices. The CGL equation looks quite similar to our equation (1.5) in the absence of injection and of Swift–Hohenberg terms. However, since it is not possible to properly define the coefficients α and β in equation (1.5), the condition for the occurrence of DMT cannot be met in equation (1.5) in the absence of injection. This is probably why DMT has never been observed experimentally in nonlinear optics and, in particular, in lasers without injection.

One interesting question that we would like to address in this article is does the external forcing due to the optical injection help or hinder DMT? In the literature, there are contradictory statements. On one side it is clear that stationary optical vortices are prevented from forming by the presence of the external forcing E_{IN} as it breaks the phase invariance $\exp(i\phi)$ of the CGL equation (1.7) [37]. On the other side, Chaté, Pikovsky and Rudzick [38] demonstrated that DMT survives when perturbing the CGL equation with the term E_{IN} in a regime where the condition $(1 + \alpha\beta) < 0$ is satisfied. The cases that we have presented in figures 1.20 and 1.21, however, do not satisfy the condition of reference [38] because our unperturbed laser equations never satisfy the phase-to-amplitude instability condition $(1 + \alpha\beta) < 0$. The original question therefore remains unanswered for the laser and SROPO cases. To answer it we need to investigate in more detail the instabilities that lead to the appearance of the turbulent regime displayed in figures 1.20(c) and (d) and 1.21(b)–(d).

As shown in [27], spatially modulated stationary structures such as those displayed in figures 1.20(b) and 1.21(a) correspond to phase-bound solutions since the variations of the phase of the field E when moving across the spatial variable are limited to a finite arc. In the case of equation (1.5), the arc is very close to the limit cycle solution of the equation [30, 39]

$$\begin{gathered} E_{LL}(t) = A_0[\cos(\Omega t + \pi) + i\sin(\Omega t + \pi)], \\ \Omega = \sqrt{\omega^2 - \omega_L^2}; \qquad \omega_L = E_{IN}/A_0; \qquad A_0 = \sqrt{\frac{3(P-1)}{P}}. \end{gathered} \tag{1.8}$$

Note that when Ω is real, i.e. in the absence of locked states, the trajectory (1.8) is the phase-drift solution of the Adler equation $d_t\phi = \omega - \omega_L \sin(\phi(t))$ [40]. When the spatially modulated stationary states become unstable, we first observe a phase instability where the new field amplitudes remain close to the limit cycle (1.8) but the phase spreads around the full circle. After this phase spreading, we observe an amplitude instability with large fluctuations away from the limit cycle $E_{LL}(t)$ (1.8). These fluctuations progressively approach points of zero-intensity where both the real and imaginary parts of the field E are simultaneously zero. It is exactly at these points that we observe the nucleation of pairs of vortex defects (see for example figure 1.21(b)). These vortices then move around the transverse space and annihilate vortices of opposite topological charge. Once the amplitude instability driven by E_{IN} takes place, the nucleation and annihilation of optical vortices keeps happening and the DMT regime sets in asymptotically.

We have now established that regimes of DMT are commonplace in our equations (1.5) and (1.6). This is somewhat surprising since we have external forcing and cannot meet the condition $(1 + \alpha\beta) < 0$ in the unperturbed systems. It is important to note that it is exactly the external forcing that induces the first phase and then amplitude instabilities for the generation and stabilization of DMT. A few important considerations are in order: (1) our DMT is not related to laser relaxation oscillations because we have considered class A instead of class B lasers. For the effect of relaxation oscillations on RWs in forced laser systems, see sections 1.2 and 1.3. Of course, when studying class B lasers with injection in 2D, our DMT may resonate with relaxation oscillations and lead to wild dynamical behavior. (2) Although DMT can have effects on the dynamics of 1D systems [39], full interaction, nucleation and annihilation of vortex couples can only be properly described in systems with 2D. (3) This mechanism of vortex turbulence is at the base of the generation of RWs in the 2D externally driven systems described by the spatiotemporal dynamics of equations (1.5) and (1.6), as demonstrated below.

Systems (1.5) and (1.6) are outside thermodynamic equilibrium, do not display relaxation oscillations, and present a delicate balance between the energy input and the losses (i.e. between the pump P, injection E_{IN}, nonlinearity, and output mirror losses). For example, during the turbulent evolution, we observe that in spite of large intensity variations locally, the total power displays only small fluctuations around an averaged value, as seen in figure 1.22. In the cases displayed in figure 1.22, the power fluctuations have a standard deviation of only 8×10^{-3} in the laser case and 1.2×10^{-2} in the SROPO case. Note also that in the laser case the average power is very close to an analytical estimate of $\pi w_0^2 A_0^2$ where w_0 is the beam width of the input pump and A_0 is given in (1.8). The fact that the turbulent state maintains an almost constant power in the presence of moving vortices of zero intensity implies the simultaneous appearance of large amplitude spikes. If the vortex density is large, multi-vortex collisions can occur with the production of large, short-lived spikes in the intensity (see figures 1.20(c) and (d)). Short-lived large intensity spikes are rare but possible events, fitting the characteristics of RWs. The particular shape and symmetry of these spikes is crucially determined by the number and position of the surrounding vortices, a feature that is unique to our particular mechanism of RW formation. Note that without spatial coupling due to diffraction, no RWs can be observed in systems described by equations (1.6) and (1.5). To characterize our spatiotemporal RWs, we use the well-known PDF of the intensity peaks [41] and focus on the shape of the rare event tail. A well-defined change in the curvature of the PDF leading to long statistical tails is an essential and clear-cut signature of the

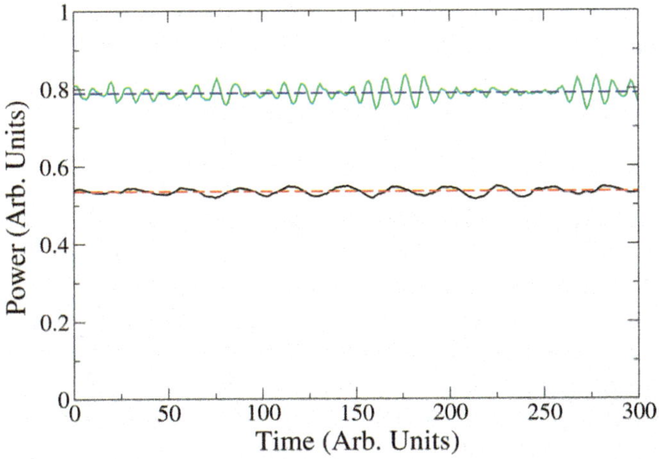

Figure 1.22. Time evolution of the transverse power for equation (1.5) (lower black curve with $\omega = 0.53$, $E_{IN} = 0.6$, $P = 4$ and $\Gamma = 0$) and equation (1.6) (upper green curve with $\omega = 1$, $E_{IN} = 1.2$, $P = 4$ and $\Gamma = 0$) during vortex turbulence. The horizontal dashed lines are the corresponding averaged values of the power.

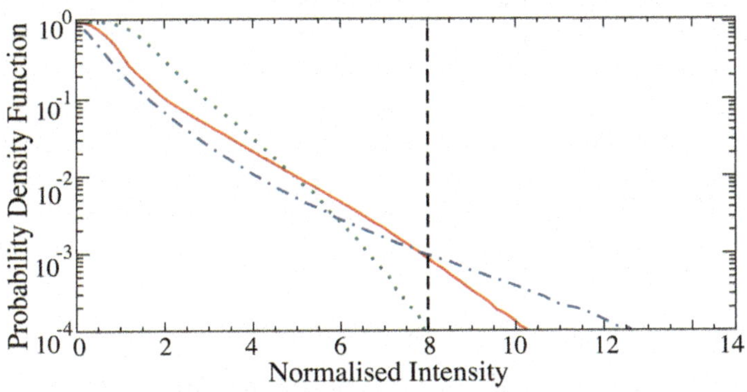

Figure 1.23. Intensity PDFs for equation (1.5) with $P = 2$, $\omega = 0.3$, $E_{IN} = 0.24$, $\Gamma = 0$ (green dotted curve) and with $P = 8$, $\omega = 2.4$, $E_{IN} = 3.40$, $\Gamma = 0$ (red solid curve) and for equation (1.6) with $P = 8$, $\omega = 1$, $E_{IN} = 1.48$, $\Gamma = 0$ (dot-dashed blue curve). The vertical black line is the standard threshold for defining waves as an extreme event [41].

presence of RWs [41]. In figure 1.23 we display the progressive change in shape of the PDF when moving to higher values of pump P, injection E_{IN}, and detuning ω. These changes are due to an increased density of vortices in the regimes of DMT. At lower values of energy input (lower values of P and E_{IN}) (see the green dotted curve in figure 1.23), the wave statistics are well-approximated by a Gaussian fit and the tail is not suitable to describe rare events with extremely large intensity peaks. At higher energy inputs (see the red solid and the dot-dashed blue curves in figure 1.23) a clear change in the shape of the PDF becomes apparent with the appearance of long non-Gaussian tails that correspond to the generation of RWs. In the case of the dot-dashed blue curve in figure 1.23, the statistics is very well approximated by a Weibull distribution [42] given by

$$f(x) = A\left(\frac{x}{b}\right)^{c-1} \exp\left(-\frac{x}{b}\right)^{c}, \tag{1.9}$$

as shown in [29].

Note that non-Gaussian PDFs cannot be replicated by superpositions of random waves typical of optical speckles [43]. It has to be stressed that in regimes of DMT, the turbulent dynamics of vortices are deterministically driven by the nonlinearity. We also note that RWs in the vortex turbulence demonstrated in figure 1.23 are different from those due to vorticity in models of inviscid fluids [44]. Equations (1.5) and (1.6) have been shown to be equivalent to the flow of a compressible and viscous fluid with density $\rho = |E|^2$ and velocity $\mathbf{v} = \nabla\phi$, where ϕ is the phase of the field [45]. In the case of our forced systems, $\nabla \times \mathbf{v}$ remains extremely close to zero in the locations where RWs are observed. We conclude that our RWs are due to the interaction of free vortices in the absence of vorticity.

In conclusion, we have shown a mechanism for producing RWs in the 2D transverse area of externally driven nonlinear optical devices via vortex turbulence. Given the universality of our models, this mechanism should be observable in a large variety of systems. Models of lasers with injected signal, where the invariance of the Adler limit cycle is well known [34, 30], can easily be extended to semiconductor media and to class B lasers, thus including the largest majority of solid-state lasers. Outside optics, vortex-mediated turbulence without driving has been observed in nematic liquid crystals [46], chemical reactions [47], and fluid dynamics [49]. In the unlocked regime of these systems with driving, vortex turbulence can excite RWs and lead to the formation of highly inhomogeneous fields with non-Gaussian statistics.

1.5 Conclusion

In this chapter, we have reviewed recent results on the statistical properties of forced oscillatory media with different numbers of spatial dimensions. Although the theoretical models differ in their exact form, they are essentially equivalent and an almost direct comparison can be performed, where the impact of spatial dimensionality can be at least partly addressed. In all three cases, heavy-tailed statistics have been observed in some specific parametric conditions and dynamical regimes. In the 0-dimensional case, noise can be a cause for rogue waves but purely deterministic dynamics can also be sufficient. The impact of noise has not been analyzed yet in the 1D and 2D cases but, as in the 0D case, rogue waves can be observed as a fully deterministic phenomenon. In the 0D case, the vicinity of an external crisis corresponding to the collision of a chaotic attractor with the stable manifold of the saddle-focus point corresponding to the laser 'low intensity' state can be at the origin of rogue events. The corresponding trajectory has also been observed experimentally, and some predictions can be realized due to the strongly dominant role of determinism in the dynamics. In the 1D system, two types of rogue events have been detected in real time and analyzed in a spatiotemporal diagram. The first type occurs in a disordered regime where the phase of the laser relative to the forcing evolves erratically but in a mostly bounded way except in some rare occurrences when the electric field completes a full 2π rotation around the locked point, implying the existence of a defect. The experimental observations clearly show a focusing phenomenon taking place over several roundtrips during the formation of a rogue event which can be interpreted as an unstable phase soliton, a feature which is obviously absent in the 0D case. Although a crisis of attractors would be very difficult to identify in this partial differential equations model, it is notable that the emergence of extreme events is related to the system approaching the saddle point corresponding to the laser 'low-intensity' state, suggesting that this point plays a similar role to the one it plays in the 0D case. The second type of extreme events, in contrast, is essentially related to the presence of spatial degrees of freedom since large statistical deviations occur when (otherwise stable) localized wave packets partially destructively collide with transient pulses. In the 2D case, rogue waves are found in a regime of defect mediated turbulence. In this case, destructive collisions between vortices of opposite charge are at the origin of the extreme events and they cannot be found in 0D or 1D systems.

Acknowledgments

CM was supported in part by the Spanish Ministerio de Ciencia e Innovación through project FIS2015-66503-C3-2-P (MINECO/FEDER) and the ICREA ACADEMIA program, Generalitat de Catalunya, Spain. The collaboration between the Nice and Strathclyde groups has been supported by the CNRS, the Royal Society (London), and the University of Strathclyde.

References

[1] Solli D R, Ropers C, Koonath P and Jalali B 2007 *Nature* **450** 1054–7

[2] Akhmediev N, Kibler B, Baronio F, Belić M, Zhong W-P, Zhang Y, Chang W, Soto-Crespo J M, Vouzas P and Grelu P *et al* 2016 *J. Opt.* **18** 063001

[3] Akhmediev N, Ankiewicz A and Soto-Crespo J M 2009 *Phys. Rev.* E **80** 026601

[4] Kibler B, Fatome J, Finot C, Millot G, Dias F, Genty G, Akhmediev N and Dudley J M 2010 *Nat. Phys.* **6** 790–5

[5] Kedziora D J, Ankiewicz A and Akhmediev N 2013 *Phys. Rev.* E **88** 013207

[6] Suret P, El Koussaifi R, Tikan A, Evain C, Randoux S, Szwaj C and Bielawski S 2016 *Nat. Commun.* **7** 13136

[7] Lecaplain C, Grelu P, Soto-Crespo J M and Akhmediev N 2012 *Phys. Rev. Lett.* **108** 233901

[8] Rimoldi C, Barland S, Prati F and Tissoni G 2017 *Phys. Rev.* A **95** 023841

[9] Coulibaly S, Clerc M G, Selmi F and Barbay S 2017 *Phys. Rev.* A **95** 023816

[10] Pierangeli D, Di Mei F, Conti C, Agranat A J and DelRe E 2015 *Phys. Rev. Lett.* **115** 093901

[11] Pierangeli D, Di Mei F, Di Domenico G, Agranat A J, Conti C and DelRe E 2016 *Phys. Rev. Lett.* **117** 183902

[12] Mussot A, Kudlinski A, Kolobov M, Louvergneaux E, Douay M and Taki M 2009 *Opt. Express* **17** 17010–5

[13] Dudley J M, Genty G, Dias, Kibler B and Akhmediev N 2009 *Opt. Express* **17** 21497–508
[14] Coullet P, Gil L and Lega J 1989 *Phys. Rev. Lett.* **62** 1619–22
[15] Longhi S 1998 *Quantum Semiclass. Opt.* **10** 617–35
[16] Lega J, Moloney J V and Newell A C 1994 *Phys. Rev. Lett.* **73** 2978–81
[17] Oppo G L, Politi A, Lippi G L and Arecchi F T 1986 *Phys. Rev.* A **34** 4000–7
[18] Coullet P, Gil L and Rocca F 1989 *Opt. Commun.* **73** 403–8
[19] Wieczorek S, Krauskopf B, Simpson T B and Lenstra D 2005 *Phys. Rep.* **216** 1–128
[20] Bonatto C, Feyereisen M, Barland S, Giudici M, Masoller C, Rios Leite J R and Tredicce J R 2011 *Phys. Rev. Lett.* **107** 053901
[21] Zamora-Munt J, Garbin B, Barland S, Giudici M, Rios Leite J R, Masoller C and Tredicce J R 2013 *Phys. Rev.* A **87** 035802
[22] Gustave F, Columbo L, Tissoni G, Brambilla M, Prati F, Kelleher B, Tykalewicz B and Barland S 2015 *Phys. Rev. Lett.* **115** 043902
[23] Gustave F, Columbo L, Tissoni G, Brambilla M, Prati F and Barland S 2016 *Phys. Rev.* A **93** 063824
[24] Gustave F 2016 Phase dynamics and dissipative solitons in semiconductor lasers *PhD Thesis* Université de Nice Sophia Antipolis
[25] Gustave F, Rimoldi C, Walczak P, Columbo L, Brambilla M, Prati F, Tissoni G and Barland S 2017 *Eur. Phys. J.* D **71** 154
[26] Gibson C J, Yao A M and Oppo G L 2016 *Phys. Rev. Lett.* **116** 043903
[27] Longhi S 1998 *J. Eur. Opt. Soc.* B **10** 617–35
[28] Selmi F, Coulibaly S, Loghmari Z, Sagnes I, Beaudoin G, Clerc M G and Barbay S 2016 *Phys. Rev. Lett.* **116** 013901
[29] Eslami M, Khanmohammadi M, Oppo G L and Kheradmand R 2017 *Phys. Rev.* A **96** 033836
[30] Mayol C, Toral R, Mirasso C R and Natiello M A 2002 *Phys. Rev.* A **66** 013808
[31] Oppo G-L, Yao A M and Cuozzo D 2013 *Phys. Rev.* A **88** 043813
[32] Lega J, Moloney J V and Newell A C 1994 *Phys. Rev. Lett.* **73** 2978
Sanchez-Morcillo V *et al* 1997 *Phys. Rev.* A 52 3237
Oppo G-L, Yao A M, Prati F and de Valcarcel G J 2009 *Phys. Rev.* A 79 033824
[33] D'Alessandro G and Oppo G L 1992 *Opt. Commun.* **88** 130
[34] Oppo G L, Politi A, Lippi G L and Arecchi F T 1994 *Phys. Rev.* A **34** 4000
Solari H and Oppo G L 1994 *Opt. Commun.* **111** 173
[35] Coullet P, Gil L and Rocca F 1989 *Opt. Commun.* **73** 403
Gil L, Emilsson K and Oppo G L 1992 *Phys. Rev.* A **45** R567
[36] Coullet P, Gil L and Lega J 1989 *Phys. Rev. Lett.* **62** 1619
Aranson I S and Kramer L 2002 *Rev. Mod. Phys.* **74** 99
[37] Coullet P and Emilsson K 1992 *Physica* A **188** 190
Coullet P and Emilsson K 1992 *Physica* D **61** 119
[38] Chaté H, Pikovsky A and Rudzick O 1999 *Physica* D **131** 17
[39] Erneux T and Glorieux P 2010 *Laser Dynamics* (Cambridge: Cambridge University Press)
[40] Adler R 1946 *Proc. IRE* **34** 351
[41] Dudley J M, Dias F, Erkintalo M and Genty G 2014 *Nat. Photonics* **8** 755
[42] Onorato M, Residori S, Bortolozzo U, Montina A and Arecchi F T 2013 *Phys. Rep.* **528** 47
[43] Shvartsman N and Freund I 1994 *Phys. Rev. Lett.* **72** 1008
[44] Abrashkin A and Soloviev A 2013 *Phys. Rev. Lett.* **110** 014501

[45] Brambilla M *et al* 1994 *Phys. Rev.* A **43** 5090
Brambilla M *et al* 1994 *Phys. Rev.* A **43** 5114
[46] Firsch T and Gilli J M 1995 *J. Physique II* **5** 561
[47] Ouyang Q and Flesselles J M 1996 *Nature* **379** 143
[48] Bodenschatz E, Pesch W and Ahlers G 2000 *Annu. Rev. Fluid Mech.* **32** 709
[49] Rimoldi C, Gustave F, Columbo L, Brambilla M, Barland S, Prati F and Tissoni G 2017 *Opt. Express* **25** 22017–31
[50] Walczak P, Rimoldi C, Gustave F, Columbo L, Brambilla M, Prati F, Tissoni G and Barland S 2017 *Opt. Lett.* **42** 3000–3

Chapter 2

Extreme waves in stimulated backscattering and frequency conversion processes

Fabio Baronio, Shihua Chen, Philippe Grelu, Jose M Soto-Crespo and Stefan Wabnitz

We present the extreme rogue wave dynamics that originate from the three-wave resonant interaction equations that describe the transient stimulated scattering processes in cubic media, and the parametric frequency-conversion processes in phase-matched weakly-dispersive quadratic media. Explicit analytical fundamental and second-order rogue wave solutions are presented and characterized. The link between the existence of rogue wave solutions and the presence of baseband modulation instability is highlighted. We also present numerical simulations that confirm the robustness and possible manifestation of these rogue waves for various parameters and initial conditions.

2.1 Introduction

A three-wave resonant interaction (TWRI) is ubiquitous in various branches of science, as it describes the interaction of three waves with different frequencies propagating in a nonlinear and dispersive medium [1]. In fact, TWRI occurs whenever the nonlinear waves can be considered as a first-order perturbation to the linear solutions of the propagation equation.

TWRI has been extensively studied in nonlinear optics, since it applies to transient stimulated Raman scattering, stimulated Brillouin scattering, parametric amplification, and frequency conversion [2–9]. In the context of plasma physics, TWRI also describes laser–plasma interactions, radio frequency heating, and plasma instabilities [10, 11]. Other important domains of application of TWRI are light–acoustic interactions, interactions of water waves, and wave–wave scattering in solid-state physics [12–14].

In the early 1970s, the integrability of the TWRI model was established, and soliton solutions were identified. The first type of TWRI solitons describes the mixing of three

doi:10.1088/978-0-7503-1460-2ch2

© IOP Publishing Ltd 2017

pulses that travel with their respective linear group velocities and interact for just a short time [15]. The second type of solitons, also known as simultons, results from the mutual trapping of pulse envelopes at three different frequencies [16]; hence the three wave packets travel, or vividly, are locked together, with a common group velocity. In the 2010s, such TWRI solitons were experimentally demonstrated in optics [17–19]. In all of the above-discussed domains of application, parametric TWRI solitons play a pivotal role because of their particle-like behavior, which enables the coherent energy transport and processing of short wave packets.

TWRIs are classified according to the group velocities of the waves and the signs of the wave energies [1]. Interactions of different types, as determined by these parameters, display very different behaviors. Depending on these parameters, TWRI may feature stimulated backscattering (SB) dynamics or soliton exchange (SE) dynamics. TWRI may also exhibit an explosive behavior in that the coupled waves may develop a singularity, or collapse, in a finite time.

Interestingly, it has been recently pointed out that, besides velocity-locked traveling solitons, the TWRI model also admits families of spatiotemporally localized solutions on a finite background [20]. Overlooked in the last decades, solitons on finite background have recently attracted significant interest as their localization dynamics have been considered for the description of extreme amplitude optical rogue waves (see [21–23] and references therein).

From a fundamental point of view, rogue waves are deterministic since they obey nonlinear partial-differential equations, and their appearance results from the evolution of a given dynamical system from well-defined initial conditions [24]. However, in most experimental optical systems [25, 26], the manifestation of rogue waves remains inherently unpredictable, up to the point where rogue waves are deemed to appear from nowhere and disappear without a trace [21, 27]. This peculiar feature comes from a large sensitivity of rogue wave evolution with respect to their initial excitation conditions. Such sensitivity is typically exacerbated in any realistic environment, where rogue waves can be triggered from noise among chaotic wave fields. To understand the span of rogue wave manifestations in a natural environment with random initial conditions, higher-order rogue waves should also be considered.

In this chapter, we show the characteristics of TWRI rogue waves that arise in the SB and SE processes [20, 28, 29]. Explicit fundamental and second-order rogue wave solutions are presented, suggesting that markedly different rogue wave behaviors are actually allowed by the different types of mixing processes. Rogue wave dynamics in the degenerate SB–SE situation are also investigated. The link between the existence of rogue wave solutions and the presence of baseband modulation instability (MI) is highlighted. Furthermore, we provide numerical evidences for the robustness of the considered rogue wave patterns.

2.2 Fundamental rogue wave solutions

The TWRI model that governs the propagation of three coupled waves, perfectly phase-matched in a weakly dispersive nonlinear medium, can be written in dimensionless form [1, 20]

$$u_{1t} + V_1 u_{1x} = u_2^* u_3^*,$$
$$u_{2t} + V_2 u_{2x} = - u_1^* u_3^*, \tag{2.1}$$
$$u_{3t} + V_3 u_{3x} = u_1^* u_2^*,$$

where $u_n(x,t)$ $(n = 1,2,3)$ are the slowly varying complex envelopes of the three fields. From a physical standpoint, these fields can model the stimulated Brillouin scattering (transient stimulated Raman scattering) process where the optical pump wave scatters off a material acoustic (optical) phonon wave to form the Stokes wave [4, 9], or the fields may denote signal, pump, and idler optical waves in the parametric mixing process that occurs in a quadratic medium [5–8]. We denote the time and space variables by t and x. Here, time t is assumed to be the evolution variable. Subscripts x ant t stand for partial derivatives, and the asterisk denotes the complex conjugation. The coefficients V_n correspond to the relative velocities of the three waves and we suppose $V_{1,2} > V_3$. Without loss of generality, we set $V_3 = 0$, which implies that equations (2.1) are written in a reference frame co-moving with u_3. The above choice of signs before the quadratic terms is indicative of the nonexplosive character of the interaction [1]. Basically, as $V_1 < V_2$, the interaction features the SB property. On the other hand, the condition $V_1 > V_2$ corresponds to the SE process. We will show that in all the situations, equations (2.1) admit exact rogue wave solutions because of their complete integrability and the nonexplosive property. The degenerate case corresponds to the condition $V_1 = V_2$.

Considering the resonant conditions for the frequencies and momenta, the plane-waves satisfying equations (2.1) can be expressed as

$$u_{10}(x,t) = a_1 \exp\left[-i(k_1 x - \omega_1 t)\right],$$
$$u_{20}(x,t) = a_2 \exp\left[i(k_2 x - \omega_2 t)\right], \tag{2.2}$$
$$u_{30}(x,t) = i a_3 \exp\left[i(k_1 - k_2)x - i(\omega_1 - \omega_2)t\right],$$

where

$$k_1 = \frac{\omega_1}{V_1} + \frac{a_2^2}{\delta V_1}, \quad k_2 = \frac{\omega_2}{V_2} + \frac{a_1^2}{\delta V_2}, \quad a_3 = \frac{a_1 a_2}{\delta}, \tag{2.3}$$

with $a_n(>0)$ being the background amplitudes, k_n the wavenumbers, and ω_n the frequencies. For convenience, we introduce $A = \Gamma_1 a_1^2 + \Gamma_2 a_2^2$, $B = \Gamma_1 a_1^2 - \Gamma_2 a_2^2$, $\Gamma_j = V_j/(V_1 - V_2)$ $(j = 1, 2)$, $\kappa = \omega_1 + \omega_2$, and $\delta = \omega_1 - \omega_2$. These coupled background fields are intrinsically unstable [30], and as a result, a tiny localized deformation may entail MI or trigger the generation of rogue waves.

By exploiting the standard Darboux transformation procedure [31, 32], we obtain the fundamental (first-order) rogue wave solutions of equations (2.1)

$$u_1^{[1]} = u_{10}\left[1 + \frac{3\Lambda\xi\theta_1^*/\alpha_1^*}{|\xi|^2 + \Gamma_1 a_1^2|\theta_1/\alpha_1|^2 + \Gamma_2 a_2^2|\theta_2/\alpha_2|^2}\right],$$
$$u_2^{[1]} = u_{20}\left[1 - \frac{3\Lambda\xi^*\theta_2/\alpha_2}{|\xi|^2 + \Gamma_1 a_1^2|\theta_1/\alpha_1|^2 + \Gamma_2 a_2^2|\theta_2/\alpha_2|^2}\right], \tag{2.4}$$
$$u_3^{[1]} = u_{30}\left[1 + \frac{3\delta\Lambda\theta_1\theta_2^*/(\alpha_1\alpha_2^*)}{|\xi|^2 + \Gamma_1 a_1^2|\theta_1/\alpha_1|^2 + \Gamma_2 a_2^2|\theta_2/\alpha_2|^2}\right],$$

where $\Lambda = \lambda_0 - \lambda_0^* \equiv 2i\mathrm{Im}(\lambda_0)$, $\xi = t + \beta_2 x$, $\alpha_j = \mu_0 + \lambda_0 - (-1)^j\delta/2$, and $\theta_j = \xi - i/\alpha_j$ $(j = 1, 2)$. Here and for later use, we define the parameters $\beta_n = \frac{1}{V_1 - V_2}\left(\frac{a_1^2}{\alpha_1^n} + \frac{a_2^2}{\alpha_2^n}\right)$, $(n = 2,3,4)$.

The specific spectral parameter λ_0 in equations (2.4) is the complex root (with a nonzero imaginary part) of the discriminant condition

$$\Delta = \sigma^3 - \rho^2 = 0, \tag{2.5}$$

under which the cubic equation

$$\mu_0^3 - 3\sigma\mu_0 + 2\rho = 0, \tag{2.6}$$

will have a double root μ_0. Here

$$\sigma = \lambda_0^2 + \frac{\delta^2}{12} + \frac{A}{3}, \quad \rho = \frac{1}{4}(\delta^2 - 2A)\lambda_0 - \lambda_0^3 + \frac{\delta B}{4}. \tag{2.7}$$

We point out that equations (2.4) apply to either the SB process or the SE process. Generally, one can solve equation (2.5) analytically for λ_0, and then solve exactly the cubic equation (2.6) for μ_0. Substitution of the values of λ_0 and μ_0 into equations (2.4) results in the fundamental rogue wave dynamics. We find that the rogue wave solutions for $V_1 < V_2$ (the SB case) only exist in the limited regime where

$$|\delta| \leqslant \left\{\frac{3}{2}(A^2 - B^2)^{1/3}[(-A - B)^{1/3} + (-A + B)^{1/3}] - A\right\}^{1/2}, \tag{2.8}$$

while rogue wave solutions for $V_1 > V_2$ (the SE case) exist without any parametric constraints. We emphasize that the vector Manakov system with normal dispersion, the Fokas–Lenells equation, and the long-wave–short-wave resonance, also entail a parametric condition similar to equation (2.8) for the existence of rogue waves [33, 34].

We confine our discussions within the simple parametric condition $B = 0$, for the clarity of basic properties, which implies

$$\frac{a_2}{a_1} = \sqrt{V_1/V_2}. \tag{2.9}$$

Under these circumstances, it follows from equations (2.5) and (2.6) that

$$\lambda_0 = \frac{i}{4}\eta + \frac{i(4A + \delta^2)}{12\eta}, \tag{2.10}$$

$$\mu_0 = \frac{i}{4}\eta - \frac{i(4A + \delta^2)}{12\eta}, \tag{2.11}$$

where the parameter η is given by

$$\eta = \pm\left[2\sqrt{A(A - 2\delta^2)} + 2A - \delta^2\right]^{1/2}. \tag{2.12}$$

It is easily seen that, as $V_1 > V_2$ (noting that $A > 0$), the value of η will be real for $|\delta| \leqslant \sqrt{A/2} \equiv \delta_h$, but it becomes complex when $|\delta| > \delta_h$. However, as $V_1 < V_2$ (which means $A < 0$), η will always be real in the regime $|\delta| \leqslant 2\sqrt{-A} \equiv \delta_m$. We can see that as $|\delta| > \delta_m$, the rogue wave solutions will be forbidden in the latter case, according to the parameter condition (2.8). Besides, we note that solutions (2.4) involve structures that are independent of the parameter κ. For this reason, we always assume $\kappa = 0$ from here on.

Therefore, by use of equations (2.10) and (2.11), the fundamental rogue wave solutions (2.4) can be reduced to their explicit forms. Specifically, in the SB situation or in the SE situation with $|\delta| \leqslant \delta_h$, rogue wave solutions can be expressed as

$$\begin{aligned}
\frac{u_1^{[1]}}{u_{10}} &= 1 + \frac{[i\delta(t - \chi x) - i\varrho x - 1]\ell + 4i\nu\delta^2 x}{\delta^2[(t - \chi x)^2 + \varepsilon^2\nu^2 x^2 + 1/\eta^2]},\\
\frac{u_2^{[1]}}{u_{20}} &= 1 + \frac{[i\delta(t - \chi x) + i\varrho x - 1]\ell - 4i\nu\delta^2 x}{\delta^2[(t - \chi x)^2 + \varepsilon^2\nu^2 x^2 + 1/\eta^2]},\\
\frac{u_3^{[1]}}{u_{30}} &= 1 - \frac{2i\delta(t - \chi x)\ell + 2(\ell - \delta^2/A)}{\delta^2[(t - \chi x)^2 + \varepsilon^2\nu^2 x^2 + 1/\eta^2]},
\end{aligned} \tag{2.13}$$

where $\varepsilon = \eta - \frac{4A+\delta^2}{\eta}$, $\chi = \frac{V_1 + V_2}{2V_1V_2}$, $\nu = \frac{V_1 - V_2}{4\delta V_1V_2}$, $\varrho = 2(2A + \delta^2)\nu$, and

$$\ell = \begin{cases} 1 + \sqrt{1 - 2\delta^2/A}, & \text{SB},\\ 1 - \sqrt{1 - 2\delta^2/A}, & \text{SE and } |\delta| \leqslant \delta_h. \end{cases} \tag{2.14}$$

Otherwise, for the SE case with a large frequency difference $|\delta| \geqslant \delta_h$, the solutions take the form

$$\begin{aligned}
\frac{u_1^{[1]}}{u_{10}} &= 1 + \frac{ie_2[\tau - 2(m + \delta)(\imath + 1)\nu x] - e_1\imath\delta/A}{\delta^2\tau^2 + 4h^2(\delta^2 - m^2)\nu^2 x^2 + \delta^2/h^2},\\
\frac{u_2^{[1]}}{u_{20}} &= 1 + \frac{ie_1[\tau - 2(m - \delta)(\imath + 1)\nu x] - e_2\imath\delta/A}{\delta^2\tau^2 + 4h^2(\delta^2 - m^2)\nu^2 x^2 + \delta^2/h^2},\\
\frac{u_3^{[1]}}{u_{30}} &= 1 - \frac{2i\delta[\tau - 2m\nu x(\imath - 1/\imath)] + 2\delta^2/A}{\delta^2\tau^2 + 4h^2(\delta^2 - m^2)\nu^2 x^2 + \delta^2/h^2},
\end{aligned} \tag{2.15}$$

where $\tau = t - (\chi - 2\imath m\nu)x$, $e_1 = \delta + m/\imath$, $e_2 = \delta - m/\imath$, $\imath = 1/2 + \sqrt{1/4 + A/\delta^2}$, and $h = \sqrt{A(1 + 1/\imath)}$ and $m = \pm\sqrt{\imath\delta^2 - A}$ are the real and imaginary parts of η, respectively. In the following, we will take the positive sign for m, unless otherwise stated.

We need to emphasize that the above analytic solutions (2.13) and (2.15) have been expressed as a ratio of second-order polynomials and they have been translated along the t-axis, so that their center is exactly on the origin. One can find directly from equations (2.13) that the relative amplitudes of the three rogue wave components at the origin are given by

$$E_1 = E_2 = \left| 1 - \frac{\eta^2\ell}{\delta^2} \right|, \qquad E_3 = \left| 1 - \frac{2\eta^2\ell}{\delta^2} + \frac{2\eta^2}{A} \right|, \tag{2.16}$$

where we define $E_n = \frac{1}{a_n}|u_n^{[1]}(0, 0)|$, $(n = 1, 2, 3)$. In the same way, we directly obtain from equations (2.15) the set of relative central amplitudes

$$E_j = \left| 1 - \frac{e_j\imath h^2}{\delta A} \right| \; (j = 1, 2), \qquad E_3 = \left| 1 - \frac{2h^2}{A} \right|. \tag{2.17}$$

Obviously, the central amplitudes given by equations (2.16) and (2.17) depend on the frequency difference δ, as seen in figure 2.1, where we assume $B = 0$ for both the SB and SE processes. In these plots, the values of δ in the abscissa axis are expressed in terms of δ_m (see panel (a)) or δ_h (see panel (b)), and hence the shown characteristic evolution curves do not depend on the specific values of amplitudes and group velocities, provided that the condition (2.9) is met. For the sake of brevity, here and in what follows, we only take the positive domain $\delta > 0$ (i.e. $\omega_1 > \omega_2$) into account. The problem in the negative domain $\delta < 0$ can be treated along similar lines. As one might see from figure 2.1(a), in the SB situation the rogue wave components u_1 and u_2 have a vanishing central amplitude at $\delta = \delta_a$, and on the other hand, the third wave amplitude u_3 vanishes at $\delta = \delta_b$, where

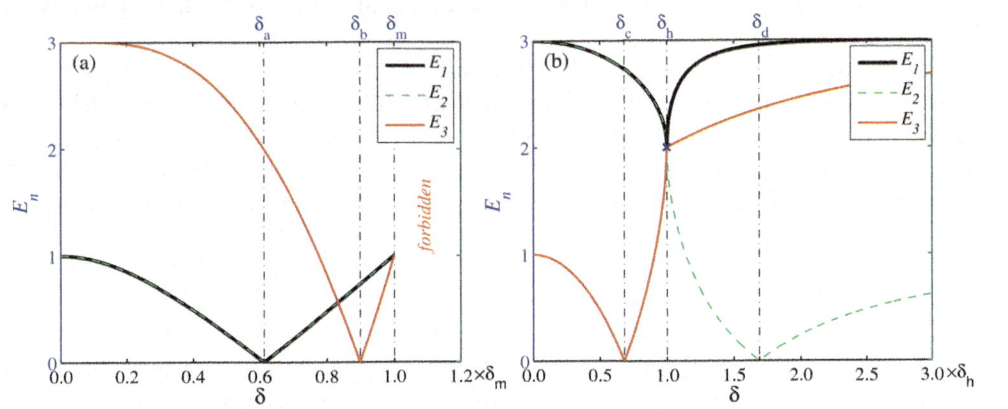

Figure 2.1. Dependence of the relative central amplitudes E_n on the frequency difference δ, expressed in terms of δ_m or δ_h. (a) SB process ($V_1 < V_2$); (b) SE process ($V_1 > V_2$).

$$\delta_a = \frac{1}{2}\sqrt{-6A} \approx 0.61\delta_m, \qquad \delta_b = \frac{1+\sqrt{3}}{2}\sqrt{-3^{1/2}A} \approx 0.90\delta_m. \tag{2.18}$$

Also, in the SE case, we find from figure 2.1(b) that the components u_2 and u_3 have a central amplitude vanishing at $\delta = \delta_d$ and δ_c, respectively, where

$$\begin{aligned} \delta_c &= \frac{-1+\sqrt{3}}{2}\sqrt{3^{1/2}A} \approx 0.68\delta_h, \\ \delta_d &= \sqrt{\left[\left(\sqrt{2}+1\right)^{4/3} + \left(\sqrt{2}-1\right)^{4/3} + 5\right]A/6} \approx 1.69\delta_h. \end{aligned} \tag{2.19}$$

It is worth noting that, if we chose a minus sign for m, then it will be the field component u_1, rather than u_2, that becomes zero at $\delta = \delta_d$. In addition, figure 2.1(a) shows further that the rogue wave states are totally forbidden to occur in the SB process as soon as $\delta > \delta_m$.

As a matter of fact, different types of fundamental rogue wave structures do appear as δ is varied, because the latter is closely related to the values of the central amplitude of each field component. We may define a rogue wave to be *bright* (Peregrine-type) whenever its peak amplitude at the center is more than twice the background height [35, 36]. We refer to rogue waves with one single hole on a nonzero background as being *dark* [33, 37]. Of particular interest is the *black* rogue wave, which corresponds to a special dark rogue wave whose intensity exactly falls to zero at the dip center. We may also loosely identify rogue wave states with a double-peak–double-dip structure as *intermediate* waves.

In the frame of the above phenomenological descriptions, in the SB case we find that the three rogue wave components u_n exhibit intermediate–intermediate–bright (IIB) structures as $\delta < \delta_a$, dark–dark–intermediate (DDI) structures as $\delta_a < \delta < \delta_b$, and finally dark–dark–dark (DDD) structures as $\delta_b \leqslant \delta < \delta_m$. Particularly, at $\delta = \delta_a$, dark–dark–bright (DDB) rogue wave structures appear, as illustrated in the upper row in figure 2.2. The first two field components exhibit a black structure, whereas the third component is of a bright nature, with its peak amplitude reaching just twice the background height.

On the other hand, in the SE case, one finds that the three rogue wave components are of the bright–bright–dark (BBD) type as $\delta \leqslant \delta_c$, or BDB as $\delta \geqslant \delta_d$, but remain bright–bright–intermediate (BBI) as $\delta_c < \delta < \delta_d$. As an illustration, we demonstrate in figure 2.2 (lower row) the rogue wave states for this SE process as $\delta = \delta_d$. It is shown that the two rogue wave components $|u_1|$ and $|u_3|$ are of a bright type, while the second one $|u_2|$ features a black structure. Particularly at $\delta = \delta_h$, which corresponds to the blue cross in figure 2.1(b), the three rogue wave components will be of the bright–bright–bright (BBB) type, all with a peak amplitude just twice the background height. We should point out that for $\delta = \delta_h$ and $B = 0$, the cubic equation (2.6) has a triple zero root. As discussed in [20], in this special SE case the fundamental rogue wave dynamics can be expressed in terms of a more general solution (involving additional structures such as rogue wave doublets) than the current one, equations (2.4). Therefore, we may state that in the SB

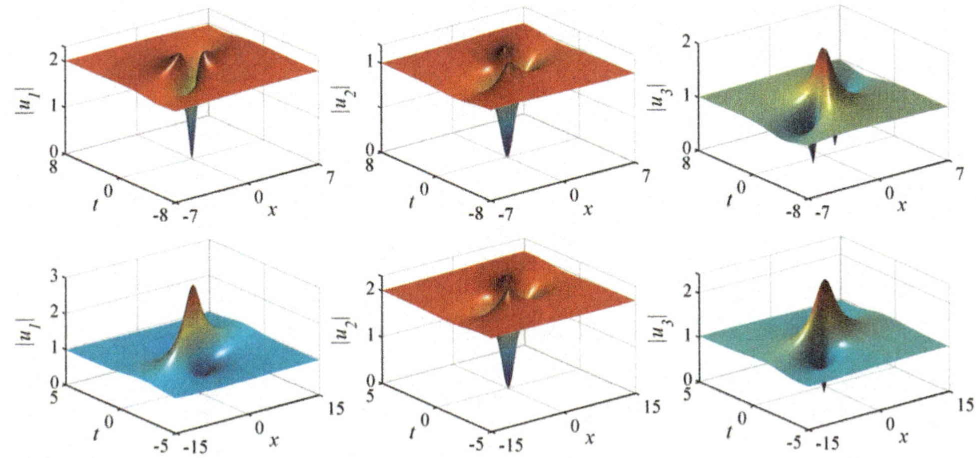

Figure 2.2. Typical fundamental rogue waves occurring in the SB or SE TWRI process. Upper row: DDB rogue waves in the SB process with initial parameters $\delta = \delta_a$, $a_1 = 2$, $V_1 = 1$, and $V_2 = 4$. Lower row: BDB rogue waves in the SE process with $\delta = \delta_d$, $a_1 = 1$, $V_1 = 4$, and $V_2 = 1$.

situation, at least two rogue wave components will be dark as $\delta \geqslant \delta_a$ (see figure 2.1(a) and 2.2(upper row)), whereas in the SE case, there are at least two rogue wave fields which are bright (see figure 2.1(b) and 2.2(lower row)).

2.3 Higher-order rogue wave solutions

To understand the complexity of rogue wave manifestations in a natural environment with random initial conditions, high-order rogue waves should also be considered. Theoretically, once the parameters λ_0 and μ_0 are known, the general second-order rogue wave solutions can be cast in the following compact form:

$$\begin{aligned}
\frac{u_1^{[2]}}{u_{10}} &= 1 + \frac{3i\Lambda[R_1^*(R_0 m_{22} - S_0 m_{21}) + S_1^*(S_0 m_{11} - R_0 m_{12})]}{a_1(m_{11}m_{22} - m_{12}m_{21})}, \\
\frac{u_2^{[2]}}{u_{20}} &= 1 + \frac{3i\Lambda[R_0^*(R_2 m_{22} - S_2 m_{21}) + S_0^*(S_2 m_{11} - R_2 m_{12})]}{a_2(m_{11}m_{22} - m_{12}m_{21})}, \qquad (2.20) \\
\frac{u_3^{[2]}}{u_{30}} &= 1 + \frac{3\Lambda[R_2^*(R_1 m_{22} - S_1 m_{21}) + S_2^*(S_1 m_{11} - R_1 m_{12})]}{a_3(m_{11}m_{22} - m_{12}m_{21})},
\end{aligned}$$

where

$$R_0 = \gamma_1 - i\gamma_2\phi\xi, \qquad R_j = \frac{a_j}{\alpha_j}(i\gamma_1 + \gamma_2\phi\theta_j), \tag{2.21}$$

$$S_0 = \gamma_1 p - i\gamma_2\phi(q\xi + gx) + \gamma_3 - i\gamma_4\phi\xi, \tag{2.22}$$

$$S_j = \frac{a_j}{\alpha_j}\left[\gamma_1\left(ip - \frac{\phi^2\theta_j + i\Lambda + i\varphi}{\alpha_j}\right) + i\gamma_3 + \gamma_4\phi\theta_j + \gamma_2\phi\left(q\theta_j + gx + \frac{\vartheta_j\alpha_j - i\phi^2}{3\alpha_j^3}\right)\right], \tag{2.23}$$

$$m_{11} = |R_0|^2 + \Gamma_1|R_1|^2 + \Gamma_2|R_2|^2, \tag{2.24}$$

$$m_{12} = R_0^*S_0 + \Gamma_1R_1^*S_1 + \Gamma_2R_2^*S_2 - m_{11} \equiv m_{21}^*, \tag{2.25}$$

$$m_{22} = |S_0|^2 + \Gamma_1|S_1|^2 + \Gamma_2|S_2|^2 - m_{12} - m_{21}, \tag{2.26}$$

with $\vartheta_j = (\theta_j\phi^2 + 3i\Lambda + 3i\varphi)(2 + i\alpha_j\xi)$, $p = -\frac{\phi^2\xi^2}{2} - i\varphi\xi - i(\Lambda\beta_2 - \beta_3\phi^2)x$, $q = p + \frac{1}{3}\phi^2\xi^2 + \frac{\Lambda - 2\varphi}{6\mu_0} + \frac{(\Lambda + \varphi)^2}{2\phi^2}$, and γ_s $(s = 1, 2, 3, 4)$ being four arbitrary complex constants (termed structural parameters). The other parameters ϕ, φ, and g in equations (2.21)–(2.23) are defined by

$$\phi = \left[\Lambda\left(\alpha_1 + \alpha_2 - \frac{A + \delta^2}{3\mu_0}\right)\right]^{1/2}, \tag{2.27}$$

$$\varphi = \frac{6\Lambda\lambda_0 - \phi^2}{6\mu_0}, \tag{2.28}$$

$$g = -2\beta_3(\Lambda + \varphi) + \beta_4\phi^2. \tag{2.29}$$

In comparison with the fundamental rogue waves of the previous section, the second-order rogue wave solutions (2.20) may exhibit a greater variety of complex patterns [28, 31, 38, 39]. Here, for brevity, we shall exclusively address the triplet dynamics (with the assumption $B = 0$ again, for the clarity of basic properties). In this case, the values of λ_0 and μ_0 are explicitly given by equations (2.10) and (2.11). We show that these rogue wave triplets will have almost the same dynamics as the corresponding fundamental rogue wave solutions. Specifically, in the SB situation, rogue triplets also exhibit IIB, DDI (including DDB), or DDD structures in the regime $\delta < \delta_a$, $\delta_a \leqslant \delta < \delta_b$, or $\delta_b \leqslant \delta < \delta_m$, respectively. Quite differently, in the SE situation these triplets will be BBD as $\delta \leqslant \delta_c$ or BDB as $\delta \geqslant \delta_d$, but BBI (including BBB) as $\delta_c < \delta < \delta_d$.

As an example, we show in figure 2.3 the DDB rogue wave triplets for the SB case by using identical background parameters as in figure 2.2(upper row), with structural parameters being specified in the caption. It is clear that the three field components have the same evolution characteristics as in figure 2.2(upper row), apart from each having a triplet feature in the present case. We point out that the

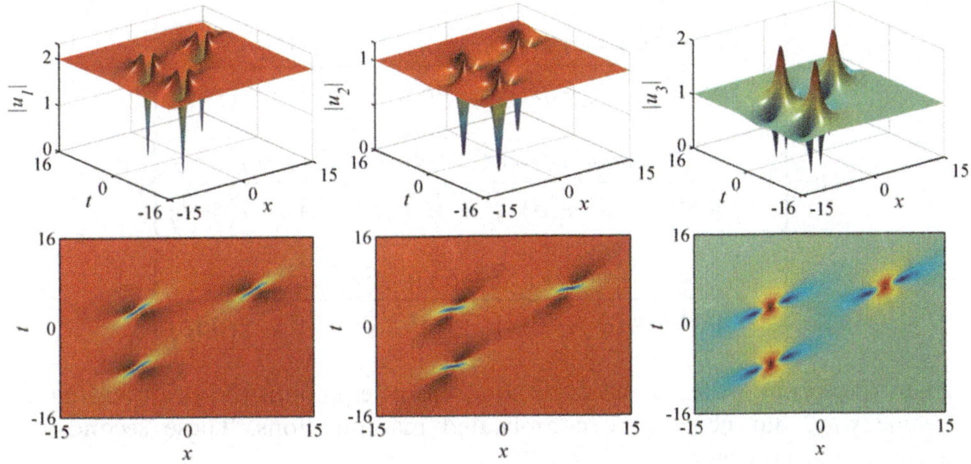

Figure 2.3. Surface and contour plots showing the DDB rogue wave triplets in the SB case with $\delta = \delta_a$, $a_1 = 2$, $V_1 = 1$, and $V_2 = 4$, the same parameters as used in figure 2.2(upper row). The four structural parameters are specified by $\gamma_1 = 5$, $\gamma_2 = 1$, and $\gamma_3 = \gamma_4 = 0$.

three constituents of each triplet are usually held in a triangle position, but their layout is not fixed, and it can change with the values of the four structural parameters γ_s.

2.4 Rogue wave solutions in the degenerate case

One may naturally ask whether the TWRI equations (2.1) admit exact rogue wave solutions in the degenerate case $V_2 = V_1 = V$, since the behavior shown in figure 2.1 exhibits different dynamical characteristics in the SE and SB cases. In the degenerate case, considering the resonant conditions for the frequencies and momenta, we can rewrite the coupled plane-wave solutions of TWRI equations (2.1)

$$\begin{aligned} u_{10}(x,t) &= a_1 \exp\left[-i(k_1 x - \omega_1 t)\right], \\ u_{20}(x,t) &= a_2 \exp\left[i(k_2 x - \omega_2 t)\right], \\ u_{30}(x,t) &= i a_3 \exp\left[i(k_1 - k_2)x - i(\omega_1 - \omega_2)t\right], \end{aligned} \tag{2.30}$$

where

$$a_3 = \frac{a_1 a_2}{\delta}, \qquad k_1 = \frac{\omega_1}{V} + \frac{a_2^2}{\delta V}, \qquad k_2 = \frac{\omega_2}{V} + \frac{a_1^2}{\delta V}, \tag{2.31}$$

with a_n being the respective background amplitudes. For the sake of convenience, we use $A = a_1^2 + a_2^2$, $B = a_1^2 - a_2^2$, $\kappa = \omega_1 + \omega_2$, and $\delta = \omega_1 - \omega_2$. Here, we assume again $\omega_1 > \omega_2$, i.e. $\delta > 0$, without loss of generality.

Based on the known nondegenerate solutions (2.4), we take the limit $V_2 \to V_1 = V$ and obtain the exact fundamental rogue wave solutions [29]

$$u_1^{[1]} = u_{10}\left\{1 + \frac{2i\delta V[\delta^2 Vt - (\delta^2 + A)x] - \delta^2 AV^2/a_1^2}{[\delta^2 Vt - (\delta^2 + B)x]^2 + 4\delta^2 a_3^2 x^2 + A^2 V^2/\left(4a_3^2\right)}\right\},$$

$$u_2^{[1]} = u_{20}\left\{1 + \frac{2i\delta V[\delta^2 Vt - (\delta^2 - A)x] - \delta^2 AV^2/a_2^2}{[\delta^2 Vt - (\delta^2 + B)x]^2 + 4\delta^2 a_3^2 x^2 + A^2 V^2/\left(4a_3^2\right)}\right\}, \tag{2.32}$$

$$u_3^{[1]} = u_{30}\left\{1 - \frac{4i\delta^3 V(Vt - x) + A^2 V^2/a_3^2}{[\delta^2 Vt - (\delta^2 + B)x]^2 + 4\delta^2 a_3^2 x^2 + A^2 V^2/\left(4a_3^2\right)}\right\}.$$

Further, the second-order rogue wave solutions of equations (2.1) can be found in the same way, but need more complicated manipulations. These second-order solutions are given explicitly below [29]:

$$u_1^{[2]} = u_{10}\left\{1 - \frac{4ia_2R_1^*[M_4(|R_2|^2 M_2^* - M_1) + R_0(|R_1|^2 M_2 + |R_2|^2 M_2^*)]}{\delta(M_1^2 + |R_1R_2M_2|^2)}\right\},$$

$$u_2^{[2]} = u_{20}\left\{1 - \frac{4ia_1R_2[M_3^*(|R_1|^2 M_2 + M_1) + R_0^*(|R_1|^2 M_2 + |R_2|^2 M_2^*)]}{\delta(M_1^2 + |R_1R_2M_2|^2)}\right\}, \tag{2.33}$$

$$u_3^{[2]} = u_{30}\left\{1 - \frac{4iR_1R_2^*[|R_1|^2(2 + M_2^*) + |R_2|^2(2 - M_2)]}{M_1^2 + |R_1R_2M_2|^2}\right\},$$

where

$$R_0 = \gamma_1 + \gamma_2\xi, \qquad R_j = \frac{a_j}{\alpha_j}(\gamma_1 + \gamma_2\theta_j), \quad \text{(here and below } j = 1,\ 2), \tag{2.34}$$

$$S_0 = -i\gamma_1\xi\varphi + \gamma_2[q\xi - 4ia_3x/(\delta V)] + \gamma_3 - i\gamma_4\xi, \tag{2.35}$$

$$S_j = \frac{\alpha_j S_0 + i\gamma_1\phi^2\left[\theta_j - A/\left(2\delta^2 a_3\right)\right] + \gamma_2\left(\phi^2\vartheta_j - iq\right) - \gamma_4}{\alpha_j\left(\gamma_1 + \gamma_2\theta_j\right)}, \tag{2.36}$$

$$\xi = t - \left[\frac{1}{V}\left(1 + \frac{B}{\delta^2}\right) + \frac{2ia_3}{\delta V}\right]x, \tag{2.37}$$

$$\alpha_j = \frac{\delta}{2A}\left[B - 2i\delta a_3 - (-1)^j A\right], \qquad \theta_j = \xi - \frac{i}{\alpha_j}, \tag{2.38}$$

$$\vartheta_j = \frac{\left[\theta_j - 3A/\left(2\delta^2 a_3\right)\right]\left(2 + i\alpha_j\xi\right)\alpha_j - i}{3\alpha_j^2}, \tag{2.39}$$

$$\phi = \frac{2\delta^2 a_3}{A^2}\sqrt{A^2 - 2B^2 + 4i\delta B a_3}, \qquad \varphi = \frac{i\phi^2 A}{2\delta^2 a_3} + \frac{2ia_3(\delta^2 - A)}{3A}, \tag{2.40}$$

$$q = -i\xi\varphi + \frac{\phi^2}{3}\xi^2 - \frac{A^2\phi^2}{4\delta^4 a_3^2} - \frac{1}{2}, \tag{2.41}$$

$$M_1 = |R_1|^2 + |R_2|^2 > 0, \qquad M_2 = S_2 - S_1, \tag{2.42}$$

$$M_3 = S_1R_0 - S_0 - 2R_0, \qquad M_4 = S_2R_0 - S_0 - 2R_0. \tag{2.43}$$

Let us take a more exhaustive look into the structure of the most simple Peregrine rogue waves defined by equations (2.32). It is shown that, while the rogue wave component u_3 possesses a peak amplitude fixedly three times its background height (significant wave height), the first two components u_1 and u_2 involve a peak amplitude varying with the relative magnitude of a_1 and a_2, given by $a_1|1 - 4a_2^2/A|$ and $a_2|1 - 4a_1^2/A|$, respectively. For example, if $a_2 = a_1$, these two rogue waves will have the same height at the origin as in the far periphery, and thus feature an intermediate rogue wave structure, for which one can refer to [28] for details. On the other hand, as $a_2 = \sqrt{3}\,a_1$, while the field u_1 takes the bright structure peaking twice as high as the average background, the field u_2 will feature a *black rogue wave* whose intensity will fall to zero in the dip center. However, no matter what situation it is in, the third field u_3 always displays a bright Peregrine structure.

The above conclusions can apply as well to the situation of higher-order rogue waves, by which one can determine whether they are bright or not. Considering the second-order rogue wave solutions (2.33), it is obvious that the dynamics are more complicated than what Peregrine solitons could exhibit, as the former ones depend not only on the starting plane-wave parameters but also on four structural parameters γ_s. To be specific, the parameter γ_2, if nonvanishing, gives rise to a second-order rogue wave skeleton, on which the spatiotemporal distribution of patterns could be influenced remarkably by the other three structural parameters. For instance, an increase of the value of any of these three parameters at a given γ_2 will enable the second-order rogue wave to evolve into a triplet pattern which consists of three Peregrine rogue waves [28, 38, 39].

More interestingly, it is revealed that the sum of the intensities of the components u_1 and u_2 will always be conserved and can be given by

$$|u_1|^2 + |u_2|^2 = a_1^2 + a_2^2 = A, \tag{2.44}$$

independently of what parameters we choose for deterministic rogue wave structures. As a matter of fact, the relation (2.44) is a natural consequence of equations (2.1), which suggest that $(\frac{\partial}{\partial t} + V\frac{\partial}{\partial x})(|u_1|^2 + |u_2|^2) = 0$ in the degenerate case. Therefore, if the field u_1 takes a bright Peregrine or a bright triplet structure, the field u_2 will take the corresponding dark counterpart so that equation (2.44) can be fulfilled. In other words, they are spatiotemporally complementary, and hence the name *complementary rogue waves*. Figure 2.4 illustrates the complementary

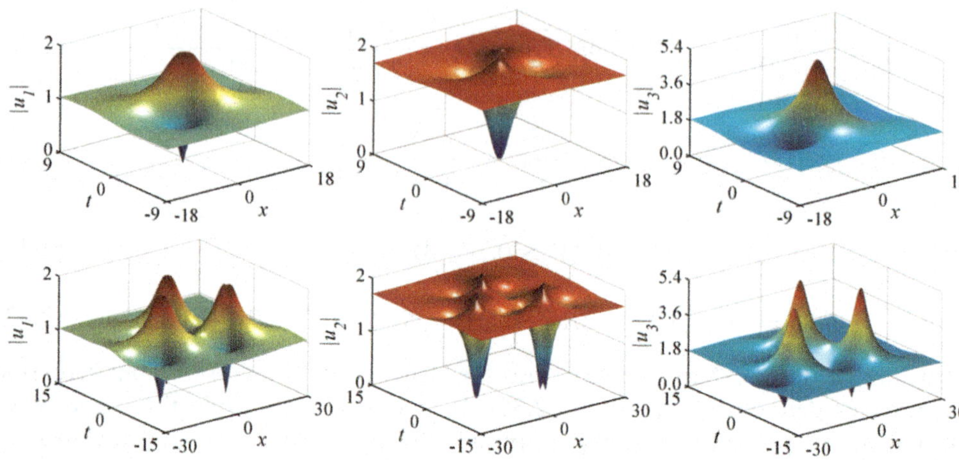

Figure 2.4. Typical (upper row) fundamental rogue waves and (lower row) rogue wave triplets occurring in the degenerate TWRI system with the same initial parameters $a_1 = 1$, $a_2 = a_3 = \sqrt{3}$, $\delta = 1$, and $V = 9$. The extra parameters for rogue wave triplets are given by $\gamma_1 = 6$, $\gamma_2 = 1$, and $\gamma_3 = \gamma_4 = 0$.

dynamics of typical Peregrine-like rogue waves and of the rogue wave triplets, by choosing the same initial plane-wave parameters for both situations and an extra set of structural parameters for the latter situation. It is clear that the three fields u_n would take the bright, black, and bright structures, respectively, but the first two are always spatiotemporally balanced according to equation (2.44), independently of whether the three rogue wave components are Peregrine-like (see upper row) or are of triplet pattern (see lower row).

2.5 Rogue wave existence and baseband MI

The parametric condition for the rogue wave existence can also be contrasted with the recently developed baseband-MI conjecture, which asserts that rogue waves can manifest only when nonzero MI gain is available for arbitrarily low modulation frequencies [33]. This conjecture has been validated for several integrable nonlinear systems [34]. To see this, we add small-amplitude Fourier modes to the plane-wave solutions (2.2) and express them as $u_n = u_{n0}\{1 + p_n \exp[-i\Omega(\mu t - x)] + q_n^* \exp[i\Omega(\mu^* t - x)]\}$ ($n = 1, 2, 3$), where p_n and q_n are small amplitudes of the Fourier modes, and the parameters Ω and μ are assumed to be positive and complex, respectively [28, 30, 33, 37]. A substitution of these perturbed plane-wave solutions into equations (2.1) followed by linearization yields the dispersion relation

$$\Omega^2\mu^2 - \frac{4\mu a_1^2}{\mu - V_2} - \left[\left(\frac{a_1^2}{\mu - V_2} + \frac{a_2^2}{\mu - V_1}\right)\frac{\mu}{\delta} - \delta\right]^2 = 0. \tag{2.45}$$

From this dispersion relation, one can then obtain the MI gain spectra which depend on the nonzero imaginary part of μ. We expect that the baseband MI, corresponding to $\Omega \to 0$, is the prerequisite of rogue wave formation [33]. Obviously, in the baseband limit, equation (2.45) will reduce to a real-coefficient quartic equation

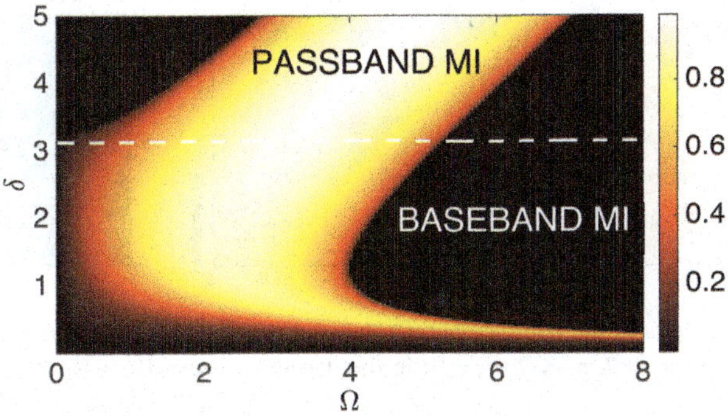

Figure 2.5. Map of the MI gain in the (Ω, δ) plane, divided into the baseband and passband parts by the dash-dotted line, calculated for given parameters $V_1 = 1$, $V_2 = 4$, $a_1 = 2$, and $a_2 = 1$.

$(B \neq 0)$ whose solution μ can be solved algebraically. It is easy to prove that as $V_1 > V_2$ (the SE case), equation (2.45) always permits a pair of complex conjugate roots in the whole regime of δ, whereas for $V_1 < V_2$ (the SB case), it only supports complex roots when the parameter condition (2.8) is fulfilled. This is completely consistent with the analytical derivation based on equations (2.5) and (2.6), confirming again the baseband-MI conjecture proposed in [33]. As a matter of fact, as shown in figure 2.5, the MI gain map, $\gamma = \Omega\,\mathrm{Im}(\mu)$, calculated from equation (2.45), consists of baseband and passband domains. However, only the baseband part contributes to the formation of rogue waves.

In the degenerate case, equation (2.45) simplifies to give

$$\Omega^2\mu^2 - \frac{4\mu a_1^2}{\mu - V} - \left[\frac{\mu A}{\delta(\mu - V)} - \delta\right]^2 = 0. \tag{2.46}$$

It is easy to show that in the baseband limit (i.e. $\Omega = 0$), equation (2.46) always permits a pair of complex conjugate roots in the whole regime of δ, as can be predicted from the analytical solutions, see equations (2.32) and (2.33). One can also refer to figure 4 in [29] for details.

2.6 Numerical simulations

The stability of these TWRI rogue waves with respect to background broadband noise perturbations (e.g. quantum noise) is of major concern to the general soliton community [40], more especially in view of their experimental investigation. For this purpose, we integrated numerically the TWRI equations using the standard split-step Fourier method and inspected whether the rogue wave generation was still observed in the presence of the MI activated by such quantum noise.

Specifically, we mimic the white noise by multiplying the real and imaginary parts of all the three field components u_n at sufficient negative times by a factor $[1 + \varepsilon r_i(x)]$ $(i = 1, \ldots, 6)$, respectively, where r_i are six uncorrelated random functions uniformly distributed in the interval $[-1, 1]$ and ε is a small parameter defining the noise level. Shown in figure 2.6 are the typical numerical results corresponding to

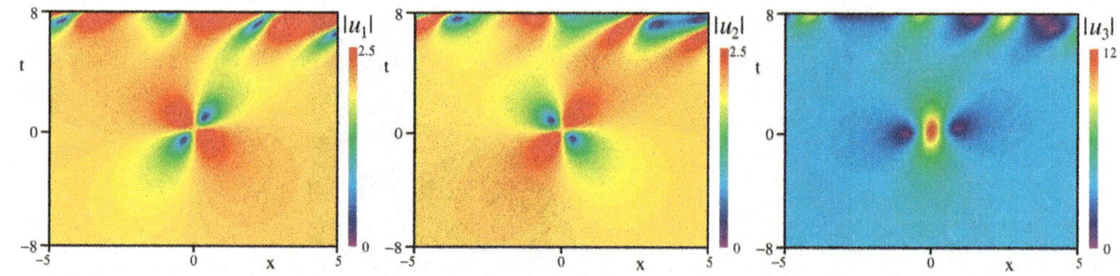

Figure 2.6. Simulation results of the complementary rogue wave states under initial white noise perturbation. The initial parameters are $a_1 = 2$, $a_2 = 2$, $a_3 = 4$; in the degenerate case $V_2 = V_1 = 4$.

the IIB rogue waves in the degenerate SB–SE situation, perturbed by a noise of strength $\varepsilon = 10^{-6}$. As seen, in the presence of a tiny perturbation, the fundamental rogue waves can still propagate very neatly for a rather long time, until eventually the spontaneous MI of the background fields grows up. Usually, when a larger noise intensity is adopted, noise-induced MI tends to interfere more strongly with the trailing edge of the rogue waves. Yet, in spite of the intrinsic instability of the background, the rogue waves themselves are typically robust enough to fully develop [28, 29]. Indeed, the rogue wave generation process may be thought of as induced MI, which is generally prevailing over spontaneous MI at least over relatively short propagation distances.

As regards the special yet interesting complementary rogue waves, we would like to put forward an experimental setup for observing them in the context of nonlinear optics. In fact, nonlinear optics has recently been seen as a fertile, reproducible, and safe ground for experimental demonstration of deterministic scalar bright [41] and vector dark rogue waves in fibers [42]. For the current TWRI issue under consideration, we may launch three phase-matched carrier waves of about 10-ps pulse duration, which can mimic quasi-continuous wave signals, in a periodically poled lithium niobate crystal of appropriate poling period to trigger a temporal type-II second harmonic interaction. With this experimental setup and by virtue of fast pulse measuring techniques, we anticipate that the MI evidence and the related complementary rogue wave dynamics could be observed with peak field intensities of hundreds of MW cm^{-2}.

2.7 Conclusions

We have presented the complex rogue wave dynamics that originate from the SB processes, the SE processes in transient stimulated scattering, and phase-matched weakly dispersive quadratic media. The explicit rogue wave solutions of the first and second order to the TWRI equations are shown, revealing that markedly different vector rogue wave behaviors are allowed by the different resonant interaction processes. Indeed, the relative magnitudes of the group velocities of the waves are of the highest importance, determining the prototypical shape of the vector rogue waves. Specifically, in the SB situation, there may exist two or three rogue wave components which can be dark, while in the SE situation, there are usually two or three rogue wave components that would be bright, each holding true for both the fundamental and the triplet dynamics. Note that in the SE situation, large group velocity differences can lead to three giant wall-like rogue wave humps that exhibit a wide spatiotemporal separation [43]. In contrast, the degenerate situation entails manifestly correlated rogue wave components. Indeed, we revealed that the sum of the intensities of the two waves sharing the same group velocity was spatiotemporally invariant. The link between the existence of rogue wave solutions and the presence of baseband MI is developed. We show once again that, although the MI consists of baseband and passband domains, only the baseband one is responsible for the generation of rogue waves. In this regard, the baseband-MI theory, whose validity has been verified for a wide range of nonlinear systems, offers an alternative method for calculating the parametric conditions for the existence of rogue waves. We finally performed numerical simulations confirming the possible manifestation of these vector rogue waves in the presence of background white noise perturbations. We have shown that the rogue waves in TWRI systems can be robust enough to develop in spite of the onset of spontaneous MI, and hence are of practical interest for experimental investigations.

Acknowledgments

This work was supported by the Italian Ministry of University and Research (MIUR, Project No. 2012BFNWZ2), by the National Natural Science Foundation of China (Grants No. 11174050 and No. 11474051), and by the European Union's Horizon 2020 research and innovation program under the Marie Sklodowska-Curie grant agreement No. 691051. Ph. G. acknowledges funding from CEFIPRA/IFCPAR (project 5104-2), and the LABEX ACTION (ANR-11-LABX-01-01). The work of J.M.S.C. was supported by MINECO under contract TEC2015-71127-C2-1-R, by C.A.M. under contract S2013/MIT-2790, and by the Volkswagen Foundation.

References

[1] Kaup D J, Reiman A and Bers A 1979 Space–time evolution of nonlinear three-wave interactions. I. Interaction in a homogeneous medium *Rev. Mod. Phys.* **51** 275–309

[2] Kim B, Blake J, Engan H and Shaw H 1986 All-fiber acousto-optic frequency shifter *Opt. Lett.* **11** 389–91

[3] Russell P, Culverhouse D and Farahi F 1991 Theory of forward stimulated Brillouin scattering in dual-mode single-core fibers *IEEE J. Quantum Electron.* **27** 836–42

[4] Picholle E, Montes C, Leycuras C, Legrand O and Botineau J 1991 Observation of dissipative superluminous solitons in a Brillouin fiber ring laser *Phys. Rev. Lett.* **66** 1454–57

[5] Ibragimov E and Struthers A 1996 Second harmonic pulse compression in the soliton regime *Opt. Lett.* **21** 1582–84

[6] Picozzi A and Haelterman M 1998 Spontaneous formation of symbiotic solitary waves in a backward quasi-phase-matched parametric oscillator *Opt. Lett.* **23** 1808–10

[7] Conforti M, Baronio F, Degasperis A and Wabnitz S 2006 Inelastic scattering and interactions of three-wave parametric solitons *Phys. Rev.* E **74** 065602

[8] Conforti M, Baronio F, Degasperis A and Wabnitz S 2007 Parametric frequency conversion of short optical pulses controlled by a CW background *Opt. Express* **15** 12246–51

[9] Abdolvand A, Nazarkin A, Chugreev A V, Kaminski C F and St. J. Russell P 2009 Solitary pulse generation by backward Raman scattering in H_2-filled photonic crystal fibers *Phys. Rev. Lett.* **103** 183902

[10] Hasegawa A 2001 *Plasma Instabilities and Nonlinear Effects* (Berlin: Springer-Verlag)

[11] Cheng W, Avitzour Y, Ping Y, Suckewer S, Fisch N, Hur M and Wurtele J 2005 Reaching the nonlinear regime of Raman amplification of ultrashort laser pulses *Phys. Rev. Lett.* **94** 045003

[12] Svaasand L 1969 Interaction between elastic surface waves in piezoelectric materials *Appl. Phys. Lett.* **15** 300–2

[13] Craik A 1985 *Wave Interactions and Fluid Flows* (Cambridge: Cambridge University Press)

[14] Lamb K 2007 Tidally generated near-resonant internal wave triads at shelf break *Geophys. Res. Lett.* **34** L18607

[15] Zakharov V E and Manakov S V 1973 Resonant interaction of wave packets in nonlinear media *JEPT Lett* **18** 243–245

[16] Nozaki K and Taniuti T 1973 Propagation of solitary pulses in interactions of plasma waves *J. Phys. Soc. Jpn.* **34** 796–800

[17] Baronio F, Conforti M, Andreana M, Couderc V, De Angelis C, Wabnitz S, Barthélémy A and Degasperis A 2009 Frequency generation and solitonic decay in three wave interactions *Opt. Express* **17** 13889

[18] Baronio F, Conforti M, De Angelis C, Degasperis A, Andreana M, Couderc V and Barthélémy A 2010 Velocity-locked solitary waves in quadratic media *Phys. Rev. Lett.* **104** 113902

[19] Baronio F, Andreana M, Conforti M, Manili G, Couderc V, De Angelis C and Barthélémy A 2011 Soliton triads ensemble in frequency conversion: from inverse scattering theory to experimental observation *Opt. Express* **19** 13192–200

[20] Baronio F, Conforti M, Degasperis A and Lombardo S 2013 Rogue waves emerging from the resonant interaction of three waves *Phys. Rev. Lett.* **111** 114101

[21] Onorato M, Residori S, Bortolozzo U, Montina A and Arecchi F T 2013 Rogue waves and their generating mechanism in different physical contexts *Phys. Rep.* **528** 47–90

[22] Onorato M, Residori S and Baronio F 2016 *Rogue and Shock Waves in Nonlinear Dispersive Media* (Berlin: Springer)

[23] Akhmediev N *et al* 2016 Roadmap on optical rogue waves and extreme events *J. Opt.* **18** 063001

[24] Bonatto C, Feyereisen M, Barland S, Giudici M, Masoller C, Leite J R R and Tredicce J R 2011 Deterministic optical rogue waves *Phys. Rev. Lett.* **107** 053901

[25] Lecaplain C, Grelu Ph, Soto-Crespo J M and Akhmediev N 2012 Dissipative rogue waves generated by chaotic pulse bunching in a mode-locked laser *Phys. Rev. Lett.* **108** 233901

[26] Dudley J M, Dias F, Erkintalo M and Genty G 2014 Instabilities, breathers and rogue waves in optics *Nat. Photon.* **8** 755–64

[27] Akhmediev N, Ankiewicz A and Taki M 2009 Waves that appear from nowhere and disappear without a trace *Phys. Lett.* A **373** 675–78

[28] Chen S, Baronio F, Soto-Crespo J M, Grelu Ph, Conforti M and Wabnitz S 2015 Optical rogue waves in parametric three-wave mixing and coherent stimulated scattering *Phys. Rev.* A **92** 033847

[29] Chen S, Cai X M, Grelu Ph, Soto-Crespo J M, Wabnitz S and Baronio F 2016 Complementary optical rogue waves in parametric three-wave mixing *Opt. Express* **24** 5886–895

[30] Conforti M, Baronio F and Degasperis A 2011 Modulational instability of dark solitons in three wave resonant interaction *Physica D* **240** 1362–9

[31] Ling L, Guo B and Zhao L C 2014 High-order rogue waves in vector nonlinear Schrödinger equations *Phys. Rev.* E **89** 041201

[32] Chen S and Mihalache D 2015 Vector rogue waves in the Manakov system: diversity and compossibility *J. Phys. A: Math. Theor.* **48** 215202

[33] Baronio F, Conforti M, Degasperis A, Lombardo S, Onorato M and Wabnitz S 2014 Vector rogue waves and baseband modulation instability in the defocusing regime *Phys. Rev. Lett.* **113** 034101

[34] Baronio F, Chen S, Grelu Ph, Wabnitz S and Conforti M 2015 Baseband modulation instability as the origin of rogue waves *Phys. Rev.* A **91** 033804

[35] Chabchoub A, Hoffmann N P and Akhmediev N 2011 Rogue wave observation in a water wave tank *Phys. Rev. Lett.* **106** 204502

[36] Baronio F, Degasperis A, Conforti M and Wabnitz S 2012 Solutions of the vector nonlinear Schrödinger equations: Evidence for deterministic rogue waves *Phys. Rev. Lett.* **109** 044102

[37] Chen S, Grelu Ph and Soto-Crespo J M 2014 Dark- and bright-rogue-wave solutions for media with long-wave–short-wave resonance *Phys. Rev.* E **89** 011201

[38] Ankiewicz A, Kedziora D J and Akhmediev N 2011 Rogue wave triplets *Phys. Lett.* A **375** 2782–5

[39] Chen S, Soto-Crespo J M and Grelu Ph 2014 Dark three-sister rogue waves in normally dispersive optical fibers with random birefringence *Opt. Express* **22** 27632–42

[40] Wabnitz S and Wetzel B 2014 Instability and noise-induced thermalization of Fermi–Pasta–Ulam recurrence in the nonlinear Schrödinger equation *Phys. Lett.* A **378** 2750–6

[41] Kibler B, Fatome J, Finot C, Millot G, Dias F, Genty G, Akhmediev N and Dudley J M 2010 The Peregrine soliton in nonlinear fibre optics *Nature Phys.* **6** 790–5

[42] Frisquet B, Kibler B, Morin P, Baronio F, Conforti M, Millot G and Wabnitz S 2016 Optical dark rogue waves *Sci. Rep.* **6** 20785

[43] Chen S, Soto-Crespo J M and Grelu Ph 2015 Watch-hand-like optical rogue waves in three wave interactions *Opt. Express* **23** 349–59

IOP Publishing

Nonlinear Guided Wave Optics
A testbed for extreme waves
Stefan Wabnitz

Chapter 3

Irreversibility and squeezing of shock waves

Maria Chiara Braidotti, Giulia Marcucci, Silvia Gentilini and Claudio Conti

3.1 Introduction

Historically, the first evidence of a shock wave can be dated back to 1887, when Ernst Mach published the first visualization of a shock front generated by the motion of a body traveling at supersonic speed. Nowadays, shock waves (SWs) are observable in many physical systems ranging from fluidodynamics to astrophysics, dispersive gas dynamics [1], plasma physics [2], granular systems [3], Bose–Einstein condensate, and nonlinear optics [4–6]. In particular, in this chapter we focus our attention on shock waves in optical nonlocal nonlinear defocusing media.

Usually, shock originates from the occurrence of a singularity in some quantity involved in the wave propagation. The way such singularity is overcome by the system allows us to distinguish between viscous and dispersive SWs. The former show a steep front associated with a jump in energy or pressure. As an example, water waves are typically characterized by dissipative effects. Conversely, dispersive shock waves (DSWs), such as those occurring in optics, regularize the singularity by the formation of characteristic fast oscillations due to the action of dispersion or diffraction.

In previous works, optical SWs have been theoretically handled by resorting to the hydrodynamic approximation that holds true in the highly nonlinear regime and allows us to reduce the nonlinear Schrödinger equation (NLSE) to a hyperbolic partial differential equation for the phase chirp [7]. Such an approach allows us to theoretically predict the shock point and the propagation velocity of the singularity at the variance of the interplay between the nonlinearity degree and the diffraction [8].

However, the exact description of the SW profiles is rarely available and techniques like the Whitham approach are limited to integrable or nearly integrable systems [7, 9]. Recently, unnormalizable wavefunctions named nonlinear Gamow vectors (GVs) have proved to be fruitful: the SWs are described by the eigensolutions of the so-called reversed harmonic oscillator (RHO) [10–12].

doi:10.1088/978-0-7503-1460-2ch3

© IOP Publishing Ltd 2017

In the first section, we briefly introduce the hydrodynamic approximation valid in the highly nonlinear regime in order to derive the main equations in which the NLSE can be reduced.

In the second section, we release the hydrodynamic approximation and we focus on taking into consideration a highly nonlocal response. This situation requires us to reformulate the equation ruling the refractive index change and consequently the NLSE accordingly. The remarkable aspect of such reformulation is that it enables us to express the NLSE in terms of Hamiltonian formalism in the rigged Hilbert space (RHS). Most important is the physical meaning of this novel mathematical approach, which implies an understanding of the optical SWs as the first physical realization of intrinsically time irreversible phenomena from their first theoretical prediction due to Gamow, in order to explain exponentially decaying states in terms of GVs [12].

Finally, in the last section, we will apply the GVs formalism to investigate the implication of the quantum nature of light in nonlinear optical propagation. In particular, we theoretically study the spontaneous squeezing of a photon fluid in a nonlocal nonlinear medium. These results open a new scenario in quantum photon fluids and may lead to novel applications in fields such as metrology and analogues of quantum gravity.

3.2 Hydrodynamic approximation of dispersive shock waves

In nature, SWs are very common in fluidodynamics when, for example, two fluids with different densities collide each other: the density jump experienced by the fluids at the discontinuity surface can be the site for the formation of a shock front. Other examples of this mechanism can be observed in clouds along a shoreline, and in sea waters at the exit of straits that connect waters at different salinity. Moreover, the occurrence of SWs can be also observed during the supersonic flights of aircraft and in space, where shock fronts are generated by the explosion of a star in a supernova.

The reason for the ubiquity of SWs can be retrieved in their mathematical description. Indeed, mathematically, they represent a singular solution of hyperbolic partial differential equations (PDEs), which are a class of equations describing a wide variety of wave-like physical phenomena:

$$a(x,y,z)\frac{\partial z}{\partial x} + b(x,y,z)\frac{\partial z}{\partial y} = c(x,y,z). \tag{3.1}$$

A singular solution of equation (3.1) develops a discontinuity in its derivative as a consequence of an abrupt jump in some of the quantities involved in the wave propagation. The point at which such a singularity arises is the shock point, and can be mathematically predicted together with its speed.

A paradigmatic example of PDEs are represented by the flux conservative equations for a velocity field u. The simplest case is represented by the Hopf equation, considered the prototype for the shock phenomenon:

$$\frac{\partial u}{\partial z} + u\frac{\partial u}{\partial x} = 0. \tag{3.2}$$

Equation (3.2) rules the dynamics of shallow water and many other phenomena in fluidodynamics. It displays a peculiar term in which the derivative of the velocity depends on the velocity itself. The Hopf equation supports a singular solution and the shock point is defined as the distance z_s at which the derivative of the velocity goes to infinity, i.e. $\partial u/\partial x \to \infty$.

Physically, the discontinuity is overcome via dissipation or dispersion mechanisms. In optics, the regularizing mechanism is played by the dispersion, or diffraction, giving rise to the emergence of a characteristic fast oscillatory front, also called *undular bores.*

The propagation of a monochromatic optical beam of wavelength λ and amplitude A, in a self-defocusing medium, is described by the NLSE:

$$2ik\frac{\partial A}{\partial Z} + \nabla^2_{XY}A + 2k^2\frac{\Delta n}{n_0}A = 0. \tag{3.3}$$

Equation (3.3) describes a wave propagating along the Z direction with wave number $k = 2\pi/\lambda$ and experiencing a nonlinear refractive index perturbation with respect to its linear refractive index n_0, given by $\Delta n = n_2 I$, where n_2 is the local Kerr coefficient and $I = |A|^2$ is the field intensity. Equation (3.3) can be conveniently rewritten in term of scaled coordinates $x,y,z = X/w_0, Y/w_0, Z/L$ and of the complex variables $\psi = A/\sqrt{I}$ and $\theta = kL_{nl}\Delta n$, where $L_{nl} = (k|n_2|I)^{-1}$ is the nonlinear length scale, $L_d = kw_0^2$ is the diffraction length, and $L \equiv \sqrt{L_{nl}L_d}$:

$$i\varepsilon\frac{\partial\psi}{\partial z} + \frac{\varepsilon^2}{2}\nabla^2_\perp\psi - \theta\psi = 0, \tag{3.4}$$

$$-\sigma^2\nabla^2_\perp\theta + \theta = |\psi|^2, \tag{3.5}$$

where the parameter $\varepsilon = \sqrt{L_{nl}/L_d}$ takes into account the competition between nonlinearity and diffraction. Equation (3.5) rules the refractive index change, taking into account the degree of nonlocality through the parameter σ. When $\varepsilon \ll 1$, it is possible to resort to the hydrodynamical approach [7], which is a decisive model used for dispersive shock waves [6, 13] and also analog gravity systems [14, 15]. By assuming that the optical field has a Gaussian symmetry $\psi(r,z) = \sqrt{\rho(r,z)}$ $\exp[i\phi(r,z)/\varepsilon]$, as it is at the waist position of a fundamental laser mode, the evolution can be studied in the framework of the Wentzel–Kramers–Brillouin (WKB) transformation, writing the field ψ in terms of a fluid-like variable. Substituting in equation (3.4) and retaining only the leading orders in ε, equation (3.4) reduces to a set of two equations, the continuity equation for hydrodynamics and the Euler equation:

$$\rho_z + (\rho u)_r = 0, \tag{3.6}$$

$$u_z + uu_r + \theta_r = 0, \tag{3.7}$$

where $u \equiv \phi_r$ is the phase chirp and we are limiting to consider the one-dimensional case. In equations (3.6) and (3.7), we can recognize the hyperbolic systems of conservative law that describe gas dynamics. In the optical case the above equations describe the development of a shock front characterized by the emergence of fast oscillations, appearing first in the phase chirp u and then in the intensity ρ, that connect the high and low sides of the wave front and expand outwards. This regularizing mechanism is due to the action of the diffraction that plays a major role in proximity of a singular point.

Experimentally, the physics of the defocusing nonlinearity can be validated by exploiting the thermal nonlinearity of a strongly absorptive medium, displaying a nonlinearity of thermal origin. A commonly used sample exhibiting such optical behavior is the aqueous solution of Rhodamine B (RhB). The shock can be excited and measured by means of the setup shown in figure 3.1(a). Figures 3.1(b)–(d) show images of the transmitted beam for increasing values of the input laser power P. The profiles display post-shock rings, with the outer ring being more intense than the inner ones, as is typical for DSWs from Gaussian beams: as expected, the number and visibility increase with P. In figures 3.2(a) and (b) are shown the top fluorescence image of the propagating beam for increasing values of the laser power P. From this image, it is possible to evaluate the shock point z_s according to the theoretical definition, as the maximum of steepness of the intensity profiles (for details, see [16]).

From a numerical point of view, the shock phenomenon can be simulated by resorting to the method of characteristics. Indeed, since the hydrodynamic approximation holds true, it is possible to apply such mathematical technique to reduce the NLSE in an ordinary differential equation, according to which equation (3.6) can be rewritten as follows:

$$\frac{d^2x}{dz^2} = -\frac{dU}{dx}. \tag{3.8}$$

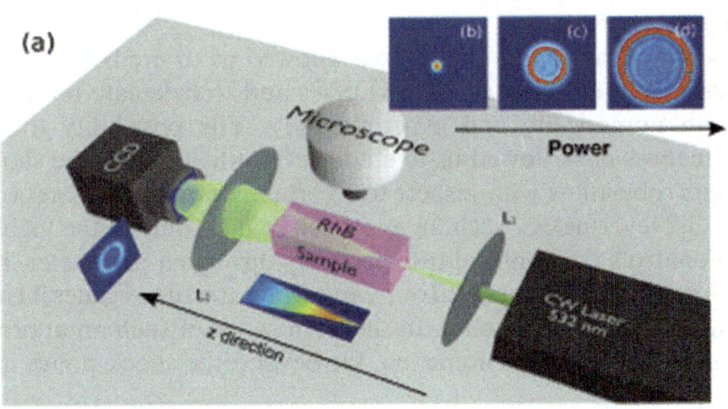

Figure 3.1. (a) Schematics of the experimental setup to obtain transmitted and top fluorescence images of the laser beam propagating in the RhB samples. The input lens, L_1, focuses the laser beam into the sample. The collecting lens, L_2, images the exit plane of the sample onto the CCD camera. The microscope placed above the sample captures top-view images of the laser beam along the propagation direction. (b–d) Images of the exit plane for increasing values of the laser power.

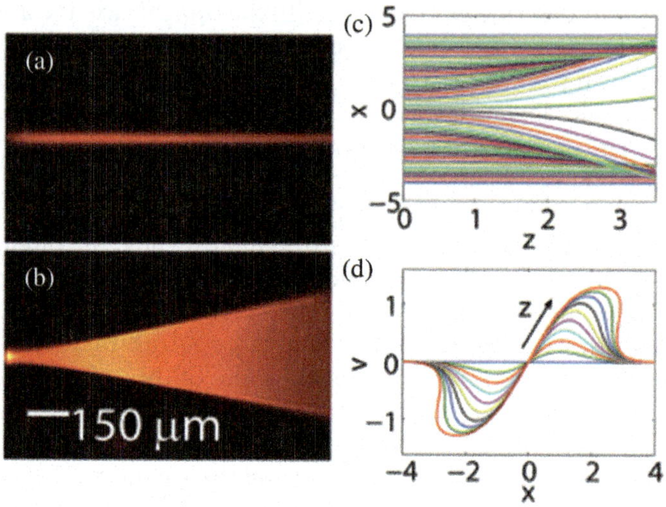

Figure 3.2. (a) and (b) Beam propagation as observed from top fluorescence emission for different input power $P = 8$ mW (a) and 450 mW (b). (c) Trajectories of colliding particles forming shock vs propagation distance z and (d) shock profile in the phase space.

Equation (3.8) resembles that ruling the trajectories of a particle experiencing a square well potential, i.e. the refractive index profile induced in a thermal media.

In figures 3.2(c) and (d), we show several of these trajectories resulting from initial uniformly distributed position in the x-axis and zero initial velocity $v = dx/dz$, as obtained by a stochastic Runge–Kutta algorithm. Upon propagation, the particles collide and the shock is signaled by the coalescence of multiple trajectories. The collision point is the way to numerically estimate the shock point (figure 3.2(c)); in the phase space of v and x (figure 3.2(d)), these correspond to the folding of the velocity profile into a multivalued function when increasing z, which induces the wave-breaking phenomenon.

The hydrodynamical approximation has allowed us to predict the appearance of the characteristic fast oscillations of the DSWs and to calculate the point of shock formation both numerically and experimentally. The possibility to carry out a quantitative analysis has allowed us to study several features of the shock phenomenon such as its robustness with respect to a certain amount of disorder [17] and the scaling law [16]. Nevertheless, such an approach is valid only in the highly nonlinear regime and it controls the nonlocal response only through a parameter. In the rest of the chapter, we enlarge the perspective by modeling the more general case of media displaying highly nonlocal response. In the framework of such an approximation, a deeper insight of the shock phenomenon, far beyond the shock point, is provided.

3.3 Highly non-local limit and irreversibility

In this section, we report on a method used to describe highly nonlocal systems. Thanks to the highly nonlocal approximation (HNA), we show that the nonlocal NLSE can be formally reduced to the linear Schrödinger equation of quantum mechanics. We show that SWs are described by the eigensolutions of the RHO. This formalism allows us to find features coming from the quantum nature of light, which helps to explain the shock irreversibility. Indeed, SWs furnish a paradigmatic example of the issues related to irreversibility in classical systems. Being singular solutions excited by smooth initial conditions, their evolution is dominated by non-regular, highly disordered waves (wave-breaking) [18]. The evolution of smooth initial conditions towards a highly disordered and incoherent regime is intuitive. However, the reverse process, albeit possible, is strongly puzzling and against well accepted ideas about entropy. Dispersive shock waves and wave-breaking in Hamiltonian models therefore question reversibility; they seem to indicate a preferential arrow of time that should be excluded from first principles, because no coupling with the environment is foreseen [1, 18–22]. We show in the following that the specific case of a nonlocal nonlinearity enables to unveil irreversibility in these systems.

3.3.1 Highly nonlocal approximation

We start by considering a Gaussian beam with wavelength λ propagating in a medium with a nonlocal nonlinear response function. The presence of the beam in the sample creates a temperature gradient which affects the sample density distribution. This translates in a variation of the refractive index

$$\Delta n = \left(\frac{\partial n}{\partial T}\right)_0 \Delta T. \tag{3.9}$$

The transport equation regulates the temperature excursion as follows:

$$-\kappa_T \nabla^2(\Delta T) = \alpha I(\boldsymbol{R}), \tag{3.10}$$

where $I(\boldsymbol{R})$ is the beam intensity and κ_T is the thermal conductance. By solving equation (3.10) and considering equation (3.9), we find the variation of the refractive index

$$\Delta n = n_2 \int d\boldsymbol{R}' G_2(\boldsymbol{R} - \boldsymbol{R}') I(\boldsymbol{R}'), \tag{3.11}$$

where n_2 is the nonlinear refractive index and $G_2(\boldsymbol{R} - \boldsymbol{R}') = \left(\frac{\partial n}{\partial T}\right)_0 G(\boldsymbol{R} - \boldsymbol{R}')$ is the kernel function normalized such that $\int G_2 d\boldsymbol{R} = 1$. $G(\boldsymbol{R} - \boldsymbol{R}')$ is the Green function of the solution of heat equation (3.10). The local Kerr effect corresponds to $G_2 = \delta(\boldsymbol{R} - \boldsymbol{R}')$. The nonlocality $G_2(\boldsymbol{R})$ is typically a bell-shaped function with a characteristic length L_{nloc}. We consider an exponential nonlocality $G_2(X, Y) = G_2(X)G_2(Y)$ such that $G_2(X) = \exp(-|X|/L_{nloc})/(2L_{nloc})$. Since the problem is separable, we consider the case of one transverse dimension X. For a defocusing nonlinearity ($n_2 < 0$), we write equation (3.3) in terms of dimensionless variables $x = X/W_0$ and $z = Z/Z_d$, with W_0 being the beam waist and $Z_d = kW_0^2$ being the diffraction length:

$$i\partial_z \psi + \frac{1}{2}\nabla_{xy}^2 \psi + P \int K(x - x')|\psi(x)|^2 dx' \psi = 0, \tag{3.12}$$

where $\psi = AW_0/\sqrt{P_{MKS}}$ and $\langle\psi|\psi\rangle = 1$, brackets denoting the Hilbert space scalar product. $P = P_{MKS}/P_{REF}$ denotes the strength of the nonlocal term, with $P_{REF} = \lambda^2/4\pi^2 n_0|n_2|$ and $K(x) = W_0 G_2(xW_0) = \exp(-|x|/\sigma)/2\sigma$, with $\sigma = L_{nloc}/W_0$.

When the nonlocal function K is much wider than the beam waist, we can apply the HNA where the convolution integral in equation (3.12) reduces to a single function of the position x: $\kappa(x) \simeq K*|\psi(x)|^2$. Hence under HNA, equation (3.12) becomes an effectively linear Schrödinger equation

$$i\partial_z\psi = \hat{H}\psi \tag{3.13}$$

with Hamiltonian

$$\hat{H} = \frac{1}{2}\hat{p}^2 + V(x) \tag{3.14}$$

being $V(x) = P\kappa(x)$ and $\hat{p} = -i\partial_x$. The nonlinear dynamics of shock waves in highly nonlocal media can be described through the effectively linear Schrödinger equation (3.13) with Hamiltonian (3.14).

3.3.2 Irreversibility and Gamow vectors

As discussed previously, Hamiltonian systems such as the NLSE describe a reversible evolution in the Hilbert space. Here, we report on a time-asymmetric formalism of quantum mechanics living in the RHS, which unveils irreversibility in Hamiltonian systems. In this theory, the exponential dynamics in Hamiltonian systems [23, 24] is explained in terms of the excitation of states that, like plane waves, are not normalizable but admit complex eigenvalues for the energy and give rise to exponential decays. These states, named GVs, were originally introduced in nuclear physics but later demonstrated to provide a theory of irreversibility in Hamiltonian systems within RHS [25–28]. In this paragraph, we first introduce the nonlinear Gamow vectors (N-GVs) from a theoretical point of view, and then we report experimental results on shock waves excited in nonlocal nonlinear media showing the notable feature of having quantized decay rates which depend on the amplitude of the wave. This property provides a clear experimental signature of the excitation of GVs and hence allows distinguishing the intrinsic nature of the irreversibility of the process from having an extrinsic cause, i.e. due to the coupling with the environment. A detailed mathematical description can be found in [29, 30].

Gamow vectors are generalized eigenvectors $\phi_n^G(x)$ of a certain Hamiltonian $\hat{H}$ with complex eigenvalues E_n with $n = 0, 1, 2, \ldots$ such that their eigenvalue equation is $\hat{H}\phi_n^G(x) = E_n\phi_n^G(x)$; they exist outside the Hilbert space of Lebesgue integrable functions and belong to the RHS which includes no normalizable states. In the RHS, they furnish a generalized basis for normalizable wavepackets,

$$\psi(x,0) = \sum_{n=0}^{\infty} \phi_n^G(x) + \phi_b(x) = \phi^G(x) + \phi_b(x). \tag{3.15}$$

Equation (3.15) is the initial beam state given by two components: ϕ_G is the discrete contribution of a finite number of vectors with complex eigenvalues $E_n = E_n^R + i\Gamma_n/2$, with real part E_n^R and imaginary part Γ_n, which corresponds to the decay rates; ϕ_b is the background component due to that part of the continuum spectrum of $\hat{H}$ that does not contain GVs. The usual spectrum of normalizable bound states with real valued eigenvalues is not present in equation (3.15). Equation (3.15) can be obtained in quantum mechanics by the analytical prolongation of the continuous spectrum eigenvectors, or equivalently of the Green function, or by the scattering S matrix [26, 31, 32]. It is worth saying that ϕ^G and ϕ_b differs in the fact that, upon evolution, the former decays exponentially while exponential decay is not allowed for ϕ_b, which decays algebraically [29, 30]. Considering the evolution of a state in the quantum mechanics formalism, $\psi(x,z) = e^{-i\hat{H}z}\psi(x,0)$, we can write the evolution in GVs as

$$\psi(x,z) = \sum_{n=0}^{\infty} e^{-iE_n^R z - \frac{\Gamma_n}{2}z}\phi_n^G(x,0) + e^{-i\hat{H}z}\phi_b(x). \tag{3.16}$$

During the beam propagation, GVs cannot emerge from the background ϕ_b, as they decay exponentially under the action of the propagator of quantum mechanics, i.e. these states form a semigroup under the action of the evolution operator $e^{-i\hat{H}z}$ for $z > 0$, which prevents the reversed evolution [29, 30, 33].

3.3.3 The reversed harmonic oscillator

Having in the system an exponential nonlocality, we write the even function $\kappa(x)$ as its second-order expansion, that is, $\kappa(x) = \kappa_0^2 - \kappa_2^2 x^2/2$, where $\kappa_0^2 = 1/2\sigma$ and $\kappa_2^2 = 1/\sqrt{\pi\sigma^2}$. Thus, the Hamiltonian (3.14) results:

$$\hat{H} = P\kappa_0^2 + \hat{H}_{\mathrm{RHO}} \qquad \text{with} \qquad \hat{H}_{\mathrm{RHO}} = \frac{\hat{p}^2}{2} - \frac{\gamma^2}{2}\hat{x}^2, \tag{3.17}$$

where the real eigenvalue of $\hat{H}$ is $E_n^R = P\kappa_0^2$ while $\gamma^2 = P\kappa_2^2$. $\hat{H}_{\mathrm{RHO}}$ is the Hamiltonian of the RHO, which describes the dynamics of a pointlike mass around an unstable equilibrium, i.e. an intrinsically irreversible evolution.

As illustrated in figure 3.3, a system with an arbitrary energy landscape V will behave as a harmonic oscillator (HO) in proximity of an equilibrium configuration, i.e. a local minimum of V. In a one-dimensional (1D) case, the HO Hamiltonian for the position momentum operators $(\hat{x}, \hat{p})$ reads as

$$H_{\mathrm{HO}} = \frac{\hat{p}^2}{2} + \frac{\omega^2}{2}\hat{x}^2, \tag{3.18}$$

where ω is the angular frequency of the corresponding classical system. $\hat{H}_{\mathrm{HO}}$ exhibits discrete energy levels $E_n = \omega(n + \frac{1}{2})$, where $n = 0, 1, 2\ldots.$ The eigenstates are $\psi_n^{HO}(x) = \frac{\sqrt[4]{\gamma}}{\sqrt{2^n n! \sqrt{\pi}}} e^{-\frac{\gamma}{2}x^2}\mathcal{H}_n(\sqrt{\gamma}x)$, with $\mathcal{H}_n(x) = (-1)^n x^2 \frac{d^n}{dx^n} e^{-x^2}$ the Hermite

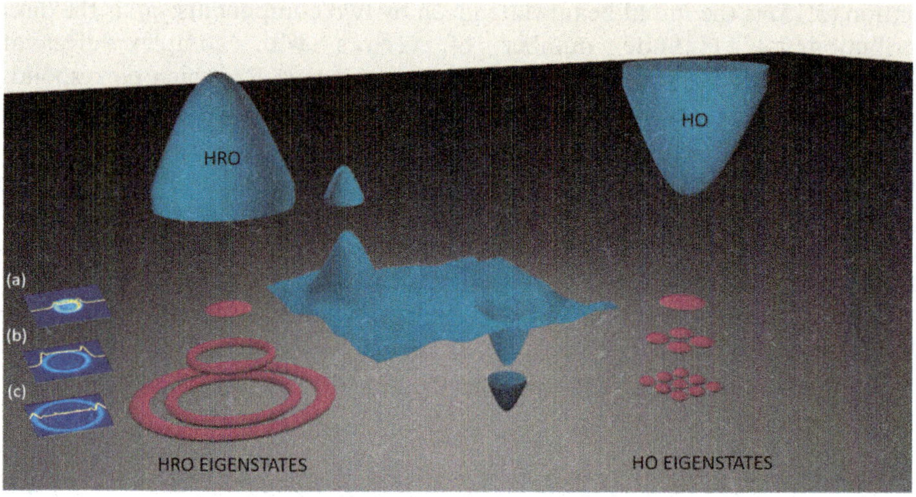

Figure 3.3. Arbitrary potential energy landscape of a system. When the system is in proximity of a local maximum, it obeys the Hamiltonian of a Harmonic reversed oscillator (HRO). In proximity of the minimum, the system obeys the Hamiltonian of a HO. The corresponding eigenstates are also reported: on the left side are sketched the eigenstates corresponding to the imaginary eigenvalue, $\Gamma_n/2$, of the HRO Hamiltonian; on the right side are shown the bounded eigenstates associated with increasing quantized energy levels of the HO Hamiltonian. (a)–(c) Transmitted profile of DSWs excited in a nonlinear optothermal medium during the experimental realization of a RHO.

polynomials. On the contrary, $\hat{H}_{HRO}$ corresponds to a system in proximity of a maximum in its potential energy V. The form of the eigenvalues and eigenvectors of $\hat{H}_{\mathrm{RHO}}$ can be derived by that of the standard harmonic oscillator. Indeed, the RHO is a complex extension of the HO.

Let's consider a family of operators [34] $\hat{V}_\nu = \exp\left[\frac{\nu}{2}(\hat{x}\hat{p} + \hat{p}\hat{x})\right]$. We have $[\hat{x}, \hat{p}] = i$, so the action of $\hat{V}_\nu$ on a state ϕ results as $\hat{V}_\nu\phi(x) = e^{-i\frac{\nu}{2}}\phi(e^{-i\nu}x)$, while $\hat{V}_\nu\hat{x}\hat{V}_\nu^{-1} = e^{-i\nu}\hat{x}$ and $\hat{V}_\nu\hat{p}\,\hat{V}_\nu^{-1} = e^{i\nu}\hat{p}$. From this, one can find $\hat{V}_{\pm\pi/4}\hat{H}_{\mathrm{RHO}}\hat{V}_{\pm\pi/4}^{-1} = \pm i\hat{H}_{\mathrm{RHO}}$, and hence one can transform the spectrum of the HO into that of the RHO by an extension to the complex plane. The eigenvalues and eigenvectors of the RHO are

$$E_n^{\mathrm{RHO}} = i\gamma\left(n + \frac{1}{2}\right), \tag{3.19}$$

$$\mathfrak{f}_n^{\pm}(x) = \frac{\sqrt[4]{\pm i\gamma}}{\sqrt{2^n n!\sqrt{\pi}}} e^{\frac{\mp i\gamma}{2}x^2}\mathcal{H}_n\left(\sqrt{\pm i\gamma}\,x\right). \tag{3.20}$$

For any n, two GVs exist: $\mathfrak{f}_n^-$ displaying an exponential decay, and $\mathfrak{f}_n^+ = (\mathfrak{f}_n^-)^*$ with exponential growth. As said before, these states are a basis for the RHS and are quasi-orthogonal and quasi-complete, namely,

$$\langle \mathfrak{f}_n^- | \mathfrak{f}_m^+ \rangle = \delta_{nm} \qquad \text{and} \qquad \sum_{n=0}^{\infty} \mathfrak{f}_n^-(x)\mathfrak{f}_n^+(x') = \delta(x - x'), \tag{3.21}$$

with δ_{nm} being the Kronecker symbol and $\mathfrak{f}_n^+$ being the time-reversed correspondent of $\mathfrak{f}_n^-$. The model just described intrinsically includes irreversibility in its formulation. In order to verify the presence of these decaying states in the propagation of DSWs, we calculate the probability $p_n(z)$ of finding our evolved state in a GV, by projecting the decaying component in equation (3.15) $\phi^G(x)$ over $\sqrt{\Gamma_n}\mathfrak{f}_n^+$. This gives the evolution of the irreversible part of the wavefunction, which can be also interpreted as the decaying process of elementary excitations with energy given by $E_n^R = \kappa_0^2 P$ and quantized decay rates $\Gamma_n = \gamma(2n + 1)$:

$$\phi_n^G(x,z) = \sum_{n=0}^{N} \left\langle \mathfrak{f}_n^+ \middle| \psi(x,0) \right\rangle \mathfrak{f}_n^- e^{-i\kappa_0^2 Pz - \frac{\Gamma_n}{2}z}. \tag{3.22}$$

The expansion in (3.22) is valid only for $z > 0$ and

$$p_n(z) = \Gamma_n \left| \left\langle \mathfrak{f}_n^+ \middle| \psi(x,0) \right\rangle \right|^2 e^{-\Gamma_n z} \tag{3.23}$$

gives the z-dependent weight of the n-order GV. If $\psi(x,0)$ is a pure GV $\psi(x,0) = \mathfrak{f}_n^-(x)$, we have $p_n(z) = \Gamma_n \exp(-\Gamma_n z)$ with normalized $\int_0^\infty p_n(z)dz = 1$. The excited GVs depend on the initial profile $\psi(x,0)$. For a Gaussian beam, $\psi(x,0) = \exp(-x^2/2)/\sqrt[4]{\pi}$, all the odd terms in equation (3.22) vanish due the x-parity and, for the first two even GVs, we have

$$p_0(z) = 2\frac{\gamma^{3/2}}{\sqrt{1+\gamma^2}}e^{-\gamma z} \qquad \text{and} \qquad p_2(z) = 5\frac{\gamma^{3/2}}{\sqrt{1+\gamma^2}}e^{-5\gamma z}. \tag{3.24}$$

The wavefunction upon evolution can be hence written as a superposition of functions that decay exponentially in the z direction. The components of this expansion can be calculated analytically or numerically, and the resulting scenario predicts quantized decay constants Γ_n [10]. We remark that this analysis is valid in the regime for the generation of photothermal shock waves, and we show in the following that it provides a comprehensive description of experimental results. We also address how the exponent Γ_n scales as the square root of the beam power, which confirms the nonlinear origin of the process, as the standard linear decay rates do not depend on the beam power. The power dependence of Γ_n is an experimentally testable prediction, which directly allows us to discriminate GVs from other damping mechanisms such as scattering or absorption.

3.3.4 Gamow vector experiments

In this section, experimental results are presented showing the presence of GVs in DSWs and proving the power of this formalism in the description of the shock phenomenon. The experiment consists of exciting a SW in a nonlinear, nonlocal optothermal medium generating a quantized inverted oscillator potential. As seen before, the heat diffusion in the photothermal liquid, belonging to the presence of the intense laser beam in the sample, causes the extension of the refractive index perturbation far beyond the beam intensity profile [32, 35–37]. Hence, light experiences a refractive index with an inverted parabolic spatial distribution [32, 38, 39]. We remark that the presence of nonlinearity in the system gives us the possibility to distinguish GVs from linear losses, which are not included in the RHO model. In figure 3.1, we illustrate the experimental setup and the top-view imaging apparatus for the experiment. A continuous wave (CW) laser beam at 532nm wavelength is focused through a spherical lens (L_1) with focal length 100 mm and a focus spot size of 10 μm into the sample. The light is collected by a spherical lens (L_2) and a charge coupled device (CCD) camera. A microscope is placed above the sample in order to capture top-view fluorescence images of the laser beam along the propagation direction z. The setup is placed having the beam propagating vertically through the sample in order to reduce convection in the water. Samples are prepared by dispersing 0.1 mM of RhB in water. The solution is placed in a cuvette 1 cm thick along the propagation direction. The measured defocusing Kerr coefficient is $|n_2| = 2 \times 10^{-12}$ m^2 W^{-1} and the absorption length $L_{abs} \simeq 1.6$ mm at the laser wavelength [17].

In figures 3.4(a) and (b), we report the CCD top images obtained for two increasing laser powers P. At low power ($P = 20$ mW), beam propagation does not manifest strong divergence and is dominated by diffraction. At high power ($P = 400$ mW), the beam displays a strongly divergent funnel shape, a signature of optical nonlinearity in the RhB solution. This regime is characterized by a fast and power-dependent beam decay along the propagation z direction.

Three transverse sections of the intensity profile at different propagation distances z = 0.2, 0.6, and 0.9 mm are displayed in figure 3.4c: the emergence of the typical shock 'M-shape' profile. For long propagation length, internal undular bores, which are the characteristic fast oscillations inside the field due to the regularization of the wave-breaking, emerge between the intensity lateral peaks. In order to observe GV decay rates Γ_n, we take slices of the intensity profile $I(x, z)$ at $x \simeq 0.1$ mm and fit the intensity versus z data with two exponential functions. The analysis is reported in figure 3.4(d) and repeated at different power levels. The signature double exponential behavior is most evident at high power. Observed and calculated double-exponential decay dynamics are found to obey the quantized spectrum scaling $\Gamma_2/\Gamma_0 = 5$ at all investigated power levels, P (see figure 3.4(c)). This demonstrates that we excite the fundamental state ($\mathfrak{f}_0^-$, $n = 0$) and the first excited state ($\mathfrak{f}_2^-$, $n = 2$). The state $\mathfrak{f}_1^-$ is not excited, as expected from the input beam symmetry. Each of the two rates is found to have a square root dependence on P (see the super imposed dashed lines in figure 3.4(e)), a signature of the underlying nonlinearity. This power dependence distinguishes RHO dynamics from linear losses, due to absorption and scattering.

In conclusion we have demonstrated that RHS Hamiltonian theories are able to describe irreversible processes and include the arrow of time as originating from

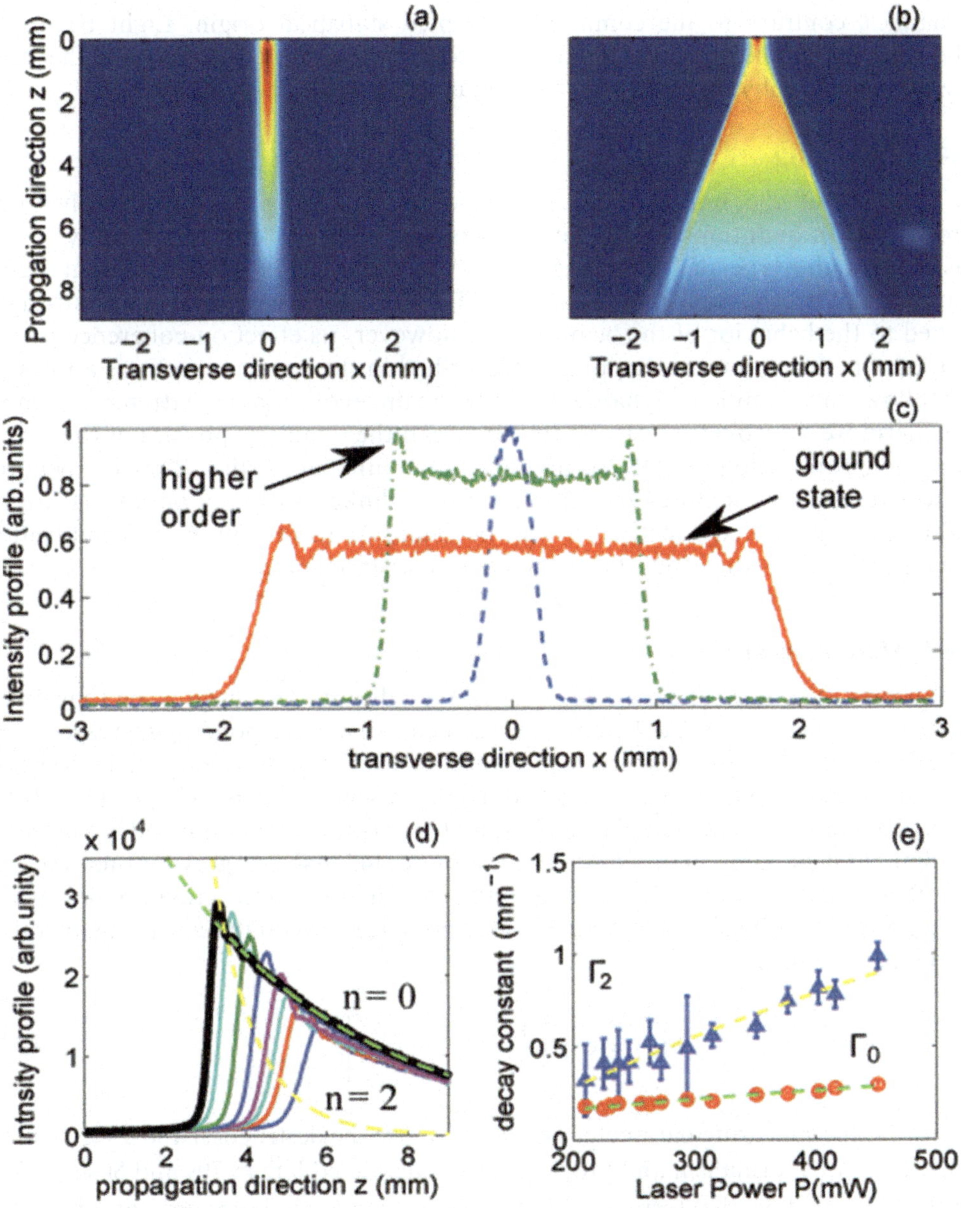

Figure 3.4. (a) and (b) Top-view fluorescence image for a 10 μm waist at $P = 20$ mW (a) and $P = 400$ mW (b). The excitation of the nonlinear response of the sample is signaled by the strongly diverging funnel shape of the beam and the emergence of the fast oscillations. (c) Transverse sections of the image in (b) taken at $z = 0.2$, 0.6, and 0.9 mm. The flat region corresponds to a ground state. (d) Observed intensity decay at different laser powers as obtained by slicing along the propagation z-direction the top-view intensity distribution shown in panel (b). (e) Decay rates vs P for the fundamental state, Γ_0 (filled circles) and the excited state, Γ_2 (triangles).

dynamics, contrary to the commonly accepted statistical origin. Light beams in photothermal liquids provide direct evidence of the main prediction of the theories: the existence of physical processes that manifest discrete decay rates.

3.4 Squeezing

A measure of the non-classical nature of a state is given by its coherence properties. In quantum optics, the study of the degree of coherence allows us to distinguish the type of light considered; for example, between coherent light, number states, and squeezed light [40, 41]. The shock phenomenon is strongly linked to the behavior of the beam phase; however, its effect on coherence is still not known. The theoretical approach described in the previous sections allows us to follow the complete dynamics of the beam evolution in extreme regimes. Furthermore, it provides a strong link between the quantum physics of the RHO and the propagation of DSWs in nonlocal nonlinear media. This connection fosters research of features in nonlinear optics linked to the quantum nature of light. An unpredicted effect which can be found thanks to this new formalism is the emergence of squeezing during the shock phenomenon.

3.4.1 Mathematical

A squeezed state is a state for which exists a quadrature Q_1, or Q_2, such that it has dispersion below that of a coherent state, at a cost of a corresponding increase of the other variable. The phase-space distribution of such a state is elliptic, in contrast with the circular phase-space distribution of a coherent beam of light [41]. First discovered in 1927 by Kennard [42], these states are purely quantum and are recently becoming outstanding in modern quantum optics [43, 44] and gravitational physics for their fundamental role in quantum computation and gravitational wave detection [45–50]. Their mathematical description is based on the squeezing operator which has the form

$$\hat{S}(\zeta) = \exp\left[\frac{1}{2}\left(\zeta^* \hat{a}^2 - \zeta \hat{a}^{\dagger 2}\right)\right], \tag{3.25}$$

where ζ is the squeezing parameter and $\hat{a}$ and $\hat{a}^\dagger$ are the destruction and construction operators. $\hat{S}(\zeta)$ formalism and properties were discussed late in 70s and 80s [51–54]. Widely debated is the form of the $\hat{S}(\zeta)$ spectrum: a community of physicists maintains that its spectrum is composed of a family of discrete eigenvalues belonging to that of Gamow; however, being a unitary operator ($\hat{S}\hat{S}^\dagger = 1$), its spectrum is usually considered continuum and covering the unitary circular in the complex plane. However, Chruscinski in [55] showed that the action of $\hat{S}$ can be completely described by its discrete spectrum.

As we showed in the previous section, the wave propagation in a defocusing nonlinear, nonlocal media can be described thanks to the eigenstates of a RHO. The wave evolution can be written in terms of the quantum mechanics propagator of the

RHO, $\hat{U} = \exp - (i\hat{H}_{RHO}z)$. By the canonical transformation $(x, p) \longrightarrow (u, v)$ we can show that $\hat{U}$ takes the form of the squeezing operator $\hat{S}$. Introducing the annihilation and creation operators,

$$\hat{a} = \frac{u + iv}{\sqrt{2}} \qquad \text{and} \qquad \hat{a}^{\dagger} = \frac{u - iv}{\sqrt{2}}, \tag{3.26}$$

we obtain

$$\hat{H}_{RHO} = \frac{\gamma i}{2}(a^2 - a^{\dagger 2}) \tag{3.27}$$

and hence

$$\hat{U} = e^{\frac{\gamma}{2}(a^2 - a^{\dagger 2})z}. \tag{3.28}$$

which is the squeezing operator $\hat{S}$ with squeezing parameter $\xi = \gamma z$. By knowing the eigenvalues of $\hat{H}_{rho}$ in the rigged Hilbert space, we find the eigenvalue equation for the squeezing operator:

$$\hat{S}\mathfrak{f}_n^{\pm} = s_n^{\pm}\mathfrak{f}_n^{\pm} \qquad \text{with} \qquad s_n^{\pm} = e^{\pm\xi\left(n+\frac{1}{2}\right)}. \tag{3.29}$$

We can hence conclude that the nonlinear defocusing dynamics in a nonlocal medium corresponds to the action of the squeezing operator on the initial state. Furthermore, it is worth noticing that $\gamma \propto \sqrt{P}$. By using this squeezing formalism, we can compute the nonlinear evolution of a Gaussian beam in a highly nonlocal medium, with initial condition $\psi_0(x) = e^{-x^2/2}/\sqrt[4]{\pi}$. The evolved state takes the form

$$\hat{S}(\xi)\psi_0(u) = e^{\xi/2}\psi_0 u e^{\xi}. \tag{3.30}$$

In order to verify that the state is a squeezed state, we have to compute the uncertainties for the squeezed quadratures u and v:

$$\begin{aligned} \Delta u^2 &= \langle S(r)\psi_0(u)|\hat{u}^2|S(r)\psi_0(u)\rangle = \frac{e^{-2r}}{4}\left(\frac{1+\gamma^2}{\gamma}\right), \\ \Delta v^2 &= \frac{e^{2r}}{4}\left(\frac{1+\gamma^2}{\gamma}\right). \end{aligned} \tag{3.31}$$

The $\hat{u}$ uncertainty decreases at the cost of the corresponding increase in Δv. The uncertainty principle reads as

$$\Delta u^2 \Delta v^2 = \left(\frac{(1+\gamma^2)^2}{16\gamma^2}\right). \tag{3.32}$$

From equation (3.32), we see that the squeezing depends on the beam power P. The physical consequences of this are that one can inject a laser beam in a thermal nonlinear medium and, by tuning the power, achieve different degrees of squeezing. Worthwhile to notice is the fact that the squeezed quadratures u and v rotate when changing the input flux, and hence the elliptical phase-space distribution also rotates in the quadratures plane.

3.4.2 Squeezing in nonlinear optics

This squeezing formalism offers new perspective into the comprehension of the nonlocal nonlinear wave propagation with respect to the hydrodynamic approach valid when the HNA holds true. As shown in the first paragraph, the nonlinear Schrödinger equation in the hydrodynamical limit can be written as

$$i\varepsilon\partial_\rho\psi + \frac{1}{2}\varepsilon^2\partial_\chi^2\psi - PK(\chi/\varepsilon)*|\psi|^2\psi = 0 \tag{3.33}$$

where the adimensional parameter $\varepsilon = \sqrt{L_{nl}/L_d} \ll 1$ takes into account the competition between diffraction and nonlinearity, and $\chi = x\varepsilon$ and $\rho = z\varepsilon$ are adimensional coordinates. We can express the squeezed variables u and v in this new coordinate system:

$$\hat{u} = \frac{\gamma x - p}{\sqrt{2\gamma}} \longrightarrow \frac{\frac{\gamma\chi}{\varepsilon} + i\varepsilon\partial_\chi}{\sqrt{2\gamma}} \quad \text{and} \quad \hat{v} = \frac{\gamma x + p}{\sqrt{2\gamma}} \longrightarrow \frac{\frac{\gamma\chi}{\varepsilon} - i\varepsilon\partial_\chi}{\sqrt{2\gamma}}. \tag{3.34}$$

By writing $\psi = Ae^{i\phi/\varepsilon}$ and assuming $\varepsilon \ll 1$, we can apply the WKB approach and calculate the uncertainties Δu^2 and Δv^2 for $\varepsilon \to 0$:

$$\Delta u^2 \to \frac{\gamma}{2}\Delta x^2 \quad \text{and} \quad \Delta v^2 \to \frac{\gamma}{2}\Delta x^2. \tag{3.35}$$

From equation (3.35), we find that the hydrodynamical approach does not predict the squeezing of the u and v variables. However, this behavior shows a three-stage propagation: at first, the quadratures start squeezing, than squeezing stops and there is a transient towards the highly nonlinear propagation in the hydrodynamic regime.

This theoretical analysis is confirmed by numerical simulations: a Gaussian beam $\psi(x, 0) = \psi_0(x)$ propagates according to equation (3.33) in a nonlocal nonlinear medium. Figure 3.5(b) reports the quadrature uncertainty Δu^2 and Δv^2 trends, showing the three predicted regions. The first stages of propagation see the squeezing of the quadrature u and an exponential increasing in Δv^2, validating the reverse harmonic oscillator approximation. Worth noting is that the indetermination product is constant in this region, as expected from equation (3.32). When the maximum squeezing degree is reached, the dynamics changes and the quadratures get the same hydrodynamical trend. By this analysis, we can conclude that the Gamow approach allows us to predict new features of wave propagation in highly nonlinear and nonlocal media which are hidden from the hydrodynamical description. The reason why the hydrodynamic approximation fails at the lowest order is due to the fact that it neglects the derivative of the density $|\psi|^2$ and only accounts for the dynamics of the phase.

A further proof of the presence of squeezing is given by the phase-space distribution of the numerically evolved beam. A squeezed state has an elliptical phase-space distribution which can be verified through the Wigner transform:

$$W(x, k) = \frac{1}{\pi}\int \psi^*(x + y)\psi(x - y)e^{2iky}dy. \tag{3.36}$$

Figure 3.6 shows the evolved beam simulation for input power $P = 100$ after equation (3.33) and the correspondent Wigner distribution $W(x, k)$ for the initial state ($z = 0$) and the evolved state at ($z = 2$). After a few steps of propagation,

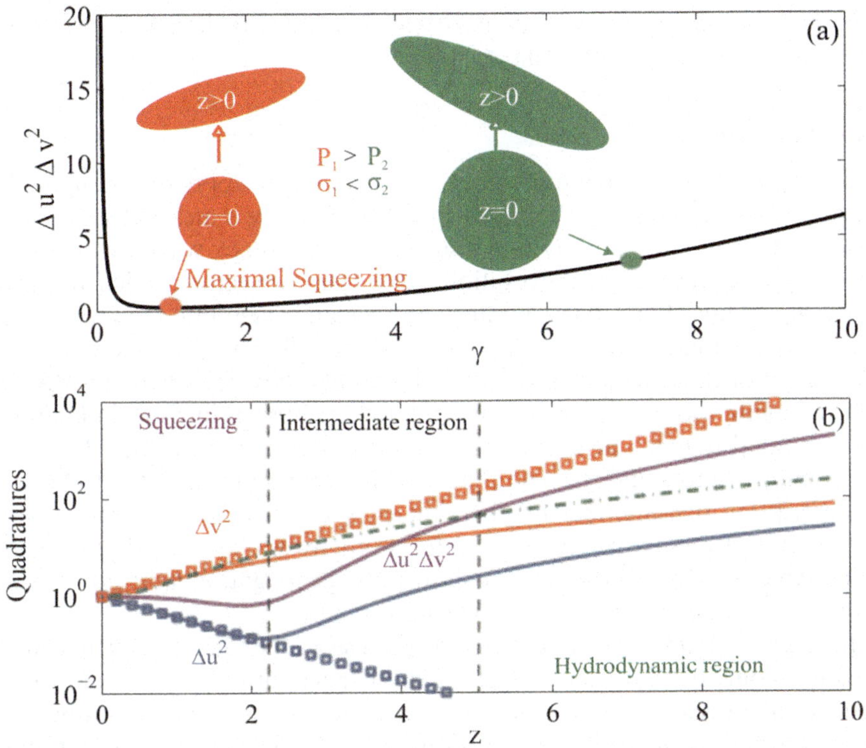

Figure 3.5. (a) Uncertainty principle $\Delta u^2 \Delta v^2$ behavior as function of γ. The inset shows that the phase-space distribution area changes with γ and hence with P and σ, while the phase-space distribution form varies from circular to elliptical during the propagation. (b) Quadrature u (blue line) and v (red line) uncertainties and uncertainty principle $\Delta u^2 \Delta v^2$ (magenta line). Squares represent the theoretical behavior for the two quadratures u (blue squares) and v (red squares) in semi-log-scale after equation (3.31), with $P = 100$ and $\sigma = 15$. Green dot-dashed curve provides the expected result in the hydrodynamical regime after equation (3.35). Dashed line evidence the 3 region limits.

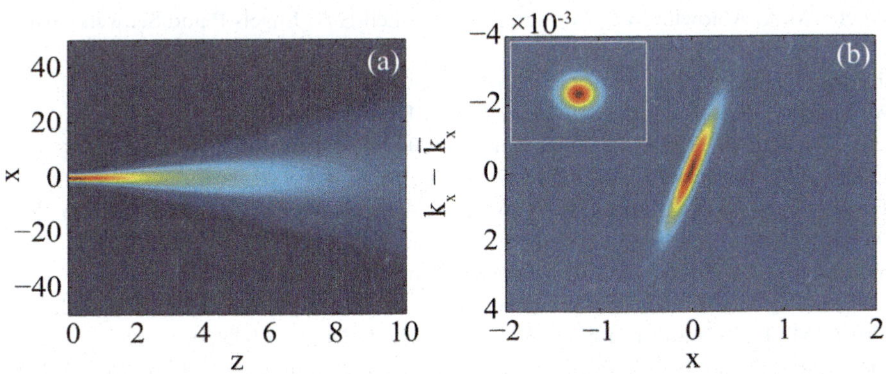

Figure 3.6. (a) Numerical solution of equation (3.33) with $P = 100$ and $\sigma = 15$; (b) Wigner function calculated after equation (3.33) of the evolved beam in panel (a) at $z = 2$, with $\bar{k}_x = 257$; the inset shows the Wigner function of the initial condition of panel (a).

$W(x, k)$ becomes elongated and compressed at a slanted direction, assuming the characteristic squeezing elliptic shape.

3.5 Conclusions

We have shown that by assuming the highly nonlocal approximation in the the nonlinear propagation regime, the excitation of the Gamow states is clearly evident by the measured decay rates, and by the occurrence of the shock waves, which is shown to be directly linked to the linear evolution of the Gamow vectors, which also gives the characteristic undular bore. These findings provide an insight into the way irreversible dynamics arise within a conservative system, providing a clear indication of the intrinsic origin of the exponential decays. The results show, in a way directly distinct from the coupling with the environment, that irreversibility is intrinsic in Hamiltonian dynamics, and also give a demonstration that decay rates are quantized.

The Gamow vectors and their nonlinear counterparts are expected to have a role in the description of a number of phenomena, such as shock and rogue waves in Hamiltonian and dissipative systems, evanescent waves, and localized resonances including Anderson localization, ultra-short pulse propagation, and sub-wavelength plasmonics and imaging at the subwavelength regime.

In the present chapter we have reported the use of the Gamow eigenvectors of a reversed harmonic oscillator in a rigged Hilbert space formalism in order to explain the squeezing process of photon fluids both in temporally and spatially nonlocal media. The proposed analysis, even if limited to the classical regime, establishes a link between the classical photon fluid description and quantum optics that may potentially surpass limits of squeezing generation for various applications in quantum information and gravitational wave detection.

Funding is recognized from the John Templeton Foundation (Grant No. 58277) and Sapienza Ateneo 2017.

References

[1] Hoefer M A, Ablowitz M J, Coddington I, Cornell E A, Engels P and Schweikhard V 2016 *Phys. Rev.* A **74** 023623

[2] Taylor R J, Baker D R and Ikezi H 1970 *Phys. Rev. Lett.* **24** 206

[3] Molinari A and Daraio C 2009 *Phys. Rev.* E **80** 056602

[4] Barsi C, Wan W, Sun C and Fleischer J W 2007 *Opt. Lett.* **32** 2930

[5] Conforti M, Baronio F and Trillo S 2014 *Phys. Rev.* A **89** 013807

[6] El G A, Gammal A, Khamis E G, Kraenkel R A and Kamchatnov A M 2007 *Phys. Rev.* A **76** 053813

[7] Whitham G B 1999 *Linear and Nonlinear Waves* (New York: Wiley)

[8] Ghofraniha N, Conti C, Ruocco G and Trillo S 2007 *Phys. Rev. Lett.* **99** 043903

[9] Crosta M, Trillo S and Fratalocchi A 2012 *New J. Phys.* **14** 093019

[10] Gentilini S, Braidotti M C, Marcucci G, DelRe E and Conti C 2015 *Phys. Rev.* A **92** 023801

[11] Gentilini S, Braidotti M C, Marcucci G, DelRe E and Conti C 2015 *Sci. Rep.* **5** 15816

[12] Glauber R J 1986 *Ann. N. Y. Acad. Sci* **480** 336

[13] Wan W, Jia S and Fleischer J W 2007 *Nat. Phys.* **3** 46

[14] Fouxon I, Farberovich O V, Bar-Ad S and Fleurov V 2010 *Euro. Lett.* **92** 14002
[15] Elazar M, Fleurov V and Bar-Ad S *Phys. Rev.* A **86** 063821
[16] Ghofraniha N, Santamaria Amato L, Folli V, DelRe E, Trillo S and Conti C 2012 *Opt. Lett.* **21** 612
[17] Ghofraniha N, Gentilini S, Folli V, DelRe E, Trillo S and Conti C 2012 *Phys. Rev. Lett.* **109** 243902
[18] Gentilini S, Ghofraniha N, DelRe E and Conti C 2013 *Phys. Rev.* A **87** 053811
[19] Fratalocch A, Armaroli A and Trillo S 2011 *Phys. Rev.* A **83** 053846
[20] Garnier J, Xu G, Ghajeri F, Ghofraniha N, DiFalco A and Conti C 2014 *Opt. Exp.* **22** 1667
[21] Lecaplain C, Soto-Crespo J M, Grelu P and Conti C *Opt. Lett.* **39** 263
[22] Moro A and Trillo S 2014 *Phys. Rev.* E **89** 023202
[23] Bohm A 1981 *J. Math. Phys.* **22** 2813
[24] Castagnino M, Diener R, Lara L and Puccini G 1997 *Int. J. Theo. Phys.* **36** 2349
[25] Chruściński D 2002 *Open Sys. Inf. Dyn.* **9** 207
[26] Gamow G 1928 *Z. Phys.* **51** 204
[27] Sudarshan E C G, Chiu C B and Gorini V 1978 *Phys. Rev.* D **18** 2914
[28] Parravicini G, Gorini V and Sudarshan E C G 1980 *J. Math. Phys.* **21** 2208
[29] Marcucci G and Conti C 2016 arXiv.1607.08109
[30] Marcucci G, Braidotti M C, Gentilini S and Conti C 2017 *Ann. Phys. Lpz.* 1600349
[31] Sun C, Jia S, Barsi C, Rica A, Picozzi A and Fleischer J W 2012 *Nat. Phys.* **8** 471
[32] Snyder A W and Mitchell D J *Science* **276** 1538
[33] Braidotti M C, Gentilini S and Conti C 2016 *Opt. Exp.* **24** 21963
[34] Chruściński D 2004 *J. Math. Phys.* **45** 841
[35] Bang O, Krolikowski W, Wyller J and Rasmussen J J 2002 *Phys. Rev.* E **66** 046619
[36] Turitsyn S K 1985 *Teor. Mat. Fiz.* **64** 226
[37] Rotschild C, Cohen O, Manela O, Segev M and Carmon T 2005 *Phys. Rev. Lett.* **95** 213904
[38] Gordon J P, Leite R C C, Moore R S, Porto S P S and Whinnery J R 1965 *J. Appl. Phys.* **36** 3
[39] Conti C, Peccianti M and Assanto G 2004 *Phys. Rev. Lett.* **92** 113902
[40] Boyd R W 2008 *Nonlinear Optics* (Academic Press)
[41] Loudon R 1983 *The Quantum Theory of Light* (Oxford: Clarendon Press)
[42] Kennard E 1927 *Z. Phys.* **44** 326
[43] Walls D 1983 *Nature* **306** 141
[44] Walls D and Milburn G 1999 *Quantum Optics* (Berlin: Springer-Verlag)
[45] Gisin N, Ribordy G, Tittel W and Zbinden H 2002 *Rev. Mod. Phys.* **74** 145
[46] Ladd T D, Jelezko F, Laflamme R, Nakamura Y, Monroe C and O'Brien J L 2010 *Nature* **464** 45
[47] Collaborationa L S 2011 *Nat. Phys.* **7** 962
[48] Chua S S Y, Slagmolen B J J, Shaddock D A and McClelland D E 2014 *Class. Quantum Grav.* **31** 83001
[49] Schnabel R, Mavalvala N, McClelland D E and Lam P K 2010 *Nat. Commun.* **1** 121
[50] Vahlbruch H, Mehmet M, Danzmann K and Schnabel R 2016 *Phys. Rev. Lett.* **117** 110801
[51] Stoler D 1970 *Phys. Rev.* D **1** 3217
[52] Stoler D 1971 *Phys. Rev.* D **4** 1925
[53] Fishe R A, Nieto M M and Sandberg V D 1984 *Phys. Rev.* D **29** 1107
[54] Ma X and Rhodes W 1990 *Phys. Rev.* A **41** 4625
[55] Chruscinski D 2004 *Phys. Lett.* A **327** 290

IOP Publishing

Nonlinear Guided Wave Optics
A testbed for extreme waves
Stefan Wabnitz

Chapter 4

Observation of the rupture of a photon dam in an optical fiber

Matteo Conforti, Gang Xu, Arnaud Mussot, Alexandre Kudlinski and Stefano Trillo

The study of the breaking of a dam is a classic problem of hydrodynamics. In the small dispersion and highly nonlinear regime, the nonlinear Schrödinger equation, which describes evolution of the optical power and chirp inside the fiber, is analogous to the 1D shallow water equations describing the temporal evolution of the water height and vertically averaged velocity. By exploiting such an analogy in a fiber optics setup, we investigate the temporal photonic analogue of the dam-breaking phenomenon in shallow water. We observe the decay of a steep input front, representing the photonic dam, into a rarefaction wave and a dispersive shock wave propagating in opposite directions. Our measurements show the clear evidence of a critical transition of the dispersive shock wave into a self-cavitating state, characterised by the presence of a null-intensity point into the shock fan. The precise measurements of the temporal waveform of the cavitating state dynamics allows for a fully quantitative test of the Whitham modulation theory applied to the defocusing nonlinear Schrödinger equation.

4.1 Introduction

Natural hazards related to water flows cause remarkable losses every year in terms of human lives, ships, buildings, and infrastructure. For example, the sudden release of water stored in an artificial lake due to a dam collapse can lead to serious damage to the downstream valley. Consequently, dam break flows have become an important practical problem in civil engineering. Field observations of the flood wave propagation present some obvious difficulties such as the danger, isolation of the phenomenon from external influences, and lack of accuracy in the measurements. A way to overcome these limitations is to perform experimental study using suitable small-scale models in controlled conditions [1, 37].

doi:10.1088/978-0-7503-1460-2ch4 © IOP Publishing Ltd 2017

A different and more elegant approach lies in exploiting the formal analogy between the equations governing the propagation of light and those ruling the evolution of water waves. Light can indeed propagate behaving as a photon fluid in many different physical situations [2, 5, 7, 13, 14, 19, 32, 42–44]. In particular, laser light propagating in nonlinear media can exhibit phenomena that characterise fluid flows such as rogue waves [32, 36], instabilities [19], transition to turbulence [42], coherent [7, 14, 44] and incoherent [13] shock waves, and superfluid flow around obstacles [5, 43]. In regimes described by the defocusing nonlinear Schrödinger equation (NLSE), a distinctive trait of the photon fluid evolution is the formation of dispersive shock waves (DSWs, or undular bores) [7, 12, 14, 28, 44, 45, 48], fast oscillating wavetrains that spontaneously emerge from the tendency to develop a gradient catastrophe [11, 21, 31]. Wave breaking regularised by dispersion, and in particular DSWs, is ubiquitous, being observed in other systems ruled by the NLSE such as cold atom condensates [8, 17] and spin waves [18], as well as in other dispersive hydrodynamic settings involving, e.g. electrons [29], water waves [40, 41], and viscous fluid conduits [26] to name a few.

A major breakthrough in the analytical description of the DSW is the Whitham modulation theory [9, 11, 15, 21, 31, 33], which, however, assumes DSWs to develop from an ideal step-like initial condition (known as Riemann problem in classical shock theory), at variance with the experiments performed to date, which have been mostly concerned with smooth or periodic initial conditions. It is therefore of crucial importance to design experiments which can give access to the study of the Riemann problem in order to advance the general understanding of dispersive hydrodynamic flows. Indeed, the modulation theory predicts for such problems a non-trivial dynamics with critical transitions in the behavior of the shock, which have not been observed to date [9, 17].

In this chapter, we exploit a fiber optics setup to investigate experimentally the specific Riemann problem associated with an initial jump in the optical power as a function of time. In the absence of any frequency chirp across the jump, such a problem is equivalent to the classic 1D dam-break problem of hydrodynamics [25, 38, 46]. Indeed, we demonstrate that the light evolves as a fluid mimicking the basic features of the dam-break according to non-dispersive shallow water equations (or Saint Venant equations), namely the decay into a pair of waves, specifically a shock and a rarefaction wave, connected by an expanding plateau. The dispersive character of the shock, however, leads to a critical transition to cavitation, which is characteristic of any system ruled by the NLSE and which can be predicted in the framework of Whitham theory [9, 17]. Above a critical height of the jump, we report evidence for the onset of self-cavitation, i.e. the appearance of point of total darkness of the optical power. Importantly, our experimental results agree well quantitatively with modulation theory [47].

4.2 Theory of classic and dispersive dam breaking

We assume that the evolution of the electric field envelope $E(T,Z)$ along the fiber distance Z is described by the NLSE

$$i\frac{\partial E}{\partial Z} - \frac{k''}{2}\frac{\partial^2 E}{\partial T^2} + \gamma |E|^2 E = 0, \tag{4.1}$$

where $T = T_{lab} - k'Z$ is the retarded time in a frame moving with group-velocity $1/k' = dk/d\omega|_{\omega_0}^{-1}$, whereas $k'' = d^2k/d\omega^2|_{\omega_0}$ and γ are the dispersion and the nonlinear coefficient of our fiber, respectively, ω_0 being the carrier frequency. In all our experiments, higher-order dispersion and nonlinearity (self-steepening and Raman) play an insignificant role. Moreover, even if the linear losses can be relevant, we managed to fully compensate for them as detailed below. We work in the normal dispersion region of the fiber $k'' > 0$, which corresponds to the defocusing regime of the NLSE. We introduce the non-dimensional variables $z = Z/Z_0$, $t = T/T_0$ with $Z_0 \equiv (\gamma P)^{-1}$ and $T_0 \equiv \sqrt{k''/\gamma P}$, $\psi(z,t) = E(Z,T)/\sqrt{P}$, with P being the peak input power. The NLSE in normalised units reads

$$i\frac{\partial \psi}{\partial z} - \frac{1}{2}\frac{\partial^2 \psi}{\partial t^2} + |\psi|^2 \psi = 0. \tag{4.2}$$

The Madelung transform $\psi(t,z) = \sqrt{\rho(t,z)}\exp(-i\int_{-\infty}^{t} u(t',z)dt')$ allows us to formulate the NLSE in the following equivalent hydrodynamical form [24]:

$$\rho_z + (\rho u)_t = 0, \tag{4.3}$$

$$u_z + \left(\frac{u^2}{2} + \rho\right)_t = \frac{1}{4}\left[\frac{\rho_{tt}}{\rho} - \frac{(\rho_t)^2}{2\rho^2}\right]_t. \tag{4.4}$$

By neglecting the RHS containing higher-order derivatives, equations (4.3) and (4.4) reduce to the one-dimensional dispersionless hyperbolic model, or vector conservation law, known as the shallow water equations (SWEs) [25], which are commonly used in hydraulics to predict 1D (i.e. along a channel) water flows in the limit of small water depth. SWEs can be derived from Euler equations by assuming a hydrostatic pressure and by reducing the velocity to a vertically averaged velocity which is directed along the pure horizontal direction. A more general 2D version of the SWEs involves two components of the horizontal velocity, whereas the fluid velocity in the vertical direction is always neglected. The role of the local water elevation and longitudinal velocity in hydrodynamics is played, in the optical problem, by the normalised power $\rho = |E|^2/P$ and the chirp u, whereas space and time in the optical setting have interchanged role with respect to hydrodynamics. Equations (4.3) and (4.4) can also be cast in the form

$$\rho_z + (\rho u)_t = 0, \tag{4.5}$$

$$(\rho u)_z + \left(\rho u^2 + \frac{1}{2}\rho\right)_t = 0, \tag{4.6}$$

of a conservation law $q_z + f_t = 0$ for mass and momentum $q = [\rho, \rho u]^T$ and the appropriate flux f, which can be easily derived [6]. Equations (4.3) and (4.4) with the RHS set to zero, or equivalently equations (4.5) and (4.6), are also called the dispersionless NLSE. While this terminology finds its justification in the fact that it neglects higher-order terms which are entirely due to the dispersion (the RHS of equation (4.4), which, in a semiclassical formulation of the NLSE, are of higher order in the overall parameter that quantifies the smallness of dispersion), it is worth pointing out that the SWEs contain terms of leading order which arise from the dispersive term in the NLSE. Indeed, fully neglecting the dispersion from the beginning would give the power profile $\rho(t)$ as a conserved quantity along the evolution.

The SWEs or dispersionless NLSE can also be written in diagonal form in terms of Riemann invariants $r^{\pm}(t,z) \equiv u(t,z) \pm 2\sqrt{\rho(t,z)}$,

$$\partial_z r^{\pm} + v^{\pm}\partial_t r^{\pm} = 0, \tag{4.7}$$

where the eigenvelocities are $v^{\pm} = v^{\pm}(r^{-},r^{+}) = (3r_{\pm} + r_{\mp})/4$.

Since SWEs are a two-dimensional vector conservation law, the dynamics of the general Riemann initial value problem, i.e. a step in ρ and u, entails the superposition of a wave pair. Each of the waves can be a smooth rarefaction wave (RW) or classical shock wave (SW), i.e. a traveling discontinuity between left ($\rho_{s,L}$, $u_{s,L}$) and right ($\rho_{s,R}$, $u_{s,R}$) values. The velocity V_c of a classical SW can be found by means of the Rankine–Hugoniot condition for a vector conservation rule [6, 17, 25]

$$V_c(q_{s,L} - q_{s,R}) = f(q_{s,L}) - f(q_{s,R}), \tag{4.8}$$

where we emphasize that V_c must be the same scalar constant for the two equations arising from the two-component field q and the relative flux f.

A classification into (i) RW–RW, (ii) SW–SW, and (iii) RW–SW can be made according to the values of the step in ρ and u. Such classification can be suitably extended to the dispersive regime ruled by the NLSE [9, 17]. In particular, the dam-break Riemann problem corresponds to a step in the variable $\rho(t,z=0)$ with $u(t,z=0) = 0$, and leads to a RW–SW pair separated by a plateau with constant values $(\rho, u) = (\rho_i, u_i)$, ρ_i being an intermediate value between the two boundary values of the jump ρ_L, ρ_R. Assuming the initial step to be in $t = 0$, when the step is increasing ($\rho_L < \rho_R$), the SW turns out to be left-going, heading towards $t < 0$ (as is the case in our experiment, see below), whereas it is right-going for a decreasing step $\rho_L > \rho_R$. In other words, the shock always points towards the direction where the initial step decreases.

Figures 4.1(a) and (b) contrast the evolutions of $\rho(t,z)$ for the dam-break process, starting from a step which is increasing from left to right, ruled either by the SWEs or dispersionless NLSE (figure 4.1(a)) or by the full NLSE (figure 4.1(b)). For positive times, the two solutions exhibit substantially the same behavior, featuring a right-going RW sitting on an intermediate constant state (the small oscillations which are visible close to the connection between the RW and the higher constant state are due to the Gibbs phenomenon, and should not be confused with the larger oscillations that characterise the DSW). For negative times, the SWEs give rise to a left-going classical shock wave whose velocity is given by the RH conditions

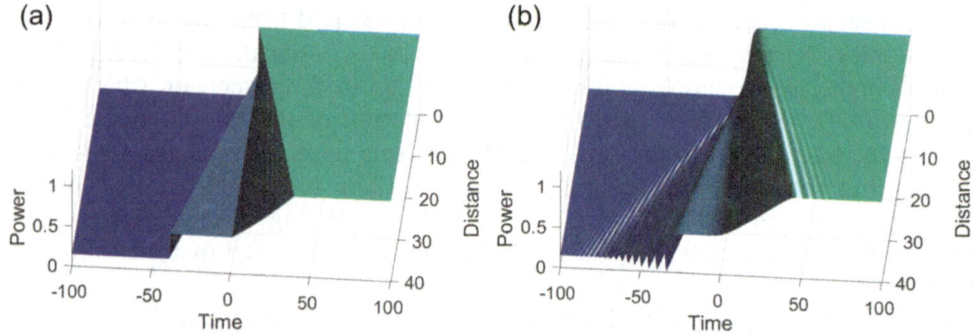

Figure 4.1. Classical versus dispersive dam break problem. (a) Exact solution of shallow water equations (4.5) and (4.6). (b) Numerical solution of NLSE equation (4.2). Here the initial step is characterised by the boundary values $\rho_R = 1$, $\rho_L = 0.15$, $u_{L,R} = 0$.

whereas, in the NLSE solution, the traveling discontinuity is replaced by an expanding fan filled by fast oscillations that form spontaneously, which constitute the undular bore or DSW. In the latter case, the velocity of the shock is not unique, as the RH velocity is replaced by the two velocities that characterise the edges of the expanding oscillatory zone. In the next section, we show how to calculate the key parameters of the two solutions, namely the values of the constant state along with the velocity of the classical shock wave in the dispersionless limit, as well as the edge velocities of the DSW in the fully dispersive regime.

4.2.1 Hydrostatic dam breaking

A solution to the dam-break problem for the SWEs with non-vanishing downstream level, i.e. $\rho_L \neq 0$ or 'wet-bed' case, was first due to Stoker [22, 38], who has suitably generalised the classical solution given by Ritter for the 'dry-bed' case [34] often reported in textbooks on water waves [20]. We point out, however, that this is a highly non-trivial generalization. In fact, in the dry-bed case, no shock can be formed and the solution of the SWEs implies a single RW smoothly connecting the upper state of the dam with the dry bed. Vice versa, in the wet-bed case, the general solution involves, besides the RW, a SW propagating downstream over the quiescent bed. The dry-bed case can be seen as the limiting case where the SW disappears due to the downstream water level that tends to vanish, i.e. $\rho_L \to 0$.

According to the Stoker approach, the Rankine–Hugoniot equations (4.8) conditions for the SW between the plateau (ρ_i, u_i, both *a priori* unknown) and the left state ρ_L, $u_L = 0$, allow us to express at the same time the SW velocity V_c and u_i as a function of the boundaries (quiescent states ρ_L, ρ_i and $u_R = 0$):

$$u_i = (\rho_L - \rho_i)\sqrt{\frac{\rho_i + \rho_L}{2\rho_i\rho_L}}, \tag{4.9}$$

$$V_c = -\rho_i\sqrt{\frac{\rho_i + \rho_L}{2\rho_i\rho_L}}. \tag{4.10}$$

Then, exploiting the constancy of the Riemann invariant $r^- = u - 2\sqrt{\rho}$ across the RW, one has the additional constraint $u_i = 2\sqrt{\rho_i} - 2\sqrt{\rho_R}$. By equating the two expressions for u_i, one obtains an equation for ρ_i which can be solved by a root-finding algorithm. Once ρ_i and u_i are known, one can fully characterise the RW and the SW. An example of the SWE solution is shown in figure 4.1(a). The shock and the rarefaction wave propagate in opposite directions, towards $t < 0$ and $t > 0$, respectively, being connected by the expanding constant state ρ_i, u_i.

A different formulation, which is more directly suited for the extension to the dispersive case ruled by the full NLSE, amounts to solving in terms of so-called *simple waves* (i.e. waves for which one of the Riemann invariants remains constant). The dam-break problem with $\rho_L < \rho_R$ implies an increasing (decreasing) step for $r^+(t, 0)$ [$r^-(t, 0)$], as displayed in figure 4.2 (steps from $r_L^+ = 2\sqrt{\rho_L}$ to $r_R^+ = 2\sqrt{\rho_R}$, and from $r_L^- = -2\sqrt{\rho_L}$ to $r_R^- = -2\sqrt{\rho_R}$, respectively). For this case, characterised by $r_R^- < r_L^- < r_L^+ < r_R^+$, we report in figure 4.2 the simple wave solutions of equations (4.7), such that one of the Riemann invariants is constant and other one varies in self-similar way, i.e. $r = r(\tau = t/z)$. The solution involves, for $\tau > 0$, a constant $r^- = r_R^-$ and a RW for r^+ described by $\tau = v^+(r_R^-, r^+)$, whereas for $\tau < 0$, $r^+ = r_L^+$ is constant and $\tau = v^-(r^-, r_L^+)$ describes the characteristic overturning (see figure 4.2). The solution can be regularised by an SW through the scalar Rankine–Hugoniot condition (not to be confused with the vector Rankine–Hugoniot equations (4.8)) imposed on the equation for r^- ($r^+ = r_L^+$ is constant), which gives a shock velocity $\tau_s = (\tau_1 + \tau_2)/2 = 3(r_L^- + r_R^-)/8 + r_L^+/4 = -(3\sqrt{\rho_R} + \sqrt{\rho_L})/4$ (cyan step in figure 4.2). The upper state of the shock is the plateau in $\rho, u = \rho_i, u_i$, which is found by exploiting continuity in $\tau = 0$ to be $u_i = (r_L^+ + r_R^-)/2 = \sqrt{\rho_L} - \sqrt{\rho_R}$ and $\rho_i = [(r_L^+ - r_R^-)/4]^2 = (\sqrt{\rho_L} + \sqrt{\rho_R})^2/4$.

The approriate limits of the solution give the relevant values of the self-similar variable τ_i, $i = 1, 2, 3, 4$, shown in figure 4.2, as follows:

$$\tau_4 = \frac{3r_R^+ + r_R^-}{4} = \sqrt{\rho_R}, \tag{4.11}$$

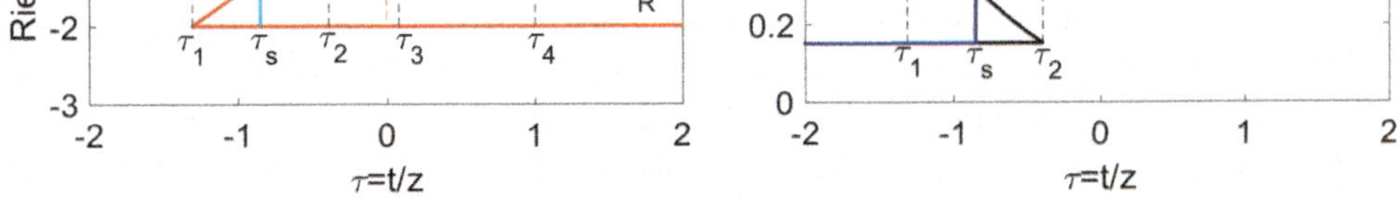

Figure 4.2. (a) Self-similar evolution of the two Riemann invariants $r^\pm$ of the SWEs in the dam-break problem. The dashed line steps are the Riemann initial values. The cyan step is the SW obeying the scalar Rankine–Hugoniot condition. (b) Corresponding plot of the density ρ. Initial step $\rho_L = 0.15$, $\rho_R = 1$.

$$\tau_3 = \frac{3r_L^+ + r_R^-}{4} = \frac{3\sqrt{\rho_L} - \sqrt{\rho_R}}{2}, \tag{4.12}$$

$$\tau_2 = \frac{3r_L^- + r_L^+}{4} = -\sqrt{\rho_L}, \tag{4.13}$$

$$\tau_1 = \frac{3r_R^- + r_L^+}{4} = \frac{\sqrt{\rho_L} - 3\sqrt{\rho_R}}{2}, \tag{4.14}$$

where we have exploited the fact that, in the dam-break, $r_{L,R}^{\pm} = \pm 2\sqrt{\rho_{L,R}}$. It is important to emphasise that τ_4 and τ_3 in equations (4.11) and (4.12) represent the edge velocities of the RW which hold valid, due to the RW smoothness, as well as when dispersive corrections are accounted for. Conversely, when dispersive corrections are applied to the SW, one needs to calculate the edge velocities by averaging over the DSW oscillation as explained below. We anticipate that this gives a result markedly different from the expression of τ_2 and τ_1 in equations (4.13) and (4.14).

4.2.2 Dispersive dam breaking

The SWEs predict a parabolic rarefaction fan and a sharp shock front. However, the nonhydrostatic pressure provokes undulations on the flow profile not predicted by equations (4.5) and (4.6). This fact has been quite recently pointed out theoretically and experimentally also in the context of hydraulics [3, 4, 23, 35, 37]. However, the details of the oscillatory regularization of the shock obviously depend on the specific form of the dispersive corrections to the hyperbolic hydrodynamic model. On this basis, one expects differences between the hydraulic case (where the dispersive evolution is described by equations such as the Korteweg–de Vries, Boussinesq, or their generalization) and the optical evolution ruled by the NLSE. In particular, in the framework of the NLSE, the classical SW is regularised into a DSW by the effect of the RHS of equation (4.4) that stems from second-order dispersion.

An example of the breaking scenario obtained from the full NLSE is reported in figure 4.1(b). The DSW is delimited by two characteristic edge velocities $\tau_{1,2}$, where the oscillations vanish (linear edge) or become deepest (soliton edge), respectively. The modulation theory based on Whitham averaging can be exploited to calculate such velocities. Indeed, by means of the Whitham theory, one can describe the DSW as a modulated cnoidal wave with parameters that vary slowly with respect to its natural periodicity. The averaged Whitham equations describe the slow variation of the parameters and allow us to compute the velocities of the linear edge (i.e. where the amplitude of the oscillation vanishes) and soliton edge (i.e. where the frequency of the oscillation vanishes) of the DSW. For the details on the method, which can be suitably extended also to non-integrable nonlinear dispersive PDEs, we refer the reader to the original literature [9, 15] and to the excellent reviews on the topic, in particular to references [10, 17]. Below, we only briefly outline the steps that lead us to obtain the velocities associated with the DSW.

The Whitham equations for the NLSE can be expressed in the following diagonal form [33], by introducing four Riemann invariants $r_i = r_i(t,z)$, $i = 1, 2, 3, 4$, ordered in such a way that $r_1 < r_2 < r_3 < r_4$, which are a suitable combination of the parameters of the cnoidal wave [9, 11, 15, 17, 33]

$$\frac{\partial r_i}{\partial z} + v_i \frac{\partial r_i}{\partial t} = 0; \quad i = 1, 2, 3, 4. \tag{4.15}$$

Here, the velocities $v_i = v_i(r_1, r_2, r_3, r_4)$ depend on combinations of $\{r_i\}$ and elliptic integrals of the first ($K(m)$) and second ($E(m)$) kind. For instance, the velocity $v_2 = v_2(r_1, r_2, r_3, r_4)$, which will be relevant in the following, reads explicitly as

$$v_2 = V + \frac{1}{2}(r_2 - r_1)\left[1 - \frac{(r_3 - r_1)E(m)}{(r_3 - r_2)K(m)}\right]^{-1}, \tag{4.16}$$

where $V = \frac{1}{4}(r_1 + r_2 + r_3 + r_4)$ and $m = \frac{(r_4 - r_3)(r_2 - r_1)}{(r_4 - r_2)(r_3 - r_1)}$.

As shown above, a left-going classical shock of the SWEs is characterised by a constant r^+ and a jump in r^- from r_L^- to r_R^- (with given left and right boundaries r_L^-, r_R^- such that $r_R^- < r_L^-$). The corresponding dispersive regularisation of the shock, i.e. the DSW, can be described by a solution of equations (4.15) where only one of the Riemann invariants varies. For the left-going DSW, this solution, sketched in figure 4.3, is generated by the initial value $r_{i0} = r_i(t, z = 0)$ with $r_{10} = r_R^-$, $r_{30} = r_L^-$, $r_{40} = r^+$, and a step r_{20} in $t = 0$ from r_R^- to r_L^- (obtained from the so-called data regularisation procedure, see [17]).

The solution is such that r_1, r_3, r_4 remain constant at any z, whereas r_2 gives rise to a smooth RW (owing to the fact that r_{20} is non-decreasing) that depends on the self-similar variable $\tau = t/z$. Indeed, all Whitham equations are formally satisfied when

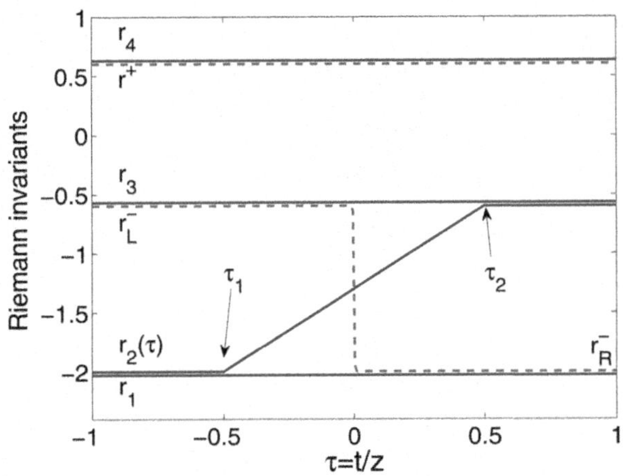

Figure 4.3. Sketch of the Riemann invariants of equations (4.15), where r_1, r_3, r_4 stay constant and $r_2 = r_2(\tau = t/z)$ varies smoothly in the range $\tau_1 \le \tau \le \tau_2$ between the values r_R^- and r_L^-. τ_1 and τ_2 correspond to the linear and soliton edges of the DSW, respectively. The dashed red lines are the Riemann invariants in the dispersionless limit. Reprinted figure with permission from [47]. Copyright 2017 by the American Physical Society.

$r_{1,3,4}(t,z) = r_{10,30,40}$ and $r_2(t,z) = r_2(\tau)$, provided the equation $(\tau - v_2)r_2' = 0$ is fulfilled. For $r_2(\tau) \neq constant$, this implies $\tau = v_2$. The latter relation, once v_2 is expressed as $v_2 = v_2(r_{10}, r_2, r_{30}, r_{40})$ according to equation (4.16), becomes a nonlinear equation in the only unknown $r_2(\tau)$, which can be solved to find the RW. The velocity $\tau_1 \equiv \tau_{linear}$ and $\tau_2 \equiv \tau_{soliton}$ of the leading and trailing edges of the DSW correspond to the edges of this rarefaction wave and can be calculated from the following limits (see figure 4.3):

$$\begin{aligned}\tau_1(r^+) &= \lim_{r_2 \to r_R^-} v_2(r_R^-, r_2, r_L^-, r^+)\\ &= \frac{2r_R^- + r^+ + r_L^-}{4} + \frac{(r^+ - r_R^-)(r_L^- - r_R^-)}{2r_R^- - r^+ - r_L^-},\end{aligned} \tag{4.17}$$

$$\tau_2(r^+) = \lim_{r_2 \to r_L^-} v_2(r_R^-, r_2, r_L^-, r^+) = \frac{r_R^- + r^+ + 2r_L^-}{4}. \tag{4.18}$$

To obtain final formulas that depend only on the quiescent states $r_{L,R}^{\pm}$, we observe that, across the shock, one has a constant $r^+ = r_L^+$ as is clear from figure 4.2(a). Hence, by expressing $r_{L,R}^{\pm}$ as a function $\rho_{L,R}$, we obtain

$$\tau_2 = -\frac{\sqrt{\rho_L} + \sqrt{\rho_R}}{2}; \qquad \tau_1 = \frac{\rho_L - 2\rho_R}{\sqrt{\rho_R}}. \tag{4.19}$$

A snapshot of the breaking scenario obtained from the full NLSE is compared with the dispersionless limit in figure 4.4(a). The snapshot corresponds to the output of our experiment, using the fiber parameters and input conditions corresponding to the experiments that will be described in the following. The DSW and the RW are delimited by the two pairs of edge velocities $\tau_{1,2}$ and $\tau_{3,4}$, respectively, which define the wedges where the DSW and RW expand, as shown in figure 4.4(b), in dimensional units, i.e. including the multiplicative factor $T_0/Z_0 = \sqrt{k''\gamma P_R}$.

4.2.3 Self-cavitating or vacuum points

Importantly, the modulation theory entails a crossover between two different regimes separated by a critical condition. In the first regime, the DSW envelopes are monotone and the DSW power never vanishes. However, below a critical value of the key parameter, namely the ratio between the quiescent (down- and up-stream) states that we henceforth denote as $r = \rho_L/\rho_R = P_L/P_R$, the DSW exhibits a self-cavitating point, i.e. zero power, corresponding to a vacuum point in gas dynamics.

Since the minimum power of the cnoidal wave turns out to be $\rho_{min} = (r_1 - r_2 - r_3 + r_4)^2/16$ [9, 17], the existence in the DSW of a cavitating or vacuum state $\rho_{min} = 0$ requires $r_2 = r_1 - r_3 + r_4$. Therefore, in analogy with equations (4.17) and (4.18), we obtain the self-similar location of the vacuum by performing the following limit

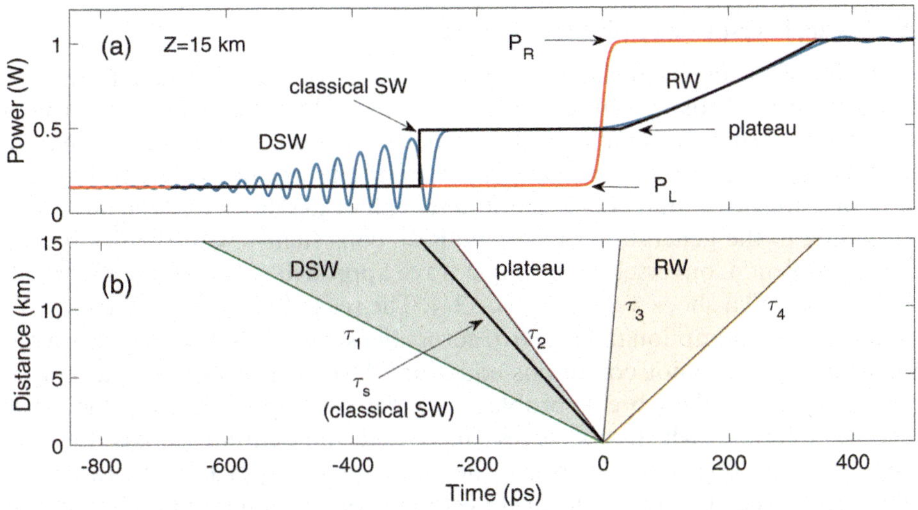

Figure 4.4. (a) Snapshots comparing the DSW–RW pair (solid blue curve) obtained from the NLSE with step-like input $E(T,0) = [P_L + (P_R - P_L)(1 + \tanh(T/T_r))/2]^{1/2}$ (solid red curve) with the ideal dispersionless solution of the SWEs (solid black). The oscillations over the top of the RW are due to the Gibbs phenomenon. (b) Wedges in (T,Z) plane corresponding to the RW (fable orange), the DSW (green), and the plateau in between. The boundaries correspond to slopes $\sqrt{k''\gamma P_R}\tau_j$, $j = 4, 3, 2, 1$. Here, $k'' = 176$ ps^2 km^{-1}, $\gamma = 3$ (W km)$^{-1}$, $P_R = 1$ W, $P_L = 150$ mW, and $T_r = 10$ ps. Reprinted figure with permission from [47]. Copyright 2017 by the American Physical Society.

$$\begin{aligned}\tau_0 &= \lim_{r_2 \to r_R^- - r_L^- + r_L^+} v_2(r_R^-, r_2, r_L^-, r^+) \\ &= \frac{r_L^+ + r_R^-}{2} - \frac{r_L^- - r_L^+}{2}\left[1 - \frac{r_R^- - r_L^-}{r_L^+ + r_R^- - 2r_L^-}\frac{E(m)}{K(m)}\right]^{-1} \\ &= \sqrt{\rho_L} - \sqrt{\rho_R} + 2\sqrt{\rho_L}\left[1 - \frac{\sqrt{\rho_R} - \sqrt{\rho_L}}{\sqrt{\rho_R} - 3\sqrt{\rho_L}}\frac{E(m)}{K(m)}\right]^{-1}\end{aligned} \tag{4.20}$$

where $m = (\frac{r_L^- - r_L^+}{r_R^- - r_L^-})^2 = 4\rho_L/(\sqrt{\rho_L} - \sqrt{\rho_R})^2$, and we expressed the boundary value of Riemann invariants in terms of ρ, i.e. $r_{L,R}^{\pm} = \pm 2\sqrt{\rho_{L,R}}$. The threshold for cavitation can be obtained by imposing $m = 1$ (cavitation on soliton edge of the DSW), or equivalently $\tau_2 = u_i$, from which we obtain

$$r_{th} \equiv \left(\frac{\rho_L}{\rho_R}\right)_{th} \equiv \left(\frac{P_L}{P_R}\right)_{th} = \frac{1}{9} \simeq 0.11. \tag{4.21}$$

For P_L/P_R below such threshold, a vacuum point always exists, which tends towards the linear edge of the DSW in the limit $P_L \to 0$. In this limit, however, both the oscillation amplitude and the plateau extension tend to vanish [9], and one recovers the limiting hydrodynamic case known as 'dry-bed' dam-break, characterised by a single RW extending to zero and no shock [24, 27, 46].

4.2.4 Classical versus dispersive shock waves

The velocity of the classical shock V_c and the values of the intermediate state ρ_i, u_i obtained from the Stoker solution differ from the velocity τ_s and the intermediate state calculated from the simple wave approach. This difference is very small for a shallow step $\rho_L \approx \rho_R$, but becomes important when the step grows larger. In the limit $\rho_L \to 0$, the Stoker solution entails the suppression of the shock and the intermediate state, leading to the generation of a single RW, consistently with the dry-bed dam-breaking problem. Conversely, the simple wave approach gives a finite intermediate value $\rho_i = 1/4$ and a shock velocity $\tau_s = -3/4$. The reason for this discrepancy is that the simple wave assumption ($r^+ = const.$ across the shock) is not compatible with the vector Rankine–Hugoniot conditions equation (4.8) for the shallow-water system. For the hydrostatic dam break problem, the Stoker approach gives the physical admissible solution, which conserves mass and momentum across the shock. However, for the dispersive problem, the simple wave approach [9] turns out to be the correct one, giving an accurate estimate of the intermediate state and of the soliton edge velocity of the DSW. Figure 4.5 reports a comparison between the numerical solution of the NLSE (blue curve) and the Stoker (red curve) and simple wave approaches (black curve) for three different relative values of the background $\rho_L = 0.15$ (a), 0.01 (b), and 0.001 (c).

For $\rho_L = 0.15$ (figure 4.5(a)) NLSE, the Stoker and simple wave solutions give approximately the same value of the intermediate state. For $\rho_L = 0.01$ (figure 4.5(b)), qualitative differences start to appear. The simple wave approach gives the correct value (i.e. in accordance with the NLSE) for the intermediate state, whereas the Stoker solution gives a substantially smaller value. For an even smaller $\rho_L = 0.001$ (figure 4.5(c)), the temporal width of the intermediate state shrinks almost to zero with an amplitude of 0.25, both for the NLSE and the simple wave solution. For the Stoker solution, the amplitude of the intermediate state tends to zero. However, the amplitude of the oscillation of the DSW tends to vanish as well, entailing that the DSW tends to the rarefaction wave predicted by the Stoker approach.

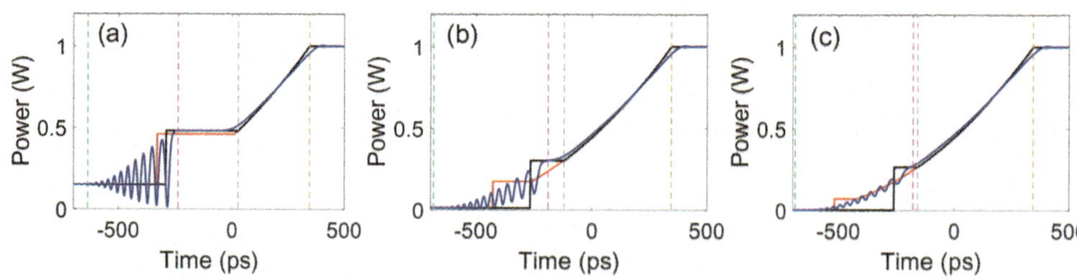

Figure 4.5. Snapshots comparing classical versus dispersive dam-breaking dynamics for different values of downstream power (elevation) ρ_L and fixed upstream value $\rho_R = 1$: (a) $\rho_L = 0.15$, (b) $\rho_L = 0.01$, and (c) $\rho_L = 0.001$. Dispersive dam breaking from numerical solution of NLSE (blue curves); Stoker solution of SWEs (red curves); simple-wave solution of SWEs (black curves). The vertical dashed lines give the predicted delays of the edges of the RW (gray and orange lines, from equations (4.11) and (4.12)) and the DSW (magenta and green lines, from equations (4.19)), respectively.

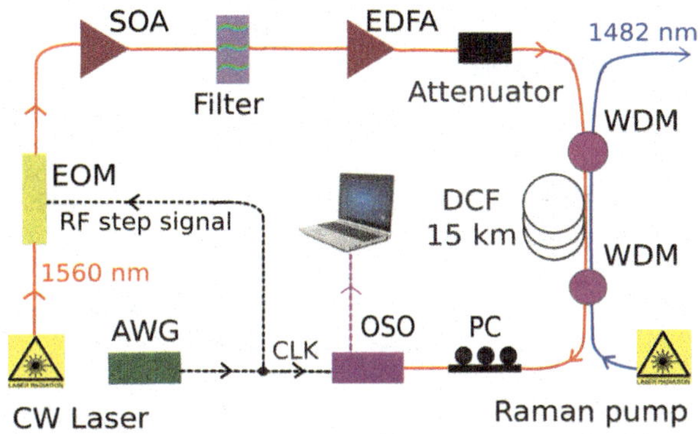

Figure 4.6. Experimental setup: EOM electro-optic modulator; EDFA, erbium-doped fiber amplifier; AWG, arbitrary wave generator; OSO, optical sampling oscilloscope; SOA, semiconductor optical amplifier; WDM, wavelength division multiplexing; DCF, dispersion compensating fiber; PC, polarization controller. Reprinted figure with permission from [47]. Copyright 2017 by the American Physical Society.

4.3 Experiment

We performed a series of experiments to provide evidence for the decay of a photonic dam (a temporal step in optical power) into the DSW-RW pair and for the critical transition to self-cavitation.

In our experimental setup sketched in figure 4.6, we make use of a continuous laser diode source emitting at $\lambda = 1560$ nm, which is intensity modulated by an electro-optic modulator driven by an arbitrary waveform generator with typical rise time $T_r \sim 25$ ps (raised tanh shape, rising from 10% to 90% in ~50 ps) and 78 MHz repetition rate. The signal is pre-amplified in a semiconductor optical amplifier, and then it passes through a spectral filter to remove amplified spontaneous emission in excess, and is finally amplified by means of an Erbium-doped fiber amplifier. This signal is launched in a 15 km long dispersion compensating fiber ($k'' = 176\ \text{ps}^2\ \text{km}^{-1}$, $\gamma = 3\ (\text{W km})^{-1}$) and analysed in the time domain with an optical sampling oscilloscope with 1.6 ps resolution.

A typical shape of the step-like signal that we obtained is shown in figure 4.7(a). Two elevating steps in power from zero to P_L and from P_L to P_R can be adjusted independently, and are followed by a descending step (trailing edge) back to zero. The advantage of using this specific signal shape is that, in the same experimental run, we can compare (i) the DSW-RW dynamics developing around the jump from P_L to P_R with (ii) the 'dry-bed' dynamics developing over the descending step. Importantly, the duration of each of the constant power states P_L, P_R was adjusted to be long, up to ~1 ns, so that the DSW-RW pair can develop without feeling the interaction with the first step and the trailing edge of the waveform. Clearly, as can be seen in figure 4.7(a), the main step (P_L to P_R) on which we mainly focus our attention in the following is not instantaneous. However, the short rising time allows us to clearly observe the dam-break phenomenon, as we will see below. Another challenging issue faced in the experiment is the loss compensation. Indeed, the DSW-RW dynamics is very sensitive to losses and even weak losses of optical fibers

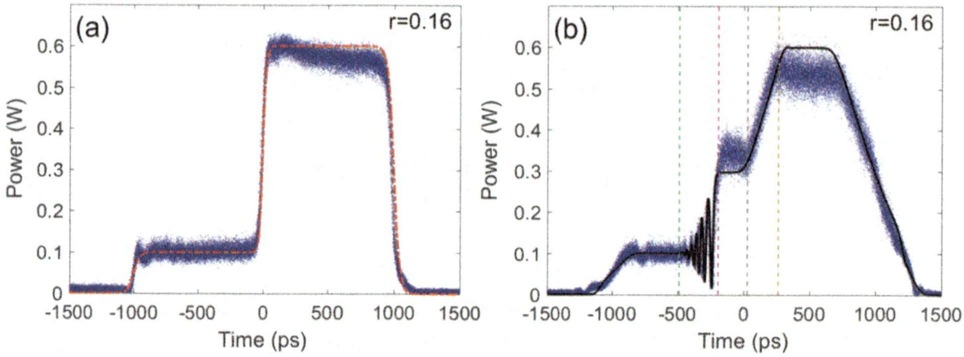

Figure 4.7. Temporal traces of the whole waveform: experiment (blue dots) versus numerics (black and red curves). The left panel (a) shows the input; the right panel (b) is the output power profile after propagation along the 15 km long fiber. The vertical dashed lines give the predicted delays of the edges of the RW (gray and orange lines, from equations (4.11) and (4.12)) and the DSW (magenta and green lines, from equations (4.19)), respectively.

(0.5 dB km^{-1}) are strongly detrimental. For instance, the plateau that connects the RW and the DSW would be completely distorted, not allowing for a quantitative comparison with theory. Inspired by transparent telecom networks, we counterbalanced linear losses by means of Raman amplification [16, 30]. To this end, a counterpropagating beam was injected in the fiber at $\lambda = 1482$ nm. In this way, we achieved significant (close to peak) Raman gain with weak relative noise intensity transfer [16, 30]. Figure 4.7(b) shows the output temporal trace obtained for the input step-like signal with $P_R = 0.6$ W, $P_L \simeq 100$ mW ($r \simeq 0.16$). As can be seen, the input optical dam at $T = 0$ breaks into a DSW-RW. The DSW is characterised by fast oscillations with ~40 ps average period (the period scales proportionally to $\sqrt{k''/(\gamma P_i)}$). The RW smoothly connects the plateau with power $P_i \simeq 0.3$ W (in agreement with the theoretical prediction $P_i = P_R\rho_i = 0.297$ W) to the peak level P_R. Conversely, over the trailing edge, no shock occurs and a smooth RW dropping to zero is observed, consistently with the case of 'dry-bed' dam-break. Overall, the data show an excellent agreement with the numerical simulations reported (black curves) based on the NLSE (4.1).

We have then proceeded to investigate in detail the breaking dynamics of the step-like input into the DSW-RW pair for different heights of the optical dam. The results shown in figure 4.8 are obtained for a fixed peak power $P_R = 0.8$ W (slightly larger than that of figure 4.7) and a variable background P_L, i.e. a variable ratio r. For a quantitative comparison with Whitham modulation theory we also report, as vertical dashed lines superimposed on the measured data, the delays $T_j = \sqrt{k''\gamma P_R}\,\tau_j L$, $j = 1, 2, 3, 4$ corresponding to the edges of the RWs in equation (4.11) and (4.12) and the DSW in equation (4.19), respectively. We remark that: (i) the experimental traces (blue dots) show a very good agreement with both simulations based on the NLSE (black curve), and the predicted delays T_j from modulation theory; (ii) the duration of the plateau connecting the DSW and the elevating RW significantly shrinks as the ratio r decreases, again in quantitative agreement with modulation theory which predicts that $|\tau_2 - \tau_3|$ reduces when

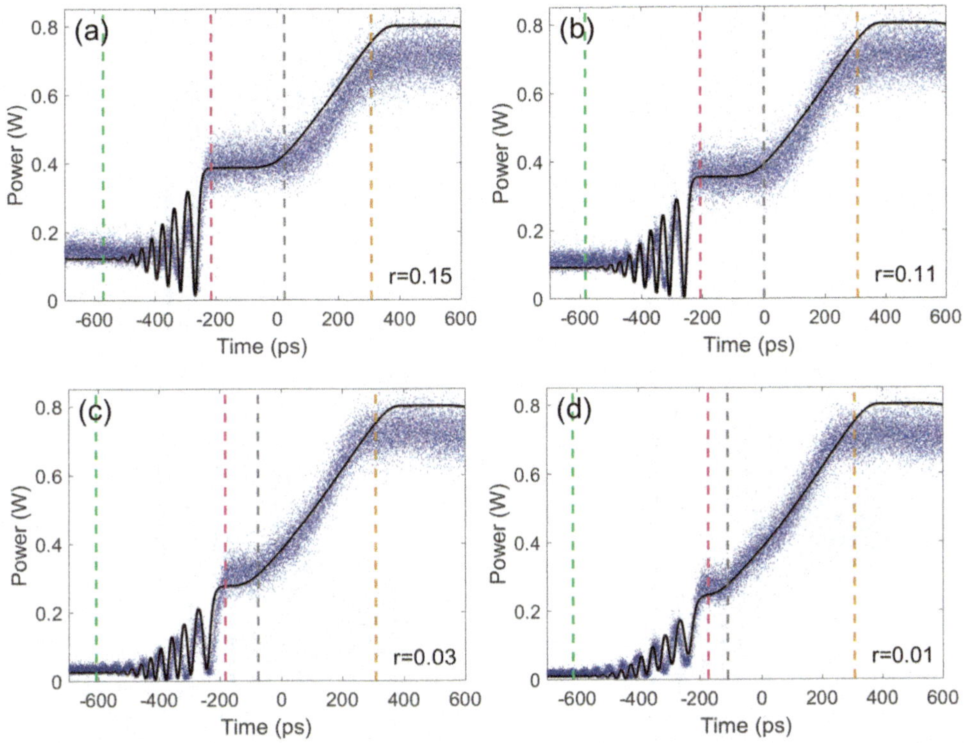

Figure 4.8. Temporal traces at the fiber output for constant peak power of the photonic dam $P_R = 0.8W$ and different background fractions r: (a) 0.15; (b) 0.11; (c) 0.03; (d) 0.01. Blue dots: experimental results; black curves: numerical simulation based on the NLSE.

r decreases; (iii) for relatively large ratios r the DSW never touches zero and the edge τ_2 of the DSW (where $m = 1$) is constituted by a gray soliton [17], as shown in figure 4.8(a). However, by decreasing r, we observe the onset of cavitation when the soliton at the trailing edge of the DSW becomes black (see figure 4.8(b), $r = 0.11$). The observed threshold value $r = 0.11$ is in excellent agreement with the theoretical prediction (equation (4.21)). While the black soliton possess a zero velocity (with respect to its background), the DSW edge maintains a finite velocity due to the chirped background (velocity u_i). Also associated with such black soliton we expect a phase jump of π which, however, cannot be measured with our setup. At even lower ratios r, the vacuum point shifts towards the left and the DSW envelope becomes non-monotone.

While the results in figure 4.8 already show the crossover to the cavitating state, a detailed quantitative study of this regime requires us to operate with a DSW possessing larger extension and shorter average period. To this end, we operate at the maximum available power in our set-up, $P_R = 1$ W. This allows us to observe a DSW exhibiting several oscillations with an average period of ~30 ps, spanning a range that exceeds 400 ps. Importantly, in this regime, the delay of the cavitating state (zero power) can be accurately identified within the DSW. Figures 4.9(a)–(d)

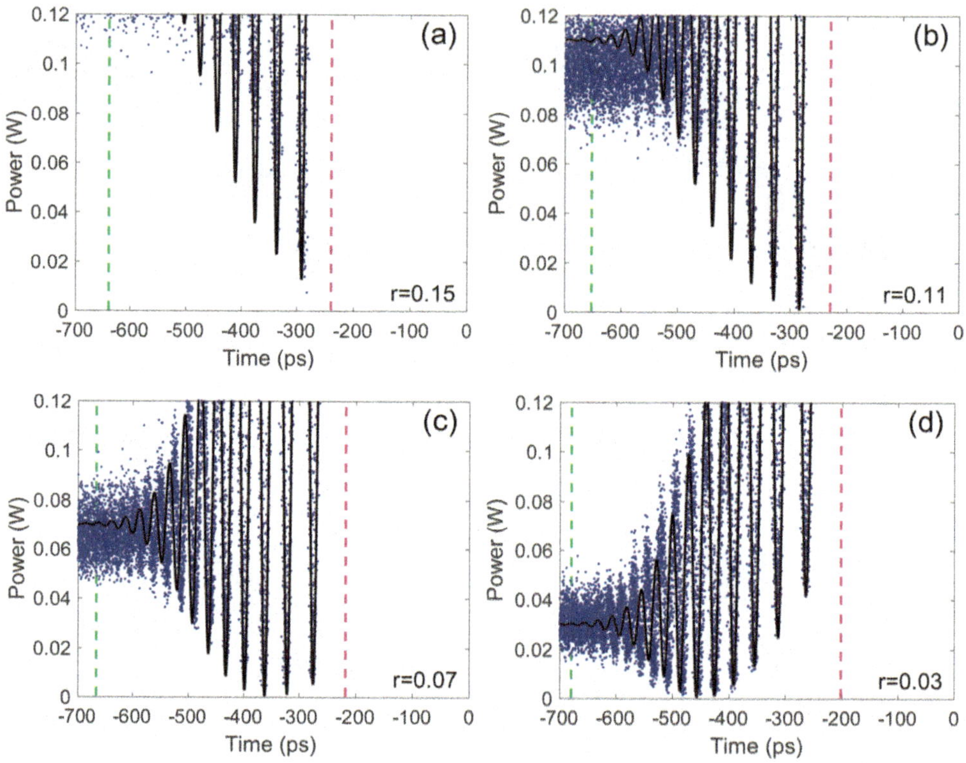

Figure 4.9. (a–d) As in figure 4.8, showing a zoom around the spontaneously cavitating state (vacuum point) of the DSW, for $P_R = 1$ W and different fractions $r = P_L/P_R$: (a) 0.15; (b) 0.11; (c) 0.07; (d) 0.03.

display a zoom over the bottom part of the DSW in order to show how the cavitating state moves when the ratio r is varied across the threshold. As shown in figure 4.9(a), when $r = 0.15$, the DSW still exhibits monotone envelopes featuring a gray soliton edge with non-vanishing dip. However, figure 4.9(b) shows very clearly the onset of cavitation (black soliton edge) at the threshold $r = 0.11$, in agreement with equation (4.21). Decreasing further r (i.e. for higher dam heights) leads the cavitating state to acquire increasingly negative delays, shifting progressively towards the linear edge of the DSW, as is clear from figure 4.9(c) for $r = 0.07$ and figure 4.9(d) for $r = 0.03$. For all cases, the numerical simulations are still in good agreement with the measured profiles.

By performing several measurements on the temporal location (delay) of the characteristic vacuum point for different values of the background ratio r, we can compare with the prediction of the Whitham modulation theory in equation (4.20). We summarise such results in figure 4.10, where we report results for the span $0.01 \leq r \leq r_{th}$, where the lower limit $r \sim 0.01$ is set by the residual noise over the bottom power level which makes it inaccurate to resolve the delay of the vacuum when this is quite close to the linear edge. The data are contrasted with both the theoretical prediction from equation (4.20) and numerical simulations from the NLSE, showing a satisfactory agreement in the whole range. In this sense, it is worth

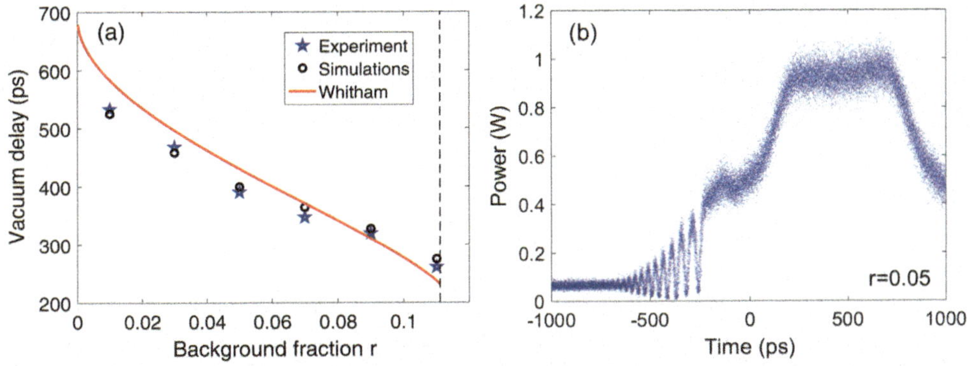

Figure 4.10. (a) Measured delay of the cavitating state versus r (blue stars), compared with prediction from equation (4.20) (red curve) and numerical simulations of the NLSE (black circles). The dashed vertical line stands for the threshold in equation (4.21). (b) Overall measured trace showing the output RW-DSW of the case $r = 0.05$.

emphasizing that it is indeed quite remarkable that the modulation theory quantitatively captures the essential traits of a complicated phenomenon, such as the development of the dispersive shock produced during the rupture of a photon dam. Minor discrepancies, in this framework, should be ascribed to the non-ideality (finite rise-time) of the input step and to the fact that the modulation theory has asymptotic character, that is, its accuracy is expected to improve as the propagation length grows very large and/or the dispersion is very small [31].

4.4 Conclusions

In summary, we have reported the fluid behavior of light in a dispersive dam-break experiment, revealing a transition to cavitation in close quantitative agreement with the predictions of modulation theory. Such behavior is expected to be universal for systems ruled by the NLSE, while qualitatively differing from other dispersive breaking scenarios observed in water [23, 35, 39]. While we addressed only the Riemann problem where the jump is in the optical power (equivalent to water elevation), our platform can be used to explore the rich scenario of critical behaviors and transitions between different regimes that one has in the general dispersive Riemann problem for the NLSE [9, 46]. This also includes the focusing regime of the NLSE, where the dam-break phenomenon presents a completely different dynamics where solitons and rogue waves are expected to play a role [11].

Acknowledgments

The present research was supported by IRCICA (USR 3380 CNRS), by the Agence Nationale de la Recherche in the framework of the Labex CEMPI (ANR-11-LABX-0007-01), Equipex FLUX (ANR-11-EQPX-0017), by the projects NoAWE (ANR-14-ACHN-0014), TOPWAVE (ANR-13-JS04-0004), and the Ministry of Higher Education and Research, Hauts de France Council and European Regional Development Fund (ERDF) through the Contrat de Projets Etat-Region (CPER

Photonics for Society P4S). ST acknowledges also the grant PRIN 2012BFNWZ2. The authors are grateful to L Bigot, E Andresen and IRCICA-TEKTRONIX European Optical and Wireless Innovation Laboratory for technical support for the electronic devices. ST acknowledges illuminating discussions with A Valiani.

References

[1] Bellos C V 2004 Experimental measurements of flood wave created by a dam break *Eur. Water* **78** 3–15

[2] Brambilla M, Lugiato L A, Penna V, Prati F, Tamm C and Weiss C O 1991 Transverse laser patterns. II. Variational principle for pattern selection, spatial multistability, and laser hydrodynamics. *Phys. Rev.* A **43** 5114–5120

[3] Cantero-Chinchilla F N, Castro-Orgaz O, Dey S and Ayuso J L 2016 Nonhydrostatic dam break flows. I: physical equations and numerical schemes *J. Hydraul. Eng.* **142** 04016068

[4] Cantero-Chinchilla F N, Castro-Orgaz O, Dey S and Ayuso-Muñoz J L 2016 Nonhydrostatic dam break flows. II: one-dimensional depth-averaged modeling for movable bed flows *J. Hydraul. Eng.* **142** 04016069

[5] Carusotto I and Ciuti C 2013 Quantum fluids of light *Rev. Mod. Phys.* **85** 299–366

[6] Conforti M, Baronio F and Trillo S 2014 Resonant radiation shed by dispersive shock waves *Phys. Rev.* A **89** 013807

[7] Conti C, Fratalocchi A, Peccianti M, Ruocco G and Trillo S 2009 Observation of a gradient catastrophe generating solitons *Phys. Rev. Lett.* **102** 083902

[8] Dutton Z, Budde M, Slowe C and Hau L V 2001 Observation of quantum shock waves created with ultra-compressed slow light pulses in a Bose–Einstein condensate *Science* **293** 663–668

[9] El G, Geogjaev V, Gurevich A and Krylov A 1995 Decay of an initial discontinuity in the defocusing NLS hydrodynamics *Physica D: Nonlinear Phenomena* **87** 186–192

[10] El G and Hoefer M 2016 Dispersive shock waves and modulation theory *Physica D* **333** 11–65

[11] El G, Khamis E G and Tovbis A 2016 Dam break problem for the focusing nonlinear Schrödinger equation and the generation of rogue waves *Nonlinearity* **29** 2798–2836

[12] Fatome J, Finot C, Millot G, Armaroli A and Trillo S 2014 Observation of optical undular bores in multiple four-wave mixing *Phys. Rev.* X **4** 021022

[13] Garnier J, Xu G, Trillo S and Picozzi A 2013 Incoherent dispersive shocks in the spectral evolution of random waves *Phys. Rev. Lett.* **111** 113902

[14] Ghofraniha N, Conti C, Ruocco G and Trillo S 2007 Shocks in nonlocal media *Phys. Rev. Lett.* **99** 043903

[15] Gurevich A V and Krylov A L 1987 Dissipationless shock waves in media with positive dispersion *Sov. Phys. JETP* **65** 944–953

[16] Headley C and Agrawal G P G P 2005 *Raman Amplification in Fiber Optical Communication Systems* (Elsevier)

[17] Hoefer M A, Ablowitz M J, Coddington I, Cornell E A, Engels P and Schweikhard V 2006 Dispersive and classical shock waves in bose–einstein condensates and gas dynamics *Phys. Rev.* A **74** 023623

[18] Janantha P A P, Sprenger P, Hoefer M A and Wu M june 2017 Observation of self-cavitating envelope dispersive shock waves in yttrium iron garnet thin films *Phys. Rev. Lett.* **119** 024101

[19] Jia S, Haataja M and Fleischer J W 2012 Rayleigh–taylor instability in nonlinear Schrödinger flow *New J. Phys.* **14** 075009

[20] Johnson R S 1997 *A Modern Introduction to the Mathematical Theory of Water Waves* (Cambridge: Cambridge University Press)

[21] Kamchatnov A M 2000 *Nonlinear Periodic Waves and Their Modulations* (Singapore: World Scientific)

[22] Kevorkian J 2000 *Partial Differential Equations: Analytical Solution Techniques* (Berlin: Springer)

[23] Kim D-H and Lynett P J 2011 Dispersive and nonhydrostatic pressure effects at the front of surge *J. Hydraul. Eng.* **137** 754–765

[24] Kodama Y and Wabnitz S 1995 Analytical theory of guiding-center nonreturn-to-zero and return-to-zero signal transmission in normally dispersive nonlinear optical fibers *Opt. Lett.* **20** 2291

[25] LeVeque R J 2002 *Finite Volume Methods for Hyperbolic Problems* (Cambridge: Cambridge University Press)

[26] Maiden M D, Lowman N K, Anderson D V, Schubert M E and Hoefer M A 2016 Observation of dispersive shock waves, solitons, and their interactions in viscous fluid conduits *Phys. Rev. Lett.* **116** 174501

[27] Martin J C and Moyce W J 1952 Part IV. an experimental study of the collapse of liquid columns on a rigid horizontal plane *Phil. Trans. R. Soc.* A **244** 312–324

[28] Millot G, Pitois S, Yan M, Hovhannisyan T, Bendahmane A, Hänsch T W and Picqué N 2015 Frequency-agile dual-comb spectroscopy *Nat. Photon.* **10** 27–30

[29] Mo Y C, Kishek R A, Feldman D, Haber I, Beaudoin B, O'Shea P G and Thangaraj J C T 2013 Experimental observations of soliton wave trains in electron beams *Phys. Rev. Lett.* **110** 084802

[30] Mollenauer L F, Islam M N and Stolen R H 1985 Experimental demonstration of soliton propagation in long fibers: loss compensated by Raman gain *Opt. Lett.* **10** 229

[31] Onorato M, Residori S and Baronio F 2016 *Rogue and Shock Waves in Nonlinear Dispersive Media* (Berlin: Springer)

[32] Onorato M, Residori S, Bortolozzo U, Montina A and Arecchi F Jul 2013 Rogue waves and their generating mechanisms in different physical contexts *Phys. Rep.* **528** 47–89

[33] Pavlov M V 1987 Nonlinear Schrödinger equation and the Bogolyubov–Whitham method of averaging *Theor. Math. Phys.* **71** 584–588

[34] Ritter A 1892 Die fortpflanzung der wasserwellen *Zeitschrift des Vereins Deutscher Ingenieure* **36** 947–954

[35] Soares Frazao S and Zech Y 2002 Undular bores and secondary waves—Experiments and hybrid finite-volume modelling *J. Hydraul. Res.* **40** 33–43

[36] Solli D R, Ropers C, Koonath P and Jalali B 2007 Optical rogue waves *Nature* **450** 1054–1057

[37] Stansby P K, Chegini A and Barnes T C D 2017 The initial stages of dam-break flow *J. Fluid Mech.* **374** 407–424

[38] Stoker J J 1992 *Water Waves* (Hoboken, NJ: Wiley)

[39] Treske A 1994 Undular bores (favre-waves) in open channels—Experimental studies *J. Hydraul. Res.* **32** 355–370

[40] Trillo S, Deng G, Biondini G, Klein M, Clauss G F, Chabchoub A and Onorato M 2016 Experimental observation and theoretical description of multisoliton fission in shallow water *Phys. Rev. Lett.* **117** 144102

[41] Trillo S, Klein M, Clauss G F and Onorato M 2016 Observation of dispersive shock waves developing from initial depressions in shallow water *Physica* D **333** 276–284

[42] Turitsyna E G, Smirnov S V, Sugavanam S, Tarasov N, Shu X, Babin S A, Podivilov E V, Churkin D V, Falkovich G and Turitsyn S K 2013 The laminar–turbulent transition in a fibre laser *Nat. Photon.* **7** 783–786

[43] Vocke D, Wilson K, Marino F, Carusotto I, Wright E M, Roger T, Anderson B P, Öhberg P and Faccio D 2016 Role of geometry in the superfluid flow of nonlocal photon fluids *Phys. Rev.* A **94** 013849

[44] Wan W, Jia S and Fleischer J W 2007 Dispersive superfluid-like shock waves in nonlinear optics *Nat. Phys.* **3** 46–51

[45] Wang J, Li J, Lu D, Guo Q and Hu W 2015 Observation of surface dispersive shock waves in a self-defocusing medium *Phys. Rev.* A **91** 063819

[46] Whitham G B 1999 *Linear and Nonlinear Waves* (Hoboken, NJ: Wiley)

[47] Xu G, Conforti M, Kudlinski A, Mussot A and Trillo S 2017 Dispersive dam-break flow of a photon fluid *Phys. Rev. Lett.* **118** 254101

[48] Xu G, Mussot A, Kudlinski A, Trillo S, Copie F and Conforti M 2016 Shock wave generation triggered by a weak background in optical fibers *Opt. Lett.* **41** 2656

IOP Publishing

Nonlinear Guided Wave Optics

A testbed for extreme waves

Stefan Wabnitz

Chapter 5

Instabilities and extreme events in all-normal dispersion mode-locked fibre lasers

Miro Erkintalo, Antoine F. J. Runge, Robert I. Woodward, Edmund J. R. Kelleher, and Neil G. R. Broderick

5.1 Introduction

As this book illustrates, the study of instabilities and extreme waves in optical systems has grown to a vibrant field of research over the last decade. A key driver behind the field's rapid growth is arguably the fact that, through analogies, optical systems can be used as convenient testbeds for the controlled investigation of extreme wave phenomena that cannot be easily investigated in their own natural environment. In addition, the study of extreme instabilities that affect optical systems is often very important in its own right, as these instabilities can impair the system's technological performance. To mitigate the impact of the instability, and thus improve the system performance, it is necessary to have a detailed understanding of the underlying instability mechanisms, which warrants their careful study [1].

Because of the underlying technological significance, the study of instabilities in optical systems is rooted in a very long history. There has been, however, a recent surge of interest in examining instabilities from the perspective of extreme nonlinear wave phenomena. Indeed, investigations of extreme waves in optical systems arguably began in 2007, when Solli *et al* showed that fluctuations in white-light fibre supercontinuum generation could display long-tailed probability distributions reminiscent of amplitude statistics characteristic of giant ocean 'rogue' waves [2]. Compounded by the fact that both pulse propagation in optical fibre and ocean wave dynamics can, under particular conditions, be modelled using the ubiquitous nonlinear Schrödinger equation [1], it was thus proposed that fibre systems could be leveraged as an analogue system to help unlock the mystery of oceanic rogue waves.

The findings of Solli *et al* immediately attracted significant attention, stimulating a number of follow-up studies aiming to better understand the physics leading to long-tailed statistics in fibre supercontinuum generation [3–6]. Furthermore, the

 © IOP Publishing Ltd 2017

investigations quickly diffused beyond the original context of fibre supercontinuum, as long-tailed probability distributions were identified across a diverse variety of optical systems [7–9]. Initially, the studies focussed on conservative (or weakly dissipative) systems, whose dynamical behaviour is not dominated by the complex interplay of gain, loss, and feedback. It was nevertheless quickly noted that extreme waves populating the tails of long-tailed probability distributions could also manifest themselves in highly *dissipative* systems, including lasers with distinct architectures and operating in a wide range of regimes. Already in 2011, three groups independently reported extreme wave statistics in three different laser systems: Bonatto *et al* in an optically injected semiconductor laser [10], Pisarchik *et al* in an erbium-doped fibre laser [11], and Kovalsky *et al* in a Kerr-lens mode-locked Ti:Sapphire laser [12].

Extreme events and instabilities have also attracted significant attention in the context of mode-locked fibre lasers. These studies began with two independent numerical investigations in 2011 [13, 14], both of which predicted abnormally large intensity spikes arising through chaotic pulse bunching effects. The first experimental observation of such 'dissipative rogue waves' was reported by Lecaplain *et al* in 2012 [15]. The authors of this paper investigated a passively mode-locked erbium-doped fibre laser operating in the so-called 'noise-like' regime [16], and observed extreme intensity fluctuations with long-tailed probability distributions. These and related results are described in detail in Chapter 11 of this book.

The study of extreme events and instabilities in mode-locked fibre lasers is particularly attractive for two reasons. First, because their 1D guided-wave character greatly simplifies theoretical analyses, mode-locked fibre lasers offer a comparatively simple system for the study of complex nonlinear dissipative dynamics [17]. Second, recent developments of new system architectures—particularly designs comprised of all-normal dispersion elements—have made mode-locked fibre lasers unprecedentedly attractive to many practical applications [18, 19], thus calling for detailed investigations of their performance limits and stability regimes.

In this chapter, we review some recent studies of extreme events and instabilities in all-normal dispersion (ANDi) mode-locked fibre lasers. First, we present a brief overview of mode-locked fibre lasers in general (for more comprehensive reviews, see [19–21]), focussing in particular on the benefits of ANDi cavity designs [21]. We then describe a particular ANDi laser implementation in greater detail, and discuss the different modes of operation supported by the system. We will focus on experimental observations of two distinct instability regimes: (i) full noise-like operation accompanied by the emission of strong Raman waves, and (ii) soliton explosions, i.e. intermittent break-up and recovery of the pulse circulating in the laser cavity. Finally, we discuss results from recent numerical simulations showing the emergence of long-lived rogue dark solitons during the build-up phase of ANDi laser mode-locking.

5.2 All-normal dispersion mode-locked fibre lasers

Passively mode-locked all-fibre lasers possess a number of attractive features compared to their bulk counterparts. These include compact size, lack of sensitivity to misalignment, superior beam quality, low manufacturing cost, and potential for environmentally stable operation. However, the performance—particularly in terms of achievable pulse energy and duration—of practical designs has until recently trailed behind bulk solid-state systems.

The first 'mainstream' mode-locked fibre lasers used erbium-doped fibres and operated around 1550 nm, where ordinary fibre components exhibit anomalous dispersion [22–24]. In these lasers, pulse shaping occurs naturally through the formation of a fundamental (bright) soliton [25]; anomalous dispersion balances self-focussing Kerr nonlinearity (and gain balances loss). Such 'soliton' lasers are in wide use today, as they are reliable and can be constructed using cheap, off-the-shelf components. Unfortunately, the generation of parasitic spectral (Kelly) sidebands [26] and the so-called soliton 'area theorem' [27] limit the pulse durations and energies that can be achieved in practical designs.

The performance deficiencies of soliton lasers were later overcome by dispersion-managed laser designs [28, 29]. In these lasers, the cavity incorporates segments of both anomalous and normal dispersion, and so the intracavity pulse 'breathes' (i.e. its duration evolves) as it completes a single cavity transit. With such dispersion-managed laser cavities, it has been possible to generate pulses with durations as short as 28 fs [30] and energies as high as 14 nJ [31]. Unfortunately, the vast majority of fibre lasers emit either around 1550nm (erbium-doped lasers) or 1030 nm (ytterbium-doped lasers), where the dispersion of standard fibres and fibre-based components is strongly anomalous (1550 nm) or normal (1030 nm). Accordingly, realising dispersion management requires the use of specialty components that increase system cost and complexity.

Because the self-focussing Kerr nonlinearity—intrinsic to silica fibres—cannot counteract temporal broadening arising from normal dispersion, it was long thought that all-normal dispersion laser cavities are unsuitable for stable mode-locking: the intracavity pulse can be anticipated to continuously broaden from roundtrip to roundtrip and never reach steady-state operation. However, in 2006, Chong *et al* remarkably demonstrated that stable mode-locking can be achieved in an all-normal dispersion configuration by inserting a spectral bandpass filter in the cavity [32]. By removing the spectral wings that have temporally separated due to dispersion, this filter effectively resets the temporal duration of the pulse after one cavity transit, thus permitting stable mode-locking.

The ANDi mode-locked fibre laser first demonstrated by Chong *et al* generated pulses with energy up to 3 nJ that could be compressed down to 170 fs outside of the laser cavity. In this context, it is worth noting that, because the pulses generated by ANDi lasers are typically highly chirped, achieving an ultrashort pulse requires extra cavity recompression. While more impressive pulse performance had been achieved using dispersion-managed cavities, the significance of the ANDi laser reported by Chong *et al* was that the cavity could be constructed from off-the-shelf components without the need for complex and expensive dispersion-control schemes. This immediately attracted wide attention, and improved designs rapidly followed this initial demonstration. To date, researchers have reported simple ANDi laser designs that are able to generate pulses with energies above 20 nJ [32] and dechirped durations down to 50 fs [33]. These lasers are comparable in output pulse performance with popular bulk lasers, while simultaneously offering the many additional benefits associated with the fibre medium. In particular, by constructing the whole ANDi laser cavity from polarisation-maintaining fibres and components, it has been possible to realise high-performance systems that are compact, low-cost, easy to operate, and stable against environmental (e.g. thermal and mechanical) fluctuations [34–36].

Because ANDi fibre lasers represent a fairly new laser technology, their dynamics and instability modes have not been extensively investigated until recently. Yet, as discussed in the Introduction, studying these instabilities is of great importance for the design of more advanced configurations with improved output pulse performance. Furthermore, as we shall see, ANDi lasers can display complex instability dynamics previously unobserved in any other fibre laser configuration, thus allowing for the study of novel nonlinear dynamical behaviours. For the sake of discussion, we next describe a particular ANDi laser implementation and subsequently describe the instability behaviours it has been observed to exhibit. However, throughout our discussion we emphasise the generality of our conclusions by drawing links to similar observations in other designs.

5.2.1 An environmentally stable design

Figure 5.1 shows a schematic illustration of the particular laser under consideration [34–36]. Similar to most ANDi laser systems, ytterbium-doped fibres act as the laser's gain media, resulting in emission around 1030 nm. The overall cavity is comprised of two loops. The *main* loop includes a segment of Yb-doped fibre to provide gain, a segment of single-mode fibre (SMF), an optical isolator to ensure unidirectional operation, an output coupler to extract part of the circulating intracavity field, and a narrow band-pass filter centred at 1030 nm to facilitate mode-locking and pulse stabilisation (as explained above). The second loop corresponds to a *nonlinear amplifying loop mirror* (NALM), which acts as the laser's (artificial) saturable absorber and enables mode-locked operation [37]. It contains a segment of Yb-doped fibre and a segment of SMF, distributed so as to impart a differential nonlinear phase shift along the two counterpropagating paths around the loop. The NALM is connected to the main laser loop using a 4-port coupler, and both Yb-doped fibre segments are pumped by identical 980 nm laser diodes. We highlight that the cavity is made entirely out of off-the-shelf polarisation-maintaining fibres and components with normal dispersion.

Compared to other ANDi lasers, the design shown in figure 5.1 is distinctive for its use of the NALM as a mode-locking element. Indeed, most other ANDi implementations rely on nonlinear polarisation evolution (NPE) [32, 38] or *real* saturable absorber devices [39–41] to initiate and sustain mode-locking. As detailed in [42], the benefit of a NALM is that it offers the flexibility of NPE while still allowing for the use of all-polarisation-maintaining fibres and components, as required for environmentally stable operation.

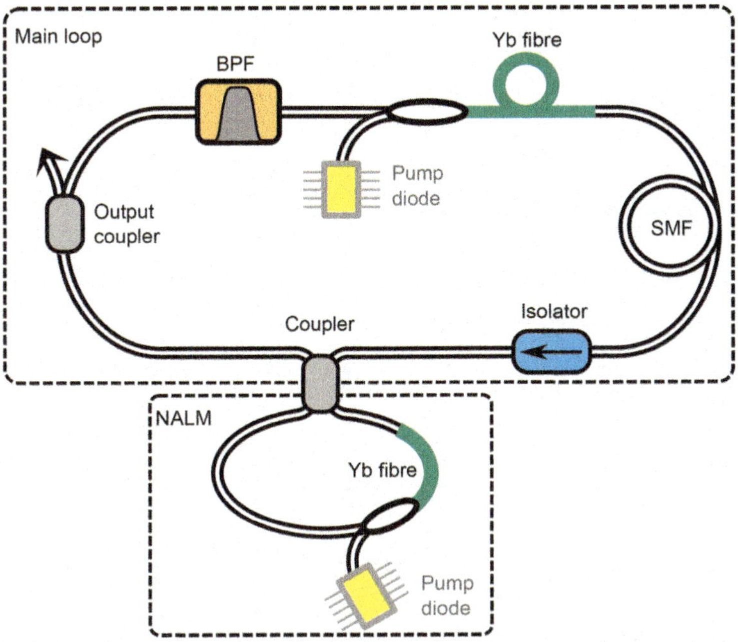

Figure 5.1. Schematic illustration of an ANDi fibre laser mode-locked with a nonlinear amplifying loop mirror (NALM). BPF: band-pass filter; SMF: single-mode fibre. Adapted from [36].

The general design shown in figure 5.1 was first demonstrated by Aguergaray *et al* in 2012 [34]. In its initial incarnation, the laser could only generate pulses with modest 0.3 nJ energy. However, the authors later demonstrated that this energy could be significantly up-scaled by increasing the length of the SMF segment separating the NALM from the active fibre in the main loop [35]. This modified configuration, with a purposefully lengthened laser cavity, can be understood to belong to a sub-class of ANDi lasers known as giant chirp oscillators (GCOs). First demonstrated by Renninger *et al* in 2008 [43], such GCOs allow for the generation of energetic (>10 nJ) pulses with low repetition rate (<10 MHz), making them ideal as seed sources for chirped-pulse amplification schemes.

For moderate cavity lengths, the pulse energy output from a GCO scales almost linearly with the cavity length [35, 43]. But of course, elongation of the cavity also enhances the accumulation of (detrimental) nonlinear effects, resulting in the eventual destabilisation of the mode-locked operation. As we shall see below, it is precisely in this long-cavity regime where extreme wave instabilities manifest themselves.

5.3 Stable mode-locking

Figure 5.2 shows typical temporal and spectral output characteristics when the laser shown in figure 5.1 is operating in the stable mode-locking regime. These results were obtained for a comparatively short laser cavity, with a total length $L_{\mathrm{tot}} \approx 20$ m

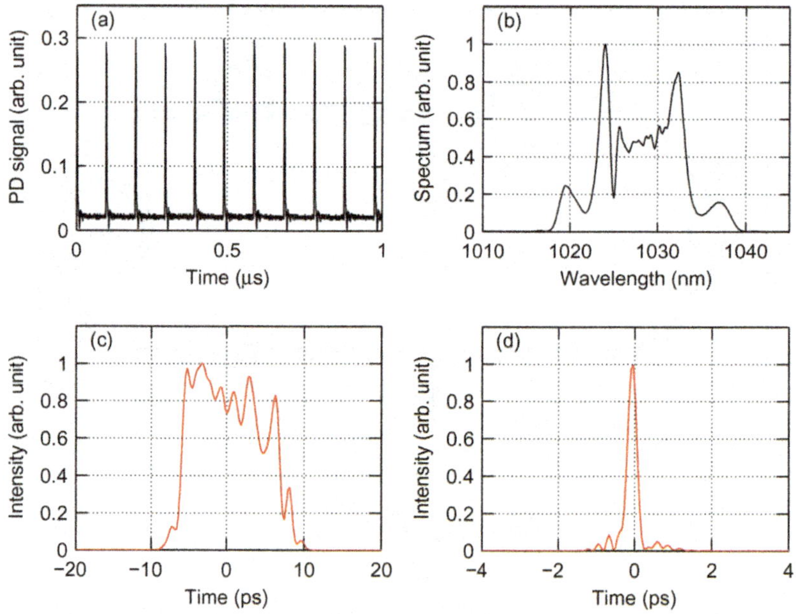

Figure 5.2. Experimentally measured output pulse characteristics for a 20-m-long ANDi laser mode-locked with a NALM [35]. (a) Pulse train measured on a slow photodetector. (b) Optical spectrum measured on an optical spectrum analyser. (c) Temporal profile of the uncompressed output pulse, extracted from a frequency-resolved optical gating measurement. (d) Corresponding temporal profile after the output pulses have been de-chirped with an external grating-pair compressor. Note the different time-axes in (c) and (d).

and a corresponding fundamental repetition rate of approximately 10 MHz (for other cavity design parameters, see [35]). Figure 5.2(a) shows the laser output as measured on a slow photodiode, and we indeed see a stable train of pulses separated by approximately 100 ns. The average spectrum, depicted in figure 5.2(b) in linear scale, is centred at 1030 nm and has a full-width at half-maximum bandwidth of about 10 nm. The spectrum exhibits a complex structure displaying strong oscillations attributed to self-phase modulation.

The pulses emitted by the laser are too short to be resolved directly on a photodetector; measurement of their temporal profile requires the use of ultrafast pulse diagnostic techniques. One such technique is known as frequency-resolved optical gating (FROG), which allows for the full measurement of the pulse amplitude and phase [44]. Figure 5.2(c) shows the temporal profile of the pulses output by the laser as extracted from a FROG measurement. As can be seen, the pulses have a temporal duration of about 13 ps, which is almost one hundred times longer than the transform limit of 155 fs. This means that the pulses are highly chirped, which is typical of ANDi lasers due to their all-normal dispersion landscape. However, because the chirp is linear, the pulses can be externally dechirped close to their transform limit. Indeed, figure 5.2(d) shows the temporal profile after the laser output is passed through a grating-pair compressor, achieving a short pulse with a duration of about 250 fs.

5.4 Noise-like emission

The pulses shown in figure 5.2 carry an energy of about 2.4 nJ. As mentioned in section 5.2.1, a conceptually simple way to obtain higher energies is to elongate the cavity by adding a long segment of SMF between the main gain section and the NALM. For example, when the total cavity length increases to 60 metres (120 metres), pulses with 10 nJ (16 nJ) energy can be achieved [35].

For intermediate cavity lengths, the temporal and spectral pulse characteristics are qualitatively similar to those shown in figure 5.2: the pulse train is stable, the spectrum highly structured, and the pulses can be re-compressed close to their transform limit [35]. However, for longer cavity lengths, it is found that a secondary operation regime can manifest itself alongside the desired, stable mode-locking regime [45]. This new regime can be easily distinguished based on its optical spectrum, as seen in figure 5.3(a) and (b), where we plot spectra measured from a 100- and 200-m-long cavity. When plotted using a linear scale [figure 5.3(a)], it is evident that the new mode of operation does not display any of the fine spectral features found in the stable mode-locking regime; the spectrum is rather smooth. An even more noticeable difference is observed when the spectra are compared on a logarithmic scale (figure 5.3(b)). The new mode of operation is clearly associated with a strong parasitic spectral component at 1070 nm that is absent in the stable operation regime. This spectral component arises from stimulated Raman scattering [27], as confirmed by the fact that its wavelength (1075 nm) coincides with the peak of the Raman Stokes gain spectrum for a pump wave centred at 1030 nm.

The new mode of operation is unstable: the pulses emitted by the laser exhibit significant fluctuations from shot to shot. The transition to a smooth spectrum is a first hint of this behaviour. Indeed, conventional optical spectrum analyser are slow, and so their measurements of mode-locked laser spectra typically correspond to ensemble averages over thousands of individual pulses. Such ensemble averaging obscures fine features when the spectra of individual pulses fluctuate, giving rise to a smooth spectral shape. A second indication of unstable operation is provided by the autocorrelation trace shown in figure 5.4. Again ensuing from a slow, ensemble averaged measurement, the autocorrelation trace consists of a sharp peak atop a

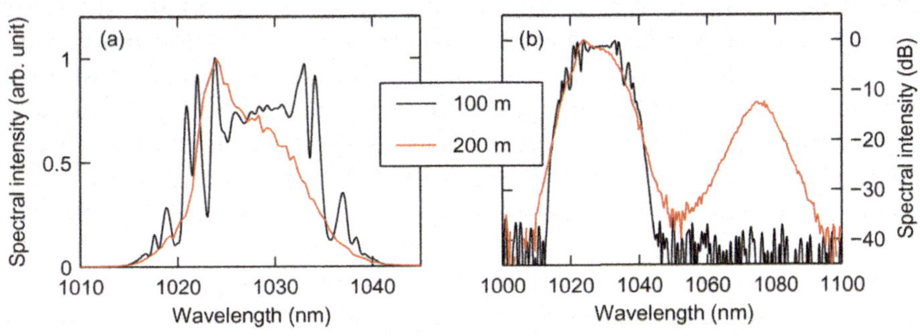

Figure 5.3. Experimentally measured output spectra for 100-m-long (black curve) and 200-m-long (red curve) ANDi lasers. (a) Shows the spectra in linear scale, while in (b) the spectra are compared in logarithmic scale and over a broader wavelength span. Adapted from [45].

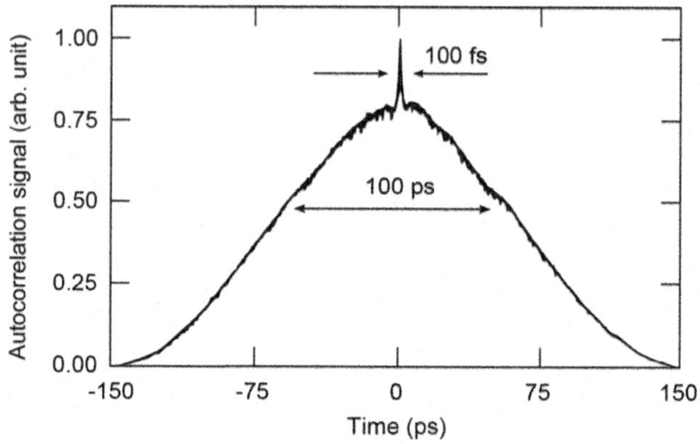

Figure 5.4. Experimentally measured autocorrelation trace for a 200-m-long ANDi laser operating in the noise-like regime. Adapted from [45].

broad pedestal. This suggests that the individual pulses comprise a broad envelope with fine internal structure. In this context, the sharp peak should not be confused as the duration of the pulse. Rather, it corresponds to a 'coherence artefact' that is indicative of the timescales of the rapidly fluctuating sub-structures within each pulse envelope [46].

The smooth spectrum and the particular shape of the autocorrelation trace suggest that the laser is generating 'noise-like' pulses, i.e. temporally localised bursts of noise that fluctuate from shot to shot [16]. First observed by Horowitz *et al* at the output of an Er-doped soliton laser in 1997, such noise-like pulses have been reported in numerous different mode-locked fibre laser cavities, including ANDi systems [47, 48]. Several different mechanisms have been proposed to explain their origins, but so far no single causative process has been identified. It appears, rather, that a number of different processes can give rise to noise-like operation.

For the laser described above (see figure 5.1), the generation of the parasitic Raman component has been proposed to lie at the origin of the noise-like regime [45]. While that component is indeed always found to accompany the noise-like pulses (in this particular laser configuration), the causative relationship between the two remains under speculation. It is worth highlighting that, while most noise-like pulses do not display a similar Raman component, the feature is not unique to the Yb-doped ANDi laser shown in figure 5.1. Indeed, similar Raman components have also been observed in Er-doped devices [49, 50]. Although in these lasers the appearance of the Raman wave appears to always be correlated with the destabilisation of uniform mode-locking, studies in other lasers have shown that Raman scattering can also have the exact opposite effect, facilitating stable mode-locking [51, 52]. This observation underlines how the precise impact of different physical phenomena is intimately tied to the specific cavity construction, and how it is unlikely that a single universal explanation can be formulated to account for the emergence of noise-like pulses in different laser cavities.

5.5 Real-time measurements and extreme Raman fluctuations

The precise characteristics of noise-like pulses remain poorly understood. This is arguably because, as highlighted previously, conventional techniques used to characterise mode-locked laser pulses rely on ensemble averaging. Given that noise-like pulses (are predicted to) exhibit strong shot-to-shot fluctuations, these techniques can provide only limited insight.

Fortuitously, the surge in studies of extreme events in optical systems has led to significant advances in the real-time characterisation of rare events [53–55]. A technique that has attracted particular attention is the dispersive Fourier transform [53, 56], which allows for convenient real-time measurements of shot-to-shot pulse spectra. This technique leverages the fact that a time-varying signal morphs into its Fourier transform when subjected to a sufficiently large quadratic spectral phase. By passing the signal through a long segment of dispersive fibre, the signal's time-domain profile is transformed into its spectrum, which can then be measured in real time using a fast oscilloscope (see figure 5.5). The technique has been instrumental in the study of extreme events in optical systems, permitting significant insights on the statistical fluctuations in e.g. fibre modulation instability [53] and supercontinuum generation [57, 58], as well as elucidating the dynamics of non-repetitive events in mode-locked lasers [59, 60].

The dispersive Fourier transformation has also been applied to gain insights on the Raman-induced noise-like pulses described previously [61]. The false colour plots in figure 5.6(a) and (b) show shot-to-shot spectra of 350 consecutive noise-like pulses emitted by the laser, obtained for two different pump power levels (see caption).

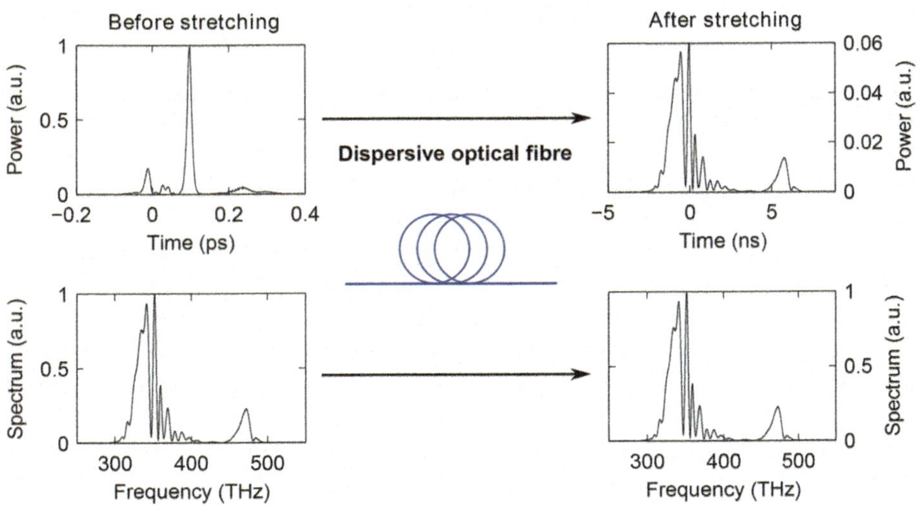

Figure 5.5. Schematic illustration of the dispersive Fourier transform method. When a time-domain signal (top left panel) is launched into a long segment of dispersive fibre, its temporal profile is stretched into a shape similar to its spectrum. The temporally broad output signal can be measured on a fast oscilloscope, thus allowing for the measurement of shot-to-shot spectra. Note the different time axes for plots before and after the dispersive fibre.

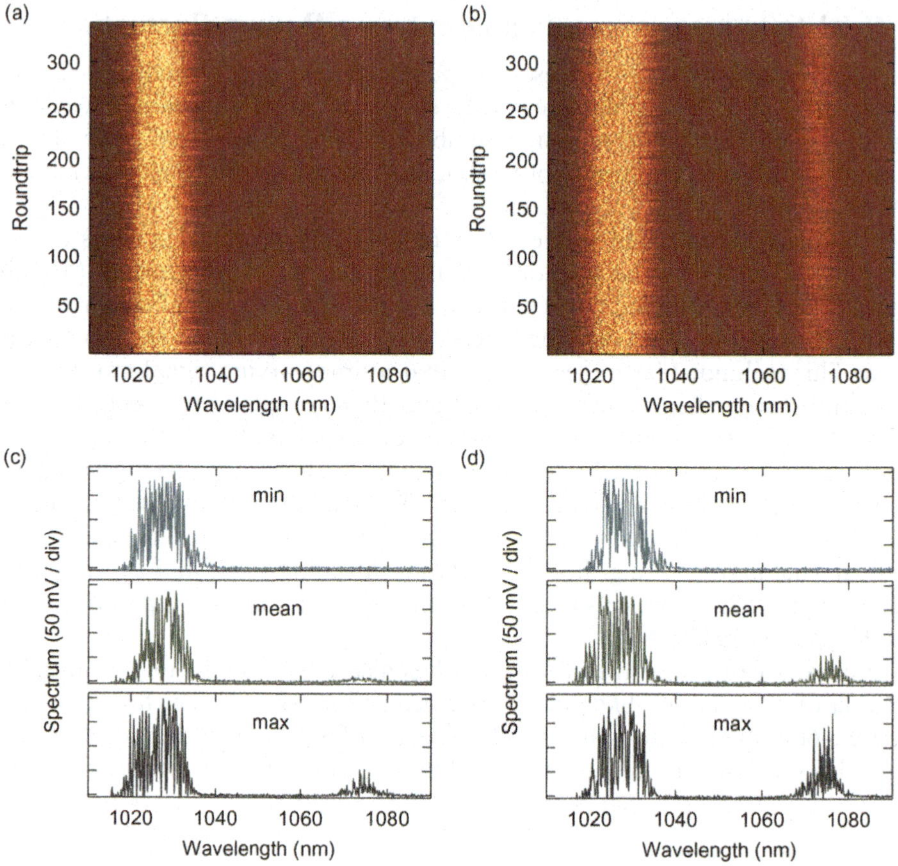

Figure 5.6. Shot-to-shot spectra measured at the output of a 200-m-long ANDi laser using the dispersive Fourier transform method. In (a), the pump power in the main loop was 190 mW, while in (b) it was 230 mW; the NALM pump power was 250 mW for both measurements. (c) and (d) show line plots of selected spectra corresponding to (a) and (b), respectively. The spectra in (c) and (d) correspond to events with the smallest (top), mean (middle), and largest (bottom) energy in the Raman band. Adapted from [61].

For clarity, line plots of selected realisations are shown in more detail in figure 5.6(c) and (d).

The measurements in figure 5.6 confirm that the pulses generated by the laser exhibit strong fluctuations from shot to shot. Furthermore, it is evident that spectra associated with individual pulses exhibit a very complex structure, thus confirming that the smoothness in the average spectral measurement (cf. figure 5.3(b)) indeed arises due to the ensemble averaging of the slow spectrum analyser. It is worth highlighting that, while these characteristics had been widely attributed to noise-like pulses for many years [16], quantitative experimental evidence was only recently reported in 2013, facilitated by the application of the dispersive Fourier transform [59].

Inspection of figure 5.6 also shows clearly that the generation probability of (strong) Raman waves depends on the pump power. Indeed, at low pump powers, strong Raman components appear only intermittently, while the vast majority of realisations display negligible Raman conversion. This is highlighted by the line plots shown in figure 5.6(c), which were selected so as to illustrate events with small, average, and large energy in the Raman band. In contrast, at large pump powers, a strong Raman wave is emitted almost every shot; nevertheless, realisations with negligible Raman conversion can be identified even in this case. Figure 5.7(a) and (b) shows the statistical distributions of shot-to-shot Raman wave energies for the two different pump power levels. For low pumping, the distribution is highly skewed, displaying a long tail, characteristic of extreme events. As detailed in [61], the most energetic events fulfil criteria often quoted for rogue waves [62]: (i) they populate the tails of highly skewed statistical distributions; (ii) they appear unpredictably; and (iii) their amplitude exceeds the significant wave height—defined as the mean amplitude of the one-third largest waves—at least by a factor of two. In contrast, as the pump power is increased, the Raman energy distribution assumes an almost Gaussian shape, showing no signatures of extreme events.

To explain the long-tailed statistics in figure 5.7(a), one first notes that the intensity of a Raman amplified signal increases exponentially with the pump intensity [27]. As a consequence, pump intensity fluctuations can be exponentially enhanced when transferred to the signal. This mechanism alone has been shown to result in extreme fluctuations in fibre Raman amplifiers [7].

In the laser just described, the main pulse at 1030 nm acts as the Raman pump; its peak intensity fluctuates from shot to shot because of the underlying noise-like operation. Thus, extreme events at the Raman signal can be anticipated solely due to the exponential amplification of the pump fluctuations. But in addition, the Raman fluctuations are further magnified by the fact that the laser output is immediately after the NALM (c.f. figure 5.1). Indeed, the NALM acts as a saturable absorber, preferentially attenuating (transmitting) the weakest (strongest) Raman waves [61].

As the pump power increases, the energy distribution transforms into a quasi-Gaussian shape for two principal reasons. First, the transmission function of the NALM is periodic with power [37]. This implies that, for sufficiently large Raman

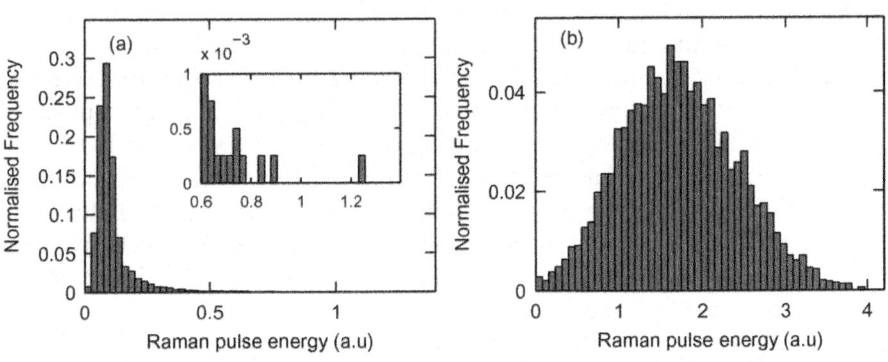

Figure 5.7. Statistical distributions of shot-to-shot Raman energy. Results in (a) and (b) were extracted from the measurements summarised in figure 5.6(a) and (b), respectively. The inset in (a) shows a zoom on the distribution tail. Adapted from [61].

fluctuations, the NALM can attenuate not only the weakest, but also the strongest waves, thus normalising the distribution around the transmission maximum. Secondly, sufficiently large Raman waves can act as pumps themselves in a second stage of Raman amplification. Such a cascaded amplification process naturally depletes energy from the first Raman components, thus normalising the distribution via selective attenuation of the strongest events [63].

The results presented in this section highlight how extreme events can arise at the output of a (partially) mode-locked laser. For the particular ANDi laser considered (cf. figure 5.1), these events manifest themselves in the form of sporadic Raman waves, but we note that many different types of extreme events have been reported in other laser configurations operating in the noise-like regime. For example, Liu *et al* have recently shown that, for a particular choice of intracavity bandpass filter, even the main pulse centred at the filter transmission maximum (as opposed to the parasitic Raman wave) can exhibit extreme-value statistics [64]. More generally, the results discussed in this section underline how real-time spectral measurements can unveil non-repetitive events at the output of (partially) mode-locked lasers, thus providing new insight that cannot be obtained from conventional ensemble-averaged measurements.

5.6 Soliton explosions

A soliton explosion is a peculiar instability phenomenon that can affect the operation of mode-locked lasers. In this regime, the pulse circulating in the laser cavity remains superficially stable for tens of roundtrips, but then suddenly undergoes an abrupt collapse (or 'explosion'), transforming itself into a complex and highly structured entity, with characteristics not dissimilar to the noise-like pulses discussed above. Remarkably, after only a few roundtrips, the intracavity field reshapes back to its original pulse-like state, maintaining this state until another collapse event occurs. The laser output can thus be understood as a quasi-coherent pulse train whose individual constituents are sporadically replaced with noise-like structures.

Soliton explosions were first identified in numerical simulations of the cubic-quintic Ginzburg–Landau equation (CQGLE) [65, 66]—a celebrated master equation that allows for the qualitative description of mode-locked laser dynamics [17]. This initial discovery was followed by numerous theoretical studies that examined in more detail how different parameters and phenomena affect the dynamics of the explosion events [67–69]. The first experimental observation was reported in 2002 by Cundiff *et al* in a Kerr-lens mode-locked Ti:Sapphire laser [70]. The authors used a diffraction grating to spatially disperse the laser output on a photodetector array, which allowed for the laser output spectrum to be examined (almost) from shot to shot. Despite the limited temporal and spectral resolution of the detection system, clear signatures of explosion events were observed.

For more than a decade, the observations of Cundiff *et al* constituted the only experimental observation of soliton explosions. However, in 2015, Runge *et al* discovered—using the dispersive Fourier transform—that qualitatively similar events could also be observed at the output of an ANDi fibre laser operating close to the boundary between stable mode-locking and noise-like emission [71]. As the laser studied by Runge *et al* (a fibre system with normal dispersion) was wholly different from that used by Cundiff *et al* (bulk laser with anomalous dispersion), these results highlighted that the soliton explosion phenomenon can manifest itself in a wide variety of different laser systems. Following the work by Runge *et al*, explosions have subsequently been observed also in a dispersion-managed Er-doped fibre laser with net normal dispersion [72, 73].

Figure 5.8 shows a sequence of shot-to-shot spectra measured for increasing levels of pump power at the output of an ANDi fibre laser operating in the soliton explosion regime. The precise laser construction is similar to that shown in figure 5.1 with a total cavity length of about 140 m. As can be seen, for low pump powers the laser is operating in the stable mode-locking regime, with all the pulses displaying

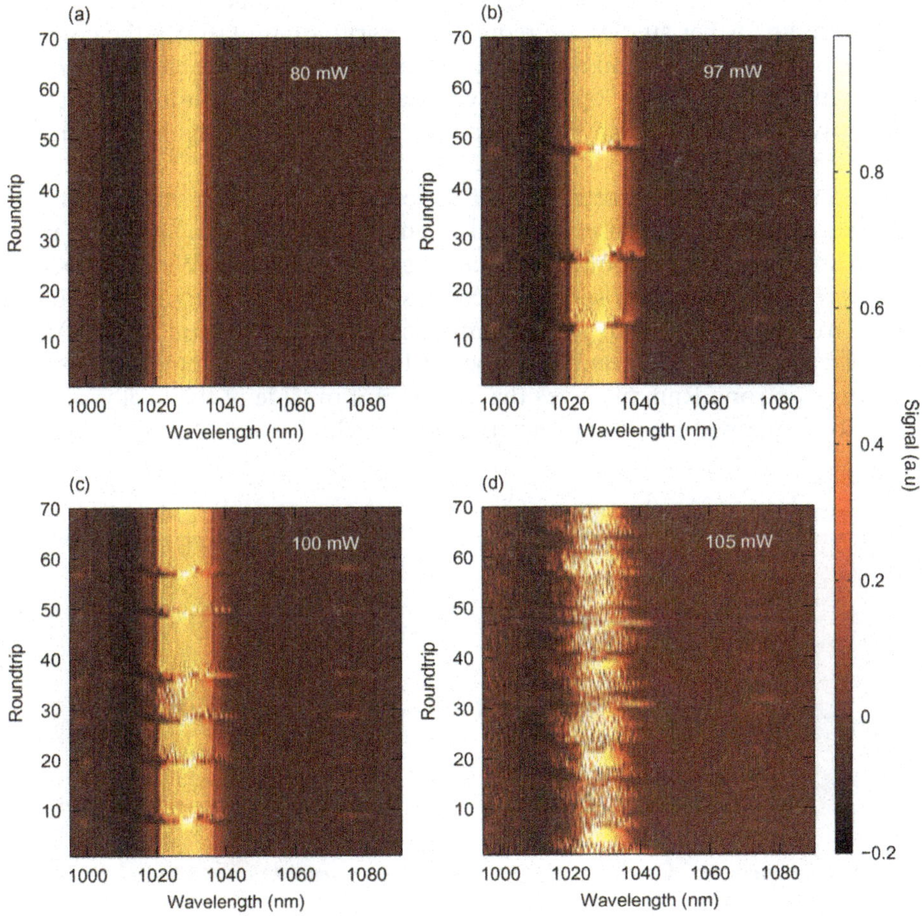

Figure 5.8. Experimentally measured shot-to-shot spectra of pulses emitted by a 140 m long ANDi fibre laser for various pump powers in the main loop as indicated (NALM pump power fixed at 150 mW). In (a), the laser is operating in the stable mode-locking regime; in (c, d) intermittent soliton explosions occur; in (d), the laser is operating in the full noise-like mode. Adapted from [74].

identical spectra (figure 5.8(a)). In contrast, for very large pump power levels, the emission is fully noise-like and the spectra of individual pulses exhibit strong shot-to-shot fluctuations (figure 5.8(d)).

Soliton explosions can be witnessed for intermediate pump powers (figure 5.8(b) and (c)). Indeed, the pulses emitted by the laser maintain (superficially) constant spectral structure for more than 10 roundtrips, but then suddenly collapse to a much narrower feature with higher amplitude. However, within just a few roundtrips, the intracavity pulse returns to its original quasi-stable state. As the pump power increases, so does the frequency of explosions; in figure 5.8(b) we can identify three clear explosions, while six events are visible in figure 5.8(c). These results suggest—even if only qualitatively—that noise-like operation could potentially be interpreted as the 'continuum limit', where explosions occur at every roundtrip, whereas explosions may be understood as transient noise-like pulses that cannot be sustained indefinitely due to insufficient gain. This interpretation should, however, be treated with caution, as further studies are required to verify quantitatively the connection between soliton explosions and noise-like pulses. We note in particular that Liu *et al* have recently discovered an operation regime where explosion-like events occur continuously, while still displaying characteristics clearly distinguishable from full noise-like operation [72].

Figure 5.9 shows the characteristics of the explosion events in more detail. Here we plot the spectra of 100 consecutive pulses emitted by the laser, and show in more detail the spectral features around particular explosion events. We see clearly how the broad, highly structured spectrum collapses into a much narrower but more intense feature, only to return to its former shape after a few roundtrips. We also note how each explosion event is associated with the concomitant emission of comparatively strong Raman waves that are absent outside of the explosion events. This can be explained by noting that the narrow, high-amplitude spectrum of an

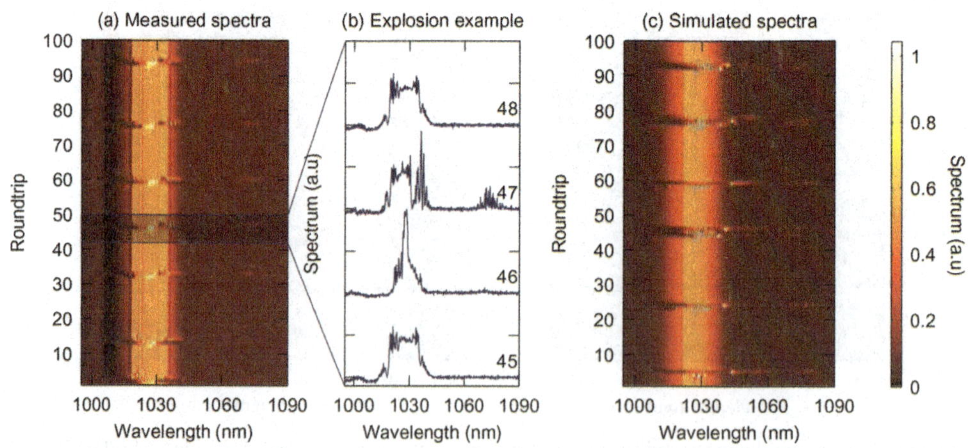

Figure 5.9. (a) Experimental measurements showing spectra of 100 consecutive pulses emitted by a 100 m long ANDi fibre laser. (b) Shows a zoom on a particular soliton explosion event. (c) Corresponding results from numerical simulations. Adapted from [71].

explosion is better suited for Raman amplification than the broad spectrum associated with stable mode-locking. These general features are also present in numerical simulations, as shown in figure 5.9(b). The simulation results were obtained from a realistic lumped cavity model, where each fibre segment (including the NALM) is rigorously modelled using generalised nonlinear Schrödinger equations (GNLSE) [42]. Overall, the simulations are in reasonably good qualitative agreement with experimental findings.

5.6.1 Temporal dynamics

The emergence of soliton explosions appears to be intimately related to the onset of multi-pulsing, at least in the ANDi fibre laser studied by Runge *et al* [71]. This can be readily appreciated from the examination of the time-domain dynamics that characterise the explosion events. To this end, figure 5.10(a) shows the shot-to-shot temporal profiles corresponding to the numerically simulated spectra shown in figure 5.9(b). We first note how each explosion event is correlated with an abrupt temporal jump, which physically implies that the cavity roundtrip time is slightly increased during an explosion event, resulting in a small local change in the laser repetition rate. In figure 5.10(b), we show the temporal dynamics around a particular explosion event in more detail, and we can clearly see how a secondary pulse develops at the trailing edge of the first pulse. The two pulses compete for a few roundtrips, but ultimately it is the new pulse that prevails, with the first one disappearing. In fact, it is the switch from the initial to the final pulse that gives rise to the abrupt temporal jump.

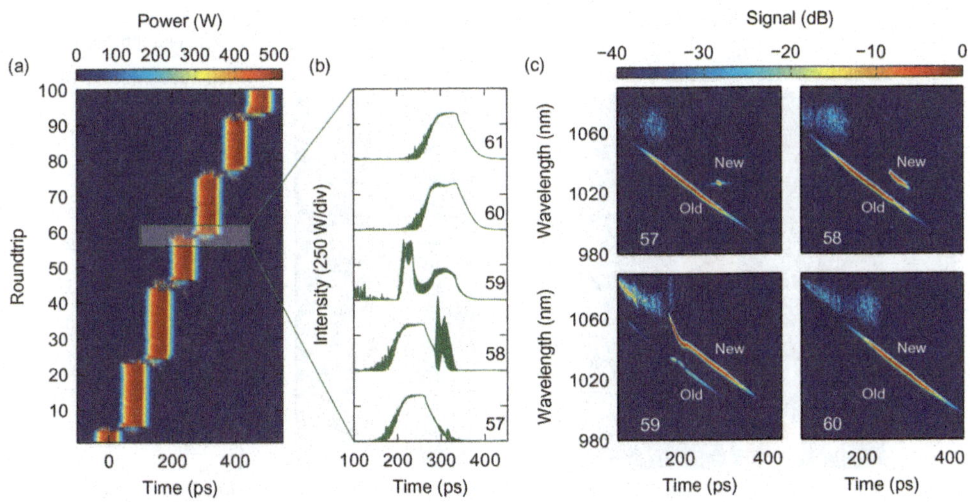

Figure 5.10. Numerical simulation results, illustrating the time-domain dynamics during soliton explosions. (a) 100 consecutive simulated pulse profiles; each explosion is associated with an abrupt temporal jump. (b) Zoom on a particular explosion event. (c) Spectrogram representations of four consecutive numerically simulated laser output pulses. (Numerical labels on bottom left corners of each panel indicate the roundtrip number.) The spectrograms show how the explosion is associated with a growth of a new pulse temporally offset from the old pulse. (a) and (b) adapted from [71]; (c) adapted from [74].

Significant new insights can be gained by simultaneously visualising the temporal and spectral features of an explosion event [74]. To this end, figure 5.10(c) shows the *spectrogram* of the explosion event whose time-domain profile is detailed in figure 5.10(b). Here, one can clearly witness how the new pulse grows and develops at the trailing edge of the old pulse, with the centre wavelengths of both pulses coinciding with the intracavity spectral filter at 1030 nm. This observation permits a straightforward explanation for the interference fringes observed at the trailing pulse edge in figure 5.10(b). Specifically, because the old pulse is highly chirped, the wavelength of its trailing edge is offset from the 1030 nm centre of the new pulse. Thus, the temporally overlapping parts of the two pulses have different wavelengths, giving rise to an interference pattern with fringe spacing $\Delta\tau = \lambda^2/(c\Delta\lambda) \sim 0.2$ ps, where $\lambda = 1030$ nm is the central wavelength of the pulses and $\Delta\lambda \sim 17$ nm is the wavelength separation between the two pulses.

Corroborating experimental evidence for the dynamics described above has been reported in [71, 74], and below we summarise the main findings. First, figure 5.11(a) shows a 'spatiotemporal' diagram of 100 consecutive pulses measured with a fast oscilloscope immediately at the laser output. The diagram was obtained by dividing a long real-time oscilloscope trace into segments whose duration equals the cavity roundtrip time, and then concatenating the traces on top of each other to form a false colour plot. As can clearly be seen, each explosion event is associated with an abrupt time shift of about 40 ps, which is in reasonable qualitative agreement with numerical simulations.

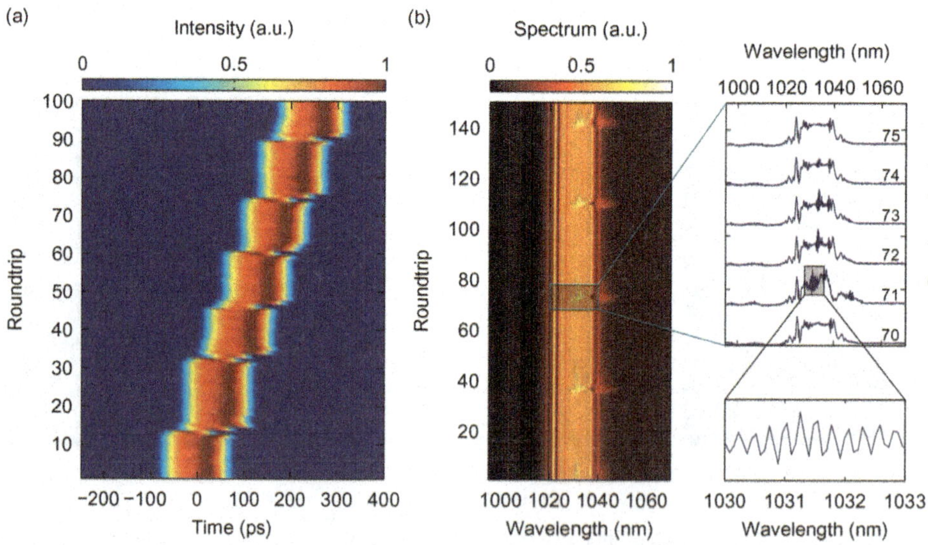

Figure 5.11. Experimental results corroborating numerically simulated time-domain dynamics. (a) Vertically concatenated oscilloscope traces corresponding to 100 consecutive pulses emitted by a 100-m-long laser. Clear evidence of abrupt time shifts is observed. (b) Real-time spectral measurements for a 50-m-long laser cavity. Spectral interference fringes observed during 'frustrated' explosions are indicative of two closely spaced pulses. (a) Adapted from [71]; (b) adapted from [74].

A second piece of corroborating evidence was obtained in a significantly shorter ANDi laser [see figure 5.11(b)]. With a total cavity length of about 50 m, that configuration predominantly operates in the stable mode-locking regime. However, it was found that the laser could sustain a new type of 'frustrated' explosions, where the intracavity pulse undergoes abrupt disturbances yet appears to avoid a full collapse [c.f. figure 5.11(b)]. Real-time spectral measurements revealed that, during these events, clear interference fringes manifest themselves around the centre of the pulse spectrum. Such fringes are indicative of the presence of two temporally separated pulses centred at the same wavelength, precisely as seen in the numerically simulated spectrograms [figure 5.10(c)].

The experimental results just summarised provide significant credence to the hypothesis that soliton explosions are linked to the onset of multi-pulsing. It is interesting to speculate whether this behaviour is universal, or unique to the ANDi fibre laser studied by Runge *et al.* Although further study is of course needed for any conclusive conclusions, we note that abrupt time shifts signalling the presence and competition between two temporally separated pulses have been noted in several theoretical studies based on the general cubic-quintic Ginzburg–Landau equation [67, 68, 75]. Such shifts have been found to arise both in the fully symmetric CQGLE [75], as well as in modified equations where time-reversal symmetry is explicitly broken by the inclusion of higher-order effects (e.g. third-order dispersion or stimulated Raman scattering) [67]. Furthermore, already in 2010, Descalzi and Brand showed very clearly that the numerically simulated time shifts could be interpreted in terms of a complex interaction between two pulse-like objects [68]. Because of the universal nature of the CQGLE, these findings certainly suggest that the connection between soliton explosions and multi-pulsing is not limited to the ANDi laser studied by Runge *et al.* Of course, it must be highlighted that the observations by Runge *et al* represent the first experimental confirmations of this type of time-domain behaviour [71], and therefore corroborate future research based on the CQGLE.

5.7 Metastable dark solitons in radiation build-up dynamics

So far we have considered extreme events and instabilities during the quasi steady-state operation of lasers. Yet, instabilities and extreme events have also been identified during radiation build-up [76]. In ANDi mode-locked lasers in particular, the radiation build-up regime has been found to exhibit a rich dynamic that supports the existence of metastable dark optical solitons. Numerical simulations of long-cavity ANDi lasers predict that these spontaneously forming dark solitons can persist for hundreds of cavity roundtrips (corresponding to tens of kilometres propagation distance), atop the bright-pulse envelope that ultimately shapes into the steady-state intracavity laser pulse. In this section, we briefly describe these observations and the underlying dynamics of the metastable dark solitons.

The existence of quasi-stationary dark solitons during the radiation build-up of mode-locking was first observed numerically through GNLSE-based simulations of ANDi lasers [14, 77, 78]. Figure 5.12 shows a typical evolution of the first 400 roundtrips of the time-domain field envelope in a 120-m-long laser cavity with a saturable absorber. The simulation initial condition corresponds to a random noise field emulating quantum shot noise. Three distinct regimes can be observed: (i) a disordered noisy phase as the field intensity initially grows, (ii) a transient regime as the saturable absorber and gain saturation act to localise the temporal field within an envelope containing internal noisy sub-structure, and (iii) the final steady-state highly chirped pulse. Of particular interest are the solitary dark structures—holes in the intensity envelope—in the transient regime which, remarkably, persist for hundreds of roundtrips before decaying. Closer inspection of these structures reveals that they are indeed dark solitons, characterised by an intensity dip and phase shift related to their 'blackness' (figure 5.12 left inset) [79].

At first glance, the appearance of dark solitons during radiation build-up in mode-locked lasers is quite unexpected. Yet, the numerical observations can be readily understood by noting that dark solitons are analytic solutions to the defocussing NLSE, which describes pulse propagation in normally dispersive optical fibres. Thus, spontaneous noise perturbations may evolve into dark solitons in the normal dispersion regime, similarly as arbitrary pulse shapes can evolve into bright solitons in the anomalous dispersion regimes [80]. Of greatest interest, however, is the long-lived behaviour of a statistically small number of these structures, resulting in extreme-valued probability distributions [77, 78].

Insight into the long-lived dynamics, and decay of dark solitons in ANDi lasers was gained using a pattern-matching algorithm to identify and track the dark solitons throughout the radiation build-up regime (figure 5.12 right inset) [78]. Ensemble simulations of varying cavity lengths, from a few metres to few kilometres, confirmed the ubiquity of existence of these structures, suggesting that this could be a universal underlying dynamic in the initiation of mode-locking in ANDi lasers. (We emphasise, however, that further work and experimental confirmation is required to verify this hypothesis.) While each repeated simulation (seeded with a different randomised shot noise field) converged to show identical pulse properties in

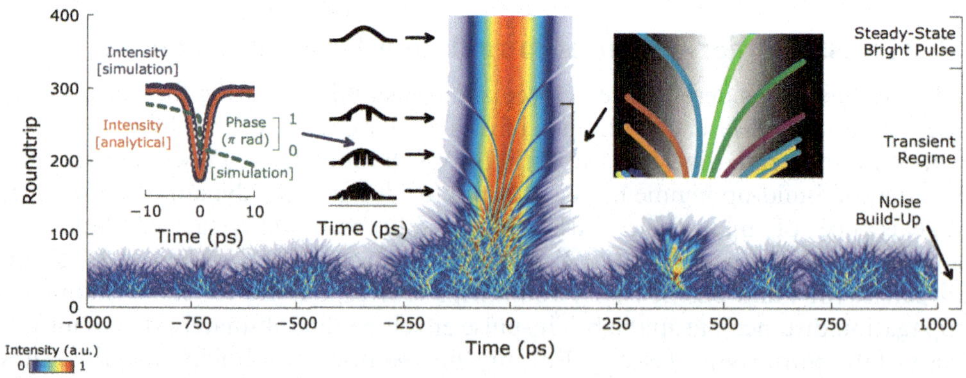

Figure 5.12. Time-domain evolution of radiation build-up in long-cavity mode-locked ANDi laser. Left inset: pulse profile at various roundtrips, with magnified region showing dark soliton well fitted by its analytic expression. Right inset: magnified region showing tracked dark solitons as coloured trajectories. Adapted from [78].

steady-state, the build-up dynamics and time taken for the decay of dark solitons in the 'transient regime' varied strongly. Briefly, the cavity design (e.g. the magnitude of the positive dispersion, component parameters, etc) defines a fixed attractor—the steady-state pulse, whose simulated properties have been verified experimentally [41, 81, 82]—towards which the system evolves along many possible trajectories from an initially random starting point in phase space.

Figure 5.13(a) shows a histogram of dark soliton lifetimes, extracted from an ensemble of 600 simulations with constant cavity parameters. (We define the term 'lifetime' as the number of roundtrips between a dark soliton spontaneously forming from noise and it accelerating (or decelerating) away from the bright pulse such that it decays.) Short-lived dark structures occur far more frequently than dark solitons with a long lifetime, yet the existence of a small number of highly persistent long-lived structures can also be identified. For solitons surviving less than 50 roundtrips, the lifetime distribution can be very well fitted by a power law. The same cannot, however, be said for the long-lifetime limit; rather, the distribution is seen to exhibit a long tail characteristic of extreme events. It is worth noting that, in contrast to optical 'rogue waves' described earlier, these dark solitons exhibit extreme statistics in their *lifetime*, rather than their peak power or energy. Nevertheless, the results clearly illustrate the fascinating and vast array of complex instabilities exhibited by the nonlinear dynamics of ANDi lasers, both during their radiation build-up and in their steady-state.

To explain the statistics, and gain further insight into the intracavity nonlinear dynamics, we first briefly review the properties of dark solitons. In the strict mathematical sense, dark solitons exist only on an infinite continuous-wave background. However, it is well known that even a *finite-width* background can sustain dark intensity dips with characteristics very similar to those expected for pure dark solitons, provided that the background width is greater than about 10 times the dark soliton duration [83]. A key parameter defining the characteristics of a dark

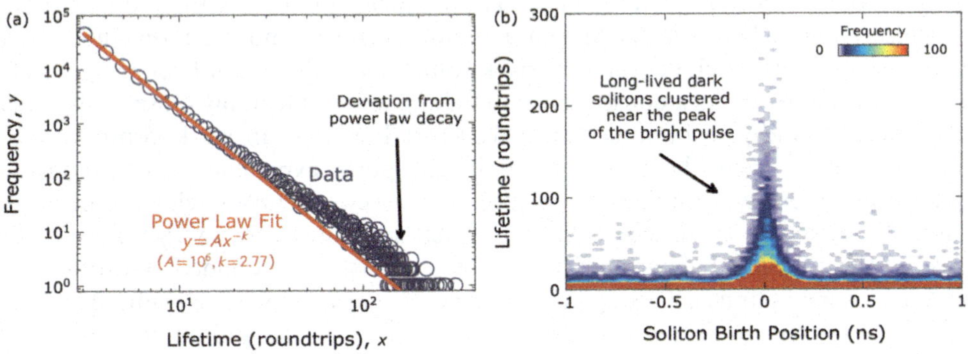

Figure 5.13. Histograms of dark soliton lifetimes: (a) one-dimensional histogram fitted with a power-law decay function, showing decreasing occurrence of long-lived dark solitons and extreme long-lived solitons that deviate from this distribution; (b) two-dimensional histogram of soliton lifetimes and birth positions in the bright pulse background (0 ns refers to the centre of the bright pulse), showing clustering of the long-lived metastable solitons near the background pulse peak.

soliton is its blackness B ($0 \leq B \leq 1$), which defines the ratio of the intensity dip minimum to the background power level: a 'black soliton' ($|B| = 1$) exhibits an intensity drop to zero with an antisymmetric π phase shift at its centre, while 'gray solitons' ($|B| < 1$) exhibit weaker intensity dips and more gradual and smaller absolute phase shifts. Importantly, in contrast to black solitons, gray solitons are not group-velocity matched with the background field. Instead, gray solitons exhibit a persistent time-domain drift (in a reference frame moving with the background), with the drift rate increasing with decreasing blackness [79]. This observation prompts the following explanation for the observed distribution of dark soliton lifetimes. The randomised initial field shapes into a sequence of dark solitons with varying blackness, and it is the gray (black) solitons that decay first (last) because of their large (small) group-velocity mismatch with the background field. Unfortunately, this explanation is unlikely to represent the full story: detailed examination of the simulation results reveal that the majority of the solitons are in fact either totally black or dark gray, indicating a small velocity mismatch [78].

A more likely cause for the observed dark soliton lifetimes stems from the shape of the background field (i.e. the bright pulse envelope). Indeed, it is well-known that even a black soliton can exhibit a group-velocity mismatch with the background field, provided that it sits at a point of non-zero background gradient [84]. This can be qualitatively understood by noting that a time-dependent intensity background imparts a time-varying nonlinear phase shift across the dark structure, resulting in a shift in the structure's instantaneous frequency and hence (because of dispersion) a change in its group velocity. Significantly, a black soliton can be expected to be group-velocity-matched with the background pulse only if located at a point where the background gradient vanishes (e.g. the peak of the pulse). If located at a point of nonzero background gradient (e.g. the leading or trailing edge of the pulse), the dark structure is expected to move away from the peak, eventually decaying.

In the simulations described above, the dark solitons evolve spontaneously from noise. Accordingly, they are initially distributed randomly across the temporal envelope of the forming bright pulse. The majority of the intensity dips exist in the pulse wings where the background is non-uniform, thus resulting in a large group-velocity mismatch between the dark solitons and the bright background pulse (as described above). As a consequence, the dark solitons forming close to the pulse wings decay rapidly, exhibiting relatively short lifetimes. In stark contrast, dark solitons forming near the peak of the bright pulse experience a near-constant background intensity, resulting in negligible group velocity mismatch with the bright pulse. It is precisely these dark solitons, which spontaneously emerge close to the peak of the forming background pulse, that persist for extended durations and exhibit extreme lifetimes. This dynamic can be seen clearly by visualising the soliton lifetime histogram in relation to the 'birth' position of the soliton relative to the bright pulse peak [figure 5.13(b)]. As can be seen, the persistent dark solitons, with extreme lifetimes, are clustered near the peak [78].

Of course, even if a purely black soliton forms exactly at the background pulse peak, it cannot persist indefinitely. This is because the peak of the pulse is an unstable equilibrium point; cavity perturbations give rise to jitter that eventually displaces the soliton, beginning its 'descent', which is accelerated by the background intensity gradient (acceleration is observed in figure 5.12 as the curvature of trajectories). An additional destabilising phenomenon, verified in more detail in reference [78], is collisions between dark solitons which results in a mutual temporal displacement [85].

The simulation results just surveyed highlight that extreme events can manifest themselves in ANDi lasers during the transition to mode-locked operation. In addition to elucidating the dynamics of ANDi lasers, improved understanding of these extreme events can also yield advances through analogies with other dispersive nonlinear systems that share an underlying mathematical framework. For example, the NLSE can describe the wavefunction of a Bose–Einstein Condensate (BEC) under conditions of 1D confinement and a dilute atomic gas [86]. Dark soliton matter waves have even been observed in BECs, where they can be a useful diagnostic for probing mesoscopic physics. Further work is required to assess the strength of this analogue, and experimental measurements are required to verify the dynamic and its proposed universality in the radiation build-up regime of ANDi lasers. While capturing these transient dark structures in the laboratory presents a significant experimental challenge, developments in the real-time shot-to-shot observations of short-time-scale dynamics continue to pave the way towards increasingly advanced measurement capabilities. Indeed, the topic of build-up dynamics is already attracting experimental attention, with initial studies offering significant insight [87]: a trend we expect will continue.

5.8 Conclusions

In this chapter, we have surveyed recent studies of dynamical instabilities and extreme events in mode-locked ANDi fibre lasers. We first presented a brief overview of the different types of passively mode-locked fibre lasers, and subsequently elaborated on the specific advantages of ANDi designs. We then introduced a particular environmentally stable configuration in more detail, and described the different modes of operation sustained by that system. In particular, we described how the laser can emit noise-like pulses associated with the generation of extreme Raman waves, how soliton explosions can manifest themselves close to the threshold of noise-like operation, and how realistic numerical simulations reveal that metastable dark solitons can form during radiation build-up. While our discussion has focussed on the environmentally stable ANDi design, we have continuously sought to draw links to other studies where similar instability phenomena have been observed. Of course, we fully acknowledge that, due to length restrictions, we have not been able to describe all the fascinating instabilities observed and studied to date [48, 76, 88].

The study of extreme events in optical systems grew from the hope of gaining insights into 'rogue' ocean waves. In contrast to some other optical systems (e.g. pulse propagation in optical fibre), lasers are not related to oceans through any direct analogies. Accordingly, it is unlikely that studies into instabilities of mode-locked lasers will yield significant insights into the generation of extreme waves in oceans [1]. However, the studies have been extremely valuable in their own right, as the improved understanding of the instability mechanisms has guided the design of more advanced laser architectures that can avoid those instabilities. For example, Bowen *et al* have recently reported on a laser design that avoids the Raman-driven instabilities described in section 5.5, thus allowing for the construction of lasers with lower repetition rates [89]. Finally, the study of non-repetitive (extreme) events has spurred the development of real-time detection schemes, which have subsequently proven very valuable in many different contexts involving the dynamics of lasers and other dissipative systems. For example, the dispersive Fourier transform has recently been applied to examine in detail the build-up dynamics of mode-locking [60], the formation of soliton molecules in lasers [87, 90], as well as the emergence of bound states in coherently driven passive cavities [91]. This highlights how the study of non-repetitive events and instabilities is impacting the practical design of lasers, as well as the fundamental study of nonlinear dissipative dynamics.

Acknowledgements

ME acknowledges support from the Rutherford Discovery Fellowships of the Royal Society of New Zealand. RIW acknowledges support from a Macquarie University MQ Research Fellowship. EJRK acknowledges support from a Royal Academy of Engineering Research Fellowship.

References

[1] Dudley J M, Dias F, Erkintalo M and Genty G 2014 Instabilities, breathers and rogue waves in optics *Nat. Photon.* **8** 755–764

[2] Solli D R, Ropers C, Koonath P and Jalali B 2007 Optical rogue waves *Nature* **450** 1054–1057

[3] Dudley J M, Genty G and Eggleton B J 2008 Harnessing and control of optical rogue waves insupercontinuum generation *Opt. Express* **16** 3644–3651

[4] Mussot A, Kudlinski A, Kolobov M, Louvergneaux E, Douay M and Taki M 2009 Observation of extreme temporal events in CW-pumped supercontinuum *Opt. Express* **17** 17 010–17 015

[5] Taki M, Mussot A, Kudlinski A, Louvergneaux E, Kolobov M and Douay M 2010 Third-order dispersion for generating optical rogue solitons *Phys. Lett.* A **374** 691–695

[6] Erkintalo M, Genty G and Dudley J M 2010 On the statistical interpretation of optical rogue waves *The European Physical Journal Special Topics* **185** 135–144

[7] Hammani K, Finot C, Dudley J M and Millot G 2008 Optical rogue-wave-like extreme value fluctuations in fiber Raman amplifiers *Opt. Express* **16** 16 467–16 474

[8] Montina A, Bortolozzo U, Residori S and Arecchi F T 2009 Non-Gaussian statistics and extreme waves in a nonlinear optical cavity *Phys. Rev. Lett.* **103** 173 901

[9] Birkholz S, Nibbering E T J, Brée C, Skupin S, Demircan A, Genty G and Steinmeyer G 2013 Spatiotemporal rogue events in optical multiple filamentation *Phys. Rev. Lett.* **111** 243 903

[10] Bonatto C, Feyereisen M, Barland S, Giudici M, Masoller C, Leite J R R and Tredicce J R 2011 Deterministic optical rogue waves *Phys. Rev. Lett.* **107** 053 901

[11] Pisarchik A N, Jaimes-Reátegui R, Sevilla-Escoboza R, Huerta-Cuellar G and Taki M 2011 Rogue waves in a multistable system *Phys. Rev. Lett.* **107** 274 101

[12] Kovalsky M G, Hnilo A A and Tredicce J R 2011 Extreme events in the Ti:sapphire laser *Opt. Lett.* **36** 4449–4451

[13] Soto-Crespo J M, Grelu P and Akhmediev N 2011 Dissipative rogue waves: extreme pulses generated by passively mode-locked lasers *Phys. Rev.* E **84** 016 604

[14] Zaviyalov A, Egorov O, Iliew R and Lederer F 2012 Rogue waves in mode-locked fiber lasers *Phys. Rev.* A **85** 013 828

[15] Lecaplain C, Grelu P, Soto-Crespo J M and Akhmediev N 2012 Dissipative rogue waves generated by chaotic pulse bunching in a mode-locked laser *Phys. Rev. Lett.* **108** 233 901

[16] Horowitz M, Barad Y and Silberberg Y 1997 Noiselike pulses with a broadband spectrum generated from an erbium-doped fiber laser *Opt. Lett.* **22** 799–801

[17] Grelu P and Akhmediev N 2012 Dissipative solitons for mode-locked lasers *Nat. Photon.* **6** 84–92

[18] Xu C and Wise F W 2013 Recent advances in fibre lasers for nonlinear microscopy *Nat. Photon.* **7** 875–882

[19] Fermann M E and Hartl I 2013 Ultrafast fibre lasers *Nat. Photon.* **7** 868–874

[20] Nelson L E, Jones D J, Tamura K, Haus H A and Ippen E P 1997 Ultrashort-pulse fiber ring lasers *Appl. Phys.* B **65** 277–294

[21] Wise F, Chong A and Renninger W 2008 High-energy femtosecond fiber lasers based on pulse propagation at normal dispersion *Laser & Photonics Reviews* **2** 58–73

[22] Richardson D J, Laming R I, Payne D N, Phillips M W and Matsas V J 1991 320 fs soliton generation with passively mode-locked erbium fibre laser *Electron. Lett.* **27** 730–732

[23] Duling I N 1991 Subpicosecond all-fibre erbium laser *Electron. Lett.* **27** 544–545

[24] Duling I N 1991 All-fiber ring soliton laser mode locked with a nonlinear mirror *Opt. Lett.* **16** 539–541

[25] Mollenauer L F, Stolen R H and Gordon J P 1980 Experimental observation of picosecond pulse narrowing and solitons in optical fibers *Phys. Rev. Lett.* **45** 1095–1098

[26] Kelly S 1992 Characteristic sideband instability of periodically amplified average soliton *Electron. Lett.* **28** 806–807

[27] Agrawal G 2001 *Nonlinear Fiber Optics* (New York: Academic Press)

[28] Fermann M E, Andrejco M J, Silberberg Y and Weiner A M 1993 Generation of pulses shorter than 200 fs from a passively mode-locked Er fiber laser *Opt. Lett.* **18** 48–50

[29] Tamura K, Ippen E P and Haus H A 1995 Pulse dynamics in stretched-pulse fiber lasers *Appl. Phys. Lett.* **67** 158–160

[30] Zhou X, Yoshitomi D, Kobayashi Y and Torizuka K 2008 Generation of 28-fs pulses from a mode-locked ytterbium fiber oscillator *Opt. Express* **16** 7055–7059

[31] Buckley J R, Wise F W, Ilday F O and Sosnowski T 2005 Femtosecond fiber lasers with pulse energies above 10nJ *Opt. Lett.* **30** 1888–1890

[32] Chong A, Renninger W H and Wise F W 2007 All-normal-dispersion femtosecond fiber laser with pulse energy above 20 nJ *Opt. Lett.* **32** 2408–2410

[33] Chichkov N B, Hausmann K, Wandt D, Morgner U, Neumann J and Kracht D 2010 50 fs pulses from an all-normal dispersion erbium fiber oscillator *Opt. Lett.* **35** 3081–3083

[34] Aguergaray C, Broderick N G R, Erkintalo M, Chen J S Y and Kruglov V 2012 Mode-locked femtosecond all-normal all-PM Yb-doped fiber laser using a nonlinear amplifying loop mirror *Opt. Express* **20** 10 545–10 551

[35] Erkintalo M, Aguergaray C, Runge A and Broderick N G R 2012 Environmentally stable all-PM all-fiber giant chirp oscillator *Opt. Express* **20** 22 669–22 674

[36] Aguergaray C, Hawker R, Runge A F J, Erkintalo M and Broderick N G R 2013 120 fs, 4.2 nJ pulses from an all-normal-dispersion, polarization-maintaining, fiber laser *Appl. Phys. Lett.* **103** 121 111

[37] Fermann M E, Haberl F, Hofer M and Hochreiter H 1990 Nonlinear amplifying loop mirror *Opt. Lett.* **15** 752–754

[38] Chong A, Wright L G and Wise F W 2015 Ultrafast fiber lasers based on self-similar pulse evolution: a review of current progress *Rep. Prog. Phys.* **78** 113 901

[39] Chong A, Renninger W H and Wise F W 2008 Environmentally stable all-normal-dispersion femtosecond fiber laser *Opt. Lett.* **33** 1071–1073

[40] Lecourt J B, Duterte C, Narbonneau F, Kinet D, Hernandez Y and Giannone D 2012 All-normal dispersion, all-fibered PM laser mode-locked by SESAM *Opt. Express* **20** 11 918–11 923

[41] Kelleher E J R, Travers J C, Ippen E P, Sun Z, Ferrari A C, Popov S V and Taylor J R 2009 Generation and direct measurement of giant chirp in a passively mode-locked laser *Opt. Lett.* **34** 3526–3528

[42] Runge A F J, Aguergaray C, Provo R, Erkintalo M and Broderick N G R 2014 All-normal dispersion fiber lasers mode-locked with a nonlinear amplifying loop mirror *Optical Fiber Technology* **20** 657–665

[43] Renninger W H, Chong A and Wise F W 2008 Giant-chirp oscillators for short-pulse fiber amplifiers *Opt. Lett.* **33** 3025–3027

[44] Trebino R 2000 *Frequency-Resolved Optical Gating: The Measurement of Ultrashort Laser Pulses* (Berlin: Springer)

[45] Aguergaray C, Runge A, Erkintalo M and Broderick N G R 2013 Raman-driven destabilization of mode-locked long cavity fiber lasers: fundamental limitations to energy scalability *Opt. Lett.* **38** 2644–2646

[46] Rhodes M, Steinmeyer G, Ratner J and Trebino R 2013 Pulse-shape instabilities and their measurement *Laser & Photonics Reviews* **7** 557–565

[47] Kobtsev S, Kukarin S, Smirnov S, Turitsyn S and Latkin A 2009 Generation of double-scale femto/pico-second optical lumps in mode-locked fiber lasers *Opt. Express* **17** 20 707–20 713

[48] Smirnov S, Kobtsev S, Kukarin S and Ivanenko A 2012 Three key regimes of single pulse generation per round trip of all-normal-dispersion fiber lasers mode-locked with nonlinear polarization rotation *Opt. Express* **20** 27 447–27 453

[49] North T and Rochette M 2013 Raman-induced noiselike pulses in a highly nonlinear and dispersive all-fiber ring laser *Opt. Lett.* **38** 890–892

[50] Santiago-Hernandez H, Pottiez O, Paez-Aguirre R, Ibarra-Villalon H E, Tenorio-Torres A, Duran-Sanchez M, Ibarra-Escamilla B, Kuzin E A and Hernandez-Garcia J C 2015 Generation and characterization of erbium-Raman noise-like pulses from a figure-eight fibre laser *Laser Phys.* **25** 045 106

[51] Bednyakova A E, Babin S A, Kharenko D S, Podivilov E V, Fedoruk M P, Kalashnikov V L and Apolonski A 2013 Evolution of dissipative solitons in a fiber laser oscillator in the presence of strong Raman scattering *Opt. Express* **21** 20 556–20 564

[52] Babin S A, Podivilov E V, Kharenko D S, Bednyakova A E, Fedoruk M P, Kalashnikov V L and Apolonski A 2014 Multicolour nonlinearly bound chirped dissipative solitons *Nat. Commun.* **5** 4653

[53] Solli D R, Herink G, Jalali B and Ropers C 2012 Fluctuations and correlations in modulation instability *Nat. Photon.* **6** 463–468
[54] Sugavanam S, Fabbri S, Le S T, Lobach I, Kablukov S, Khorev S and Churkin D 2016 Real-time high-resolution heterodyne-based measurements of spectral dynamics in fibre lasers. *Sci. Rep.* **6** 23 152
[55] Närhi M, Wetzel B, Billet C, Toenger S, Sylvestre T, Merolla J M, Morandotti R, Dias F, Genty G and Dudley J M 2016 Real-time measurements of spontaneous breathers and rogue wave events in optical fibre modulation instability *Nat. Commun.* **7** 13 675
[56] Goda K and Jalali B 2013 Dispersive Fourier transformation for fast continuous single-shot measurements *Nat. Photon.* **7** 102–112
[57] Wetzel B *et al* 2012 Real-time full bandwidth measurement of spectral noise in supercontinuum generation *Sci. Rep.* **2** 882
[58] Godin T *et al* 2013 Real time noise and wavelength correlations in octave-spanning supercontinuum generation *Opt. Express* **21** 18 452–18 460
[59] Runge A F J, Aguergaray C, Broderick N G R and Erkintalo M 2013 Coherence and shot-to-shot spectral fluctuations in noise-like ultrafast fiber lasers *Opt. Lett.* **38** 4327–4330
[60] Herink G, Jalali B, Ropers C and Solli D R 2016 Resolving the build-up of femtosecond mode-locking with single-shot spectroscopy at 90 MHz frame rate *Nat. Photon.* **10** 321–326
[61] Runge A F J, Aguergaray C, Broderick N G R and Erkintalo M 2014 Raman rogue waves in a partially mode-locked fiber laser *Opt. Lett.* **39** 319–322
[62] Lecaplain C, Grelu P, Soto-Crespo J M and Akhmediev N 2013 Dissipative rogue wave generation in multiple-pulsing mode-locked fiber laser *J. Opt.* **15** 064 005
[63] Aalto A, Genty G and Toivonen J 2010 Extreme-value statistics in supercontinuum generation by cascaded stimulated Raman scattering *Opt. Express* **18** 1234–1239
[64] Liu Z, Zhang S and Wise F W 2015 Rogue waves in a normal-dispersion fiber laser *Opt. Lett.* **40** 1366–1369
[65] Soto-Crespo J M, Akhmediev N and Ankiewicz A 2000 Pulsating, creeping, and erupting solitons in dissipative systems *Phys. Rev. Lett.* **85** 2937–2940
[66] Akhmediev N, Soto-Crespo J M and Town G 2001 Pulsating solitons, chaotic solitons, period doubling, and pulse coexistence in mode-locked lasers: complex Ginzburg-Landau equation approach *Phys. Rev.* E **63** 056 602
[67] Latas S C V and Ferreira M F S 2010 Soliton explosion control by higher-order effects *Opt. Lett.* **35** 1771–1773
[68] Descalzi O and Brand H R 2010 Transition from modulated to exploding dissipative solitons: hysteresis, dynamics, and analytic aspects *Phys. Rev.* E **82** 026 203
[69] Cartes C, Descalzi O and Brand H R 2012 Noise can induce explosions for dissipative solitons *Phys. Rev.* E **85** 015 205
[70] Cundiff S T, Soto-Crespo J M and Akhmediev N 2002 Experimental evidence for soliton explosions *Phys. Rev. Lett.* **88** 073 903
[71] Runge A F J, Broderick N G R and Erkintalo M 2015 Observation of soliton explosions in a passively mode-locked fiber laser *Optica* **2** 36–39
[72] Liu M, Luo A P, Yan Y R, Hu S, Liu Y C, Cui H, Luo Z C and Xu W C 2016 Successive soliton explosions in an ultrafast fiber laser *Opt. Lett.* **41** 1181–1184
[73] Liu M, Luo A P, Xu W C and Luo Z C 2016 Dissipative rogue waves induced by soliton explosions in an ultrafast fiber laser *Opt. Lett.* **41** 3912–3915

[74] Runge A F J, Broderick N G R and Erkintalo M 2016 Dynamics of soliton explosions in passively mode-locked fiber lasers *J. Opt. Soc. Am.* B **33** 46–53

[75] Akhmediev N and Soto-Crespo J M 2004 Strongly asymmetric soliton explosions *Phys. Rev.* E **70** 036 613

[76] Churkin D V, Sugavanam S, Tarasov N, Khorev S, Smirnov S V, Kobtsev S M and Turitsyn S K 2015 Stochasticity, periodicity and localized light structures in partially mode-locked fibre lasers *Nat. Commun.* **6** 7004

[77] Kelleher E J R and Travers J C 2014 Chirped pulse formation dynamics in ultra-long mode-locked fiber lasers *Opt. Lett.* **39** 1398

[78] Woodward R I and Kelleher E J R 2016 Dark solitons in laser radiation build-up dynamics *Phys. Rev.* E **93** 032 221

[79] Kivshar Y and Luther-Davies B 1998 Dark optical solitons: physics and applications *Phys. Rep.* **298** 81–197

[80] Gouveia-Neto A S and Taylor J R 1989 Soliton evolution from noise bursts *Electron. Lett.* **25** 736

[81] Woodward R I, Kelleher E J R, Popa D, Hasan T, Bonaccorso F, Ferrari A C, Popov S V and Taylor J R 2014 Scalar nanosecond pulse generation in a nanotube mode-locked environmentally stable fiber laser *IEEE Photon. Technol. Lett* **26** 1672–1675

[82] Woodward R I, Kelleher E J R, Runcorn T H, Loranger S, Popa D, Wittwer V J, Ferrari A C, Popov S V, Kashyap R and Taylor J R 2015 Fiber grating compression of giant-chirped nanosecond pulses from an ultra-long nanotube mode-locked fiber laser *Opt. Lett.* **40** 387–390

[83] Tomlinson W J, Hawkins R J, Weiner A M, Heritage J P and Thurston R N 1989 Dark optical solitons with finite-width background pulses *J. Opt. Soc. Am.* B **6** 329–334

[84] Kivshar Y S and Yang X 1994 Dark solitons on backgrounds of finite extent *Opt. Commun.* **107** 93–98

[85] Thurston R N and Weiner A M 1991 Collisions of dark solitons in optical fibers *J. Opt. Soc. Am.* B **8** 471

[86] Proukakis N P, Parker N G, Frantzeskakis D J and Adams C S 2004 Analogies between dark solitons in atomic Bose–Einstein condensates and optical systems *J. Opt.* B **6** S380–S391

[87] Herink G, Kurtz F, Jalali B, Solli D R and Ropers C 2017 Real-time spectral interferometry probes the internal dynamics of femtosecond soliton molecules *Science* **356** 50–54

[88] Gao L, Zhu T, Wabnitz S, Liu M and Huang W 2016 Coherence loss of partially mode-locked fibre laser *Scientific Reports* **6** 24995

[89] Bowen P, Erkintalo M, Provo R, Harvey J D and Broderick N G R 2016 Mode-locked Yb-doped fiber laser emitting broadband pulses at ultralow repetition rates *Opt. Lett.* **41** 5270–5273

[90] Krupa K, Nithyanandan K, Andral U, Tchofo-Dinda P and Grelu P 2017 Real-time observation of dissipative optical soliton molecular motions. arXiv:1702.01161 [physics]

[91] Wang Y, Leo F, Fatome J, Erkintalo M, Murdoch S G and Coen S 2017 Universal mechanism for the binding of temporal cavity solitons. arXiv:1703.10604 [physics]

IOP Publishing

Nonlinear Guided Wave Optics
A testbed for extreme waves
Stefan Wabnitz

Chapter 6

Extreme wave dynamics from incoherent dissipative solitons in fiber laser cavities

Philippe Grelu, José M Soto-Crespo, Katarzyna Krupa, K Nithyanandan, Saïd Hamdi and Aurélien Coillet

6.1 Introduction: the notion of incoherent dissipative solitons

The major sought-after virtues of dissipative solitons are their robustness and stability, as far as the development of coherent pulsed laser sources is concerned [1–3]. These advantageous properties arise from the existence of focus or fixed-point attractors that can be found in a wide range of laser cavity parameters. Thus, provided that a suitable balance between nonlinear gain and loss propagation effects is achieved, stable and coherent ultrashort optical pulses can form within diverse laser cavity architectures and for various saturable absorption solutions. In that respect, dissipative solitons are an efficient concept for understanding and designing mode-locked lasers [4, 5]. Nevertheless, beyond the pulsed regimes having a stationary amplitude that are based on fixed-point attractors, dissipative systems are known to host a wealth of complex nonlinear dynamics, including pulsating and chaotic ones, which can be accessed by altering one or several cavity parameters [4]. Adjusting for instance the laser pumping power or the cavity dispersion, the transitions between stationary and oscillating-amplitude short-pulse dynamics can be observed experimentally and understood by numerical modeling. Such possible transitions include Hopf-type [6, 7] and multiple-period bifurcations [8]. As a common observation from laser physicists, by crossing the boundary of the coherent pulse-stability domain, regular to irregular pulsations can take place and will quickly lead to the complete disruption of the mode-locked pulse operation. Characterized by a sharp spectral shrinking, such disruption will end up in a multimode incoherent emission dubbed as quasi-continuous-wave (quasi-cw), since a noisy optical field extending all along the cavity can be observed. This transition from mode locking to quasi-cw is understood as strongly imbalanced physical effects, e.g. nonlinearity, dispersion, gain, etc, can no longer sustain the field confinement in a short wavepacket traveling around the cavity.

doi:10.1088/978-0-7503-1460-2ch6

© IOP Publishing Ltd 2017

However, several experiments have underlined the existence of paradoxical pulse regimes that do not fall into the category of laser mode locking: they manifest as incoherent and chaotic, yet strongly confined, wavepackets which nevertheless exist in a wide range of cavity parameters. Starting from a coherent single-pulse operation, we can typically draw two roadmaps that lead to incoherent, self-generated laser short pulses.

The first of these is widely represented among normally dispersive laser cavities. In this roadmap, the initial coherent single pulse will feature an important frequency chirp. By increasing one or several parameters (the cavity length, typically), destabilizing physical effects such as birefringence, nonlinear polarization rotation overdrive, and/or Raman nonlinearity will entail pulse disruption. However, a subsequent balance can allow the localization of an incoherent pulse to be maintained one roundtrip after another. Consistently, this balance will involve nonlinear dissipation, such as gain spectral filtering, following the wide spectral extension of the incoherent pulse. The first comprehensive report of this peculiar pulsed laser regime was made in 1997 by Horowitz *et al* and referred to as *noiselike pulse* (NLP) generation [9]. The reader will find a detailed presentation of this roadmap in chapter 5 from Erkintalo *et al* in this volume, with a highlight on the real-time characterization challenge as well as on the connected dynamics of soliton explosion and metastable dark solitons.

The second roadmap involves multiplying the number of soliton pulses by increasing considerably the pumping power, and selecting a regime of chaotic bound states [10], where the multiple-pulse interactions entail a confinement of the pulse set—namely, pulse bunching—yet the individual pulses are perpetually set in chaotic relative motions. Such a combination of chaotic motion and bunching makes the multi-soliton structure comparable to a liquid soliton state [11, 12]. This roadmap applies typically to anomalously dispersive laser cavities, where each pulse will have a low energy and frequency chirping, as compared to the pulse features in the normal-dispersion case. According to the cavity parameters and depending on the number of pulse constituents involved, the temporal extension of the chaotic pulse bunch can vary over a large range, typically from tens of picoseconds to tens of nanoseconds. The actual temporal extension will become crucial when it comes to real-time characterization issues. Compared to the first roadmap case, the level of coherence or incoherence of the pulse bunch can be, in principle, varied, following the adjustment of pulse interactions. Also, the temporal granularity, or temporal short-length scale, is expected to remain comparable to a single soliton duration, though for the most chaotic states this may not hold true [13].

These two roadmaps are not exclusive. For instance, it is possible to observe incoherent short pulses of extended spectral widths from the path-averaged anomalous dispersion regime as well [14]. This reflects the general complexity and multiplicity of the mechanisms that can lead to the generation of incoherent pulses. In our view, such diversity justifies the use of the *incoherent dissipative soliton* general terminology, which encompasses noiselike pulse emission and soliton liquid dynamics. The suggested terminology presents a valuable parallel with the notion of incoherent solitons of conservative systems. The latter involves the self-trapping of incoherent light in either the spatial or temporal domain, using a nonlinear medium characterized by a slow response [15, 16] or nonlocality [17]. Whereas several physical mechanisms can lead to the formation of incoherent dissipative solitons, dissipation itself will always take a leading role, enabling the pulse localization in both the temporal and spectral domains in contrast, for instance, to the spectrally accelerated temporal incoherent solitons [16].

We emphasize the fact that there is not a single incoherent dissipative soliton regime. Instead, the latter is representative of a wide range of dynamics that can be differentiated by using several quantitative tools, such as the level of fluctuations in the temporal and spectral domains, from which histograms and statistics can be made. In turn, statistics link to the search of extreme wave and rogue wave dynamics, which constitute the core of the present book. Obviously, the possibilities will differ between numerical and experimental investigations, owing to the accessible simulation window of the former and the limitations of real-time measurements in the latter. Fortunately, recent advances in ultrafast characterization [7, 18] have allowed for an improved experimental characterization of incoherent dissipative solitons [14, 19].

To illustrate these aspects, this chapter presents a variety of incoherent pulse regimes that have been observed or simulated in laser cavities endowed with anomalous chromatic dispersion, following mainly the second road map. The next section will show how a chaotic multiple pulse regime can lead, through strong pulse interactions and inelastic collisions, to the generation of extreme wave dynamics. This section starts from the modeling perspective and then compares with fiber laser experiments that are analyzed in the temporal domain. By changing the nature of the saturable absorber, the extension to vector dynamics is proposed in the following section. By implementing a dispersive Fourier-transform spectro-temporal imaging technique, it enables a shot-to-shot experimental study in the optical spectral domain, entailing the notion of vector spectral rogue waves. Vector incoherent dissipative solitons are studied experimentally, highlighting the existence of vector spectral rogue waves that connect to explosive-type dynamics, which are compared to soliton explosions.

6.2 Dissipative rogue waves from chaotic pulse bunching

With similar features as those required to obtain mode locking, the fiber laser architecture needs a high-gain amplifying section to compensate for the significant cavity roundtrip losses, and a fast saturable absorber to favor high peak-intensity pulse emission. The anomalous path-averaged dispersion can be obtained by combining different fibers, using for instance anomalous passive fibers and normal gain fibers with appropriate lengths or anomalous fibers exclusively, as typical for 1.5 and 2 micron fiber lasers, respectively. To reach the multiple dissipative soliton regime from which chaotic pulse bunching will be searched, a relatively high pumping power is required, typically in the range of 0.5 W or above.

6.2.1 Numerical modeling and investigations

As stated in the Introduction, chaotic pulse bunching, noise-like pulse emission, and optical rogue waves are related nonlinear phenomena. These dynamics can either appear simultaneously in the vicinity of passive mode-locking, or they can show up separately [20].

We shall consider here a typical lumped model of a passively mode-locked laser operating in the anomalous dispersion regime, which combines these three possible aspects for a wide range of cavity parameters; it is shown schematically in figure 6.1. Namely, the ring cavity is composed of three basic components: (A) an erbium-doped fiber (EDF), (B) a single-mode fiber (SMF) with anomalous dispersion, and (C) a saturable absorber, which is followed by an output coupler that splits out a certain amount of the energy accumulated during the roundtrip.

The evolution of the field envelope, $\psi(z, t)$, in the SMF is governed by the nonlinear Schrödinger equation (NLSE):

$$i\psi_z + \frac{D}{2}\psi_{tt} + |\psi|^2\psi = 0, \tag{6.1}$$

where D is the dispersion parameter, z is the propagation distance, and t is the time in a frame of reference moving with the group velocity, all in normalized dimensionless units.

Pulse propagation in the EDF is modeled by

$$i\psi_z + \frac{D_2}{2}\psi_{tt} + \Gamma_2|\psi|^2\psi = \frac{ig_o}{1 + Q/Q_{\text{sat}}}(\psi + \beta_2\psi_{tt}), \tag{6.2}$$

where the total energy Q is

$$Q = \int_{-\infty}^{\infty} |\psi|^2 dt,$$

Q_{sat} is the parameter representing the saturation energy, g_o is the small signal gain, and β_2 represents the spectral width of the gain. Γ_2 is the ratio between the nonlinear coefficients in the EDF and SMF; more precisely, it corresponds to the ratio between the effective mode areas of the EDF and SMF, when both fibers are made of the same glass material, e.g. silica.

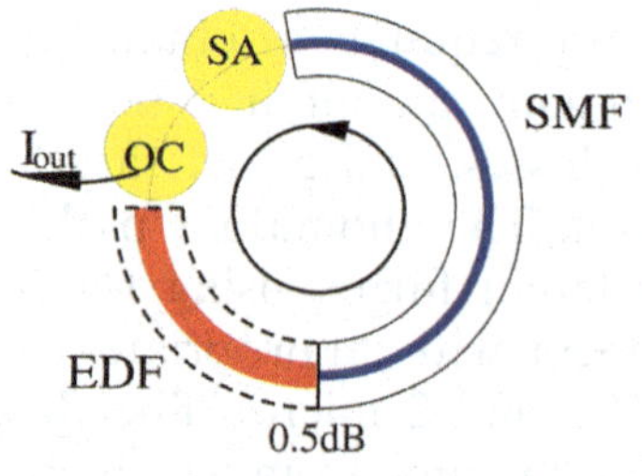

Figure 6.1. Model of the fiber laser used in our simulations. SMF, single-mode fiber; SA, saturable absorber; OC, output coupler; EDF, erbium-doped fiber. From [27].

The instantaneous saturable absorber (SA) is modeled by the following transfer function:

$$T = T_o + \Delta T \frac{I(t)}{I_{\text{sat}} + I(t)}, \tag{6.3}$$

where $I(t) = |\psi(t)|^2$, T_o is the transmission level at low intensities of the optical field, ΔT is the transmission contrast, and I_{sat} is the saturation intensity. The output coupler is characterized by the transmission coefficient T_r. The latter is the coefficient representing power transfer between the two sections of the cavity. A typical 0.5 dB coupling loss is added at the EDF–SMF inhomogeneous splice junction.

The choice of the values of the parameters determine the operation regime of the laser. The gain is the parameter that can be most easily changed in the experiments; it can be controlled either by g_o or by the saturation parameter Q_{sat}. We have chosen Q_{sat} as the control parameter, as it can be expressed directly in units of the energy pumped into the cavity when operating well above the lasing threshold.

A stationary regime of single pulse operation can be reached by choosing a realistic set of the model parameters (given in the caption of figure 6.2) and for small values of the saturation parameter Q_{sat}. By increasing its value, we have constructed the bifurcation diagrams shown in figure 6.2. The laser generates a single pulse with well-defined amplitude at Q_{sat} between 1.25 and 2.34. We then have a sequence of

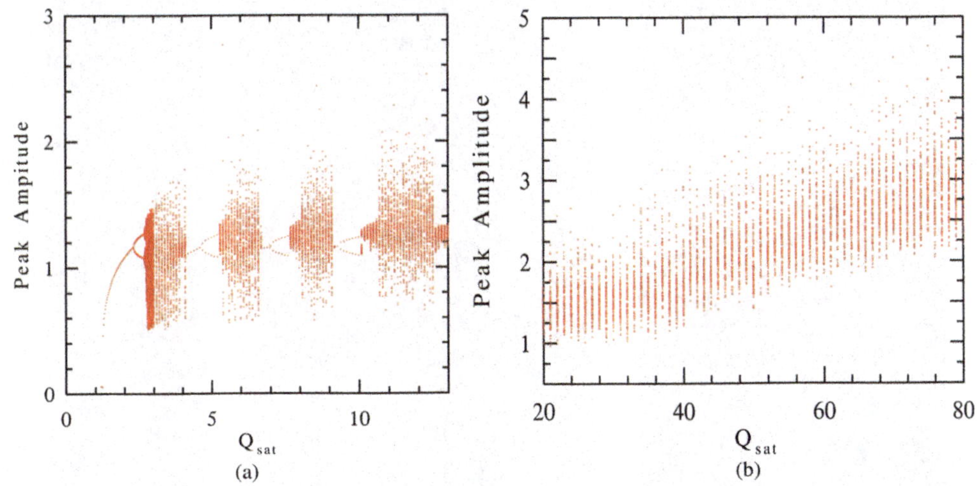

Figure 6.2. Bifurcation diagram constructed by varying Q_{sat} while keeping fixed the rest of the parameters. The plot is split into two sections (a) and (b) for convenience. Each (red) dot on the diagram represents the absolute maximum of the field amplitude observed at each roundtrip. The regions with a single maximum for a given Q_{sat} in (a) correspond to a stable single-pulse operation. The regions with a wide vertical band represent a chaotic output. These regions appear repeatedly when increasing Q_{sat}. Each of them occupies a finite interval along the Q_{sat}-axis, being a potential area for the appearance of rogue waves. Parameters of the system are: $D_2 = -0.735$, $\Gamma_2 = 3$, $g_0 = 4$, $\beta_2 = 0.008$. $L_{\text{SMF}} = 1.1$, $L_{\text{EDF}} = 0.22$, $T_r = 70\%$, $T_0 = 0.3$, $\Delta T = 0.7$, and $I_{\text{sat}} = 0.6$. From [27].

period doubling bifurcations driving the pulse to a chaotic regime [25]. There are several such transitions in figure 6.2(a) with an increasing number of pulses inside each roundtrip. Namely, the laser generates two pulses when Q_{sat} is around 4.5. As expected, the number of pulses in the cavity increases as Q_{sat} gets higher. It generates three pulses when Q_{sat} is around 7, and so on.

For higher values of Q_{sat}, the number of pulses in the cavity becomes legion; the interaction between them produces significantly more complex field profiles. Depending on the exact value of Q_{sat}, the pulses may repel or attract each other. Signal evolution inside the fiber cavity for three cases of Q_{sat} are presented in figure 6.3.

Figure 6.3(a) shows the chaotic evolution of a mostly single pulse for low values of the energy inside the cavity, namely for $Q_{sat} = 3$. The pulse was stationary and stable for $Q_{sat} \leqslant 2$, but now evolves erratically with a randomly changing profile. When Q_{sat} is considerably increased up to 30, figure 6.3(b) shows that a number of pulses coexist inside the cavity, which mainly repel each other. They tend to be distributed more or less uniformly within the cavity. The dynamics reveal their chaotic motion and collisions.

As Q_{sat} is further increased the pulses have the tendency to group into a bunch that has a nearly stable envelope but fluctuating internal structure. This behavior is that of a noise-like pulse [9, 21–24], which entails a drastic loss of the pulse coherence as compared to mode-locked laser operation [14, 19]; it is illustrated in figure 6.3(c)) for $Q_{sat} = 60$. Sporadically, the bunch may expel a single pulse that extinguishes as it

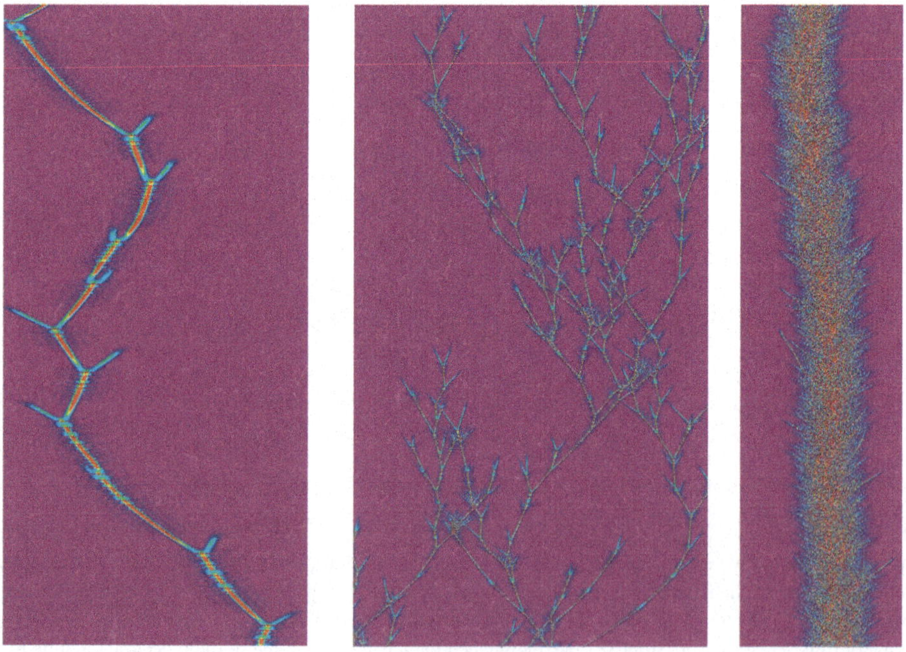

(a) (b) (c)

Figure 6.3. Signal evolution for Q_{sat} = (a) 3, (b) 30, and (c) 60. The rest of the parameters are the same as in figure 6.2. In all cases, the propagation distance is 500 roundtrips and the time intervals are 80, 500, and 230, respectively.

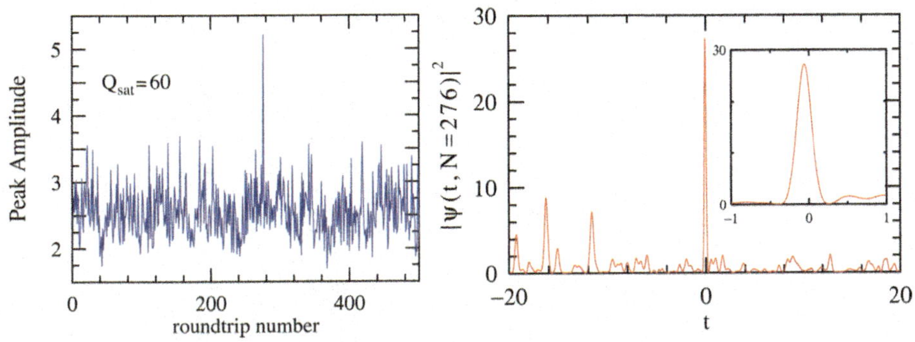

Figure 6.4. (a) Evolution of the absolute maximum amplitude of the field in t with the number of roundtrips, N, for $Q_{\text{sat}} = 60$ (solid blue line). The highest maximum in this run is observed when $N = 276$. (b) Main segment of the transverse field profile of the laser radiation at the roundtrip number $N = 276$. A large number of pulses occupies the larger temporal interval $(-50,\ 50)$. The extreme pulse in the middle can be considered as a rogue wave. The inset shows the same pulse in higher resolution. From [27].

separates from the bunch. This emission occurs more likely for less powerful bunches, i.e. for $Q_{\text{sat}} \leqslant 60$. In addition, they travel farther before extinction.

This regime looks ideal for finding rogue waves. Pulse amplitudes experience large variations, as seen in figure 6.2(b). For this purpose, at each roundtrip, we trace the pulse with the highest amplitude in the t domain. In such a way, we can detect the tallest pulse in any particular simulation. A typical example is shown in figure 6.4(a). A huge peak intensity is located at a given roundtrip number, N, namely at $N = 276$. A section of the optical field profile containing the tallest pulse for this special value of roundtrip is shown in figure 6.4(b). Indeed, the pulse, intentionally positioned in the middle of the interval, $t = 0$, can be considered as a rogue wave as its maximum intensity is almost three times higher than the maximum intensity of the pulse that is next in height. This example shows clearly the random appearance of extreme waves from our laser model. Higher peak amplitudes can also be observed but the probability of their appearance is lower. These extremely-high-amplitude pulses are good candidates to be named dissipative rogue waves.

The statistical properties of radiation can be studied by calculating the probability density function (PDF) of the peak amplitude of the pulses in the bunch for many different realizations. For $Q_{\text{sat}} = 60$, the probability density function is shown in figure 6.5.

This figure demonstrates that the high-intensity parts of the PDF are indeed elevated. The PDF curve in figure 6.5 is neither a convex parabola nor a straight line as one would expect from the classical distributions. Thus, for the sake of comparison, in figure 6.5, we also draw a straight dashed (green) line fitted to approximate the low- to medium-intensity part of the curves. It can be seen that the data at the high-intensity part of the curves are orders of magnitude above this line. Thus, high-intensity events occur more often than they should according to the Gaussian or Rayleigh distributions. The existence of a myriad of low-intensity pulses makes the significant wave height (SWH) small enough that it is easy to find pulses whose amplitude exceeds 2.2 times this SWH. In fact, even neglecting pulses with

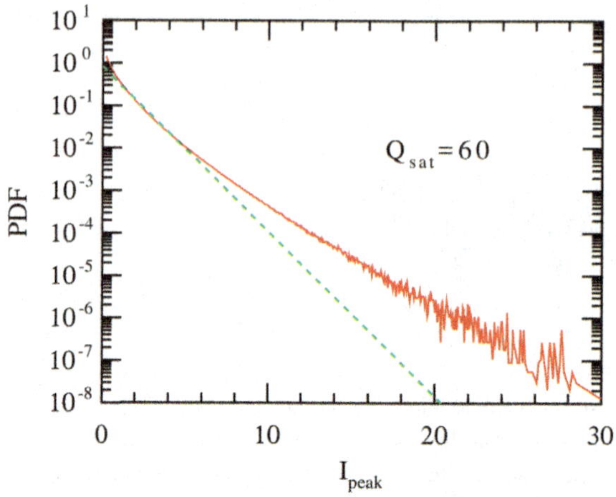

Figure 6.5. Probability density function for the maxima of intensity (red solid curve). For comparison, the dashed (green) line is drawn for a pure exponential decay of the probability. From [27].

amplitudes lesser than 0.1, the number of rogue waves is elevated [26]. From these numerical investigations, we conclude that the probability of producing extreme pulses in such a laser system is higher than in most other known nonlinear systems.

6.2.2 Fiber laser experiments

To explore the diversity of incoherent dissipative soliton dynamics in anomalous-dispersion fiber lasers, it is important to design a laser cavity with several accessible, widely tunable parameters. The pumping power level controls the energy of the pulse bunch through the number of sub-pulse constituents that also affects the temporal extension of the pulse bunch. The use of a virtual saturable absorber obtained by nonlinear interference offers at the same time a tunable nonlinear transfer function and an ultrafast response. We use the scheme of nonlinear polarization evolution (NPE), along with polarization discrimination, to shape the nonlinear transfer function. Relying on the crossed Kerr nonlinearity of the glass fiber, NPE-based mode locking is one of the most popular fiber laser mode-locking schemes, as it is widely adjustable, ultrafast, cost-effective, and can be implemented at all lasing wavelengths. Note that although NPE involves the two polarization components, the overall laser dynamics are equivalent to a scalar one, since the polarization degree of freedom is frozen by the intracavity polarizer element.

The 1.5 micron fiber laser experimental setup is sketched in figure 6.6(a). It is an all-fiber ring cavity comprised of several optical fibers and fiber pigtailed integrated components. Unidirectional laser propagation is ensured by an optical isolator. A 2 m long EDF constitutes the gain medium, pumped by two counterpropagating laser diodes operating at 980 nm that couple up to 1 W of total pump power. Pieces of passive standard (SMF 28) and dispersion-compensating (DCF) fibers can be used to control the amount of chromatic dispersion per cavity roundtrip. For analysis, the laser output intensity is sent to a fast photodiode that is connected to a real-time oscilloscope, as well as to an optical spectrum analyzer (OSA, resolution

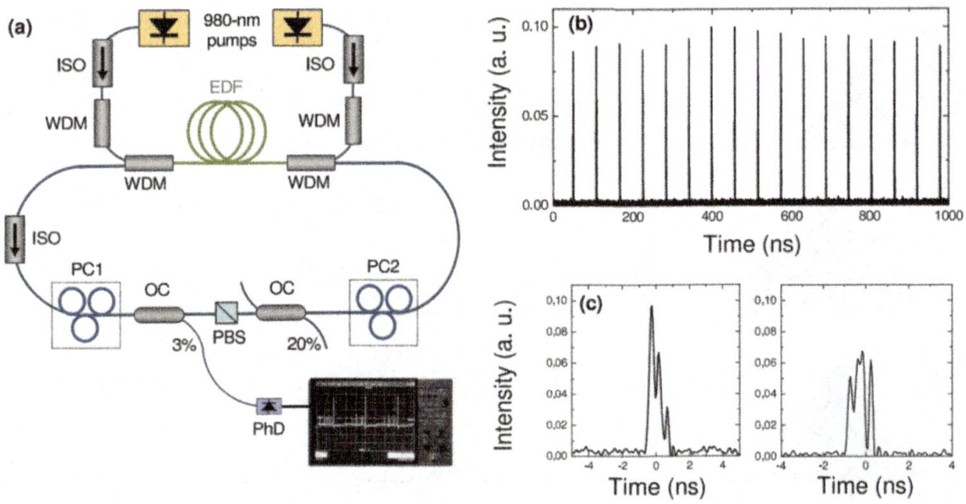

Figure 6.6. (a) Sketch of the 1.5 micron fiber laser experimental setup. EDF: erbium-doped fiber, WDM: wavelength-division multiplexer, ISO: polarization-insensitive optical isolator, OC: optical coupler, PC1,2: polarization controllers, PBS: integrated polarization beam-splitter, PhD: fast photodiode. (b) Weakly mode-locked pulse train, comprising a chaotic *soliton liquid* bunch of pulses, whose temporal internal structure is partly resolved in (c).

65 pm) and to a multishot second-order autocorrelator. The combination of a polarizing beam splitter (PBS) and polarization controllers (PCs) implements the NPE virtual saturable absorber mechanism. Transmission through the PBS transforms the polarization modulation into an amplitude modulation. The specific nonlinear transfer function is defined by the setting of the PCs located before and after the PBS. Such precise tailoring of the nonlinear transfer function drastically affects the pulsed regime of operation as well as the nature and the reach of pulse-to-pulse interactions [27]. Here, each PC consists of three fiber loops whose plane can be manually oriented: each of these loops acts as an orientable waveplate with a fixed phase delay. A wide range of pulsed regimes becomes accessible, besides the standard stationary-amplitude mode locking, by simply altering the orientation of the PCs.

A first illustration of the chaotic bunch operation is illustrated by the temporal traces shown in figures 6.6(b) and (c). Although figure 6.6(b) displays a pulse that repeats at roundtrip time intervals, a closer inspection allowed by the electronic temporal resolution of 65 ps reveals a pulse bunch whose duration is around 2 ns, with an internal structure that strongly fluctuates from one roundtrip to another (see figure 6.6(c)). Though this internal structure is not fully resolved, the dynamics of the chaotic pulse can be further investigated. We also note that the total field energy fluctuations from one roundtrip to another are minor, in contrast to the intensity fluctuations resulting from the interactions that take place among the chaotic pulse constituents of the bunch. A typical illustration of intracavity manipulation of the multiple pulse dynamics is shown in figure 6.7. By a mere adjustment of the

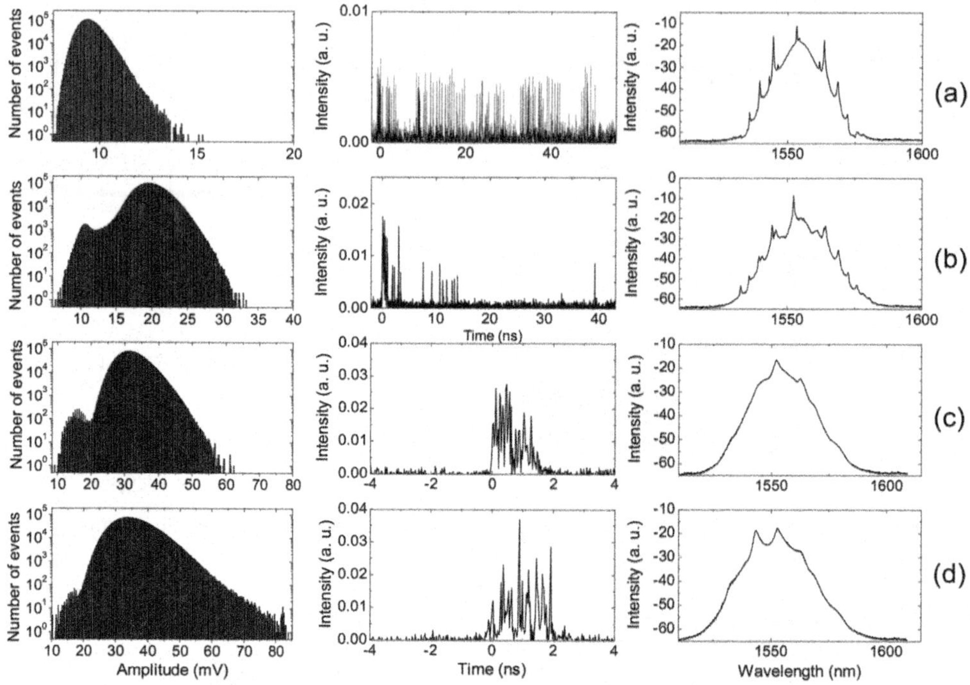

Figure 6.7. Illustration of the various multiple pulse dynamics that can be accessed by changing the orientation of one polarization controller. Starting from the uneven and non-stationary pulse distribution shown in (a) after a transition favoring the attractive interactions (b), chaotic pulse bunching is obtained (c) with an increased number of collisions that are able to generate rogue wave events (d). Left column: histogram of the peak-intensity recordings; middle: temporal snapshot of the output intensity using a 20 GHz real-time oscilloscope; right: optical spectrum. Note the scale changes in histograms and temporal snapshots between successive lines. Adapted from [14].

orientation of one PC, the transfer function is readily altered and the net interaction among pulses is changed. This way, chaotic pulse bunching takes place, and can lead to an L-shaped distribution of detected optical intensity maxima.

To detect rogue wave events in such an experimental scheme, it is crucial to have a sufficient detection bandwidth to reveal and analyze the fluctuating internal structure. Considering the same dynamics as in figure 6.6, figure 6.8 shows that for a bandwidth of 13 and 20 GHz, the L-shaped PDF will highlight dissipative rogue wave candidates, as expected from the numerical simulations developed in the former section. However, as the bandwidth reduces below 10 GHz, by loosening the resolution of the pulse bunch structure, no extreme event will be found. Even if this seems obvious, it is nevertheless a key distinction of the present rogue wave mechanism, namely through the chaotic interactions that take place inside the pulse bunch. This contrasts with another type of dynamics found in a short pulse laser, where total energy fluctuations were responsible for extreme waves which could then be detected with conventional, lower-bandwidth, electronics [28].

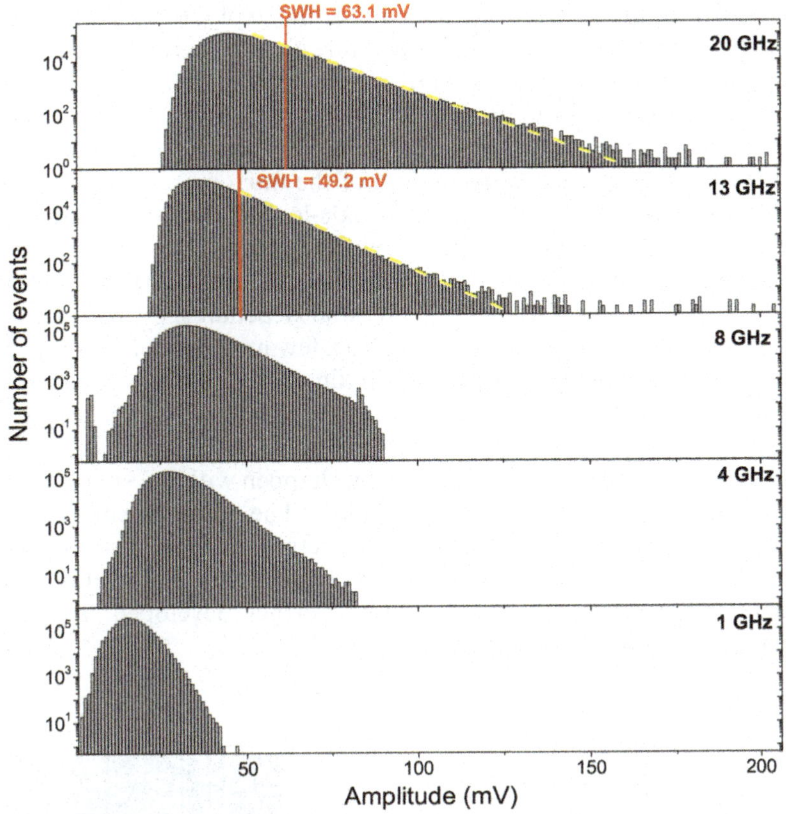

Figure 6.8. Influence of the detection bandwidth on the probability distribution function. The same chaotic pulse bunching dynamics is analyzed with an electronic bandwidth decreasing from 20 to 1 GHz, from top to bottom. From [14].

6.2.3 Incoherent solitons around the 2-micron wavelength

The dynamics and interactions of large numbers of solitons are also a subject of current research around the wavelength of 1.9 μm, in fiber lasers based on the amplification from thulium ions [29–31]. Compared to erbium doped fibers, thulium-doped fibers (TDFs) have a wider gain bandwidth, ranging from 1.8 μm to 2.1 μm [32], and the relaxation time of the excited electronic level of the thulium ions is much shorter than that of erbium. This affects quantitatively the various soliton interaction schemes within fiber cavities, and can even lead to new types of dynamics. Despite these differences, standard telecommunication fibers and components can be made compatible with the 1.9 μm wavelength, and virtual saturable absorbers in particular can be obtained with the same techniques as those used for conventional 1.55 μm mode-locked fiber lasers.

To illustrate the universality of chaotic pulse interactions in general, we present some preliminary observations made with a TDF laser that uses another type of virtual saturable absorber, namely, a nonlinear amplified loop mirror (NALM). Our fiber laser uses a 3 m long TDF as gain medium, and exploits the NALM to generate pulses (see figure 6.9(a)). This effective saturable absorber architecture consists of a Sagnac loop mirror that includes a TDF amplifier located at one extremity of the loop [33]. The imbalance brought by the TDF gain leads to the accumulation of a different nonlinear phase shift depending on the propagation direction, and therefore to an intensity-dependent transmission of the mirror.

With this effective saturable absorber, mode-locked operation can be obtained around 1.9 μm, and dissipative solitons are easily observed [33]. In order to generate a large number of low-energy solitons, we increased the length of the nonlinear loop to reach 100 m. At a pump power of 1 W, and depending on the position of the polarization controllers, we could obtain up to a few hundreds of pulses propagating within the cavity [34]. The different time-domain evolution maps displayed in figure 6.9 show a wide variety of dynamics: collisions (elastic and inelastic), molecule formation or splitting [31], velocity locking, spontaneous appearance, or extinction. In the case of figure 6.9(d), all these interactions happen within a small portion of the cavity roundtrip, with a few hundreds of pulses. The confinement in a short time interval of such a large number of objects interacting through various and complex mechanisms could easily lead to chaotic or extreme events. No doubt the investigation of extreme wave generation will be further developed in the 2-micron wavelength region in the coming years.

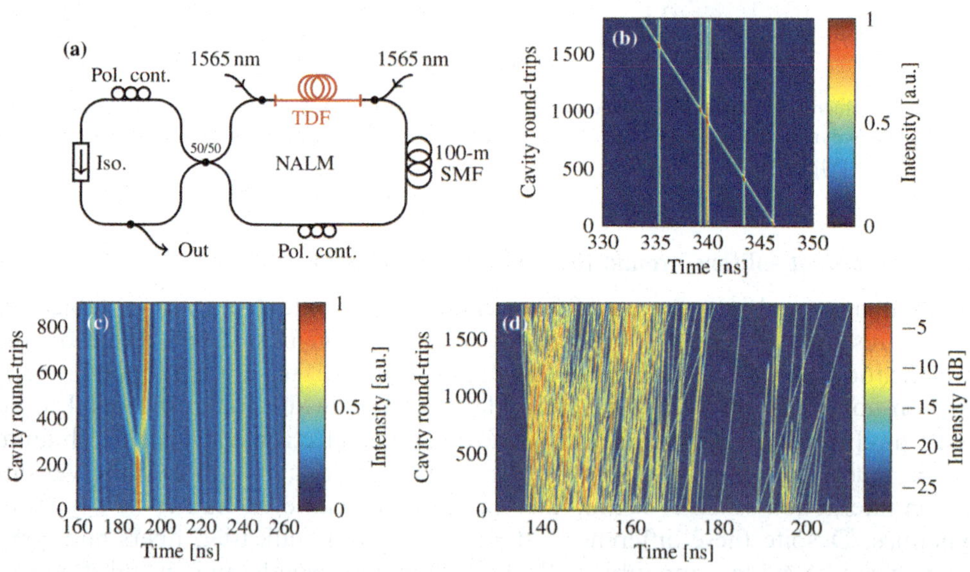

Figure 6.9. (a) Experimental setup for the TDF laser with an effective saturable absorber based on a nonlinear amplified loop mirror (NALM). TDF stands for thulium-doped fiber and SMF is a standard single-mode fiber at the telecom wavelength. (b), (c) Experimental time-domain evolution maps of two types of interactions between solitons and soliton molecules in a thulium fiber cavity. (d) For high pump power, the amount of soliton in the cavity can reach the few hundred range, and their interactions (elastic and inelastic collisions, molecule formation or splitting, velocity locking, etc) can readily become chaotic.

6.3 Extreme vector waves

Standard single-mode fiber supports two orthogonal polarization modes due to the presence of small amounts of random birefringence, inevitable in practice, which is induced by manufacturing imperfections and externally applied random strain or bending. In cavities without polarization discrimination components, a wide range of polarization dynamics can manifest, from polarization self-modulated and mode-locked dual-polarization lasers to polarization switching and chaos [35, 36]. In fiber lasers, the variety of complex polarization dynamics can be combined with that of dissipative soliton dynamics. By simply tuning the overall cavity birefringence through an intra-cavity polarization controller, dynamics with locked, switching, and rotating states of polarization have recently been investigated experimentally and demonstrated in the regimes of fundamental soliton, harmonic mode-locking, soliton molecules, and different multi-pulse operations [37–45]. In this section, we will focus on the more general case of vector-incoherent dissipative solitons including in particular a regime characterized by transient spectro-temporal ordering among the incoherent pulses. We will show that in the vicinity of polarization-switching dynamics, intermittent behaviors can manifest, and that they can be associated with the generation of vector rogue waves [46].

6.3.1 Fiber laser setup

Here, we present the results that were obtained with the experimental setup illustrated in figure 6.10. The laser configuration is based on EDF emitting at 1.55 μm, which is mode-locked using a semiconductor saturable absorber mirror (SESAM) with a relaxation time of 10 ps, inserted in the cavity through a circulator.

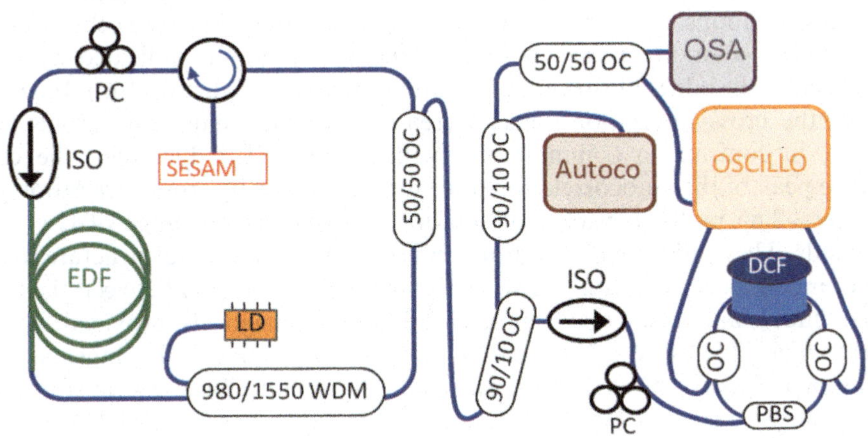

Figure 6.10. Experimental setup of the SESAM mode-locked fiber ring laser (on the left), and its ultrafast characterization line (on the right). EDF—Erbium-doped fiber, SMF—standard single-mode fiber (blue lines), DCF—dispersion compensation fiber, OC—optical coupler, WDM—multiplexer, PBS—fiber-based polarization beam-splitter, ISO—polarization-independent optical isolator, PC—polarization controller, LD—laser diode at 980 nm, Autoco—autocorrelator, OSCILLO—oscilloscope, OSA—optical spectrum analyzer.

The laser cavity is further composed of a 0.55 m long EDF (110 dB m^{-1} absorption at 1530 nm), backward-pumped by a 980 nm laser diode (LD), and of SMF-28 with a total length of several meters, which entails a fundamental repetition rate of a few tens of MHz. The group velocity dispersion at 1.55 μm is +13.5 ps^2 km^{-1} for the EDF and −22.9 ps^2 km^{-1} for the SMF, providing anomalous path-averaged cavity dispersion. A polarization-insensitive optical isolator ensures the unidirectional light propagation. An intra-cavity polarization controller allows us instead to achieve different vector dynamics. The experimental characterization was performed through the 50/50 fiber coupler with an optical spectrum analyzer (OSA) and a multi-shot second-order optical autocorrelator, whereas an additional extra-cavity polarization controller followed by a fiber-based polarization beam splitter were used to separate the two orthogonal polarization components. Moreover, we used a real-time dispersive Fourier-transform (DFT) technique [18] to unveil polarization-resolved shot-to-shot spectral and temporal features of the observed ultrafast internal dynamics. The DFT was implemented by temporally stretching the laser output in a 6 km long DCF. This mapped the spectrum of each optical pulse into a temporal waveform that was recorded via a high-speed 45GHz photodiode by a 6 GHz, 40 GSa s^{-1} real-time oscilloscope, entailing an effective spectral resolution of a fraction of nanometers.

6.3.2 Transient spectro-temporal ordering within incoherent dissipative solitons

The above-described fiber laser has allowed us to induce the regime of incoherent dissipative solitons, which is characterized by chaotic yet temporally localized dynamics similar to NLP emission, as illustrated in figure 6.11. The absence of modulation in the average OSA spectrum (see figure 6.11(d), blue curve), as well as the variability of consecutive single-shot spectra (see figure 6.11(a)) suggest the lack of stable phase relationship among the pulses from one roundtrip to the other, while the lack of precise temporal positions of the generated pulses is indicated by a large background at the edges of the autocorrelation traces (see figure 6.11(e)). Indeed, the width of the cross-correlation peaks increases when scanning time grows larger. However, in contrast to common observation from NLP dynamics, the central coherence peak of the autocorrelation trace is here flanked by cross-correlation peaks spaced by ~2.65 ps for a packet of solitons localized in the range of few tens of picoseconds. These additional peaks are reminiscent of an internal structure made of coherent mode-locked multiple pulses with a defined separation between them [4, 7, 47]. To understand such contradiction, additional insight is provided in figure 6.11(b), which illustrates a 2D contour plot of consecutive 8000 single-shot first-order autocorrelation traces that are calculated through the Fourier-transform of the corresponding single-shot DFT spectra of figure 6.11(a). We can clearly observe the existence of some partial organization occurring inside the incoherent pulses, which vanish and reappear, within time scales of 100–500 cavity roundtrips. In other words, the incoherent dissipative soliton observed here features transient spectro-temporal ordering, in analogy to the coalescence of well-separated droplets in liquids. We underline that this regime is by certain aspects different from the regime of soliton

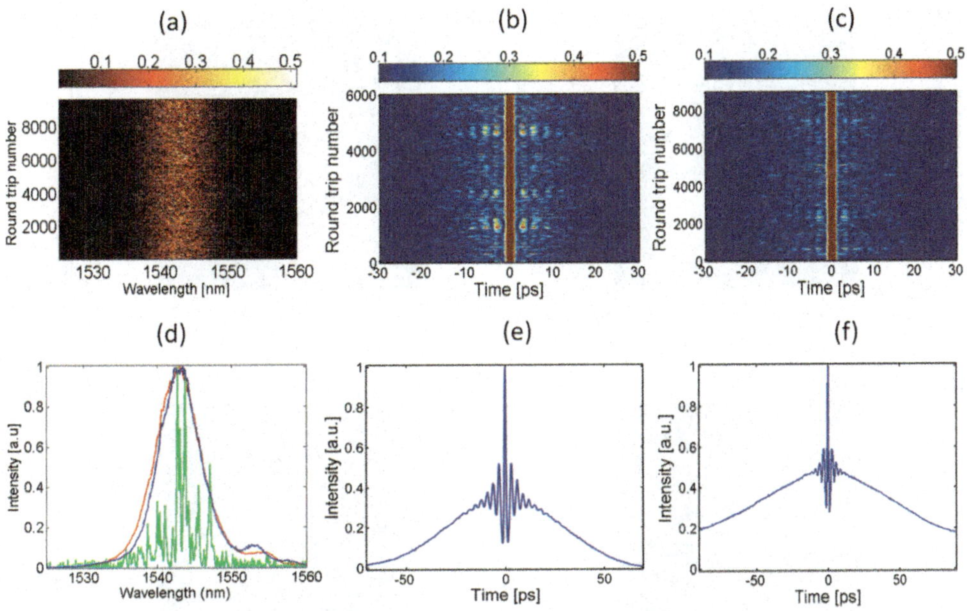

Figure 6.11. (a) 2D contour plot of 8000 consecutive single-shot DFT spectra; (b), (c) first-order autocorrelation traces obtained by Fourier transform of corresponding DFT spectra; (d) Average OSA spectrum (blue curve), spectrum averaged over 8000 DFT spectra (red curve), and an example of a single-shot spectrum (green curve); (d), (e) Multi-shot second-order autocorrelation traces. Left and middle panels: experimental results obtained for pump power of 200 mW. Right panels: experimental results obtained for pump power of 350 mW.

liquid [11, 12], whose internal structure composed of chaotically moving dissipative solitons could, in principle, be followed over arbitrary long times. The partial recurrent ordering of the present incoherent dynamics can be further modified by the level of pumping; its internal structure becomes more and more chaotic when pump power grows larger, as shown by figures 6.11(c) and (f).

6.3.3 Polarization-switching dynamics

Remarkably, the preceding regime of incoherent dissipative solitons can also feature some polarization ordering. By adjusting the intra-cavity polarization controller, we obtain two different vector dynamics, which are illustrated in figure 6.12 through the polarization-resolved DFT spectral characterization. Figure 6.12(a) demonstrates the existence of polarization-locked vector-incoherent solitons, where, by adjusting the extra-cavity PC, the output laser field could be projected onto one of the orthogonal polarization components that remains confined on the particular polarization axis. The second vector dynamics is more striking since it consists of the polarization switching of chaotic incoherent solitons, with temporally alternating x- and y-polarizations over a long quasi-period of the order of a thousand cavity roundtrips. An example of polarization switching occurring over ~2000 roundtrips with a 0.5 duty cycle ratio is shown in figure 6.12(b). However, once fixing the pump power, further control of the duty cycle ratio is possible by additional fine-tuning of

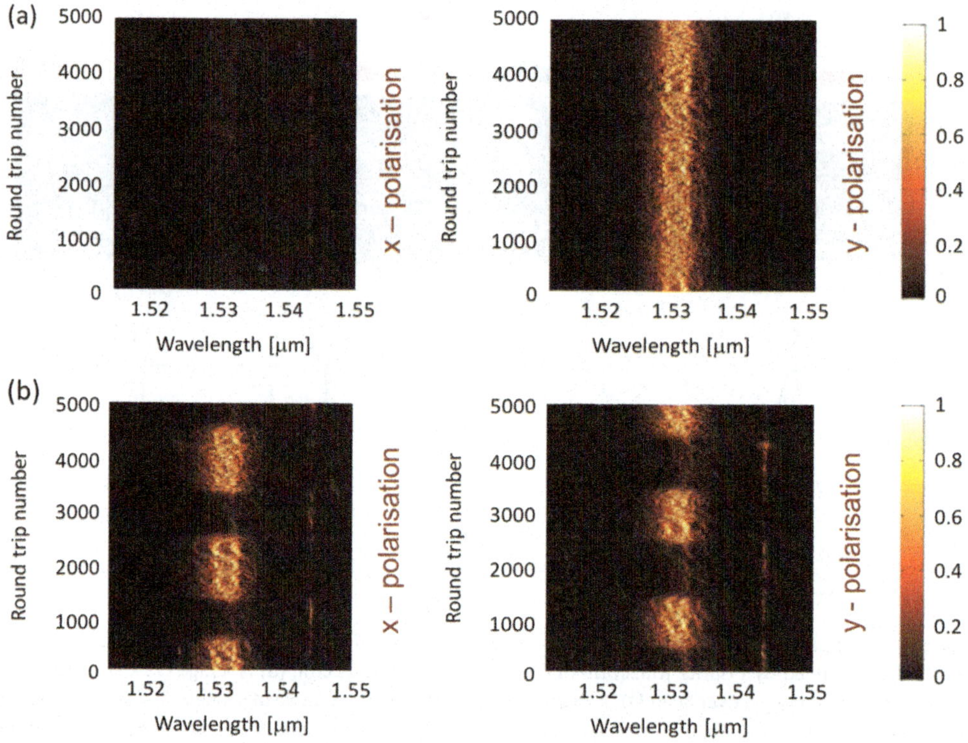

Figure 6.12. An example of experimental results of (a) polarization-locked vector solitons and (b) polarization-switching vector solitons. Both panels display 2D contour plots of consecutive DFT single-shot laser spectra. From [47].

the intra-cavity PC, whereas, for the same PC settings, we could change the value of the switching period by playing with the pump power, as shown in figure 6.13.

6.3.4 Explosive dynamics and rogue waves

In this section, we further explore the vicinity of the transition between the two previously described polarization dynamics through careful adjustment of the intra-cavity PC. We observe a peculiar instability, which shares common features with the soliton explosion dynamics. Soliton explosion can be considered as a critical regime between stable soliton regime and NLP emission, where a solitary pulse intermittently suffers an explosive instability and then recovers its original stable state [48–50]. Interestingly, the regime of soliton explosion may be associated with dissipative rogue waves (DRWs), as was predicted and experimentally demonstrated in scalar normally dispersive fiber lasers [49, 51]. Here, we will extend the concept of this extreme-wave dynamics by showing that the spectral explosive instabilities linked with DRW generation can also take place in the more general situation of vector-incoherent dissipative solitons in an anomalous dispersion regime [46].

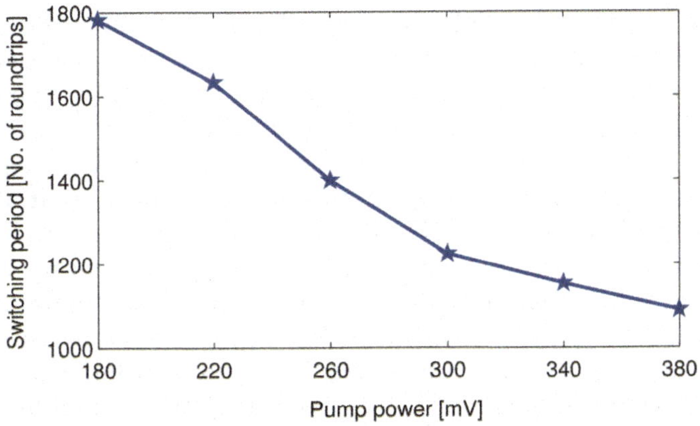

Figure 6.13. Switching period as a function of pump power in the regime of polarization-switching vector dissipative solitons.

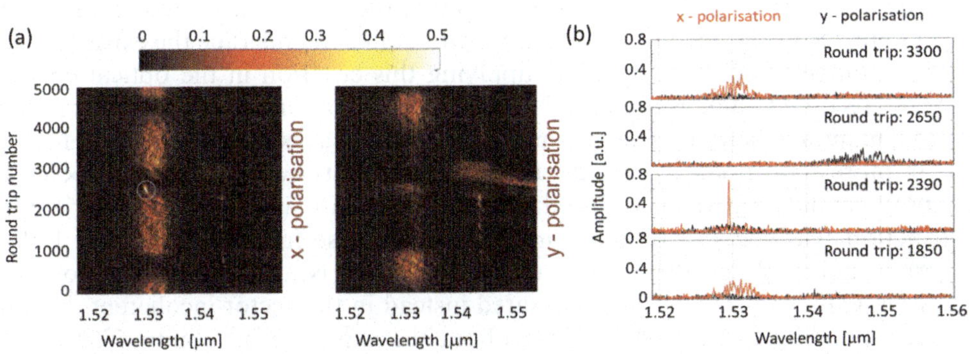

Figure 6.14. (a) Polarization-resolved 2D contour plots of 8000 consecutive single-shot spectra measured in the vector-incoherent soliton explosion regime; (b) examples of four single-shot spectra measured at the indicated roundtrip numbers, around the explosion event. From [47].

An example of spectral real-time vector analysis is presented in figure 6.14, when the laser operates in the regime of polarization-switching dynamics. Around the explosive event, four sets of single-shot DFT spectra are displayed (see figure 6.14(b)). Indeed, we can see that a train of incoherent optical pulses, circulating in the cavity with alternating polarization, suddenly experiences an abrupt structural collapse at x-polarization at roundtrip number ~3390, triggering a strong frequency down-conversion at y-polarization and then returning to its initial form. The association of the spectral collapse with a red-shifted development was also observed in the case of scalar soliton explosions. We also noticed a sudden temporal shift, which is typically attributed to the soliton explosion events, as well [48, 50, 52]. Thus, although an important difference between the present observation and the previously found soliton explosions relies on the nature of the central dynamics, which is here that of incoherent dissipative solitons, both cases share common features. The main similarity resides in the manifestation of an abrupt spectral expansion followed by recovery. The other important specificity of the present dynamics is represented by its vector expression associated with polarization switching. Moreover, by adjusting the orientation of the PC, we may increase the rate of explosive instabilities, such as in previous observations of soliton explosions [51].

Next, we will present the connection between the observed explosive instabilities and DRWs. In order to search the signature of vector spectral rogue waves, we performed a statistical analysis. We recorded the maximum spectral intensity, as voltage amplitude expressed in millivolts, for each single-shot DFT spectrum for both x- and y-polarizations, and we added it to histograms that we build from 10^6 events. Figure 6.15 illustrates an example of such a set of histograms measured in the regime of successive explosive instabilities. As we can see, both histograms exhibit non-Gaussian statistics showing long-tailed L-shaped distributions, which indicate the existence of extreme events. The highest recorded voltage amplitude corresponds to ~7 and ~3.3 times the SWH for x- and y-polarization, respectively. We note that in the present vector dynamics, one polarization is more prone to explosive instabilities, which is in agreement with the observations related to figure 6.14. The typical criterion for DRW generation is that the highest measured amplitude is larger than $2 \times$ SWH [13, 14]. However, we would like to underline the existence of a possible source of controversy while applying this criterion in the optical context since the recorded value is an optical intensity, and not a field amplitude as it is in hydrodynamics, where the scientific notion of rogue waves originates. Therefore, it may be more appropriate to consider the optical intensity events exceeding $4 \times$ SWH as DRW candidates, which is found in the present case among the x-polarization distribution. The statistics obtained in time-domain also fulfill the $2 \times$ SWH DRW criterion, but with smaller maximum amplitudes. Note also that histograms following Gaussian statistics are measured instead in the vector-incoherent dynamics corresponding to figure 6.12, i.e. without the explosive instability. The experimental results reported in this section confirm that there is a link between the DRWs and the explosive behaviors in ultrafast fiber laser operating in the vector-incoherent dynamics regime.

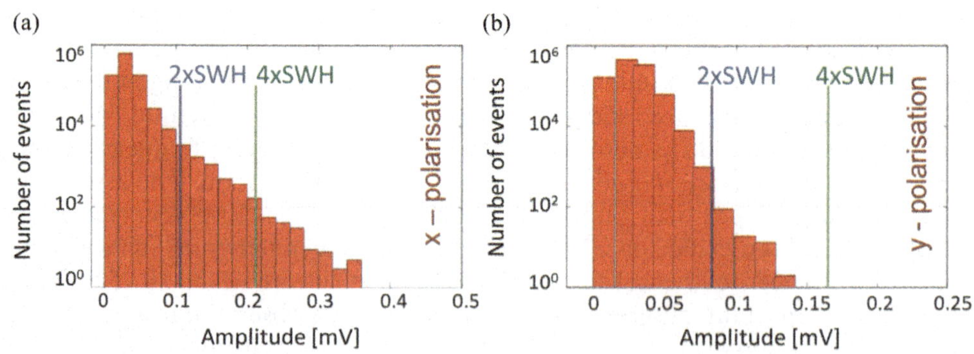

Figure 6.15. Histogram on log scale showing the statistical distribution of the maximal spectral intensity fluctuations in the explosive regime, measured as a voltage amplitude (in mV) on the oscilloscope for (a) x-polarization and (b) y-polarization components. SWH–significant wave height. From [47].

6.4 Conclusions

Chaotic pulse bunches appear ubiquitously in ultrafast fiber laser experiments as the energy inside the cavity is increased. This transition from mode-locked to chaotic bunching is particularly salient in the anomalous dispersion regime, where the pump power scaling favors the interaction among large pulse numbers per cavity roundtrip. When the internal structure of a chaotic pulse bunch becomes complicated, with a high level of intensity fluctuations, it complies with the original denomination of noise-like pulse emission. We recall that the notion of noise-like pulses was originally developed in the context of average experimental measurements. Nowadays, with improved real-time measurement, we can distinguish, albeit in a limited way, a great variety of noise-like pulse dynamics. We have explored experimentally and theoretically several of these dynamics, which we include under the more general terminology of *incoherent dissipative solitons.* The interaction of many sub-pulses inside the incoherent pulse bunch produces strong pulsations that can be categorized as rogue waves. Other transient phenomena, such as exploding solitons, can be also observed. The inclusion in the experiment of birefringent transparent elements enriches the dynamics, which can yield vector rogue wave dynamics, a topic of much recent interest from both experimental and theoretical points of view.

Acknowledgments

This work was supported by the Indo-French Center for the Promotion of Advanced Research (CEFIPRA/IFCPAR, project 5104-2); the Centre National de la Recherche Scientifique (CNRS); the Agence Nationale de la Recherche through projects SOLICRISTAL (ANR-2010-BLANC-0417-01), OPTIROC (ANR-12-BS04-0011), and LabEx ACTION (ANR-11-LABX-01-01); the Région Bourgogne; and the Fonds Européen de Développement Régional (FEDER). Ph.G. acknowledges the important contribution of C Lecaplain in the early experimental explorations of dissipative rogue waves. The work of JMSC is supported by the Volkswagen Stiftung, the MINECO under contract TEC2015-71127-C2-1-R, and by C.A.M. under contract S2013/MIT-2790.

References

[1] Akhmediev N and Ankiewicz A (ed) 2005 *Dissipative Solitons* (Berlin–Heidelberg: Springer)

[2] Grelu Ph (ed) 2016 *Nonlinear Optical Cavity Dynamics* (Weinheim: Wiley-VCH)

[3] Okhotnikov O G (ed) 2012 *Fiber Lasers* (Weinheim: Wiley-VCH)

[4] Grelu Ph and Akhmediev N 2012 Dissipative solitons for mode-locked lasers *Nat. Photonics* **6** 84

[5] Delavaux J M, Grelu Ph, Pu W and Ilday F O 2014 Special issue on short pulse fiber lasers *Opt. Fib. Technol.* **20**

[6] Soto-Crespo J M, Grelu Ph, Akhmediev N and Devine 2007 Soliton complexes in dissipative systems: vibrating, shaking, and mixed soliton pairs *Phys. Rev.* E **75** 016613

[7] Krupa K, Nithyanandan K, Andral U, Tchofo-Dinda P and Grelu Ph 2017 Real-time Observation of Internal motion within ultrafast dissipative optical soliton molecules *Phys. Rev. Lett.* **118** 243901

[8] Soto-Crespo J M, Grapinet M, Grelu Ph and Akhmediev N 2004 Bifurcations and multiple-period soliton pulsations in a passively mode-locked fiber laser *Phys. Rev.* E **70** 066612

[9] Horowitz M, Barad Y and Silberberg Y 1997 Noiselike pulses with a broadband spectrum generated from an erbium-doped fiber laser *Opt. Lett.* **22** 799

[10] Turaev D, Vladimirov A G and Zelik S 2007 Chaotic bound state of localized structures in the complex ginzburg–landau equation *Phys. Rev.* E **75** 045601 (R)

[11] Chouli S and Grelu Ph 2009 Rains of solitons in a fiber laser *Opt. Express* **17** 11776
Chouli S and Grelu Ph 2010 Soliton rains in a fiber laser: an experimental study *Phys. Rev.* A **81** 063829

[12] Amrani F, Haboucha A, Salhi M, Leblond H, Komarov A and Sanchez F 2010 Dissipative solitons compounds in a fiber laser. analogy with the states of the matter *Appl. Phys.* B **99** 107

[13] Lecaplain C, Grelu Ph, Soto-Crespo J M and Akhmediev N 2013 Dissipative rogue wave generation in multiple-pulsing mode-locked fiber laser *J. Opt.* **15** 064005

[14] Lecaplain C and Grelu Ph 2014 Rogue waves among noiselike-pulse laser emission: an experimental investigation *Phys. Rev.* A **90** 013805

[15] Mitchell M and Segev M 1997 Self-trapping of incoherent white light *Nature (London)* **387** 880

[16] Michel C, Kibler B, Garnier J and Picozzi A 2012 Temporal incoherent solitons supported by a defocusing nonlinearity with anomalous dispersion *Phys. Rev.* A **86** 041801 (R)

[17] Cohen O, Buljan H, Schwartz T, Fleischer J W and Segev M 2006 Incoherent solitons in instantaneous nonlocal nonlinear media *Phys. Rev.* E **73** 015601 (R)

[18] Goda K and Jalali B 2013 Dispersive Fourier transformation for fast continuous single-shot measurements *Nat. Photonics* **7** 102

[19] Runge A F J, Aguergaray C, Broderick N G R and Erkintalo M 2013 Coherence and shot-to-shot spectral fluctuations in noise-like ultrafast fiber lasers *Opt. Lett.* **38** 4327

[20] Donovan G M 2015 Dynamics and statistics of noise-like pulses in modelocked lasers *Physica* D **309** 1–8

[21] Kang J and Posey R 1998 Demonstration of supercontinuum generation in a long-cavity fiber ring laser *Opt. Lett.* **23** 1375

[22] Kobtsev S, Kukarin S, Smirnov S, Turitsyn S and Latkin A 2009 Generation of double-scale femto/pico-second optical lumps in mode-locked fiber lasers *Opt. Express* **17** 20710

[23] Pottiez O, Grajales-Coutino R, Ibarra-Escamilla B, Kuzin E A and Hernandez-Garcia J C 2011 Adjustable noiselike pulses from a figure-eight fiber laser *Appl. Opt.* **50** 24

[24] North T and Rochette M 2013 Raman-induced noiselike pulses in a highly nonlinear and dispersive all-fiber ring laser *Opt. Lett.* **38** 890–2

[25] Soto-Crespo J M, Grapinet M, Grelu Ph and Akhmediev N 2004 Bifurcations and multiple-period soliton pulsations in a passively mode-locked fiber laser *Phys. Rev.* E **70** 066612

[26] Soto-Crespo J M, Grelu Ph and Akhmediev N 2011 Dissipative rogue waves: extreme pulses generated by passively mode-locked lasers *Phys. Rev.* E **84** 016604

[27] Soto-Crespo J M and Grelu Ph 2005 Temporal multi-solitons complexes generated by passively mode-locked lasers *Dissipative Solitons* ed N Akhmediev and A Ankiewicz (Berlin-Heidelberg: Springer)

[28] Kovalsky M, Hnilo A and Tredicce J 2011 Extreme events in the Ti: sapphire laser *Opt. Lett.* **36** 4449

[29] Nelson L E, Ippen E P and Haus H A 1995 Broadly tunable sub-500 fs pulses from an additive-pulse mode-locked thulium-doped fiber ring laser *Appl. Phys. Lett.* **67** 19–21

[30] Wang X, Zhou P, Wang X, Xiao H and Liu Z 2014 Pulse bundles and passive harmonic mode-locked pulses in tm-doped fiber laser based on nonlinear polarization rotation *Opt. Express* **22** 6147–53

[31] Wang P, Bao C, Fu B, Xiao X, Grelu P and Yang C 2016 Generation of wavelength-tunable soliton molecules in a 2-μm ultrafast all-fiber laser based on nonlinear polarization evolution *Opt. Lett.* **41** 2254–7

[32] Li Z, Heidt A M, Daniel J M O, Jung Y, Alam S U and Richardson D J 2013 Thulium-doped fiber amplifier for optical communications at 2 m *Opt. Express* **21** 9289–97

[33] Rudy C W, Urbanek K E, Digonnet M J F and Byer R L 2013 Amplified 2-μm thulium-doped all-fiber mode-locked figure-eight laser *J. Lightwave Technol.* **31** 1809–12

[34] Xu Y, Li Song Y, Guo Du G, Guang Yan P, Yu Guo C, Liang Zheng G and Chen Ruan S 2015 Soliton dynamic patterns of a passively mode-locked fiber laser operating in a 2 μm region *Laser Phys. Lett.* **12** 045 108

[35] Brunel M, Romanelli M and Vallet P 2016 Synchronization in vectorial solid-state lasers *Nonlinear Optical Cavity Dynamics* ed Ph Grelu (Weinheim: Wiley) ch 13

[36] Wabnitz S, Lecaplain C and Grelu Ph 2016 Vector patterns and dynamics in fiber laser cavities *Nonlinear Optical Cavity Dynamics* ed Ph Grelu (Weinheim: Wiley) ch 14

[37] Cundiff S T, Collings B C, Akhmediev N N, Soto-Crespo J N, Bergman K and Knox W H 1999 Observation of polarization-locked vector solitons in an optical fiber *Phys. Rev. Lett.* **82** 3988

[38] Collings B C, Cundiff S T, Akhmediev N N, Soto-Crespo J M, Bergman K and Knox W H 2000 Polarization-locked temporal vector solitons in a fiber laser: experiment *J. Opt. Soc. Am.* B **17** 354

[39] Liu M, Luo A P, Luo Z-Ch and Xu W-Ch 2017 Dynamic trapping of a polarization rotation vector soliton in a fiber laser *Opt. Lett.* **42** 330

[40] Sergeyev S V, Mou Ch, Rozhin A and Turitsyn S K 2012 Vector solitons with locked and precessing states of polarization *Opt. Express* **20** 27434

[41] Tsatourian V, Sergeyev S V, Ch Mou, Rozhin A, Mikhailov V, Rabin B, Westbrook P S and Turitsyn S K 2013 Polarisation dynamics of vector soliton molecules in mode locked fibre laser *Sci. Rep.* **3** 3154

[42] Habruseva T, Mou Ch, Rozhin A and Sergeyev S V 2014 Polarization attractors in harmonic mode-locked fiber laser *Opt. Express* **22** 15211

[43] Zhao L M, Tang D Y, Zhang H and Wu X 2009 Bunch of restless vector solitons in a fiber laser with SESAM *Opt. Express* **17** 8103

[44] Luo A-P, Luo Z-Ch, Liu H, Zheng X-W, Ning Q-Y, Zhao N, Chen W-Ch and Xu W-Ch 2015 Noise-like pulse trapping in a figure-eight fiber laser *Opt. Express* **23** 10422

[45] Ning Q-Y, Liu H, Zheng X-W, Yu W, Luo A-P, Huang X-G, Luo Z-Ch, Xu W-Ch, Xu S-H and Yang Z-M 2014 Vector nature of multi-soliton patterns in a passively mode-locked figure-eight fiber laser *Opt. Express* **22** 11900

[46] Krupa K, Nithyanandan K and Grelu Ph 2017 Vector dynamics of incoherent dissipative optical solitons *Optica* **4** 1239

[47] Stratmann M, Pagel T and Mitschke F 2005 Experimental observation of temporal soliton molecules *Phys. Rev. Lett.* **95** 143902

[48] Cundiff S T, Soto-Crespo J M and Akhmediev N 2002 Experimental evidence for soliton explosions *Phys. Rev. Lett.* **88** 073903
[49] Chang W and Akhmediev N 2016 Exploding solitons and rogue waves in optical cavities *Nonlinear Optical Cavity Dynamics* ed Ph Grelu (Weinheim: Wiley) ch 11
[50] Runge A F J, Broderick N G R and Erkintalo M 2015 Observation of soliton explosions in a passively mode-locked fiber laser *Optica* **2** 36
[51] Liu M, Luo A-P, Xu W-Ch and Luo Z-Ch 2016 Dissipative rogue waves induced by soliton explosions in an ultrafast fiber laser *Opt. Lett.* **41** 3912
[52] Runge A F J, Broderick N G R and Erkintalo M 2016 Dynamics of soliton explosions in passively mode-locked fiber lasers *J. Opt. Soc. Am.* B **33** 46
[53] Lecaplain C, Grelu Ph, Soto-Crespo J M and Akhmediev N 2012 Dissipative rogue waves generated by a modelocked fiber laser *Phys. Rev. Lett.* **108** 233901
[54] Lecaplain C, Grelu Ph and Wabnitz S 2014 Dynamics of the transition from polarization disorder to antiphase polarization domains in vector fiber lasers *Phys. Rev.* A **89** 063812
[55] Liu M, Luo A-P, Yan Y-R, Hu S, Liu Y-C, Cui H, Luo Z-Ch and Xu W-Ch 2016 Successive soliton explosions in an ultrafast fiber laser *Opt. Lett.* **41** 1181
[56] Akhmediev N *et al* 2016 Roadmap on optical rogue waves and extreme events *J. Opt.* **18** 063001

IOP Publishing

Nonlinear Guided Wave Optics
A testbed for extreme waves
Stefan Wabnitz

Chapter 7

Ubiquitous nature of modulation instability: from periodic to localized perturbations

B Kibler, A Chabchoub, A Gelash, N Akhmediev and V E Zakharov

7.1 Introduction

Modulation instability (MI), also known as Benjamin–Feir instability, is considered to be the self-modulation of the continuous 'envelope waves' (i.e. regular wave trains in the context of deep water, or plane waves in optics) in the presence of small periodic or irregular noisy perturbations [1]. Since the 1960s, the propagation of weakly nonlinear dispersive wave packets affected by MI has been theoretically studied and experimentally observed in various fields of physics such as water waves, plasmas, laser light, electrical lines and Bose–Einstein condensates [2–16]. Their dynamical evolution can be easily modeled using the one-dimensional focusing nonlinear Schrödinger (NLS) equation [1–5]. The resulting dynamics consists of an initial linear stage and a subsequent nonlinear stage. For the former, a simplistic linear stability analysis can be applied; the effect of modulation instability is the result of interaction between a strong carrier harmonic wave ω_0 and weak amplitude sidebands $\omega_0 \pm \Omega$, i.e. a degenerate four-wave mixing process. The linearization of the NLS equation allows us to identify the instability criterion for the wave system under study [9], whereby one can determine the frequency threshold and the initial growth rate of perturbations. A typical exponential growth of the sidebands during wave propagation depicts the amplification of the initial weak perturbation imposed on the harmonic wave. Such a linear approach of MI ceases to be valid as soon as the perturbation reaches a comparable size to the carrier wave. One may expect that the rapid amplification of the perturbation favors nonlinear four-wave mixing processes, thus extending the energy cascade towards higher-order sidebands. Indeed, the linear stage of MI precedes a more complex evolution of wave packets, usually associated with the emergence of highly localized wave structures, the so-called nonlinear stage of MI [17]. The latter has received renewed attention since exact pulsating solutions of the complete MI evolution have been linked to the dynamics of rogue waves [18–21].

doi:10.1088/978-0-7503-1460-2ch7

© IOP Publishing Ltd 2017

To address the comprehensive MI problem, some exact solutions on finite background called breathers were derived during the 1970s and 80s [2, 22–27], as the NLS equation belongs to the remarkable class of integrable systems [4]. Such NLS breathers only derived for periodic initial modulations have provided, as a first step, a powerful framework for interpretation of a specific range of MI related dynamics that play a fundamental role in the theory of freak waves. The elementary solutions are the Akhmediev breather (AB) [2], which describes the simplest MI scenario with only one growth-return cycle evolved from a weak periodic perturbation, and the Peregrine breather (PB) [25], also known as the lowest-order rational solution to the NLS equation, which is the single-cycle limiting case of the AB with infinite period (i.e. the doubly localized breather). These restrictions can be overcome when one goes beyond the elementary instability by considering such first-order solutions as building blocks to construct higher-order solutions (i.e. nonlinear superpositions of breathers) to describe MI dynamics of more complex initial wave systems towards the irregular noisy perturbation. For instance, second-order AB solutions describe a higher-order MI namely, starting from five-wave systems [2]. A complete classification of the hierarchy of higher-order rational solutions of the NLS equation was reported in reference [28] by taking into account independent degrees of freedom associated with physical translations of component breathers.

Only recently, some unique features of these first- or higher-order periodic solutions have been confirmed experimentally for various nonlinear media [29–41]. Some works have confirmed that breather dynamics appear even with initial conditions that do not correspond to the mathematical ideal [29, 37, 42, 43]. As a result, such coherent structures may emerge locally in nonlinear wave systems driven by noise or chaotic states. In that configuration, interaction between many unstable modes takes place, so that rogue breather waves may appear intermittently or randomly in space and time with associated long-tailed statistics [19, 44–46]. Until recently, such characterization has remained a considerable challenge in optics. However, advanced pulse metrology techniques such as specially designed 'time-microscope' or 'time-lens' systems allow us now to analyze temporal 'snapshots' of the nonlinear propagation of ultrafast random light. Real-time optical measurements have characterized the spontaneous emergence of breathers from the unstable breakup of a continuous wave induced by low-amplitude noise [47]. Moreover, similar experiments have demonstrated the central role played by 'breather-like' structures such as the Peregrine soliton in the emergence of heavy-tailed statistics in integrable turbulence [48].

In order to describe the general physical picture of MI before going further into the problems of turbulence in integrable systems and rogue wave occurrence [49–51], one has to extend the MI analysis to *localized* weak perturbations of the plane wave, with a larger physical significance than periodic perturbations requiring the whole infinite space. To this end, a particular subset of higher-order breather solutions, known as superregular breathers, was recently derived [52, 53]. These solutions are defined as a nonlinear superposition of N pairs (where N is an integer) of quasi-Akhmediev breathers propagating in opposite directions. They can start with infinitesimally small localized perturbations and exhibit a combination of unique physical features in terms of propagation behavior. In addition to the development of perturbation into special pairs of breathers, the reverse process is another important scenario, for which N-pairs of quasi-Akhmediev breathers are almost annihilated to a small localized perturbation as a result of their collision. Corresponding experimental confirmations were simultaneously reported in optics and hydrodynamics [54]. Such N-pair breather solutions are able to describe different configurations of the nonlinear stage of MI such as the amplification, annihilation, or even ghost-interaction of nearly any individual or ensemble of localized perturbations on a plane wave [52–54]. Recently, analogs of superregular breathers were found in the higher-order generalized NLS equation [55, 56], which demonstrates the general character of these solutions.

In this chapter, we consider the nonlinear stage of MI in the case of localized weak perturbations and, by using a unique mathematical formalism, we review the recent theoretical and experimental advances, more specifically, the dynamics of superregular breathers. Complementary numerical simulations of noise-driven MI are analyzed to give an advanced comprehensive approach of this 'localized' MI problem. In section 7.2, we provide the theoretical description for NLS superregular breathers. Section 7.3 is devoted to the description of the distinct experimental configurations for the sake of observation of such breather waves in optics and hydrodynamics. Section 7.4 presents some supplementary numerical studies of MI induced by time-localized noise that confirm predictions from the nonlinear wave theory. Finally, section 7.5 provides conclusions and an outlook on novel research directions.

7.2 Breather formalism

In the following, we use the breather formalism of reference [54] within the framework of the self-focusing NLS equation in dimensionless form that describes wave propagation in space:

$$i\psi_{\xi} + \frac{1}{2}\psi_{\tau\tau} + |\psi|^2\,\psi = 0. \tag{7.1}$$

Here, the subscripted variables stand for partial differentiations. The variable ψ denotes the wave envelope which is a function of ξ (a scaled propagation distance or longitudinal variable) and τ (a co-moving time, or transverse variable, moving with the wave-group velocity). This equation usually describes the nonlinear wave evolution in various media, in particular, surface gravity waves in deep-water conditions and light waves in standard optical fibers for telecommunications.

We study the MI of the plane wave solution, written as $\psi_0 = Be^{iB^2\xi}$, where B is the finite amplitude of the plane wave. For simplicity, the latter amplitude of the background is scaled to 1. It is well-known that this solution is unstable with respect to small perturbations [9], with (angular) frequencies located within the range $0 < |\Omega| < 2$. The corresponding growth rate of instability, given by the linear stability analysis, is $\Gamma(\Omega) = \Omega(1 - \Omega^2/4)^{1/2}$. The maximum growth rate value $\Gamma = 1$ is reached for $\Omega = \pm\sqrt{2}$. In the presence of a background plane wave, the self-focusing NLS equation (7.1) has an exact N-breather solution, which can be constructed by the appropriate integration technique. This requires study of the auxiliary linear

Zakharov–Shabat system [27] with the complex spectral parameter $\lambda = \lambda_{\text{Re}} + i\lambda_{\text{Im}}$. The wave function solution of this system has a cut on the real axis where $-1 < \lambda_{\text{Re}} < 1$. Each breather adds a simple pole to the wave function at some point of λ-plane, including possibly the cut (i.e. Akhmediev breather solution). Coordinates of the pole in λ-plane describe the main breather properties (amplitude, group velocity, period of oscillation, etc). In the following, N-breather solutions were found using the dressing method as an integration technique; theoretical details can be found in references [52, 53, 57]. Note that the Jukowsky transform of the spectral parameter $\lambda = (\rho + \rho^{-1})/2$ with $\rho = R\,e^{i\alpha}$ was used [58]. This allows the mapping of the cut onto the circle of unit radius and λ-plane onto the outer part of this circle, so that the coordinates of the pole are described by the radial and angle coordinates R and α, as depicted in figure 7.1. Furthermore, each breather has two additional parameters μ and θ, which describe phase shifts in space and time and between different breathers.

The general one-breather solution has four real parameters R, α, μ, θ and can be described by the following expression [52–54, 59–61]:

$$\psi = \left(1 + 2\left(R + \frac{1}{R}\right)\cos\alpha \frac{q_1^* q_2}{|q_1|^2 + |q_2|^2}\right)e^{i\xi}, \tag{7.2}$$

where

$$q_1 = e^{-\phi} - \frac{e^{\phi - i\alpha}}{R}, \quad q_2 = e^{\phi} - \frac{e^{-\phi - i\alpha}}{R},$$

$$\phi = \eta\tau + \gamma\xi + \tfrac{\mu}{2} + i\left(k\tau + \delta\xi - \tfrac{\theta}{2}\right),$$

$$\eta = -\tfrac{1}{2}\left(R - \tfrac{1}{R}\right)\cos\alpha, \quad k = -\tfrac{1}{2}\left(R + \tfrac{1}{R}\right)\sin\alpha,$$

$$\gamma = -\tfrac{1}{4}\left(R^2 + \tfrac{1}{R^2}\right)\sin 2\alpha, \quad \delta = \tfrac{1}{4}\left(R^2 - \tfrac{1}{R^2}\right)\cos 2\alpha.$$

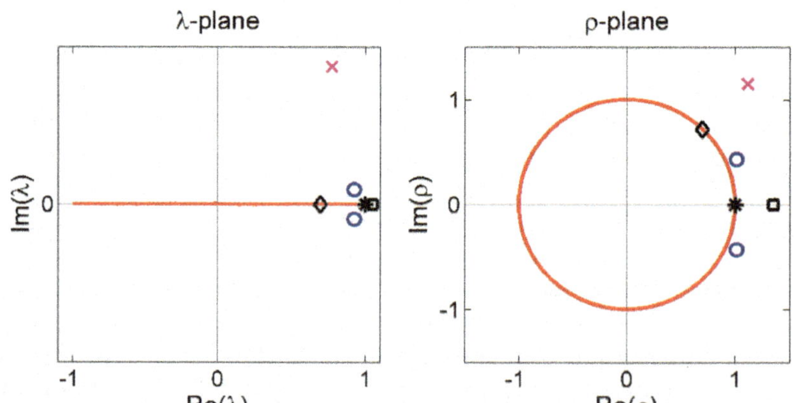

Figure 7.1. Application of the Jukowski transform to the plane of the complex spectral parameter λ. Poles of fundamental one-breather solutions are shown with markers: (◇) Akhmediev breather, (□) Kuznetsov–Ma breather, (□) Peregrine breather, (○) quasi-Akhmediev breather, (×) general one-breather solution. The superregular one-pair solution is formed by the two quasi-AB with opposite group velocities (○-○).

The parameters R, α are the polar coordinates of the point in the ρ-plane (see figure 7.1, $R \geqslant 1$, $\frac{\pi}{2} > \alpha > -\frac{\pi}{2}$), and they control the main breather properties. In general, the one-breather solution is a localized object with characteristic size $\tau_0 = 1/((R - 1/R)\cos\alpha)$, moving on top of the continuous wave with group velocity $V_g = -\gamma/\eta = -\sin\alpha(R^4 + 1)/(R(R^2 - 1))$, and oscillating (in ξ variable) with period $T = 4\pi R^2/((R^4 - 1)\cos 2\alpha)$. Parameters R and α control the main breather properties, whereas μ, θ define the location as well as phase of the breather and vary in the range $[0, 2\pi]$. When $R = 1$, $\alpha \neq 0$, the solution is an Akhmediev breather (see figure 7.2(a)) that becomes the Peregrine soliton in the limit $\alpha \to 0$. For low values of V_g ($R \neq 1$, $\alpha \to 0$), we approach the limit of Kuznetsov–Ma breathers [62], whereas for high values of V_g and τ_0 ($R = 1 + \varepsilon$, where ε is a small parameter), we obtain the class of quasi-Akhmediev breathers (see figures 7.2(b) and (c)).

A general two-breather solution can be also obtained [52, 53]. It has four main parameters $R_{1,2}$, $\alpha_{1,2}$ and four additional phase-shift parameters $\mu_{1,2}$, $\theta_{1,2}$ (subscripts 1 and 2 correspond to the first and second breather). In the general N-breather solution, each i-th breather has similar four parameters R_i, α_i, μ_i, θ_i, so that the total solution is described by $4N$ parameters. The cut in the λ-plane allows construction of such a configuration of poles (when the poles are paired near the cut on the opposite sides) that the corresponding solution must be a small perturbation of the initial plane wave at the moment of breather collision. These solutions are referred to as superregular breathers. To construct the simplest one-pair superregular breather, the phases of the solution have to be the same at $\pm\infty$ by the τ coordinate at any fixed ξ, since a small perturbation of the plane wave background cannot change the phase of the whole solution at infinity. This happens when breathers move in opposite directions while $\alpha_1 = -\alpha_2 = \alpha$ (see figure 7.1). In the case $R_1 = R_2 = 1$, the solution is

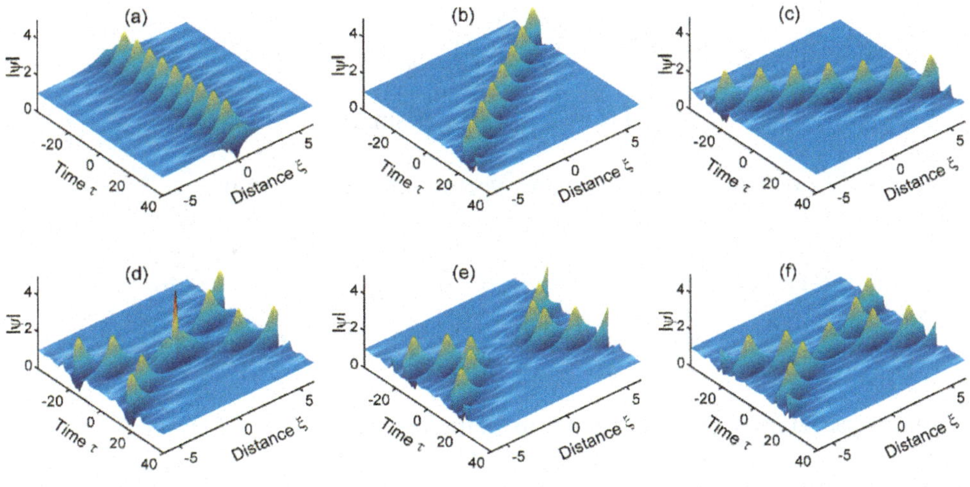

Figure 7.2. Examples of one-breather solutions and one-pair breather solutions. (a) Akhmediev breather ($R = 1$, $\alpha = 0.4$, $\theta = 0$, $\mu = 0$). (b) and (c) Quasi-Akhmediev breathers ($R = 1.1$, $\alpha = \pm 0.4$, $\theta = 0$, $\mu = 0$). (d)–(f) One-pair breather solution ($R = 1.1$, $\alpha = 0.4$) with the following phase shifts: $\theta_{1,2} = 0$, $\theta_{1,2} = \pi/2$, (superregular case), and $\theta_{1,2} = \pi$, respectively ($\mu_{1,2} = 0$ for all cases).

a pure plain wave (except the special case of degenerate breather solutions, which we do not discuss here), so that two Akhmediev breathers with opposite values of angular parameter completely annihilate each other. Consequently, when R_1 and R_2 are close to 1, two quasi-Akhmediev breathers almost annihilate at the moment of collision with the formation of a small perturbation of the plane wave. Thus, the amplitude of the perturbation is controlled by the difference of $R_{1,2}$ from the unity. The small parameter $\varepsilon = (R - 1)$ characterizes the amplitude of the perturbation at the moment of breather quasi-annihilation. In principle, quasi-annihilation can be observed at wide range of phase parameters $\mu_{1,2}$, $\theta_{1,2}$, however, the most effective annihilation occurs when $\theta_1 + \theta_2$ is close to π. Indeed, phase shifts affect the shape and the amplitude of the perturbation. The difference between $\theta_1 + \theta_2$ and π defines the degree of complexity of the wave profile at the area of collision. The solution is superregular when the phases are adjusted in a way that the perturbation is sufficiently small. The detailed description of the impact of parameters μ and θ can be found in references [52, 53], as well as some examples of complicated two and three-pair breather solutions and discussion of a general N-pair solution.

In the following, we study the simplest one-pair breather solution. It can be obtained from a general two-breather solution by setting $R_1 = R_2 = R = 1 + \varepsilon$, $\alpha_1 = -\alpha_2 = \alpha$, and can be written in the following form:

$$\psi = \left(1 + \left(R^2 - \frac{1}{R^2}\right)\frac{N}{\Delta} \sin 2\alpha\right)e^{i\xi} \tag{7.3}$$

$$N = \left(R - \tfrac{1}{R}\right) \sin \alpha\left(\left|\boldsymbol{q}_1\right|^2 q_{21}^* q_{22} + \left|\boldsymbol{q}_2\right|^2 q_{11}^* q_{12}\right) - i\left(R + \tfrac{1}{R}\right)\cos \alpha\left[\left(\boldsymbol{q}_1^* \cdot \boldsymbol{q}_2\right) q_{21}^* q_{12} - \left(\boldsymbol{q}_1 \cdot \boldsymbol{q}_2^*\right) q_{11}^* q_{22}\right],$$

$$\Delta = \left(R + \tfrac{1}{R}\right)^2 \cos^2 \alpha \left|q_{11} q_{22} - q_{12} q_{21}\right|^2 + \left(R - \tfrac{1}{R}\right)^2 \left|\boldsymbol{q}_1\right|^2 \left|\boldsymbol{q}_2\right|^2 \sin^2 \alpha,$$

where $\boldsymbol{q}_i = (q_{i1}, q_{i2})$ is a two-component vector function, so that

$$q_{11} = e^{-\phi_1} - \frac{e^{\phi_1 - i\alpha}}{R}, \quad q_{12} = e^{\phi_1} - \frac{e^{-\phi_1 - i\alpha}}{R}, \quad q_{21} = e^{-\phi_2} - \frac{e^{\phi_2 + i\alpha}}{R}, \quad q_{22} = e^{\phi_2} - \frac{e^{-\phi_2 + i\alpha}}{R},$$

$$\phi_1 = \eta\tau + \gamma\xi + \frac{\mu_1}{2} + i\left(k\tau + \delta\xi - \frac{\theta_1}{2}\right), \quad \phi_2 = \eta\tau - \gamma\xi + \frac{\mu_2}{2} - i\left(k\tau - \delta\xi + \frac{\theta_2}{2}\right),$$

and the parameters η, k, γ, and δ are as defined previously.

Figures 7.2(d)–(f) show some particular arrangements of the just-discussed one-pair breather solution ($R = 1.1$, $\alpha = 0.4$) with the following values of phase shifts: $\theta_{1,2} = 0$, $\theta_{1,2} = \pi/2$, and $\theta_{1,2} = \pi$, respectively ($\mu_{1,2} = 0$ for all cases). The synchronization of the collision is obtained for $\theta_{1,2} = 0$, thus leading to the formation of the high-amplitude peak at the origin ($\xi = 0$, $\tau = 0$) as shown in figure 7.2(d) [37, 60, 61]. Figure 7.2(e) shows the quasi-annihilation of breathers at the origin when $\theta_{1,2} = \pi/2$. This corresponds to a small localized perturbation of the continuous wave on the line $\xi = 0$ (superregular case) [51–53]. Another particular interaction is obtained for $\theta_{1,2} = \pi$ (see figure 7.2(f)). Then, the collision point is just another maximum of either breather solution. Each breather then appears seemingly without influence of the collision process; this particular case is here called a ghost-interaction of breathers. Quasi-Akhmediev breathers shown in figures 7.2(b) and (c) are neither periodic in time nor space. They are periodic along the line connecting the peak maxima given by the group velocity V_g (see figures 7.3(a) and (b)), and breathers propagate in opposite directions to infinity with a constant velocity before or after the collision area. Note that for the one-pair solution, shown in figures 7.2(d)–(f), there is only one collision point in the plane (ξ, τ).

In the following, we focus on the novel scenario of quasi-annihilation of breathers. When $\theta_1 + \theta_2$ is exactly equal to π (for example, $\theta_{1,2} = \pi/2$ in figure 7.2(e)), the quasi-annihilation is the most effective. In this case, at the moment of breather collision ($\xi \sim 0$), the perturbation $\delta\psi$ on the continuous wave (i.e. $\psi = 1 + \delta\psi$) can be approximated by the following formula:

$$\delta\psi \approx \frac{4i\varepsilon \cosh(i\alpha - \xi \sin 2\alpha)\cos\left(2\tau\sin\alpha - \dfrac{\theta_1 - \theta_2}{2}\right)}{\cosh(2\varepsilon\tau \cos \alpha)}. \tag{7.4}$$

The perturbation is localized in time and can contain many oscillations, but their number decreases by decreasing α. The number of oscillations also defines the main perturbation frequency of the wave envelope. Note that R and α also govern the

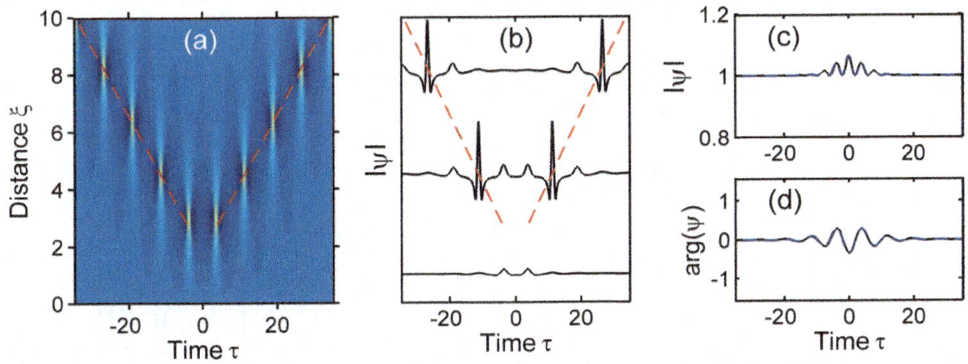

Figure 7.3. (a) Top view of figure 7.2(e), focused on the perturbation development into a pair of quasi-Akhmediev breathers propagating in opposite directions. Dashed red lines show the group velocity $V_g = \gamma/\eta$ calculated for each breather. (b) Corresponding amplitude profiles of the one-pair superregular breather at $\xi = 1$, 4.5, and 8.1. (c) and (d) Amplitude and phase profiles of the approximated perturbation (blue dashed line) calculated with equation (7.4) and compared to the exact profile (black solid line) of the quasi-annihilated breathers at $\xi = 0$ observed in figure 7.2(e).

angle that separates quasi-AB trajectories in the plane (ξ, τ). Moreover, the temporal width of the initial perturbation increases with decreasing ε, while its amplitude decreases. Figures 7.3(c) and (d) show the comparison between the approximated perturbation and the exact profile of the quasi-annihilated breathers at $\xi = 0$ from figure 7.2(e). We observe that the simplified approach of $\delta\psi$ gives an excellent fit of the small localized perturbation even here with $\varepsilon = 0.1$.

An initially small localized perturbation of the continuous wave grows at first exponentially. It is described by the well-known equations for the linear stage of modulation instability. It should be stressed that the Fourier spectrum of the superregular-type perturbations overlaps with the MI growth region $0 < |\Omega| < 2$. Similar to the Akhmediev breather solution (see, for instance reference [63]), the main frequency of superregular perturbation corresponds to the maximum of the MI growth rate ($\Omega_{max} = \pm\sqrt{2}$) for $\alpha = \pi/4$, as illustrated in figure 7.4.

Then it develops into a pair of breathers propagating in opposite directions, as shown in figure 7.3. We note that such solutions leave after the propagation of breathers a continuous wave with a different phase (the difference with the phase at infinity is 4α) [53]. The perturbed part of the plane wave almost returns to the initial state, and this can be treated as a local Fermi–Pasta–Ulam (FPU) recurrence. Recall that the FPU recurrence is described by Akhmediev breathers for periodic perturbations and the presence of a similar phase shift [64]. Moreover, the pure homoclinic Akhmediev two-breather solution was studied in reference [53]. Namely, it was shown that when angular parameters of the solution are linked by $|\alpha_1| + |\alpha_2| = \pi/2$, the pure homoclinic (i.e. with zero phase shift) FPU recurrence occurs. This idea can be generalized to the multiple Akhmediev breather case and used to describe the complex periodic scenario of MI development leading to the FPU recurrence to exactly the same initial condition. To conclude this section, it is worth noting that the reverse process, implying the quasi-annihilation of a pair of quasi-Akhmediev breathers, is simply obtained by applying the time-reversal operation, which corresponds to $\theta_{1,2} = \frac{\pi}{2} + \pi$ (i.e. ξ becomes $-\xi$ in figures 7.2 and 7.3).

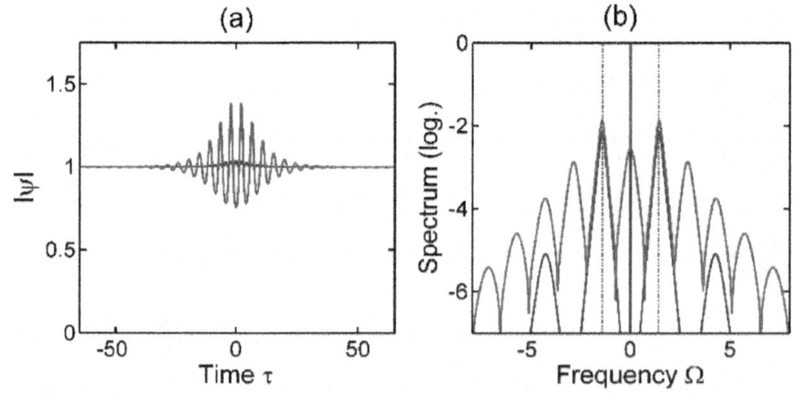

Figure 7.4. Modulation instability development for the one-pair superregular breather with $R = 1.1$ and $\alpha = \pi/4$. (a) Temporal profiles $|\psi(\tau)|$ at different propagation distances $\xi = 0$ (blue lines) and $\xi = 1$ (green lines). (b) Corresponding Fourier spectra $|\tilde{\psi}(\Omega)|$ in log scale. The dotted red lines indicate the positions of maximum MI growth rate Ω_{max} and coincide with the positions of maximal amplitudes of the perturbation spectrum.

7.3 Experimental demonstrations

The current technical capability to manipulate light and water waves allows us to develop experimental setups to demonstrate the existence of superregular breathers in real nearly conservative physical systems [53], namely a water wave tank and an optical fiber workbench, as shown in figure 7.5. Such systems were specially designed to ensure as small a contribution as possible from higher-order effects beyond the second-order dispersion and the cubic nonlinearity. The parameters were chosen to highlight the creation and annihilation dynamics related to superregular breathers, as shown in figure 7.3. Note that a low value of the parameter α and considerable value of ε were chosen due to some technical restrictions of the experimental setups.

Each system is capable of synthesizing non-trivial wave profiles in the temporal domain according to equations (7.2) and (7.3). Such ideal excitation of the MI process is a prerequisite to exactly observe the dynamics shown in figure 7.3 instead of more complex behaviors [33]. As breather solutions are unstable waves, a non-ideal condition will induce the emergence of asymmetric wave profiles or/and introduce some complex spatial recurrence phenomena.

7.3.1 Water waves

In water wave experiments (see figure 7.5(a)), the initial wave profiles are generated with a single-flap paddle located at one end of a tank. An electric signal, calculated using the exact mathematical expression describing the water surface elevation, controls a hydraulic cylinder that drives the paddle to directly modulate the water surface height in the temporal domain. To this end, one has to fix the initial amplitude a_0 and the wavenumber of the carrier wave k_0, thus defining the steepness a_0k_0 of the quasi-regular wave train. We set the amplitude to be $a_0 =$ 0.01 m and the wavenumber $k_0 = 10$ rad m^{-1} while the water depth is 1 m to satisfy deep-water conditions. The frequency of the carrier wave is given by the dispersion relation of linear deep-water wave theory, $\omega_0 = (gk_0)^{\frac{1}{2}}$, where g is the gravitational acceleration. The surface elevation $\eta(z,t)$ is related to the NLS solution ψ to second order in steepness as follows:

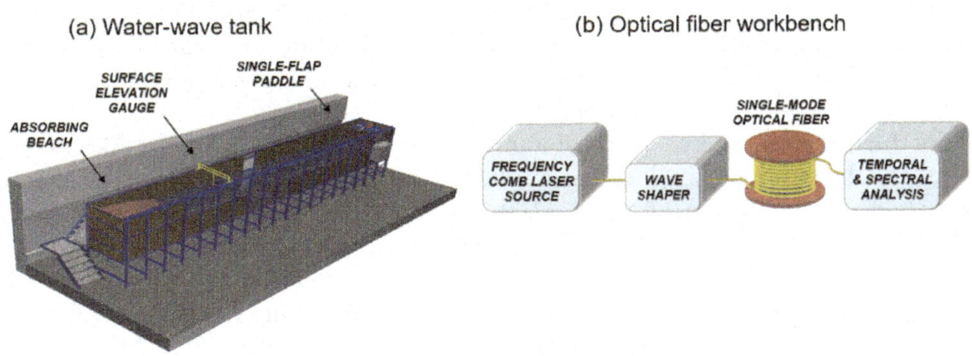

Figure 7.5. Experimental setups used to demonstrate the existence of superregular breather waves based on (a) a water wave tank and (b) an optical fiber workbench. (Adapted from [54].)

$$\eta(z,\,t) = \mathrm{Re}\left\{\psi(z,\,t)e^{i(k_0 z-\omega_0 t)} + \frac{1}{2}k_0\psi^2(z,\,t)e^{2i(k_0 z-\omega_0 t)}\right\}. \tag{7.5}$$

The tank dimensions are 15 m × 1.6 m × 1.5 m. In order to avoid wave reflections in the wave flume, an absorbing beach is installed at the opposite end. To overcome the limited length of the wave flume, the propagation was split into several stages and we reconstructed the full wave dynamics afterwards. After starting the breather wave generation repetitively with different initial conditions given from theory, the wave profiles were measured at 4.5 and 9 m from the flap with wave gauges. Measurements at 9 m served as new initial conditions for the wave maker in the next stage of the experiments. This process was repeated four times to reach the propagation distance of 31.5 m.

The correspondence between the theory and experiment can be retrieved by recalling that dimensional distance z (m) and time t (s) are related to the previous rescaled variables by $z = \tau/(\sqrt{2}k_0^2 a_0) + c_g t$ and $t = 2\xi/(k_0^2 a_0^2 \omega_0)$, with $c_g = \omega_0/(2k_0)$. Using these latter relations and inverting the scaling, the analytical solution, multiplied by the carrier amplitude a_0, is then written in dimensional form, satisfying the hydrodynamic NLS equation [5]. We emphasize that the hydrodynamic time and the space NLS frameworks provide different wave profile evolutions for Akhmediev-type breathers [65]. At a specific distance, the temporal variation of envelope can be easily reconstructed from the surface measurements by use of the Hilbert transform [21].

Figure 7.6(a) shows the results of experiments carried out in the water wave tank by considering an initial localized perturbation centered at $t = 0$, and by using the parameters $R = 1.15$, $\alpha = 0.4$, $\theta_{1,2} = \pi/2$, $\mu_{1,2} = 0$. After 18 m of propagation, we clearly observe that the localized perturbation develops into a pair of quasi-Akhmediev breathers that begin to move in the opposite directions, leaving behind (i.e. at the center $t = 0$) a non-perturbed plane wave. Note that the envelope asymmetry between the two breathers has to be inserted into initial conditions in order to take into account breather propagation along the tank during the excitation. For each time series, we also calculated the envelope profile corresponding to the surface elevation of the wave trains to facilitate the comparison with the theory shown in figure 7.6(b). A good agreement is obtained for both amplitude and phase of the wave trains. Theoretical profiles were based on the calculation of the free surface elevation from breather solutions and variable ψ given by equations (7.3) and (7.5) to second order in steepness.

For the reverse process of quasi-annihilation of one pair of quasi-Akhmediev breathers, the parameters $\mu_{2,1}$ of the solution are slightly shifted due to experimental restrictions ($R = 1.15$, $\alpha = 0.4$, $\theta_{1,2} = \pi/2 + \pi$, $\mu_{2,1} = \pm 2$). We experimentally confirm, in figure 7.6(c), the quasi-annihilation dynamics of the initial pair of quasi-Akhmediev breathers into a localized perturbation centered at $t = 0$, after 31.5 m of propagation. The agreement with theoretical predictions (see figure 7.6(d)) is again excellent, thus demonstrating the existence of superregular breathers in hydrodynamics, in particular for describing modulation instability in the framework

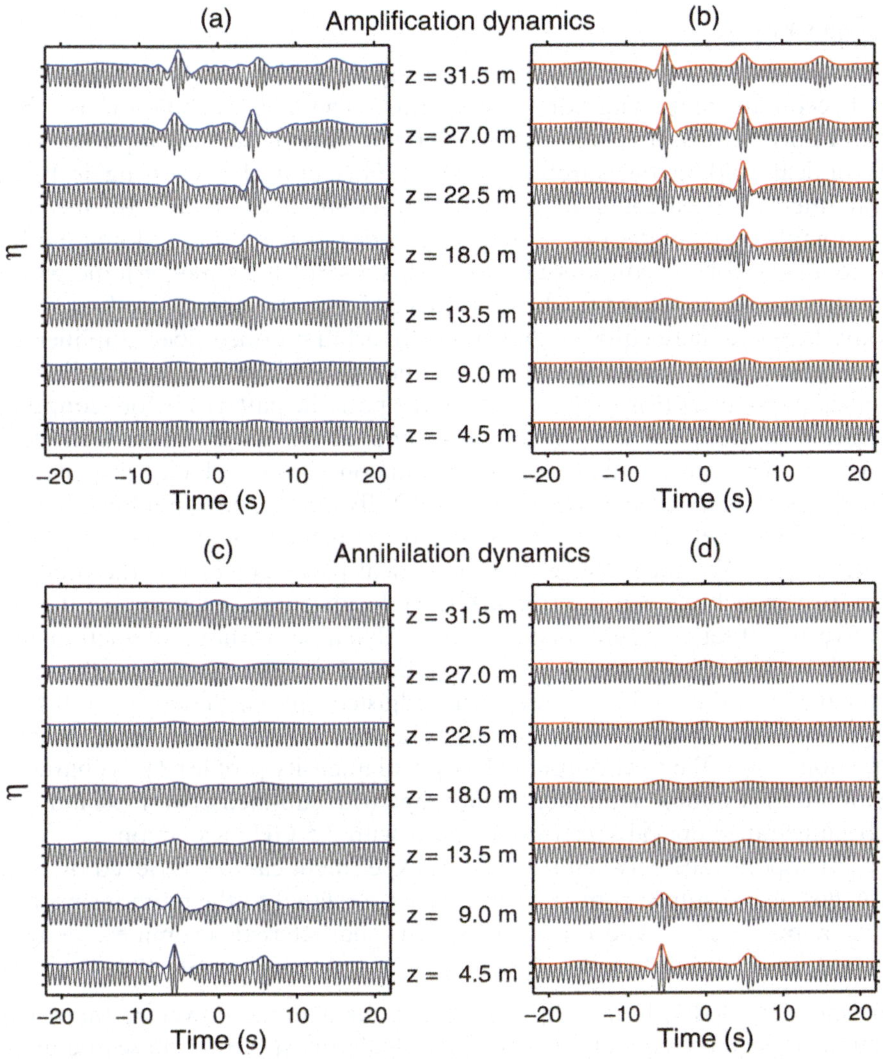

Figure 7.6. (a) and (c) Evolution of time series of surface elevation measurement as a function of propagation distance for amplification and annihilation dynamics, respectively. The surface elevation is kept in normalized units so that the amplitude of the unperturbed wave train is 1 (in dimensional units, it corresponds to 1 cm). The corresponding envelopes of the wave trains (blue lines) were calculated using the Hilbert transform. (b) and (d) Theoretical predictions of time series of surface elevation as a function of propagation distance for amplification and annihilation dynamics, respectively. The envelope of the analytical solution is shown by the red line. Adapted from [54].

of localized perturbations. Note that amplitudes are slightly lower in the experiments than theoretical prediction, for instance, due to the dissipation naturally existing when performing experiments in water wave facilities [66]. Other deviations can be also attributed to higher-order effects that cannot be captured within the standard NLS equation [67–69].

7.3.2 Light waves

For light waves, the experimental setup (see figure 7.5(b)) is mainly based on high-speed telecommunications-grade components. The ideal excitation is obtained through Fourier-transform optical pulse shaping with phase and amplitude controls of the optical field in the spectral domain. This optical processing is based on spectral line-by-line shaping of a frequency comb laser source by means of a programmable optical filter (wave shaper) in order to synthesize tailored modulated initial conditions of the continuous wave. This results in a time-periodic pattern of the perturbed continuous wave whose frequency is equal to the spectral separation of the comb lines (i.e. here equal to 20 GHz). An erbium-doped fiber amplifier is used to amplify and fix the average power of the synthesized wave according to the theoretical parameters (for each element of the periodic pattern) before coupling and propagation into 0.76 or 1.5 km long segments of the same standard single-mode optical fiber. The optical fiber exhibits anomalous group-velocity dispersion $\beta_2 = -21.1\ \mathrm{ps}^2\ \mathrm{km}^{-1}$, low linear losses equal to 0.2 dB km^{-1}, and a nonlinearity $\gamma = 1.2\ \mathrm{W}^{-1}\ \mathrm{km}^{-1}$. More details about the fiber-based frequency comb source can be found in reference [38]. We then restrict the theoretical parameters to fit the limited time window over which we can inscribe arbitrary localized perturbation on the continuous wave and observe the complete nonlinear dynamics without interaction between neighboring elements of the periodic pattern. The parameters of the one-pair superregular breather solution are then adjusted, in particular $\varepsilon = 0.4$. After propagation, wave profiles are then simultaneously measured in temporal and spectral domains at the fiber output. The optical intensity profile $|A|^2$ is characterized using an ultrafast optical sampling oscilloscope with sub-picosecond resolution and a high-dynamic-range optical spectrum analyzer with 2.5 GHz resolution.

The correspondence between theory and experiment can be retrieved by recalling that dimensional distance z [m] and time t [s] are related to the previous normalized parameters by $z = \xi L_{\mathrm{NL}}$ and $t = \tau t_0$, where the characteristic (nonlinear) length and time scales are $L_{\mathrm{NL}} = (\gamma P_0)^{-1}$ and $t_0 = (|\beta_2| L_{\mathrm{NL}})^{\frac{1}{2}}$, respectively [70]. The dimensional field $A(z, t)[W^{\frac{1}{2}}]$ is $A = P_0^{\frac{1}{2}}\psi$, P_0 being the average power of the perturbed continuous wave, here equal to 0.74 W. By calculating spectral (phase and intensity) profile differences between the initial frequency comb and the corresponding Fourier transform of initial condition at fiber input, one can apply the resulting (phase and intensity) corrections as a pulse-shaping mask to the wave shaper.

Figure 7.7(a) reports the development of a small localized perturbation into a pair of quasi-Akhmediev breathers measured for two propagation distances. The initial condition at the fiber input was taken from $\psi(R = 1.4,\ \alpha = 0.4,\ \theta_{1,2} = \pi/2,\ \mu_{2,1} = \pm 2,\ \xi = -1,\ \tau)$. By comparing with figure 7.7(b), we clearly reveal that each element of the measured periodic wave (i.e. a 50 ps time window) perfectly agrees with theory. Our experimental characterization is completed with direct spectral measurements in figure 7.7(c), since no phase information was available in the time domain. The strong spectral broadening of the initial perturbed continuous wave confirms again the nonlinear focusing of perturbations, in agreement with theoretical predictions over a 50 dB dynamic range.

The reverse process corresponding to the quasi-annihilation of a pair of breathers into a small localized perturbation of the continuous wave was measured using the initial condition $\psi(R = 1.4,\ \alpha = 0.4,\ \theta_{1,2} = \pi/2 + \pi,\ \mu_{2,1} = \pm 2,\ \xi = -2.85,\ \tau)$, as shown in figure 7.7(d). Here, we evidence the quasi-annihilation dynamics after 760 m of propagation. For a longer propagation (i.e. 1500 m) we note the emergence of a single localized peak, in agreement with theory, in figure 7.7(e). This typical space-time behavior was also presented in figure 7.2(e) when looking at the reverse propagation direction. The corresponding spectral signatures are reported in figure 7.7(f). In particular, the quasi-

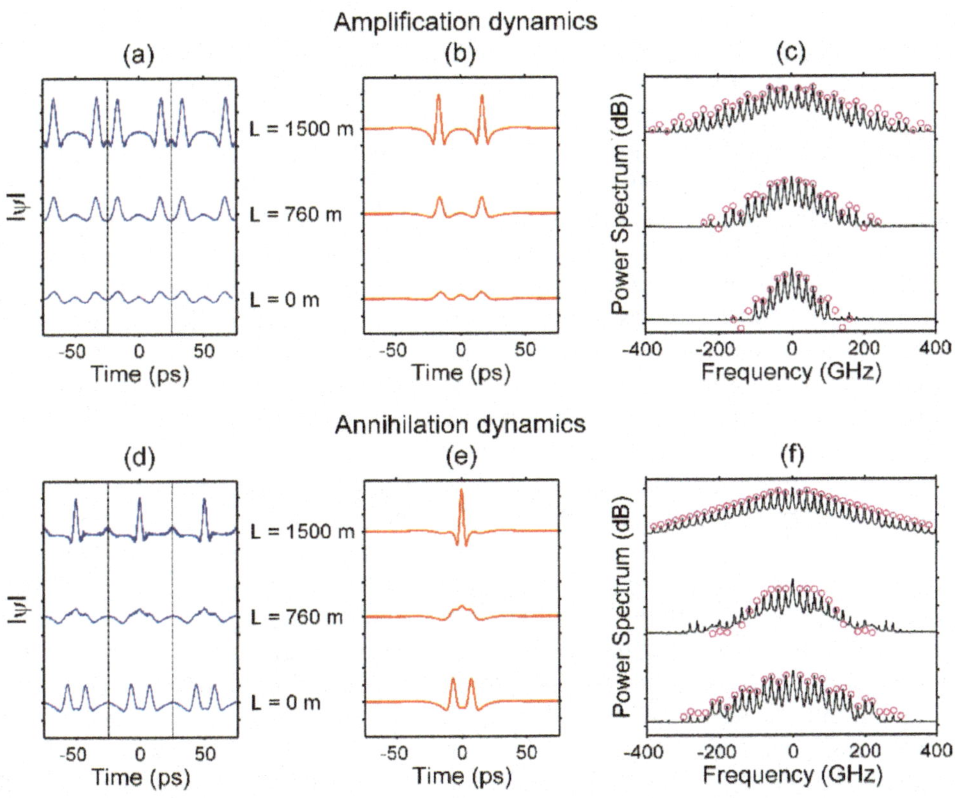

Figure 7.7. Light wave measurements for both amplification and annihilation dynamics of superregular breathers compared to theoretical predictions. Evolution of the square root of intensity profiles measured in the temporal domain as a function of propagation distance (a) and (d) and corresponding amplitude profiles from theory (b) and (e). The wave profiles are kept in normalized units so that the amplitude of the unperturbed wave train is 1. Recall that a periodic pattern of the exact solution was generated in experiments; each element of the pattern is delimited by dashed lines. (c) and (f) Comparisons of the corresponding power spectra (black solid lines for experiments, and magenta open circles for theory). Note that the theoretical spectra were sampled with the repetition rate of the experimental pattern and the delta-function component at the carrier wavelength is not shown. Adapted from [54].

annihilation stage is associated with significant spectral narrowing. The energy distributed in the multiple initial sidebands comes back to the pump and its adjacent first harmonics before being redistributed during the next nonlinear focusing.

Our experiments, which have been conducted on timescales of seconds in the water wave tank and picoseconds in the optical fiber, confirm analytical predictions of nonlinear wave theory developed recently. Demonstrating simultaneous experimental results at drastically different scales, in two different domains, are extremely rare and prove again the universal nature of MI dynamics.

7.4 Localized noise-driven modulation instability

As reported, NLS breather solutions have been generally observed following a careful control of initial excitation of the MI process. A few numerical studies have analyzed the fundamental question of how closely such analytic solutions of the NLS equation describe spontaneous emergence of localized structures in noise-driven and chaotic MI [19, 71, 72]. As suggested, some elementary and higher-order periodic breathers have been actually confirmed to exist in chaotic MI experiments [47]. A quantitative agreement can be easily obtained with numerical simulations based on the NLS equation, starting from a fully noise-perturbed plane wave. The noise perturbation is usually computed from the inverse Fourier transform of a broadband noise of one-photon per mode spectral density with random phase [71]. As a result, the characteristics of the noise-induced high-amplitude localized waves were found to cluster closely around NLS solutions. In this section, we extend the analysis to the nonlinear stage of MI emerging from time-localized noise. We address the following question: Do superregular breather solutions manifest in a genuinely chaotic and localized MI process?

In order to investigate the space-time MI dynamics excited from small localized noise, we performed simulations of MI within the framework of the NLS equation (7.1) with a locally noise-perturbed plane wave as initial condition, i.e. $\psi(\tau, \xi = 0) = 1 + \zeta(\tau)\mathrm{sec}(\tau/\Delta)$. Here, $\zeta(\tau)$ is computed from the inverse Fourier transform of a low-amplitude broadband noise with random phase. We restrict the noise perturbation in time using a sech function with variable width Δ. Figure 7.8(a) reports the evolution map of $\psi(\tau, \xi)$ when $\Delta = 4$. We clearly observe the emergence of high-amplitude temporal peaks developing as a result of the amplification of localized noisy perturbation of the input plane wave. The MI takes place in a delimited zone of the plane (ξ, τ), which increases with propagation distance. The edges of the MI area appear to be driven by wave structures similar to quasi-Akhmediev breathers that propagate with opposite group velocities. These velocities are partially related to the frequencies of maximum MI growth rate $\Omega_{\mathrm{max}} = \pm\sqrt{2}$. The corresponding spectral evolution confirms the preferential amplification of such frequencies in the initial stage of MI (i.e. $\xi < 5$), as shown in figure 7.8(c).

We can simply describe the temporal edges of MI using the one-pair superregular breather solution (see equation (7.3), with $\theta_{1,2} = \pi/2$, $\mu_{1,2} = 0$). The parameter $\alpha = \pm\pi/4$ is fixed by the value of Ω_{max} (also defining the 'time' period T of quasi-

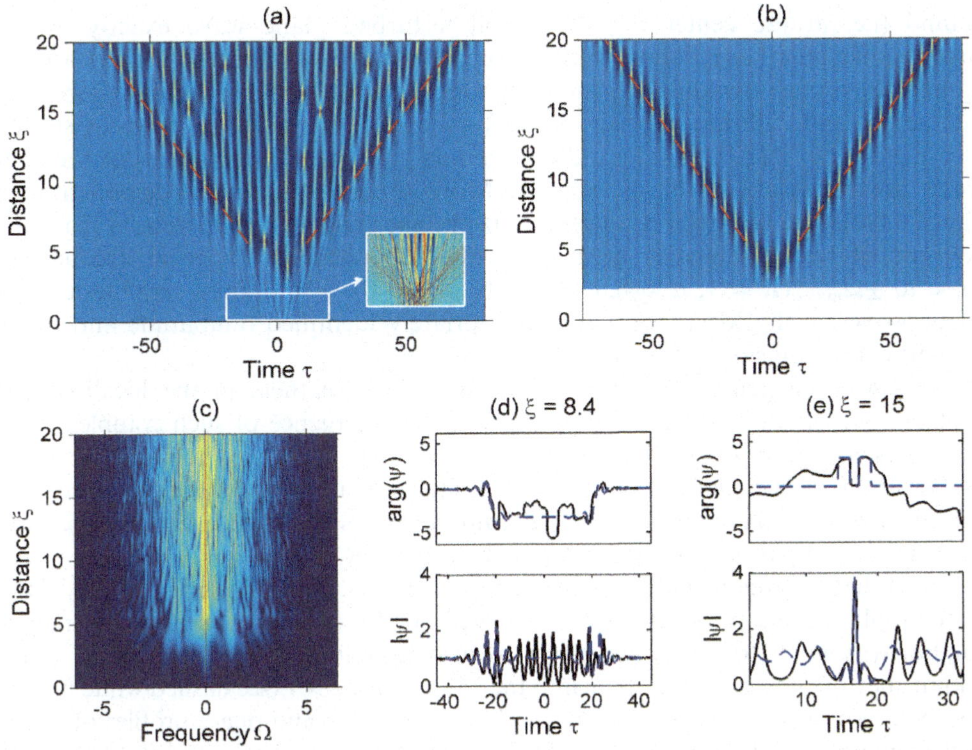

Figure 7.8. (a) NLS simulation of the temporal evolution of a small localized ($\Delta = 4$) noisy perturbation of the plane wave as a function of propagation distance. Dashed red lines indicate the temporal edges of the developing MI area. The inset (false color) zooms in on the initial transient MI stage. (b) Corresponding one-pair superregular breather solution that fits the MI edges in subfigure (a). Dashed red lines highlight opposite group velocities of the two quasi-Akhmediev breathers. (c) Spectral evolution $|\tilde{\psi}(\Omega, \xi)|$ (in log. scale) corresponding to evolution map (a). (d) Temporal profiles of both phase (top) and amplitude (bottom) obtained at $\xi = 8.4$ with NLS simulation (black lines) and breather theory (dashed blue lines), which evidence the particular properties of temporal edges of MI area. (e) Temporal phase (top) and amplitude (bottom) profiles of the highest amplitude peak detected in subfigure (a) compared to the theoretical synchronized collision of the one-pair breather solution (dashed blue lines).

Akhmediev breathers). Finally, R was adjusted to 1.2 in order to fit the absolute group velocity of MI edges highlighted in figure 7.8(a) with red dashed lines. The resulting solution is depicted in figure 7.8(b) and represents an excellent fit to the simulated MI area. Note that the origin of analytic solution was slightly shifted to $\xi_0 = 2$ to fit the numerical simulation; this depends on the energy contained in each frequency mode related to MI. Indeed, an initial transient phase (when $\xi < 2$) of MI is evidenced in the inset of figure 7.8(a), wherein a vast range of noise frequencies ($\Omega > 2$) are not amplified and then linearly dispersed from the localized perturbation. In figure 7.8(d), we confirm that temporal edges of the MI area take the form of the one-pair superregular breather solution; both amplitude and phase are well fitted by theory. Given the random initial conditions, the localized MI process cannot leave behind (i.e. at the center $\tau = 0$) a non-perturbed plane wave exactly as the superregular breather solution. When looking at the evolution into the MI area in figure 7.8(a), the dynamics is more chaotic and we observe various forms of interaction and collision between elementary breathers similar to previous studies of noise-driven MI [71, 73]. Periodic behaviors and spatial recurrences of breather waves are clearly observable as expected. More importantly, we also demonstrate in figure 7.8(e) that the highest amplitude peak emerging at $\xi = 15$ results from the collision of two Akhmediev breathers with their particular opposite frequencies close to $\pm\Omega_{\max}$ ($R = 1$, $\alpha = 0.8$, $\theta_{1,2} = 0$, $\mu_{1,2} = 0$). The highest amplitude NLS rogue waves in the MI process are appropriately identified (amplitude and phase properties) as collisions of breathers [37].

To assess the general key role of superregular breathers in the localized MI problem, we provide additional signatures of the emergence of such complex wave structures in figure 7.9(a). The NLS evolution map of $\psi(\tau, \xi)$ for another noisy perturbation of the plane wave ($\Delta = 12$) shows again that MI takes place in a delimited zone of the plane (ξ, τ). The temporal edges of the localized MI area can be formed by a pair of quasi-Akhmediev breathers ($R = 1.07$, $\alpha = \pm\pi/4$, $\theta_{1,2} = \pi/2$, $\mu_{1,2} = 0$) that propagate with opposite group velocities, as confirmed in figure 7.9(b). Both amplitude and phase at $\xi = 11.4$ are well fitted by theory (the origin of analytic solution was shifted to $\xi_0 = 2$). Moreover, we also distinguish that another pair of quasi-Akhmediev breathers is developing into the MI area for $\xi > 7$ (see dashed white lines in figure 7.9(a)). Figure 7.9(c) depicts again both amplitude and phase profiles obtained with numerical simulation at $\xi = 11.4$, however, in this case, we are indeed able to fit some wave structures in the central zone of MI with the same one-pair superregular breather solution and its origin shifted to $\xi_0 = 6.2$. Another similar pair of quasi-Akhmediev breathers appears to emerge for $\xi \sim 12$. This spatial recurrence phenomenon can be

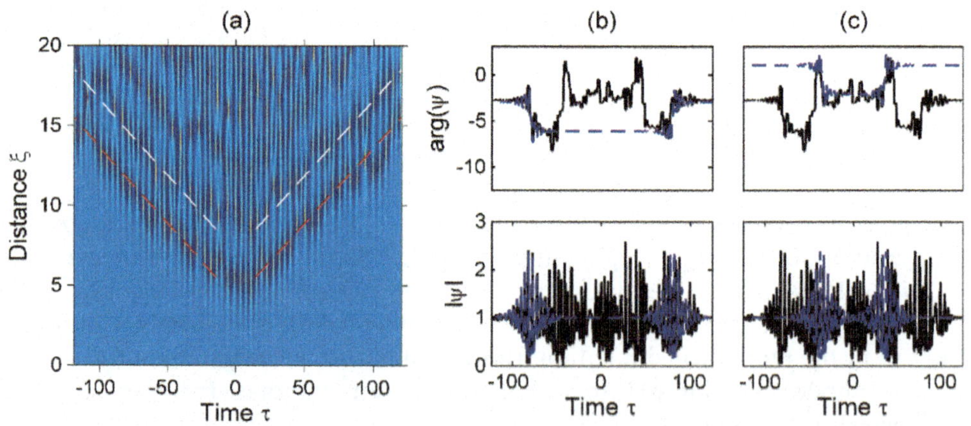

Figure 7.9. (a) NLS simulation of the temporal evolution of another small localized ($\Delta = 12$) noisy perturbation of the plane wave as a function of propagation distance. Dashed red and white lines indicate the possible position of two pairs of quasi-Akhmediev breathers in the developing MI area. (b) and (c) Corresponding temporal profiles of both phase (top) and amplitude (bottom) obtained at $\xi = 11.4$ compared with superregular breather theory (dashed blue lines).

related to the random initial conditions that are far from the exact simple perturbation described by the single one-pair superregular breather. The long-term evolution of MI dynamics from random localized perturbations still requires further studies that are beyond the scope of this chapter. Future works could also take into account the more complex theoretical framework of N-pair breather solutions.

7.5 Conclusions

This chapter reported the main physical features of NLS superregular breathers, with particular emphasis on the simplest one-pair superregular solution. We also described the first experimental evidence of such breathers in two different branches of wave physics, namely in optics and hydrodynamics. Based on the common NLS equation framework [74], the multidisciplinary approach confirmed that superregular breathers are of fundamental importance since they contribute to fully describe both amplification and annihilation stages of extreme localized waves in the modulation instability phenomenon. Here, our investigations have been restricted to an essentially reduced class of N-pair breather solutions and in the absence of complex and random dynamics. However, even the simplest one-pair breathers may describe a vast range of unstable nonlinear dynamics (for example, see figure 7.2), and contribute to solve MI-related problems such as rogue waves and turbulence in integrable systems [51, 75].

To this end, some preliminary numerical simulations were performed to analyze the localized noise-driven MI problem. These numerical results clearly highlighted that the resulting nonlinear stage of spatiotemporal MI involves the development of superregular breathers. In order to explore the rich dynamics induced by random noisy perturbation, further detailed studies are obviously still required. Note that the long-term evolution of the nonlinear stage of MI is still an open problem depending on the input perturbation considered. In particular, as was shown in references [76, 77], a broad class of modulationally unstable localized initial conditions can be described by so-called continuous spectrum solutions of the NLS equation. Besides, the impact of non-exact initial excitation of NLS breathers as well as the influence of non-ideal NLS propagation (i.e. quasi-integrable systems) are of fundamental importance to better understanding of natural phenomena and experimental systems. Any kind of disturbance of the NLS propagation induces a deviation from the theoretical breather. However, breathers may be considered as 'robust solutions' [78] in the sense that their main pulsating features occur even with non-ideal conditions. Note that recent efforts have been firstly made from both theoretical and experimental points of view for elementary breather solutions. For instance, breather solutions were derived of the integrable nonlinear Schrödinger equation hierarchy with an infinite number of higher-order terms [79, 80]. Approximate polynomial solutions can be also found for an extended NLS equation, taking into account higher-order odd terms or even-parity perturbations that break the integrability [81, 82]. Preliminary numerical studies recently investigated how some metamaterial properties (such as the magneto-optic effect) can affect the propagation of NLS breathers [83]. Nevertheless, experimental characteristics are generally chosen in accordance with NLS theory requirements in order to avoid the potential impact of higher-order dispersive or nonlinear effects. Only a few experimental studies in optics and hydrodynamics reported the impact of higher-order dispersion, propagation defect or losses, or forcing on some breather dynamics and related spatial recurrence phenomena [84–88]. Finally, it is worth mentioning that major experimental developments were also achieved in the observation of breather solutions, which open new perspectives particularly in the field of experimental physics. First, recall that real-time single-shot measurements of spontaneous breathers were reported in optics from the nonlinear propagation of initially coherent or incoherent waves [47, 48]. Similar detection analysis was also confirmed in a water wave tank in which the MI process was seeded by random noise added to a Stokes wave [89]. Secondly, the evidence of polarization modulation instability and related vector breather solutions is expected to be a crucial progress in describing spontaneous wave pattern formation in multicomponent systems [90, 91].

Acknowledgments

BK acknowledges funding from the French 'Investissements d'Avenir' program (project PIA2/ISITE-BFC, contract ANR-15-IDEX-03) and the Conseil Régional de Bourgogne. AC acknowledges the support of the Japan Society for the Promotion of Science (JSPS). The work on theoretical breather formalism linked to hydrodynamics was supported by the Russian Science Foundation (Grant No. 14-22-00174 to AG and VZ). The work on theoretical breather formalism linked to optics was supported by the Russian Science Foundation (Grant No. 17-71-10128 to AG). NA acknowledges the support of the Australian Research Council (Discovery Project numbers DP140100265 and DP150102057) and support from the Volkswagen Stiftung.

References

[1] Zakharov V E and Ostrovsky L A 2009 Modulation instability: The beginning *Physica D* **238** 540–8

[2] Akhmediev N and Ankiewicz A 1997 *Solitons, Nonlinear Pulses and Beams* (London: Chapman and Hall)

[3] Remoissenet M 2010 *Waves Called Solitons: Concepts and Experiments* (Heidelberg: Springer)

[4] Zakharov V E and Shabat A B 1971 Exact theory of two-dimensional self-focusing and one-dimensional self-modulation of waves in nonlinear media *Zh. Eksp. Teor. Fiz.* **61** 118 [Sov. Phys. JETP 34 62 (1972)]

[5] Zakharov V E 1968 Stability of periodic waves of finite amplitude on a surface of deep fluid *J. Appl. Mech. Tech. Phys.* **9** 190–4

[6] Lighthill M J 1965 Contribution to the theory of waves in non-linear dispersive systems *J. Inst. Math. Appl* **1** 269–306

[7] Whitham G B 1965 A general approach to linear and nonlinear dispersive waves using a Lagrangian *J. Fluid Mech.* **22** 273–83

[8] Bespalov V I and Talanov V J 1966 Filamentary structure of light beams in nonlinear liquids. *JETP Lett.* **3** 307–10

[9] Benjamin T B and Feir J E 1967 The disintegration of wave trains on deep water. Part 1: Theory *J. Fluid Mech.* **27** 417–30
[10] Benjamin T B 1967 Instability of periodic wavetrains in nonlinear dispersive systems *Proc. Roy. Soc.* A **299** 59–75
[11] Taniuti T and Washimi H 1968 Self-trapping and instability of hydromagnetic waves along the magnetic field in a cold plasma *Phys. Rev. Lett.* **21** 209
[12] Watanabe S 1977 Self-modulation of a nonlinear ion wave packet *J. Plasma Phys.* **17** 487–501
[13] Bailung H and Nakamura Y 1993 Observation of modulational instability in a multi-component plasma with negative ions *J. Plasma Phys.* **50** 231–42
[14] Tai K, Hasegawa A and Tomita A 1986 Observation of modulational instability in optical fibers *Phys. Rev. Lett.* **56** 135
[15] Marquie P, Bilbault J M and Remoissenet M 1994 Generation of envelope and hole solitons in an experimental transmission line *Phys. Rev.* E **49** 828
[16] Salasnich L, Parola A and Modulational Reatto L 2003 Instability and complex dynamics of confined matter-wave solitons *Phys. Rev. Lett.* **91** 080405
[17] Lake B M, Yuen H C, Rungaldier H and Ferguson W E 1977 Nonlinear deep-water waves: theory and experiment. Part 2. Evolution of a continuous wave train *J. Fluid. Mech.* **83** 49–74
[18] Akhmediev N, Ankiewicz A and Taki M 2009 Waves that appear from nowhere and disappear without a trace *Phys. Lett.* A **373** 675–8
[19] Akhmediev N, Soto-Crespo J M and Ankiewicz A 2009 Extreme waves that appear from nowhere: On the nature of rogue waves *Phys. Lett.* A **373** 2137–45
[20] Kharif C, Pelinovsky E and Slunyaev A 2009 *Rogue Waves in the Ocean* (Berlin–Heidelberg: Springer)
[21] Osborne A R 2010 *Nonlinear Ocean Waves and the Inverse Scattering Transform* (San Diego: Academic)
[22] Kuznetsov E 1977 Solitons in a parametrically unstable plasma *Sov. Phys. Dokl* **22** 507–8
[23] Kawata T and Inoue H 1978 Inverse scattering method for the nonlinear evolution equations under nonvanishing conditions *J. Phys. Soc. Jpn* **44** 1722–9
[24] Ma Y C 1979 The perturbed plane-wave solutions of the cubic Schrödinger equation *Stud. Appl. Math.* **60** 43–58
[25] Peregrine D H 1983 Water waves, nonlinear Schrödinger equations and their solutions *J. Austral. Math. Soc. Ser.* B **25** 16–43
[26] Akhmediev N and Korneev V I 1986 Modulation instability and periodic solutions of the nonlinear Schrödinger equation *Theor. Math. Phys.* **69** 1089–93
[27] Akhmediev N, Eleonskii V M and Kulagin N E 1987 Exact first-order solutions of the nonlinear Schrödinger equation *Theor. Math. Phys.* **72** 809–18
[28] Kedziora D J, Ankiewicz A and Akhmediev N 2013 Classifying the hierarchy of nonlinear-Schrödinger-equation rogue-wave solutions *Phys. Rev.* E **88** 013207
[29] Kibler B, Fatome J, Finot C, Millot G, Dias F, Genty G, Akhmediev N and Dudley J M 2010 The Peregrine soliton in nonlinear fibre optics *Nat. Phys.* **6** 790–5
[30] Hammani K, Wetzel B, Kibler B, Fatome J, Finot C, Millot G, Akhmediev N and Dudley J M 2011 The spectral dynamics of modulation instability described using Akhmediev breather theory *Opt. Lett.* **36** 2140–2
[31] Chabchoub A, Hoffmann N P and Akhmediev N 2011 Rogue wave observation in a water wave tank *Phys. Rev. Lett.* **106** 204502

[32] Bailung H, Sharma S K and Nakamura Y 2011 Observation of Peregrine solitons in a multicomponent plasma with negative ions *Phys. Rev. Lett.* **107** 255005

[33] Erkintalo M, Hammani K, Kibler B, Finot C, Akhmediev N, Dudley J M and Genty G 2011 Higher order modulation instability in nonlinear fiber optics *Phys. Rev. Lett.* **107** 253901

[34] Chabchoub A, Hoffmann N, Onorato M and Akhmediev N 2012 Super rogue waves: observation of a higher-order breather in water waves *Phys. Rev.* X **2** 011015

[35] Chabchoub A, Neumann S, Hoffmann N P and Akhmediev N 2012 Spectral properties of the Peregrine soliton observed in a water wave tank *J. Geophys. Res.* **117** C00J02

[36] Chabchoub A and Akhmediev N 2013 Observation of rogue wave triplets in water waves *Phys. Lett.* A **377** 2590–3

[37] Frisquet B, Kibler B and Millot G 2013 Collision of Akhmediev breathers in nonlinear fiber optics *Phys. Rev.* X **3** 041032

[38] Frisquet B, Chabchoub A, Fatome J, Finot C, Kibler B and Millot G 2014 Two-stage linear-nonlinear shaping of an optical frequency comb as rogue nonlinear-Schrödinger-equation-solution generator *Phys. Rev.* A **89** 023821

[39] Chabchoub A, Kibler B, Dudley J M and Akhmediev N 2014 Hydrodynamics of periodic breathers *Phil. Trans. R. Soc.* A **372** 20140005

[40] Pallabi P, Sharma S K, Nakamura Y and Bailung H 2016 Observation of second order ion acoustic Peregrine breather in multicomponent plasma with negative ions *Phys. Plasmas* **23** 022107

[41] Chabchoub A, Waseda T, Kibler B and Akhmediev N 2017 Experiments on higher-order and degenerate akhmediev breather-type rogue water waves *J. Ocean Eng. Mar. Energy* **3** 385–94

[42] Akhmediev N, Kibler B and Baronio F *et al* 2016 Roadmap on optical rogue waves and extreme events *J. Opt.* **18** 063001

[43] Chabchoub A 2016 Tracking breather dynamics in irregular sea state conditions *Phys. Rev. Lett.* **117** 144103

[44] Calini A and Schober C M 2002 Homoclinic chaos increases the likelihood of rogue wave formation *Phys. Lett.* A **298** 335–49

[45] Onorato M 2003 Landau damping and coherent structures in narrow-banded 1 + 1 deep water gravity waves *Phys. Rev.* E **67** 046305

[46] Kibler B, Hammani K, Michel C, Finot C and Picozzi A 2011 Rogue waves, rational solitons and wave turbulence theory *Phys. Lett.* A **375** 3149–55

[47] Närhi M, Wetzel B, Billet C, Toenger S, Sylvestre T, Merolla J-M, Morandotti R, Dias F, Genty G and Dudley J M 2016 Real-time measurements of spontaneous breathers and rogue wave events in optical fibre modulation instability *Nat. Commun.* **7** 13675

[48] Suret P, El Koussaifi R, Tikan A, Evain C, Randoux S, Szwaj C and Bielawski S 2016 Single-shot observation of optical rogue waves in integrable turbulence using time microscopy *Nat. Commun.* **7** 13136

[49] Zakharov V E 2009 Turbulence in Integrable Systems *Stud. Appl. Math.* **122** 219–34

[50] Agafontsev D S and Zakharov V E 2015 Integrable turbulence and formation of rogue waves *Nonlinearity* **28** 2791–821

[51] Soto-Crespo J M, Devine N and Akhmediev N 2016 Integrable turbulence and rogue waves: breathers or solitons? *Phys. Rev. Lett.* **116** 103901

[52] Zakharov V E and Gelash A A 2013 Nonlinear stage of modulation instability *Phys. Rev. Lett.* **111** 054101

[53] Gelash A A and Zakharov V E 2014 Superregular solitonic solutions: a novel scenario for the nonlinear stage of modulation instability *Nonlinearity* **27** R1–39

[54] Kibler B, Chabchoub A, Gelash A A, Akhmediev N and Zakharov V E 2015 Superregular breathers in optics and hydrodynamics: omnipresent modulation instability beyond simple periodicity *Phys. Rev.* X **5** 041026

[55] Zhang J-H, Wang L and Liu C 2017 Superregular breathers, characteristics of nonlinear stage of modulation instability induced by higher-order effects *Proc. R. Soc.* A **473** 20160681

[56] Liu C, Ren Y, Yang Z Y and Yang W L 2017 Superregular breathers in a complex modified Korteweg-de Vries system *Chaos* **27** 083120

[57] Zakharov V E and Shabat A B 1979 Integration of nonlinear equations of mathematical physics by the method of inverse scattering II *Funct. Anal. Appl.* **13** 166–74

[58] Faddeev L D and Takhtajan L A 2007 *Hamiltonian Methods in the Theory of Solitons* (Heidelberg: Springer)

[59] Tajiri M and Watanabe Y 1998 Breather solutions to the focusing nonlinear Schrödinger equation *Phys. Rev.* E **57** 3510

[60] Akhmediev N, Ankiewicz A and Soto-Crespo J M 2009 Rogue waves and rational solutions of the nonlinear Schrödinger equation *Phys. Rev.* E **80** 026601

[61] Akhmediev N, Soto-Crespo J M and Ankiewicz A 2009 How to excite a rogue wave *Phys. Rev.* A **80** 043818

[62] Kibler B, Fatome J, Finot C, Millot G, Genty G, Wetzel B, Akhmediev N, Dias F and Dudley J M 2012 Observation of Kuznetsov--Ma soliton dynamics in optical fibre *Sci. Rep.* **2** 463

[63] Akhmediev N, Ankiewicz A, Soto-Crespo J M and Dudley J M 2011 Universal triangular spectra in parametrically-driven systems *Phys. Lett.* A **375** 775–9

[64] Devine N, Ankiewicz A, Genty G, Dudley J M and Akhmediev N 2011 Recurrence phase shift in Fermi–Pasta–Ulam nonlinear dynamics *Phys. Lett.* A **375** 4158–61

[65] Chabchoub A and Grimshaw R H J 2016 The hydrodynamic nonlinear Schrödinger equation: Space and time *Fluids* **1** 23

[66] Henderson D M and Segur H 2013 The role of dissipation in the evolution of ocean swell *J. Geophys. Res.* **118** 5074–91

[67] Dysthe K B 1979 Note on the modification of the nonlinear Schrödinger equation for application to deep water waves *Proc. R. Soc.* A **369** 105–14

[68] Trulsen K and Dysthe K B 1996 A modified nonlinear Schrödinger for broader bandwidth gravity waves on deep water *Wave Motion* **24** 281–9

[69] Craig W, Guyenne P and Sulem C 2010 Hamiltonian approach to nonlinear modulation of surface water waves *Wave Motion* **47** 552–63

[70] Agrawal G P 2013 *Nonlinear Fiber Optics* 5th edition (Oxford: Academic)

[71] Toenger S, Godin T, Billet C, Dias F, Erkintalo M, Genty G and Dudley J M 2015 Emergent rogue wave structures and statistics in spontaneous modulation instability *Sci. Rep.* **5** 10380

[72] Agafontsev D S and Zakharov V E 2016 Integrable turbulence generated from modulational instability of cnoidal waves *Nonlinearity* **29** 3551–78

[73] Dudley J M, Genty G, Dias F, Kibler B and Akhmediev N 2009 Modulation instability, Akhmediev breathers and continuous wave supercontinuum generation *Opt. Exp.* **17** 21497–508

[74] Chabchoub A, Kibler B, Finot C, Millot G, Onorato M, Dudley J M and Babanin A V 2015 The nonlinear Schrödinger equation and the propagation of weakly nonlinear waves in optical fibers and on the water surface *Ann. Phys.* **361** 490–500
[75] Akhmediev N, Soto-Crespo J M and Devine N 2016 Breather turbulence versus soliton turbulence: Rogue waves, probability density functions, and spectral features *Phys. Rev.* E **94** 022212
[76] Biondini G and Mantzavinos D 2016 Universal nature of the nonlinear stage of modulational instability *Phys. Rev. Lett.* **116** 043902
[77] Biondini G, Li S and Mantzavinos D 2016 Oscillation structure of localized perturbations in modulationally unstable media *Phys. Rev.* E **94** 060201
[78] Ankiewicz A, Devine N and Akhmediev N 2009 Are rogue waves robust against perturbations? *Phys. Lett.* A **373** 3997–4000
[79] Ankiewicz A, Kedziora D J, Chowdury A, Bandelow U and Akhmediev N 2016 Infinite hierarchy of nonlinear Schrödinger equations and their solutions *Phys. Rev.* E **93** 012206
[80] Ankiewicz A and Akhmediev N 2017 Rogue wave solutions for the infinite integrable nonlinear Schrödinger equation hierarchy *Phys. Rev.* E **96** 012219
[81] Ankiewicz A, Soto-Crespo J M, Chowdhury M A and Akhmediev N 2013 Rogue waves in optical fibers in presence of third-order dispersion, self-steepening, and self-frequency shift *J. Opt. Soc. Am.* B **30** 87–94
[82] Ankiewicz A, Chowdhury A, Devine N and Akhmediev N 2013 Rogue waves of the nonlinear Schrödinger equation with even symmetric perturbations *J. Opt.* **15** 064007
[83] Boardman A, Alberucci A, Assanto G, Grimalsky V, Kibler B, McNiff J, Nefedov I, Rapoport Y and Valagiannopoulos C 2017 Waves in hyperbolic and double negative metamaterials including rogues and solitons *Nanotechnology* **28** 444001
[84] Chabchoub A, Hoffmann N, Branger H, Kharif C and Akhmediev N 2013 Experiments on wind-perturbed rogue wave hydrodynamics using the Peregrine breather model *Phys. Fluids* **25** 101704
[85] Mussot A, Kudlinski A, Droques M, Szriftgiser P and Akhmediev N 2014 Fermi–Pasta–Ulam recurrence in nonlinear fiber optics: The role of reversible and irreversible losses *Phys. Rev.* X **4** 011054
[86] Kimmoun O *et al* 2016 Modulation instability and phase-shifted Fermi–Pasta–Ulam recurrence *Sci. Rep.* **6** 28516
[87] Kimmoun O, Hsu H C, Kibler B and Chabchoub A 2017 Non-conservative higher-order hydrodynamic modulation instability *Phys. Rev.* E **96** 022219
[88] Bendahmane A, Mussot A, Szriftgiser P, Zerkak O, Genty G, Dudley J M and Kudlinski A 2014 Experimental dynamics of Akhmediev breathers in a dispersion varying optical fiber *Opt. Lett.* **39** 4490–3
[89] Chabchoub A, Genty G, Dudley J M, Kibler B and Waseda T 2017 Experiments on spontaneous modulation instability in hydrodynamics *27th Int. Ocean and Polar Engineering Con.* (ISOPE-2017)
[90] Frisquet B, Kibler B, Morin P, Baronio F, Conforti M, Millot G and Wabnitz S 2016 Optical dark rogue wave *Sci. Rep.* **6** 20785
[91] Frisquet B, Kibler B, Fatome F, Morin P, Baronio F, Conforti M, Millot G and Wabnitz S 2015 Polarization modulation instability in a Manakov fiber system *Phys. Rev.* A **92** 053854

IOP Publishing

Nonlinear Guided Wave Optics
A testbed for extreme waves
Stefan Wabnitz

Chapter 8

Rogue waves in photorefractive media

Davide Pierangeli, Claudio Conti, Eugenio DelRe

8.1 Introduction

Phenomena involving long-tail statistics are of great interest in physics because they allow the observation of extreme events that would otherwise be truly rare in systems that follow Gaussian distributions. This typically occurs in out-of-equilibrium processes that manifest complex dynamics, such as in the appearance of earthquakes in plate motion or in large-scale breakdowns in networks [1]. Long-tail statistics are also observed in wave-supporting systems in which case giant perturbations, known as rogue waves, form out of noise, reach abnormal amplitudes, and rapidly decay back into noise. Originally studied in ocean dynamics, where their origin and properties are still debated [2–5], they have now been observed in a variety of different contexts and have been shown to present common features also when different physical mechanisms specific for each system are involved in their generation [6]. Universal statistical traits and model equations based on the generalized nonlinear Schrödinger equation (NLSE) have fuelled particular research efforts in optics [7, 8]. Here, extreme waves can be exploited both for applications and as test benches to study the origin and properties of their not-well-understood oceanic counterpart. In particular, long-tail statistics have been observed in various optical systems, from nonlinear pulse propagation in optical fibers [9] to beam filamentation [10] and dissipative resonators [11]. Several physical ingredients underlying the occurrence of extreme events have been identified, such as interacting coherent structures emerging from stochastic instabilities [12–15], interference of inhomogeneous random wave fields [16–19], wave turbulence in incoherent nonlinear propagation [20–22], and spatiotemporal chaos in cavity dynamics [23, 24]. In spatially extended systems, instabilities have been shown to be key in leading to rogue waves in the transverse plane of the optical light beam [25]. Here, the presence of blocked or induced disorder or, generally speaking, of an interplay between a non-equilibrium state and a nonlinear mechanism, appears to be a crucial ingredient to extreme event generation, providing complex interaction and coupling of different spatial regions [26].

doi:10.1088/978-0-7503-1460-2ch8 © IOP Publishing Ltd 2017

Optical waves in photorefractive media has been the arena of choice for experimental nonlinear physics, allowing the investigation of general phenomena such as spatial solitons [27], shock-waves [28], wave localization [29], wave condensation [30], scale-free propagation [31], anti-diffraction [32], and negative-mass dynamics [33], to name a few. Unfortunately, the central role of disorder seemed to rule out rogue waves for beam propagation in photorefractive media, where stochastic components generally play a minor part [34]. Recently, exploiting the large photorefractive nonlinearity of disordered ferroelectric crystals in proximity to their structural phase transition has allowed the opening of this important platform to extreme wave experiments. In these conditions, the presence of long-range fluctuations leads to out-of-equilibrium states characterized by reorienting polar nanoregions that can affect light through a disordered and strongly enhanced electro-optic response [35]. This combination of disorder and nonlinearity has allowed the first observation of rogue events in photorefractive wave propagation [36]. Photorefractive rogue waves offer unique opportunities for understanding the effect of various physical ingredients on extreme event generation, since they form a rich experimental setting in which specific components can be finely tuned. Several peculiarities make this system attractive as a test bench for real-world extreme waves. At variance with optical fibers, it represents a spatially extended and high-dimensional system in which wave dynamics can be far from integrable. At the same time, the absence of gain and dissipation mechanisms makes nonlinear evolution admit a Hamiltonian wave-type description; it allows us to investigate how saturation in the nonlinearity affects rogue phenomena, possibly bridging the gap between extreme waves of linear and nonlinear origin. In another interesting approach, photorefractive media under feedback can also be used to investigate extreme events in pattern-forming systems [37].

In this chapter, we discuss the emergence of optical rogue waves as light beams that undergo highly nonlinear regimes of propagation in photorefractive ferroelectric crystals. Basic experimental findings corroborated by numerical simulations of the generalized NLSE model are reported in section 8.2. In section 8.3, we report experiments unveiling how rogue wave appearance arises from a transition to a strong turbulent scenario supported by the interplay between wave instability, high nonlinearity, and medium fluctuations. The role of the coherence length of the optical field, along with high-resolution analysis of the extreme waveforms, is investigated in section 8.4 using partially incoherent beams as disordered input fields. Here, through a picture based on non-stationary spatial quasi-solitons and saturation of the nonlinearity, we lay the foundations for attempting a full understanding of rogue waves in photorefractive media and beyond. Open points and future research directions are briefly discussed in section 8.4.

8.2 Spatial rogue waves in photorefractive ferroelectrics

Generally speaking, we are here considering nonlinear wave propagation in the spatial domain, a physical situation schematically shown in figure 8.1(a). The optical field evolves along a spatial coordinate, the propagation direction z, and the

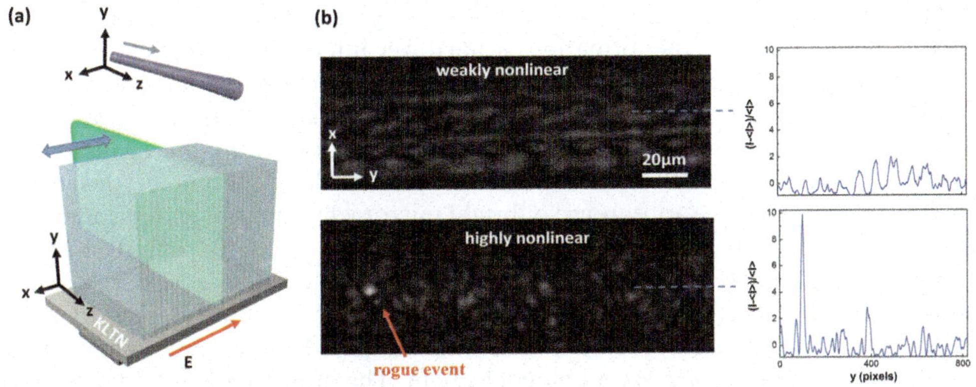

Figure 8.1. Nonlinear beam propagation in photorefractive KLTN in proximity of the ferroelectric phase transition. (a) Sketch of the physical configuration for nonlinear waves in the spatial domain and experimental setup. (b) Light intensity distribution transmitted through the medium showing (top) a disordered pattern in the weakly nonlinear regime, with spatial intensity fluctuations along the blue dashed y-line, and (bottom) a speckle-like pattern in the highly nonlinear regime including an extreme intensity spot, specified as rogue event. Figure adapted with permission from [36].

dynamics, resulting from the competition between diffraction and nonlinearity, occurs in the transverse x–y plane of the beam. We focus on optical waves propagating in a photorefractive ferroelectric crystal in proximity of its structural phase transition. In fact, as hereafter reported, we find that the transmitted light assumes a disordered spatial distribution containing bright localized spots of anomalously large intensity. The intensity distribution follows a signature long-tail statistics that disappears as the nonlinearity is weakened, suggesting that extreme events are intimately related to the enhanced nonlinear response of the medium provided by its critical conditions.

8.2.1 Observing optical extreme waves

The typical experimental geometry is sketched in figure 8.1(a). Laser light propagation ($\lambda = 532$ nm, P = 0.1 mW) is investigated focusing stripe beams, spatially extended and quasi-homogeneous in the y-direction and with full-width-at-half-maximum (FWHM) $\Delta x = 8$ μm along x, on a photorefractive crystal of KLTN, $K_{1-\alpha}Li_{\alpha}Ta_{1-\beta}Nb_{\beta}O_3$, with $\alpha = 0.04$ and $\beta = 0.38$. The $2.4^{(x)} \times 2.0^{(y)} \times 1.7^{(z)}$ mm ($l_x \times l_y \times l_z$) sample is kept at $T = T_C + 1$ K by means of a Peltier junction, where $T_C = 294$ K is the room-temperature Curie point that signals the paraelectric to ferroelectric phase transition. An external electric field ($V = 500$ V applied to the sample facets 2.4 mm apart) is applied along the x-direction, parallel to the optical field polarization. The nonlinearity can be controlled through temperature and applied voltage, and nonlinear beam propagation is explored detecting the transmitted intensity distribution through an imaging system and a CCD camera. What emerges is a disordered output intensity distribution $I(x, y)$ with micrometric spots of various intensity that are 'speckle-like' [38]. In fact, due to the critical conditions of the medium, a disordered index of refraction pattern forms over spatial scales comparable with λ, introducing scattering and fragmentation in the unstable nonlinear propagation, with the formation of narrower interacting filaments.

In order to study the statistical properties of the speckle-like intensity pattern, we first quantify the amplitude of the intensity fluctuations in the spatial light distribution, introducing the deviation $(I(x, y) - \langle I \rangle)/\langle I \rangle$ from the spatial mean value $\langle I \rangle$. Large deviations are observed in the disordered output of this highly nonlinear regime with the appearance of bright micrometric spots of extreme intensity. In figure 8.1(b) (bottom) we report, as an example, a measured output with the associated fluctuations along a y-direction in the inset; we note an extreme event, spatially localized, with a peak intensity approximately 20 times larger than the average intensity. The measured transverse width (FWHM) of this exceedingly bright micro-beam, a rogue wave, is approximately 3 μm. The disordered pattern with rogue fluctuations is observed at steady state after a transient stage, whose dynamics are the subject of section 8.3. During the initial evolution, the beam diffracts in an inhomogeneous setting, associated with linear sample disorder, as it reaches the crystal output facet. The typical spatial distribution in this 'weakly' nonlinear regime is shown in the top of figure 8.1(b), with the inset showing the smaller intensity fluctuations in this condition. This regime is superseded at subsequent times by the appearance of modulational instability and break-up, where disorder appears to have a dynamic and strongly amplified effect on light, which then leads to the final steady-state regime where rogue waves are detected. This transient nature is related to the physics of the photorefractive nonlinearity, which involves a build-up of a photo-generated space-charge field, so that the beam-induced index of refraction modulation accumulates in time [39]. As will be discussed in section 8.3, this property can be exploited to study in detail the whole process as a function of the nonlinearity.

The intensity probability distribution function (PDF) is obtained acquiring a large set of uncorrelated images (approximately 10^3) in the same experimental conditions. Results are shown in figure 8.2, where the intensity statistics observed in the highly and weakly nonlinear regime is compared. In both cases, the so-called incoherent part, which is related to random correlations between independent measurements [38], is of the order of the background intensity and is not shown. Long-tail statistics characterize highly nonlinear conditions, as reported in figure 8.2(a). The experimental behavior strongly deviates from a Gaussian distribution, that implies a decay according to $P(I) = \exp(-I/\langle I \rangle)/\langle I \rangle$, which is evaluated with the measured value of $\langle I \rangle$ (red line). In particular, the observed intensity distribution is well fitted by a stretched exponential decay (blue line) in the form of $P(I) = \exp(-cI^b - a)$. The stretching parameter b quantifies the deviation from Gaussianity, and long-tailed behavior is indicated by $b < 1$. The value obtained is $b = 0.65 \pm 0.02$, demonstrating that abnormal values of intensity occur with large probability. To further analyze the extreme events, we calculate the significant wave height, defined as the mean amplitude of the highest one-third of detected waves. Events of amplitude exceeding this value at least by a factor of two are commonly referred to as rogue waves in hydrodynamics. Setting the threshold in figure 8.2 (green dotted line), we identify a large number of events as rogue waves, among which is the one reported in figure 8.1(b). The extension of this oceanographic

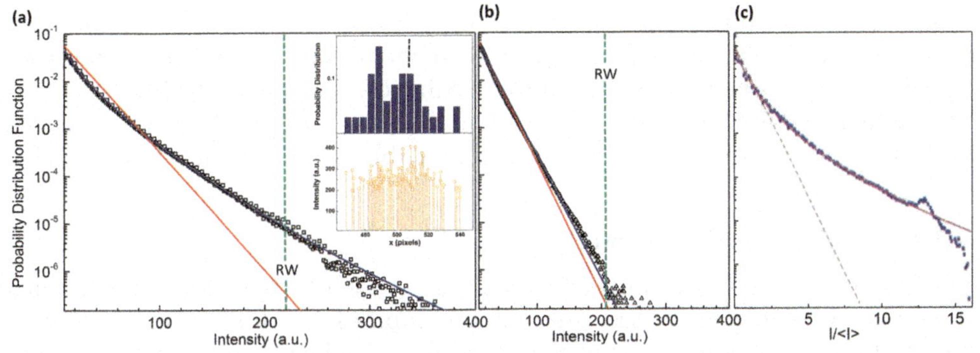

Figure 8.2. Intensity statistics. (a) Long-tail statistics in the highly nonlinear regime: experimentally observed (black dots), fitting function (blue line), and consistent Gaussian distribution for comparison (red line). The inset shows the spatial x-position of several observed rogue events (bottom) with the corresponding count histogram (top). (b) Normal statistics in the weakly nonlinear regime: experimentally observed (black dots), fitting function (blue line), and Gaussian distribution (red line). The vertical green dotted lines indicate the hydrodynamic rogue wave thresholds. (c) Analogous to panel (a) for data detected in a KLTN slab waveguide. Figure adapted with permission from [36].

criterion, which is associated with the wave amplitude, to the present optical system where the envelope intensity is measured, is here validated by the fact that the threshold approximately coincides with the extremity of the tail of the normal statistics. Taking several recorded extreme events, the spatial point at which they appear can be considered to obtain information on the role of anisotropic field effects in their generation. Counts as a function of the peak transverse x-position are reported in the inset of figure 8.2(a); the distribution is qualitatively centered near the beam-averaged midpoint (black reference line) without a prevailing shifted component. This means that self-bending in our system, which is associated with nonlocal responses [40], plays a negligible role in the emergence of rogue waves. This finding is intriguing since the effect is a spatial counterpart of the Raman shift in optical fiber propagation and temporal optical rogues have been usually associated with the most red-shifted soliton in supercontinuum generation [7]. Here, the spatial lateral shift (along x) is observed for the whole beam profile and amounts to approximately 20 μm at the crystal output. Results in the weakly nonlinear regime are reported in figure 8.2(b). The PDF loses its long tail in this case, showing only a small deviation from the Gaussian behavior (evaluated as previously, red line). The stretching exponent is now close to one, $b = 0.99 \pm 0.07$, confirming the normal scenario. This fact is quite interesting since it implies that, through the strength of the nonlinear interaction, long-tail statistics can be generated and controlled. It also assigns a fundamental role to the nonlinear response in the appearance of rogue waves. Moreover, the emergence of extreme events in photorefractive propagation is found to be almost independent of the system dimensionality. Specifically, different experiments in comparable conditions have been performed in a KLTN slab waveguide obtained through ion implantation [33]. In this case, transverse dynamics occur strictly along the y-direction since wave evolution along the x-axis is bounded by the integrated optical potential rather than by the line geometry of the input beam. The resulting long-tail statistics is reported in figure 8.2(c), and gives $b = 0.45 \pm 0.01$. Rogue waves thus appear enhanced as dimensionality is reduced to an actual two-dimensional case, a point that may be related to an enhanced interaction between localized modes as the available spatial extension decreases.

8.2.2 Generalized NLSE framework

In looking for a first understanding of the origin of each single rogue wave and how the amount of nonlinearity can affect microscopically its formation, we present numerical (2 + 1)D simulations of the generalized nonlinear Schrödinger equation describing the paraxial spatial evolution of the normalized optical field envelope $\psi(x, y, z)$ in centrosymmetric photorefractive media [39]. We consider the two leading terms in the nonlinear optical response of the material, which is a nonlinearity containing a Kerr-saturated component and a saturated nonlocal component due to the charge diffusive field. The generalized NLSE in this case reads as

$$i\partial_z\psi = -\frac{1}{2k}\nabla_\perp^2\psi + \frac{k\Delta n_0}{n}\left[\frac{1}{\left(1 + |\psi|^2\right)^2} - 2\alpha\left(\frac{\nabla_\perp|\psi|^2}{1 + |\psi|^2}\right)^2\right]\psi, \tag{8.1}$$

where $k = 2\pi n/\lambda$ is the wavenumber, n the refractive index, $\Delta n_0 = \frac{1}{2}n^3 g\varepsilon_0^2\chi^2 E^2$ the nonlinear strength, with $E\ (V/l_x)$ the bias field and χ the dielectric susceptibility, coupled to the electric field via the effective electro-optic coefficient g. The value of the diffusive nonlinear parameter $\alpha < 1$ sets the weakly nonlocal regime of the model. A split-step Fourier method analysis of equation (8.1) is performed adding a spectrally random seed noise to the input Gaussian beam. The values of the parameters are selected so as to match those for the KLTN sample and optical setup, except for χ, that is, the parameter through which we fix the strength of the nonlinearity. Scattering associated with index fluctuations of the medium is not included in the model and this constitutes the main discrepancy between experiments and numerical simulations. However, sample disorder is expected to only weakly affect the high-intensity statistical properties of the field. In fact, results are in good agreement with the experimental observation of intense light spots and show how, due to the strong nonlinear response, instabilities build up during propagation, ultimately leading to the breaking of the beam waveform. Two distinct regimes are identified increasing the nonlinearity: the first condition precedes the onset of modulational instability, while the second is characterized by randomic quasi-soliton fusion. In figure 8.3, we report a simulated beam propagation in both cases, respectively with $\chi = 5 \times 10^4$ and $\chi = 10^5$. We find multi-soliton formation in the highly nonlinear condition (figure 8.3(b)); here, collisions and mergers dynamically lead to an intensity distribution with larger spatial fluctuations (figure 8.3(c)), and localized structures with giant amplitudes emerge. As shown in figure 8.3(d), the output PDF reveals a characteristic long-tailed behavior in good agreement with the experimental statistics and confirms the rogue wave scenario. Consistent with

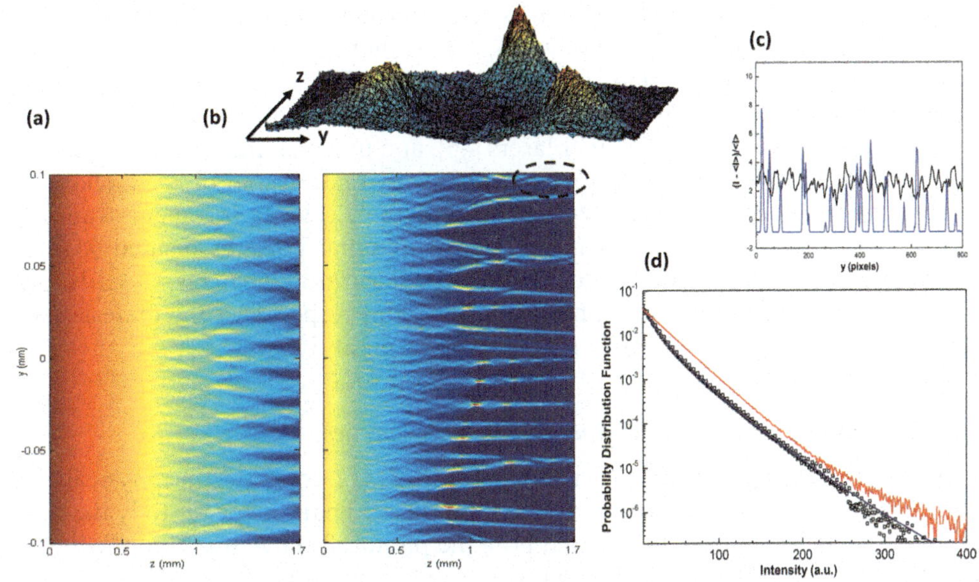

Figure 8.3. Numerical results. Typical wave dynamics in the y–z section in (a) nonlinear and (b) highly nonlinear conditions, where soliton collisions and mergers take place, leading to giant peak intensities (detail in the inset). (c) Comparison between output deviations in (a) and (b) along y. (d) Long-tail PDF in the highly nonlinear regime (red line) compared with the experimental data (black dots). Figure adapted with permission from [36].

experiments, this scenario of nonlinear origin is weakly affected by the nonlocal component in equation (8.1). In particular, the simultaneous occurrence of interaction processes and extreme intensity fluctuations suggest that the dynamics of quasi-soliton mergers may be an important microscopic mechanism at the basis of extreme intensity waves. This places our results in a general context that may relate rogue waves to soliton collisions in the nonlinear Schrödinger model. However, in the following we do not give much emphasis to collisions between soliton structures in the wake of extreme waves simply because, up to now, this process has lacked direct experimental observation. The obstacle also applies to other systems where this mechanism is thought to have a leading role [10, 15].

8.3 Optical instabilities and strong wave turbulence

We now discuss in more detail the physical state underlying the formation of rogue waves in photorefractive ferroelectrics. Exploiting the cumulative property of the nonlinearity, we report experiments proving that this state is triggered by a fully developed transition to strong optical turbulence [41]. It is shown how an increasing wave interaction causes a modulational unstable quasi-homogeneous optical field to be superseded by a chaotic and spatially incoherent one with the emergence of concomitant extreme events. At variance with weak wave turbulence that usually refers to weakly nonlinear wave systems dominated by external noise and having negligible intrinsic disorder [42, 43], the strong-turbulence scenario is characterized by several emerging and collapsing wave packets that interplay with an incoherent wave field. The reader can refer to [44] for a general review. In nonlinear optics, strong turbulence has been observed, for example, in multiple filamentation of intense femtosecond pulses [45], fiber lasers [46], and nonlocal media [47]; the picture is here extended to disordered photorefractive propagation and directly related to extreme wave appearance [41]. We stress that the detected turbulent transition involves the coupling of two stochastic effects: external noise associated with the initial condition and internal fluctuations of the linear/nonlinear response. In fact, local intensity fluctuations are amplified by the giant response of the material, and inhomogeneity in the nonlinearity strongly affects light dynamics.

8.3.1 Evidence of turbulent transitions in beam dynamics

With the experimental setup previously described, we consider the physical time ruling light evolution at the crystal output as a physical parameter to study the beam dynamics. In fact, we have mentioned how the photorefractive nonlinearity has the peculiar property of being noninstantaneous and cumulative in time. In this way, observations at different times correspond to beam propagation for increasing nonlinearity up to saturation. A typical time scale τ for beam dynamics is fixed through its symmetry breaking into periodic coherent structures [48, 49], a stage that can be accurately identified and characterized varying the accessible experimental parameters (see [41] for details). The dimensionless continuous control parameter of the nonlinearity is thus t/τ, where t is the evolving time.

Direct evidence of the onset of strong turbulence is shown in figure 8.4, where the y-profile of the output intensity distribution is reported as the nonlinearity increases and saturates. Once the quasi-homogeneous input beam has experienced symmetry-breaking via modulational instability, a sharp transition into a chaotic state with pseudo-recurrent patterns occurs for $t/\tau \gtrsim 1$ (figure 8.4(a)). Some of these structures can have an extremely large intensity and are identified as rogue waves. The single extreme event is found to suddenly disappear as t/τ further varies, so that no traces remain after its passing. Globally, this transition corresponds to the loss of spatial coherence that persists only on small scales (speckle-like). Specifically, the width of the spatial Fourier spectrum sharply increases by almost one order of magnitude and the long-range intensity autocorrelation abruptly decreases, as shown in figure 8.4(b). We stress that detecting the onset of turbulence as a sharp transition signaling departure from coherence is a method that goes beyond the stability properties of the input flux. In fact, an analogous transition has been reported in conditions where the homogeneous state of the dynamics is stable and unstable with respect to perturbations [50]. We note that a behavior similar to figure 8.4(a) and (b) has been numerically observed studying the nonlinear stage of modulation instability in the NLSE framework. Here, the incoherent state generated during wave evolution is referred to as integrable turbulence [51]. However, results here presented depart from this scenario since the presence of a saturable nonlinearity makes wave dynamics non-integrable. This means that the observed turbulent transition weakly depends on

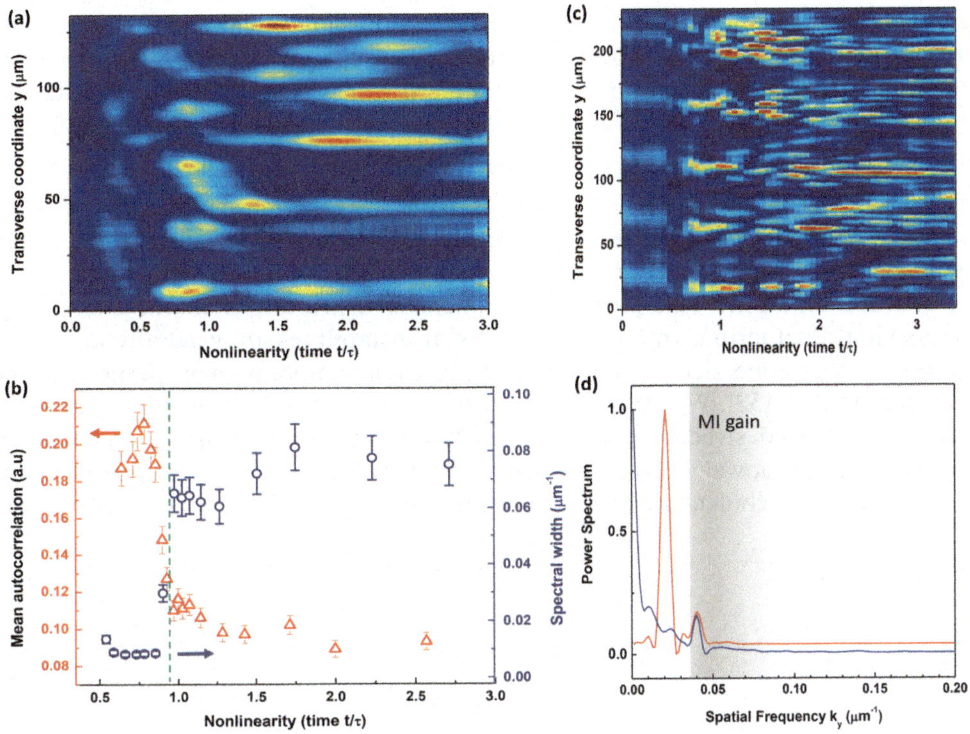

Figure 8.4. Transitions to strong turbulence in beam dynamics. (a) Detected output spatial intensity distributions as a function of the nonlinearity expressed through the continuous dimensionless control parameter t/τ (time). (b) Corresponding evolution for the width of the spatial Fourier spectrum and mean intensity autocorrelation. Green line serves as a guide at the sharp transition signaling the onset of optical turbulence. (c) Onset of turbulence for spatially-modulated input waves. (d) Input power spectrum for spatially-modulated (red line) and quasi-homogeneous beams (blue line) with the gain region of the modulational instability process (gray area). Figure adapted with permission from [41].

the input wave and can occur also without a leading role of the modulational instability process [44]. This fact is demonstrated repeating the experiments with an inhomogeneously tailored coherent input wave; using a spatial light modulator (SLM), the input field is modulated along the y-direction with a periodic component. As shown in figure 8.4(c) a transition to turbulence is observed at $t/\tau \approx 1$. In this case, beam breaking is dominated by the input spatial frequency $k_y = 0.02\ \mu\text{m}^{-1}$ and modulation instability is only weakly involved, as noise experiences small amplification on this scale. In figure 8.4(d), the input power spectrum is shown in comparison with the quasi-homogeneous case. The picture can be easily extended to generically modulated input waves.

8.3.2 Spectral and statistical evolution of the wave field

The spectral and statistical properties of the optical state significantly change before and after the transition to turbulence. An ensemble spectral analysis at moderate nonlinearity preceding the transition reveals the modulationally unstable regime. The peak in figure 8.5(a) shows that the typical spatial frequency experiencing maximum gain in the experiments is $\bar{k}_y = 0.05\ \mu\text{m}^{-1}$. This well agrees with the modulational instability gain function $G(k_y)$ numerically computed from equation (8.1), evaluating the amplitude of the Fourier spectrum $\hat{\Psi}(k_y, z)$ at the input, $z = 0$, and output, $z = l_z$, as $G(k_y) = 1/l_z \ln[\hat{\Psi}(k_y, 0)/\hat{\Psi}(k_y, l_z)]$. The broadening of $G(k_y)$ around the maximum amplified spatial frequency indicates that, during the single-shot dynamics, higher/lower frequencies can also emerge easily and compete with the characteristic one. This gives rise to intensity distributions characterized by a large spatial fluctuation. In fact, modulational instabilities are generally known to possess a strong dependence on the specific noise realization, with properties varying from shot to shot [52]. We observe that when the frequency $\bar{k}_y$ is mainly excited, the emerging localized structures have a weakly varying peak intensity, and there is equipartition of power across the generated mode. Completely different is the scenario in the turbulent regime, where intensity distributions vary stochastically

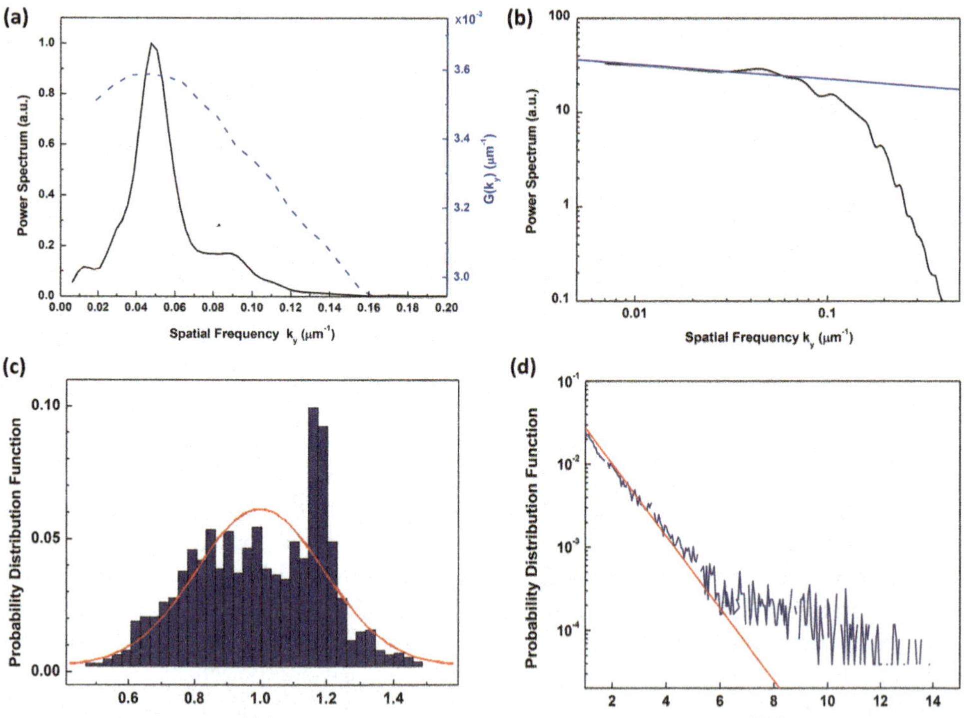

Figure 8.5. Spectral and statistical properties of the wave field before and after the onset of turbulence. (a) Spatial spectrum at $t/\tau \approx 1$ with the instability gain function (blue-dashed line). (b) Measured wave-turbulent power spectrum at $t/\tau \simeq 2.5$ (black line) and fitting power law scaling on large spatial scales (blue line). (c) Peak-intensity PDF of localized structures emerging from instability at $t/\tau \approx 1$, experimental counts (blue bars), and Gaussian trend of the distribution tails (red line). (d) Measured long-tail statistics in optical turbulence at $t/\tau \approx 2.5$ (blue line) and consistent Gaussian exponential scaling (red line), showing rogue wave generation in the turbulent regime. Figure adapted with permission from [41].

from shot to shot. Waves appear to be characterized by random phases, although from the long-tail statistics we realize that some correlations between spatial modes actually exist. In figure 8.5(b) the power spectrum is reported at $t/\tau \simeq 2.5$, where we expect the nonlinearity to be fully saturated; it is extremely broad and without specific resonances, with the peak associated with the amplification of $\bar{k}_y$ before the transition that results in being fully relaxed towards lower spatial frequencies. The spectrum is well-fitted at low frequencies by a power law behavior $\propto k_y^{-\gamma}$, with the scaling exponent $\gamma = 0.15 \pm 0.01$.

Statistical analysis demonstrates that rogue waves are strictly connected to the turbulent regime. Specifically, we first consider the PDF of peak-intensity values of localized structures emerging from instabilities before the turbulent transition. The histogram, reported in figure 8.5(c), contains a high-intensity peak embedded into a broad distribution. The peak, at $I/\langle I\rangle \simeq 1.2$, is deterministic and closely related to the peak in the gain spectrum of figure 8.5(a), as it arises from structures belonging to the maximally amplified frequency $\bar{k}_y$. Random shot-to-shot fluctuations in this stage populate the rest of the distribution, with tails compatible with a Gaussian decay, implying that extreme events occur here with low probability. On the contrary, as reported in figure 8.5(d), the PDF measured deep into the turbulent regime presents the long-tail behavior defining rogue wave phenomena, in close analogy with results reported in section 8.2.1.

8.4 Incoherence, saturation, and solitons in extreme waves

In this section, we address fundamental issues such as the role of input field incoherence and solitons in rogue waves, points recognized as crucial also in the more general case of the integrable NLSE dynamics [53]. In particular, we discuss how the emergence of rogue waves strongly depends on the correlation length of the input field distribution [54]. Using partially incoherent excitations, we report a scale-dependent behavior of the long-tail statistics, in which extreme event appearance can be suppressed or strongly enhanced for a specific scale of the spatial incoherence. Remarkably, high-resolution measurements of the rogue waveforms identify this scale with a characteristic peak-intensity-independent wave size coinciding with the minimum soliton size permitted by the saturation in the response. These findings suggest that, in photorefractive nonlinear beam propagation, the emergence, properties, and control of rogue waves are intimately connected to non-stationary quasi-solitons typical of a saturable cumulative nonlinearity.

8.4.1 Input wave disorder and extreme event generation

To investigate the effect of the spatial coherence scale of the field on extreme events, the experimental setup detailed in section 8.2.1 is extended to partially incoherent input beams [55], whose spatial autocorrelation length can be finely tailored. The photorefractive ferroelectric crystal is kept in the same conditions in which turbulent states with rogue waves have been observed starting from coherent and quasi-homogeneous stripe input excitations. Setup and methods for partially incoherent beam generation are shown schematically in figure 8.6(a). Light at a wavelength

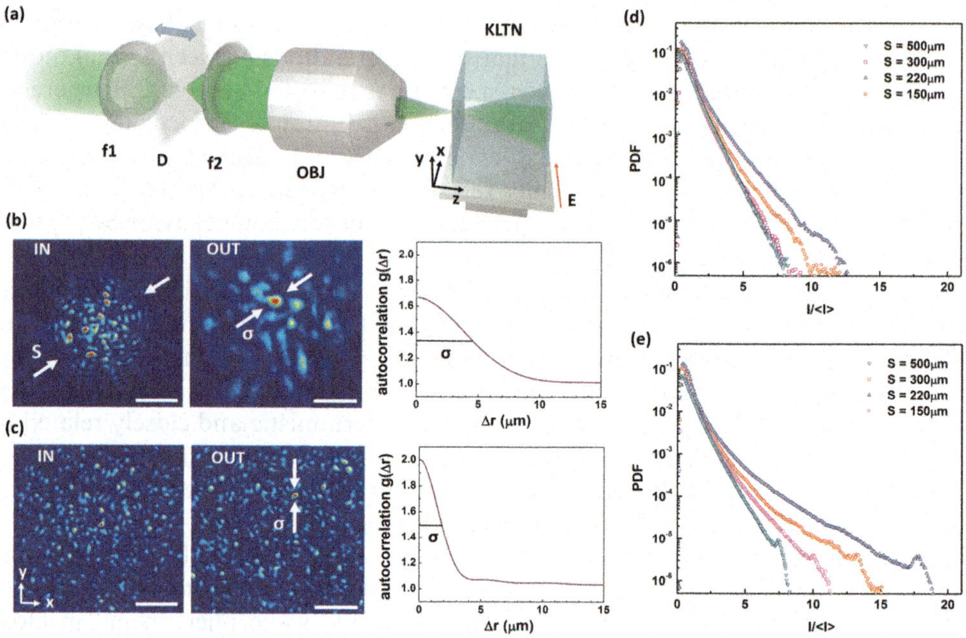

Figure 8.6. Partially incoherent beams and scale-dependent behavior of the intensity statistics. (a) Sketch of the experimental setup with lenses ($f1 = f2 = 50$ mm), adjustable glass diffuser D (average particle size of 2 μm), long-working-distance objective OBJ (NA = 0.55), and KLTN sample. (b) and (c) Input and output intensities with the corresponding spatial autocorrelation function $g(\Delta r)$ for two different positions of the scatterer. σ indicates the output autocorrelation length and S is the corresponding input source size. Scale bars correspond to 30 μm. (d) PDF measured in linear conditions for beams with different coherence length expressed through the input parameter S. (e) Corresponding distributions for nonlinear propagation, showing a long-tail behavior depending on the specific correlation length. Figure adapted with permission from [54].

$\lambda = 532$ nm from a 150 mW continuous-wave laser is expanded and focused on a glass diffuser plate, where transmitted radiation is collected producing a collimated speckle field. A long-working-distance objective (NA = 0.55) launches this field at the input facet of the KLTN crystal. The partially incoherent beam, linearly polarized in the experimental plane, copropagates with a background intensity along the z-axis and is detected at the output facet through a high-resolution imaging system (NA = 0.50) and a CCD camera. The incoherence properties of the input beam are achieved placing the diffuser between two confocal lenses ($f1$ and $f2$), its exact position fixing the spatial scale of the incoherence while small tilts and rotations on it generate different disordered realizations of the optical field. Examples of partially incoherent beams at the crystal input and output are reported in figure 8.6(b) and (c) for two positions of the scatterer along the propagation axis. For the output intensity distribution, we consider the spatial autocorrelation function

$$g(\Delta r) = \frac{\left\langle \int d^2 r I(r) I(r + \Delta r) \right\rangle}{\int d^2 r \langle I(r) \rangle \langle I(r + \Delta r) \rangle}, \tag{8.2}$$

whose width defines the spatial correlation length σ, i.e. the average speckle size. Since σ varies as nonlinear effects are involved in wave dynamics, we use the input source size S as a parameter characterizing the spatial incoherence of the input beam. We have $S \simeq 2\lambda l_z/g(0)\sigma$, which generalizes to nonlinear conditions optical speckle propagation [56].

In figures 8.6(d) and (e), the detected PDF of the output intensity is reported varying the beam incoherence both in the linear and nonlinear cases. In linear conditions, where no external field is applied, we observe in figure 8.6(d) no significant deviations from the Gaussian statistics as expected for completely random interfering waves. The exponential scaling is well verified in particular for beams presenting spatial coherence only on small scales ($S \approx 500$ μm, $S \approx 300$ μm). For more correlated beams ($S \approx 220$ μm, $S \approx 150$ μm), the PDF slightly deviates at large intensities, consistent with the presence of weak inhomogeneities in the phases of the elementary interfering waves [16]. Rogue waves occur as the nonlinearity is activated by means of the external field $E = 2\text{kV cm}^{-1}$. In the nonlinear case, the incoherent field experiences strong self-interaction and spatiotemporal fluctuations, that is, the speckle intensity dynamically varies in a turbulent fashion. Results for PDF as a function of the coherence length are shown in figure 8.6(e) and demonstrate how extreme events strongly depend on this parameter. We found long-tail statistics defining rogue waves and a peculiar scale-dependent behavior. Specifically, the spatial correlation scale of the optical field strongly affects its PDF, with a large enhancement in extreme wave appearance that occurs for incoherent beams of size $S \approx 220$ μm and their complete suppression at $S \approx 500$ μm. We note that the effect is approximately independent of the input power and of the value of the applied field, provided that both are above a certain threshold ensuring the highly nonlinear dynamics. Therefore, small-scale random-intensity fluctuations inhibit rogue wave generation, whereas a peculiar increase in their probability is triggered by a specific input disorder scale.

8.4.2 The characteristic scale of rogue waveforms

To investigate the mechanism underlying the correlation between abnormal wave statistics and incoherence scale, we use our ability to resolve the spatial waveform of each event with 0.3 μm resolution. The possibility of direct access to the waveform is a remarkable advantage offered by beam propagation in the spatial domain; the analogous time measurements in optical fiber are challenging [57, 58]. We first consider the case with incoherence corresponding to the maximum statistical-tail enhancement and, in particular, the rogue wave peak intensity I_P and its FWHM ΔX are analyzed. Examples of spatially resolved rogue waveforms emerging from partially incoherent intensity distributions are shown in figure 8.7(b) as giant pulses. In figure 8.7(a), we report an interesting behavior that is found for the two analyzed parameters: even though the abnormal waves span different peak intensities, their width is almost constant. Localized waves appear with the same transverse size irrespective of the fact that they populate the Gaussian portion of the PDF or the

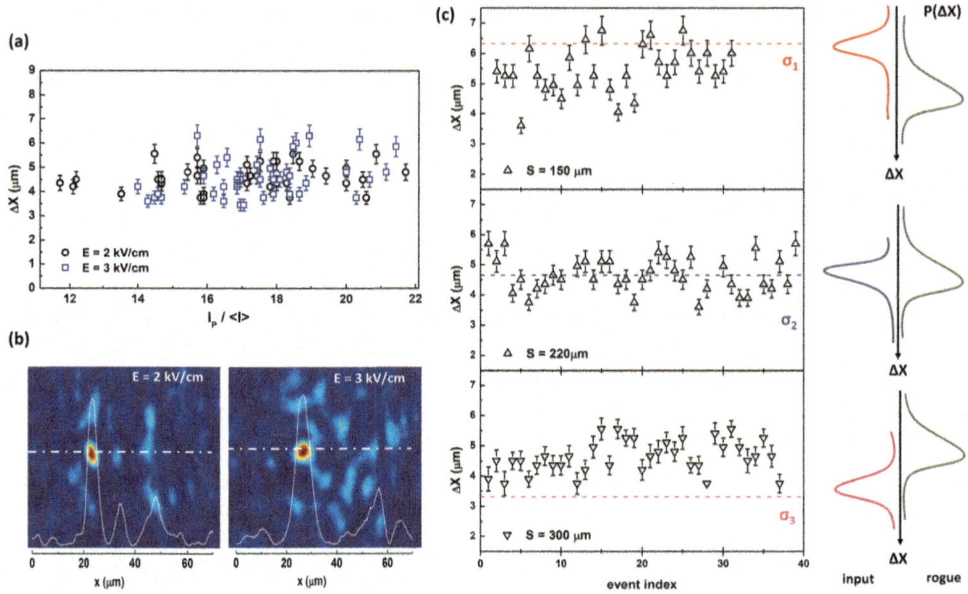

Figure 8.7. Evidence of a typical scale in rogue waveforms (a) Detected transverse width ΔX and peak intensity I_P of extreme events for data at different applied fields. (b) High-resolution spatial intensity distributions containing rogues waves. White curves are x-profiles along the dotted lines. (c) Measured extreme event widths at different coherence length σ (dashed lines). The two scales are resonant for $S = 220$ μm, that corresponds to the large enhancement in the probability of rogue wave appearance. The diagram on the right illustrates how results imply the existence of a characteristic scale for rogue waves. Figure adapted with permission from [54].

extreme one of its abnormal tail. This feature persists also at different bias fields (nonlinearity) and, as further shown in the following, it amounts to a general property of rogue waves in the saturable nonlinearity. Therefore, the analysis can be extended taking into account the width of the extreme waves as a function of the degree of incoherence of the corresponding optical field. Specifically, the typical scale ΔX is compared with the correlation scale σ of its entire intensity distribution, obtained in the nonlinear regime according to equation (8.2). This allows us to inspect whether the rogue wave has a size determined by the mean autocorrelation length of the speckle beam or an intrinsic property is involved. The whole picture is presented in figure 8.7(c). Extreme events are found to emerge on a typical scale that is significantly lower or higher than the coherence one, respectively, for beams of size $S = 150$ μm and $S = 300$ μm. Moreover, matching between these two scales is evident at $S = 220$ μm, which is exactly the case in which the large enhancement in the long-tail statistics is detected. These findings prove that the key feature providing the optimal input disorder conditions for the emergence of long-tail statistics is the existence of an intrinsic scale for rogue waves. We estimate it to be approximately $\overline{\Delta X} = 4.5$ μm. A schematic explanation is illustrated in figure 8.7(c); the coherence length distribution of the input beam acts as a probe for the probability $P(\Delta X)$ of finding rogue waves with a certain width ΔX. Their overlap, in terms of sizes, sets the amount of emerging extreme events so that the enhancement for $S = 220$ μm appears as a resonant interaction.

8.4.3 A soliton-based mechanism for the origin of rogue waves

The existence of a preferential size for extreme waves provides key information to understand statistically their appearance. Moreover, once the nonlinear propagation conditions are fixed, the spatial correlation of the optical field can be tuned to arbitrarily modify the intensity distribution tail. The generality of this mechanism relies on the physical basis that leads to a typical size for rogue events. We address the fundamental question on its origin starting from the consideration that the main properties of the photorefractive nonlinearity underlying our optical dynamics is its saturable character. Since saturation turns out in the response of any real system for large excitations, the finding may represent a universal trait in abnormal wave events, at least in this limited condition. For photorefractive media, we here provide a physical picture that not only explains the presence of a peculiar spatial scale, but also the observed insensibility to wave intensity. The framework is based on spatial solitons in saturable nonlinearities, whose interaction properties are suggested by numerical simulations in section 8.2.2 to have a leading role for rogue waves.

We begin considering the one-dimensional stationary case, where soliton solutions have the form $\psi(x, z) = u(x)e^{i\Gamma z}$ and must satisfy specific existence conditions that can be expressed in terms of the normalized input peak amplitude u_0 and width $\Delta\xi$. These constitute the soliton existence curve in the phase space of the nonlinear wave. In highly saturated conditions ($u_0 \gg 1$), solitons are known to manifest self-similarity in the form of a scale-invariance relationship $\psi(x, z) \to q^{-1}\psi(qx, q^2z)$ [59]. Experimental results in figure 8.7 show that the scale-invariance relationship is not well verified for rogue waves. This discrepancy can be explained as an unsuitability of the time-independent soliton scenario. Indeed, non-stationary dynamics is a fundamental aspect in conditions dominated by spatiotemporal fluctuations as in the turbulent regime. We thus consider the non-equilibrium counterpart of standard photorefractive solitons, i.e. transient self-trapping waves in non-stationary conditions [60]. As detailed in reference [61], in the present case, transverse localization occurs on a size

$$\Delta x \simeq \frac{3\lambda}{2\pi n^2 a_{eo}} E^{-2}, \tag{8.3}$$

where a_{eo} is a parameter quantifying the electro-optical response of the supporting material. For the presented experimental realization, from equation (8.3) we have $\Delta x = (5 \pm 1)$ μm, where the uncertainty is related to the uncertainty in a_{eo} in proximity to the ferroelectric phase transition for the biased condition. This value of Δx is consistent with the typical scale of rogue waveforms $\overline{\Delta X}$ we have found. Moreover, equation (8.3) possesses the fundamental property of being independent of the wave intensity, in agreement with observations of extreme waves in figure 8.7. Physically, unconstrained intensities occur because, for a highly saturated response, wave dynamics are locally linear. Dependence on the external electric field E is

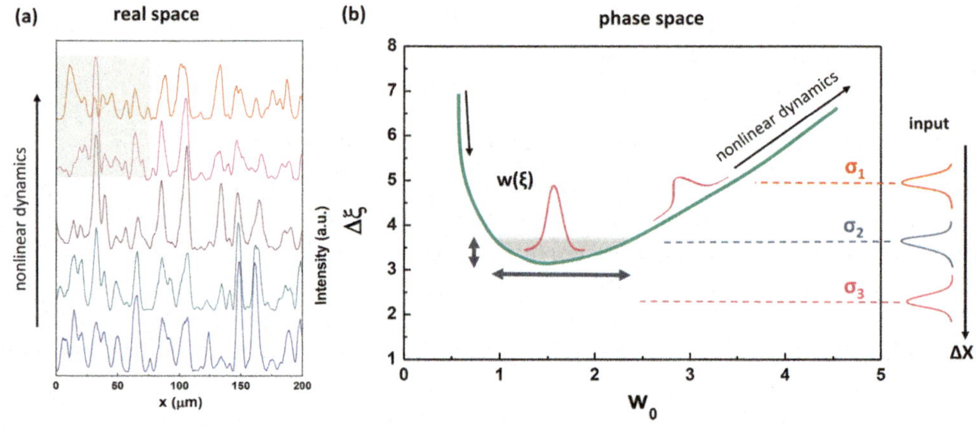

Figure 8.8. Possible mechanism underlying the appearance of rogue waves. (a) Nonlinear dynamics in the real space: experimental traces showing emergence and decaying of a rogue waves (shaded area). (b) Corresponding dynamics in the phase space of the nonlinear wave. Existence curve of non-equilibrium solitons in saturable nonlinearities (green line) on which the waveform $w(\xi)$ is schematically shown. Arrows indicate the magnitude of width and amplitude fluctuations in the gray region that is in proximity to the localized, self-trapped wave solution. For comparison with figure 8.6 and figure 8.7, input correlation lengths used in experiments are also reported. Figure adapted with permission from [54].

predicted as very weak at high values, and the result with different bias fields also verifies this feature. Therefore, observations strongly suggest that the spontaneous emergence of optical rogue waves in strong wave turbulence, along with the scale dependence of long-tail statistics on the spatial incoherence, can be understood with the mechanism illustrated in figure 8.8. When in turbulent regimes nonlinear dynamics is locally triggered by fast fluctuations of the beam intensity distribution, which are mediated by fluctuations in the nonlinear response of the material, a rogue wave may be observed to emerge and disappear in the real space (figure 8.8(a)). In the spatiotemporal phase-space of the nonlinear waves, recalled in figure 8.8(b) in terms of normalized amplitude w_0 and width $\Delta\xi$, non-equilibrium self-trapped waves form across the minimum of the existence curve according to equation (8.3). Here, a small variation in $\Delta\xi$ can lead to large fluctuations of the wave amplitude, with peak intensities reaching the giant values that populate the extreme regions of the total PDF. As for the dependence on the input coherence scale of the field, extreme events are enhanced when the input coherence scale falls in this region, whereas their excitations and suppression implies, respectively, that matching with the input autocorrelation is partial or does not occur at all.

8.5 Future developments

The experimental results presented in this chapter clearly indicate how the study of optical rogue waves in photorefractive media points out general physical properties and ideas that may be a fundamental tassel in solving the extreme event puzzle in less accessible contexts, especially from the experimental perspective. Moreover, several points remain open and represent future research directions. Among these, the extent to which rogue waves in experiments are related to a persistent out-of-equilibrium wave state and whether deviations from Gaussianity can disappear as the equilibrium state is reached should be clarified [51]. The role of the departure from integrability also deserves a deep investigation. This may allow us to understand how the evolution of exact solutions provides basic information that persists also in the dynamics of non-integrable physical systems [62]. Nonlinear waves in photorefractive media are especially promising for this purpose since quasi-integrable conditions can be experimentally obtained using a background amplitude that is much larger than the evolving wave field [39].

Another important issue is the development of a theoretical framework providing a statistical mechanical understanding of the transition to turbulence and unveiling the main properties of the turbulent regime. Nonlinear wave models taking into account the weak disorder embedded in the medium and fluctuations of its response are suggestive candidates [63]. It is relevant to note that disorder in the wave-supporting media does not prevent the dynamics of quasi-solitons [64]. Conversely, the effect of disorder has been proposed to favor and spatially constrain rogue formation [65], a suggestion that can be verified in photorefractive propagation, where soliton modes can form also as self-sustained defects in a generically nonlinear optical potential [66]. The experimental investigation of shot-to-shot correlations may provide an interesting contribution to this effort, eventually revealing the presence of a phase-locking of different spontaneous modes in the turbulent regime [67].

That said, the central point in the study of extreme phenomena is their predictability, especially for those we would like to avoid, such as ocean rogue waves. Optical rogue events can be useful for this purpose if common and universal traits exist in their underlying dynamics. Comparing the dynamical features of various extreme events in optics and hydrodynamics, such as the chaotic nature and the predictability of the process, has shown them to be generally different [68]. A behavior analogous to hydrodynamic and oceanic rogue waves in terms of correlations and statistical mechanics properties has been found for rogue waves in photorefractive media [69]; unfortunately, both consist of high-dimensional correlation dynamics whose complexity hampers long-range prediction. This makes useful extreme-event prediction based on standard dynamical methods just as difficult in the optical case as in related contexts. However, the existence of intrinsic correlations in the process, which makes it different from a stochastic evolution, is a crucial point as it justifies efforts aimed at developing a probabilistic approach to prediction in which large datasets are needed [70, 71]. At the same time, this leaves open the possibility of controlling extreme events and exploiting them in applications in optics and beyond.

Funding is recognized from the *Sapienza Ateneo 2016* and *Sapienza Ateneo 2017* project.

References

[1] Clauset A, Shalizi C R and Newman M E 2009 Power-law distributions in empirical data *SIAM Rev.* **51** 661

[2] Kharif C, Pelinovsky E and Slunyaev A 2006 *Rogue Waves in the Ocean* (Berlin: Springer)
[3] Adcock T A and Taylor P H 2014 The physics of anomalous ('rogue') ocean waves *Rep. Prog. Phys.* **77** 105901
[4] Fedele F, Brennan J, de Leon S P, Dudley J and Dias F 2016 Real world ocean rogue waves explained without the modulational instability *Sci. Rep.* **6** 27715
[5] Toffoli A, Proment D, Salman H, Monbaliu J, Frascoli F, Dafilis M, Stramignoni E, Forza R, Manfrin M and Onorato M 2017 Wind generated rogue waves in an annular wave flume *Phys. Rev. Lett.* **118** 144503
[6] Onorato M, Residori S, Bortolozzo U, Montina A and Arecchi F T 2013 Rogue waves and their generating mechanisms in different physical contexts *Phys. Rep.* **528** 47
[7] Dudley J M, Dias F, Erkintalo M and Genty G 2014 Instabilities, breathers and rogue waves in optics *Nat. Photonics* **8** 775
[8] Akhmediev N *et al* 2016 Roadmap on optical rogue waves and extreme events *J. Opt.* **18** 063001
[9] Solli D R, Ropers C, Koonath P and Jalali B 2007 Optical rogue waves *Nature* **450** 1054
[10] Birkholz S, Nibbering E T J, Bree C, Skupin S, Demircan A, Genty G and Steinmeyer G 2013 Spatiotemporal rogue events in optical multiple filamentation *Phys. Rev. Lett.* **111** 243903
[11] Lecaplain C, Grelu Ph, Soto-Crespo J M and Akhmediev N 2012 Dissipative rogue waves generated by chaotic pulse bunching in a mode-locked laser *Phys. Rev. Lett.* **108** 233901
[12] Onorato M *et al* 2009 Statistical properties of directional ocean waves: the role of the modulational instability in the formation of extreme events *Phys. Rev. Lett.* **102** 114502
[13] Shats M, Punzmann H and Xia H 2010 Capillary rogue waves *Phys. Rev. Lett.* **104** 104503
[14] Armaroli A, Conti C and Biancalana F 2015 Rogue solitons in optical fibers: a dynamical process in a complex energy landscape? *Optica* **2** 497
[15] Toenger S, Godin T, Billet C, Dias F, Erkintalo M, Genty G and Dudley J M 2016 Emergent rogue wave structures and statistics in spontaneous modulation instability *Sci. Rep.* **5** 0380
[16] Arecchi F T, Bortolozzo U, Montina A and Residori S 2011 Granularity and inhomogeneity are the joint generators of optical rogue waves *Phys. Rev. Lett.* **106** 153901
[17] Hohhmann R, Kuhl U, Stickmann H-J, Kaplan L and Heller E J 2010 Freak waves in the linear regime: a microwave study *Phys. Rev. Lett.* **104** 093901
[18] Liu C, van derWel R E C, Rotenberg N, Kuipers L, Krauss T F, Di Falco A and Fratalocchi A 2015 Triggering extreme events at the nanoscale in photonic seas *Nat. Physics* **11** 358
[19] Leonetti M and Conti C 2015 Observation of three dimensional optical rogue waves through obstacles *Appl. Phys. Lett.* **106** 254103
[20] Ganshin A N, Efimov V B, Kolmakov G V, Mezhov-Deglin L V and McClintock P V 2008 Observation of an inverse energy cascade in developed acoustic turbulence in superfluid helium *Phys. Rev. Lett.* **101** 065303
[21] Hammani K, Kibler B, Finot C and Picozzi A 2010 Emergence of rogue waves from optical turbulence *Phys. Lett.* A **374** 3585
[22] Walczak P, Randoux S and Suret P 2015 Optical rogue waves in integrable turbulence *Phys. Rev. Lett.* **114** 143903
[23] Bonatto C, Feyereisen M, Barland S, Giudici M, Masoller C, Rios Leite J R and Tredicce J R 2011 Deterministic optical rogue waves *Phys. Rev. Lett.* **107** 053901

[24] Selmi F, Coulibaly S, Loghmari Z, Sagnes I, Beaudoin G, Clerc M G and Barbay S 2016 Spatiotemporal chaos induces extreme events in an extended microcavity laser *Phys. Rev. Lett.* **116** 013901

[25] Safari A, Fickler R, Padgett M J and Boyd R W 2017 Generation of caustics and rogue waves from nonlinear instability *Phys. Rev. Lett.* **119** 203901

[26] Montina A, Bortolozzo U, Residori S and Arecchi F T Non-gaussian statistics and extreme waves in a nonlinear optical cavity *Phys. Rev. Lett.* **103** 173901

[27] Chen Z, Segev M and Christodoulides D N 2012 Optical spatial solitons: historical overview and recent advances *Rep. Prog. Phys.* **75** 086401

[28] Wan W, Jia S and Fleischer J W Dispersive superfluid-like shock waves in nonlinear optics *Nat. Phys.* **3** 46

[29] Schwartz T, Bartal G, Fishman S and Segev M 2007 Transport and Anderson localization in disordered two-dimensional photonic lattices *Nature* **446** 52

[30] Sun C, Jia S, Barsi C, Rica S, Picozzi A and Fleischer J W 2012 Observation of the kinetic condensation of classical waves *Nat. Phys.* **8** 470

[31] DelRe E, Spinozzi E, Agranat A J and Conti C 2011 Scale-free optics and diffractionless waves in nanodisordered ferroelectrics *Nat. Photonics* **5** 39

[32] DelRe E, DiMei F, Parravicini J, Parravicini G B, Agranat A J and Conti C 2015 Subwavelength anti-diffracting beams propagating over more than 1,000 Rayleigh lengths *Nat. Photonics* **9** 228

[33] DiMei F, Caramazza P, Pierangeli D, Di Domenico G, Ilan H, Agranat A J, Di Porto P and DelRe E 2016 Intrinsic negative mass from nonlinearity *Phys. Rev. Lett.* **116** 153902

[34] Günter P and Hiugnard J P 2006 *Photorefractive Materials and Their Application 1* (Berlin: Springer)

[35] Pierangeli D, Parravicini J, DiMei F, Parravicini G B, Agranat A J and DelRe E 2014 Photorefractive light needles in glassy nanodisordered KNTN *Opt. Lett.* **39** 1657

[36] Pierangeli D, Di Mei F, Conti C, Agranat A J and DelRe E 2015 Spatial rogue waves in photorefractive ferroelectrics *Phys. Rev. Lett.* **115** 093901

[37] Marsal N, Caullet V, Wolfersberger D and Sciamanna M 2014 Spatial rogue waves in a photorefractive pattern-forming system *Opt. Lett.* **39** 3690

[38] Goodman J W 1975 *Statistical Properties of Laser Speckle Patterns* (Berlin–Heidelberg: Springer)

[39] DelRe E, Crosignani B and Di Porto P 2009 Photorefractive solitons and their underlying nonlocal physics *Prog. Opt.* **53** 153

[40] DelRe E, D'Ercole A and Palange E 2005 Mechanisms supporting long propagation regimes of photorefractive solitons *Phys. Rev.* E **71** 036610

[41] Pierangeli D, Di Mei F, Di Domenico G, Agranat A J, Conti C and DelRe E 2016 Turbulent transitions in optical wave propagation *Phys. Rev. Lett.* **117** 183902

[42] Picozzi A, Garnier J, Hansson T, Suret P, Randoux S, Millot G and Christodoulides D N 2014 Optical wave turbulence: towards a unified nonequilibrium thermodynamic formulation of statistical nonlinear optics *Phys. Rep.* **542** 1–132

[43] Laurie J, Bortolozzo U, Nazarenko S and Residori S 2012 One-dimensional optical wave turbulence: Experiment and theory *Phys. Rep.* **514** 121

[44] Robinson P A 1997 Nonlinear wave collapse and strong turbulence *Rev. Mod. Phys.* **69** 509

[45] Mlejnek M, Kolesik M, Moloney J V and Wright E M 1999 Optically turbulent femtosecond light guide in air *Phys. Rev. Lett.* **83** 2938

[46] Turitsyna E G, Smirnov S V, Sugavanam S, Tarasov N, Shu S, Babin S A, Podivilov E V, Churkin D V, Falkovich G and Turitsyn S K 2013 The laminar-turbulent transition in a fibre laser *Nat. Photon.* **7** 783–6

[47] Xu G, Vocke D, Faccio D, Garnier J, Roger T, Trillo S and Picozzi A 2015 From coherent shocklets to giant collective incoherent shock waves in nonlocal turbulent flows *Nat. Commun.* **6** 8131

[48] Mamaev A V, Saffman M, Anderson D Z and Zozulya A A 1996 Propagation of light beams in anisotropic nonlinear media: From symmetry breaking to spatial turbulence *Phys. Rev.* A **54** 870

[49] Shih M F, Jeng C C, Sheu F W and Lin C Y 2002 Spatiotemporal optical modulation instability of coherent light in noninstantaneous nonlinear media *Phys. Rev. Lett.* **88** 133902

[50] Wabnitz S 2014 Optical turbulence in fiber lasers *Opt. Lett.* **39** 1362

[51] Agafontsev D S and Zakharov V E 2015 Integrable turbulence and formation of rogue waves *Nonlinearity* **28** 2791

[52] Solli D R, Herink G, Jalali B and Ropers C 2012 Fluctuations and correlations in modulation instability *Nat. Photon.* **6** 463

[53] Soto-Crespo J M, Devine N and Akhmediev N 2016 Integrable turbulence and rogue waves: breathers or solitons? *Phys. Rev. Lett.* **116** 103901

[54] Pierangeli D, Musarra G, Di Mei F, Di Domenico G, Agranat A J, Conti C and DelRe E 2016 Enhancing optical extreme events through input wave disorder *Phys. Rev.* A **94** 063833

[55] Mitchell M, Chen Z, Shih M F and Segev M 1996 Self-trapping of partially spatially incoherent light *Phys. Rev. Lett.* **77** 490

[56] Bromberg Y, Lahini Y, Small E and Silberberg Y 2010 Hanbury Brown and Twiss interferometry with interacting photons *Nat. Photon.* **4** 721–6

[57] Suret P, El Koussaifi R, Tikan A, Evain C, Randoux S, Szwaj C and Bielawski S 2016 Single-shot observation of optical rogue waves in integrable turbulence using time microscopy *Nat. Commun.* **7** 13136

[58] Narhi M, Wetzel B, Billet C, Toenger S, Sylvestre T, Merolla J-M, Morandotti R, Dias F, Genty G and Dudley J M 2016 Real-time measurements of spontaneous breathers and rogue wave events in optical fibre modulation instability *Nat. Commun.* **7** 13675

[59] DelRe E, D'Ercole A and Agranat A J 2003 Emergence of linear wave segments and predictable traits in saturated nonlinear media *Opt. Lett.* **28** 260

[60] Zozulya A and Anderson D 1995 Nonstationary self-focusing in photorefractive media *Opt. Lett.* **22** 837

[61] DelRe E and Palange E 2006 Optical nonlinearity and existence conditions for quasi-steady-state photorefractive solitons *J. Opt. Soc. Am.* B **23** 2323

[62] Grinevich P G and Santini P M 2017 The finite gap method and the analytic description of the exact rogue wave recurrence in the periodic NLS Cauchy problem arXiv:1707.05659v2

[63] Conti C and Leuzzi L 2011 Complexity of waves in nonlinear disordered media *Phys. Rev.* B **83** 134204

[64] Conti C 2005 Complex light: Dynamic phase transitions of a light beam in a nonlinear nonlocal disordered medium *Phys. Rev.* E **72** 066620

[65] Saleh M F, Conti C and Biancalana F 2017 Anderson localisation and optical-event horizons in rogue-soliton generation *Opt. Express* **25** 5457

[66] Pierangeli D, Flammini M, Di Mei F, Parravicini J, de Oliveira C E M, Agranat A J and DelRe E 2015 Continuous solitons in a lattice nonlinearity *Phys. Rev. Lett.* **114** 203901

[67] Pierangeli D, Tavani A, Di Mei F, Agranat A J, Conti C and DelRe E 2017 Observation of replica symmetry breaking in disordered nonlinear wave propagation *Nat. Commun.* **8** 1501

[68] Birkholz S, Bree C, Demircan A and Steinmeyer G 2015 Predictability of Rogue Events *Phys. Rev. Lett.* **114** 213901

[69] Pierangeli D, Di Mei F, Conti C and DelRe E 2015 Evidence of the universal dynamics of rogue waves arXiv:1511.01390

[70] Birkholz S, Bree C, Veselic I, Demircan A and Steinmeyer G 2016 Ocean rogue waves and their phase space dynamics in the limit of a linear interference model *Sci. Rep.* **6** 35207

[71] Hadjihosseini A, Wachter M, Hoffmann N P and Peinke J 2016 Capturing rogue waves by multi-point statistics *New J. Phys.* **18** 013017

Chapter 9

Vector rogue waves driven by polarisation instabilities

S V Sergeyev, H Kbashi, C Mou, A Martinez, S Kolpakov, V Kalashnikov

9.1 Introduction

Over the last two decades, waves with anomalous amplitude (rogue waves) have been a focus of researchers from different fields ranging from oceanography to laser physics [1–23]. In the scientific and social context, revealing underlying, hidden mechanisms of rogue wave (RW) emergence may enable forecasting and localisation of the source of the extreme events and so reduce their detrimental impact on society. To quantify RWs, the following criteria have been suggested: (i) probability distribution functions for RW amplitudes should have 'fat' tails significantly deviating from Gaussian or Rayleigh distributions; and (ii) RWs should have amplitudes more than twice as large as the mean amplitude of the highest third of the waves (significant wave height), or RW amplitude should exceed the standard deviation of the ocean's surface variations by more than eight times [3].

The major obstacle to studying rogue waves in the context of prediction and mitigation are the scarcity of these rare events and the inability to perform full-scale experiments on RW emergence in real-world scenarios. Mode-locked fibre lasers, as a source of pico- and femtosecond pulses with MHz repetition rates, provide a unique opportunity to observe more data on rogue waves in the short term (compared to the timescale of RWs in other systems such as in the ocean and financial markets [1–3]) and under laboratory-controlled conditions [7–12, 15–23]. Previously it has been found that RWs can be generated in mode-locked lasers in the timescale of a roundtrip [8–12] due to soliton–soliton interaction through the overlapping of their tails or soliton–dispersive wave interaction [24, 25]. The result of these interactions is coupling enhancement that leads to chaotic pulse bunching (soliton rain) in the timescale of a roundtrip [7–12]. This makes the scenario of the RW emergence similar to the dynamics of the coupled nonlinear oscillators, where

doi:10.1088/978-0-7503-1460-2ch9

© IOP Publishing Ltd 2017

desynchronisation causes random relative phase slips leading to the appearance of rogue waves [26–28].

Models of networks of coupled oscillators are widely used to study different synchronisation scenarios, from phase locking to chaotic phase drift, in the context of applications from biomedicine to laser physics [26–28]. For the phase-coupled oscillators, evolution can be considered based on the Kuramoto equation:

$$\frac{d\varphi_i}{dt} = \Omega_i + \frac{K}{N} \sum_{i=1, i \neq j}^{N} G(\varphi_i - \varphi_j). \tag{9.1}$$

Here, φ_i and Ω_i are the phase and the frequency of oscillator, K is the coupling strength, and G is the coupling function [26–28]. In the simplest case of two coupled oscillators with $G = \sin(\varphi_i - \varphi_j)$, equation (9.1) is called the Adler equation [26–28]. For a comprehensive study of the synchronisation scenarios in lasers, the amplitude dynamics along with dynamics of the population inversion have to be included in consideration, which results in a new class of coupled system, i.e. coupled oscillators with dynamic coupling strength [27]. With accounting for the vector nature of pump and lasing field interaction with active medium, orthogonal states of polarisation can be treated as coupled oscillators as well [29–35]. This provides a very good test bed for the study of synchronisation and desynchronisation of the longitudinal modes and the orthogonal states of polarisation (SOPs) in a fibre laser as a function of the laser parameters (power and ellipticity of the pump wave, and in-cavity birefringence and twist) tunable by the pump laser current driver, polarisation controller for the pump wave, and in-cavity polarisation controller [29–35]. Here, for erbium-doped lasers mode-locked with carbon nanotubes, we review our recent experimental and theoretical results and demonstrate that coupling of orthogonal SOPs provides an additional soliton–soliton and soliton–dispersive wave interaction mechanism leading to the RW emergence [16, 17]. To isolate such interaction from the contribution of the soliton–soliton and soliton–dissipative wave interactions, we used a low-pass filter with 1 MHz bandwidth. As a result, we found a new type of RW emerging at a slow timescale (tens of thousands of roundtrips) in the form of slow bright and dark rogue waves (BDRWs). With the help of a new vector model, we obtained theoretical results which correspond to the experimental data. Our theoretical analysis demonstrates that orthogonally polarised SOPs can be considered as quasi-equilibrium points, where a typical trajectory traverses the neighbourhood of one of the orthogonal SOPs with further switching to the other SOP. The dwelling time for the trajectory near each orthogonal SOP is defined by the cavity anisotropy, i.e. by the pump wave's ellipticity adjustable with the help of the pump polarisation controller [17]. The escape from the neighbourhood of each SOP is driven by the in-cavity birefringence, tunable with the help of the in-cavity polarisation controller, and leads to the output power distribution satisfying the BDRW criteria. In addition to the aforementioned results, we demonstrate that, for a long fibre laser without generic mode-locking mechanism, the tuning of in-cavity fibre birefringence and twist can result in different synchronisation scenarios ranging from mode locking to random multi-pulsing satisfying rogue wave criteria [18–22].

The chapter is organised as follows. Section 9.2 is dedicated to the experimental and theoretical study of rogue waves driven by polarisation instability in Er-doped fibre lasers mode-locked with carbon nanotubes. In section 9.3, for a long-cavity (615 m) Er-doped fibre laser, we provide analysis of the longitudinal mode synchronisation and desynchronisation caused by tuning of in-cavity fibre birefringence and twist.

9.2 Bright and dark rogue waves in mode-locked fibre laser

9.2.1 Experimental results

The laser setup is shown in figure 9.1. Details of the setup specification are given in annex I. The fast dynamics comprises 11 oscillograms with 20 K samples in each, and a 25 ps sample interval recorded with help of a fast photodetector and oscilloscope. Slow dynamics includes 20 traces with 1024 samples in each and 1 μs sample interval obtained with the help of a polarimeter. The polarisation controllers (POC1 and POC2) have been tuned to find conditions for RW emergence. The additional results obtained, with the help of autocorrelator and optical spectrum analyser, are discussed in annex I.

The results for the fast dynamics are shown in figures 9.2(a1) and (b1) for the pump currents J_p = 140 mA and figures 9.2(a2) and (b2) for 170 mA [16]. As shown in figures 9.2(a1) and (a2), the voltage offset in the oscilloscope results in minimum voltage taking the negative values. The multi-pulse dynamics in figures 9.2(a1) and (a2) are similar to the soliton rain, with the main difference in low pump power of 67 mW (I_p = 140 mA) used here as compared to the 800 mW used by the other authors [8–12]. When pump current is increased to 170 mA, the number of pulses in a bunch increases (figure 9.2(a2)). For both cases, the dynamics are stable within each frame, changing from frame to frame. Probability distribution histograms are shown in figures 9.2(b1) and (b2). Accounting for output amplitude normalisation, RW criterion shows as $V_n > 8$, and so both cases shown in figures 9.2(b1) and (b2) satisfy such criteria. Soliton–soliton interaction leading to RW appearance in the form of soliton rain on the scale of a few roundtrips is discussed in annex I [16]. As the total number of roundtrips measured by oscilloscope is 171, a low-pass filter

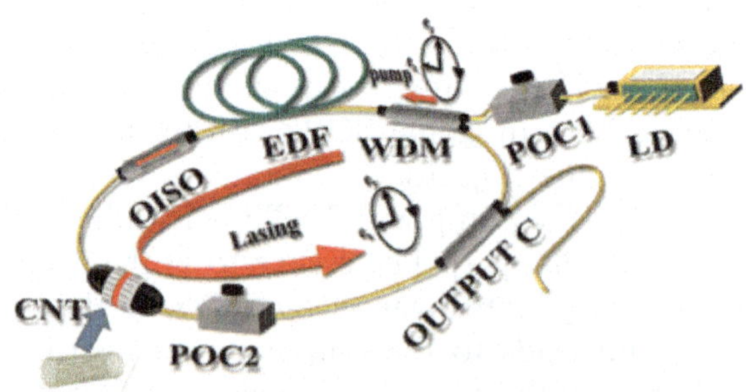

Figure 9.1. Schematic of a fibre laser mode-locked with carbon nanotubes: erbium-doped fibre (EDF); optical isolator (OISO); laser diode (LD); 980/1550 wavelength division multiplexer (WDM); carbon-nanotube-based mode locker (CNT); polarisation controllers (POC1 and POC2); output coupler (output C). Reproduced from [17].

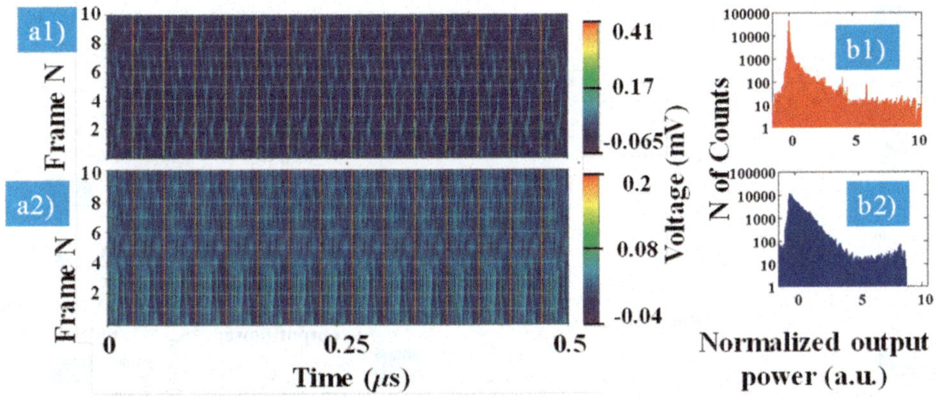

Figure 9.2. Data recorded by the oscilloscope [15]: (a1)–(a2) 11 slices of the oscillogram (25 ps sample interval, 16 roundtrips per slice); (b1)–(b2) probability distribution histograms (each output voltage V is normalised as $V_n = (V - \text{median}(V))/\sigma(V)$, where $\sigma(V)$ is the standard deviation; RW criterion is $V_n > 8$. Parameters: (a1), (b1) J_p = 140 mA; (a2), (b2) J_p = 170 mA.

cannot be applied to filter out such interactions to reveal a slow dynamics at the timescale of 30 roundtrips.

To reveal the RW emergence at the slow timescales of 1 μs–20 ms (660 K roundtrips), we used a polarimeter to record 20 traces with 1024 samples in each and 1 μs sample interval. As a result, we found the degree of polarisation (DOP), output power, and normalised Stokes parameters s_1, s_2, s_3, which are related to the output powers of two linearly cross-polarised SOPs I_x and I_y, and the phase difference between them $\Delta\phi$ as follows:

$$S_0 = I_x + I_y, \quad S_1 = I_x - I_y, \quad S_2 = 2\sqrt{I_x I_y}\cos\varphi, \quad S_3 = 2\sqrt{I_x I_y}\sin\Delta\varphi,$$
$$s_i = \frac{S_i}{\sqrt{S_1^2 + S_2^2 + S_3^2}}, \qquad DOP = \frac{\sqrt{S_1^2 + S_2^2 + S_3^2}}{S_0}, \quad (i = 1, 2, 3). \tag{9.2}$$

The obtained results are shown in figure 9.3. As follows from figure 9.3(a1) for J_p = 140 mA, the spikes in the output power correspond to the spikes in DOP and the phase difference jumps in π. This means that the spikes are result of the random switching between orthogonally polarised SOPs as shown in figures 9.3(b1) and (c1). By increasing pump current to J_p = 170 mA and tuning POC2, we obtained results whereby the output power oscillated without the presence of RW events (figure 9.3(a2)). Thus, a high degree of polarisation and oscillating (between 0 and π) rather than randomly evolving phase difference can be considered to be a signature of the absence of RW events.

Probability distribution histograms (PDH) for the total output power $I_x + I_y$, output powers I_x and I_y for the pump currents of 140 and 170 mA, are shown in figures 9.3(d1)–(f1) and (d2)–(f2). As follows from figures 9.3(d1)–(f1), the PDHs demonstrate the presence of so-called 'bright' (spikes in figure 9.3(a1)) and 'dark' (dips in figure 9.3(a1)) [13, 14] RWs driven by the interaction of two orthogonal

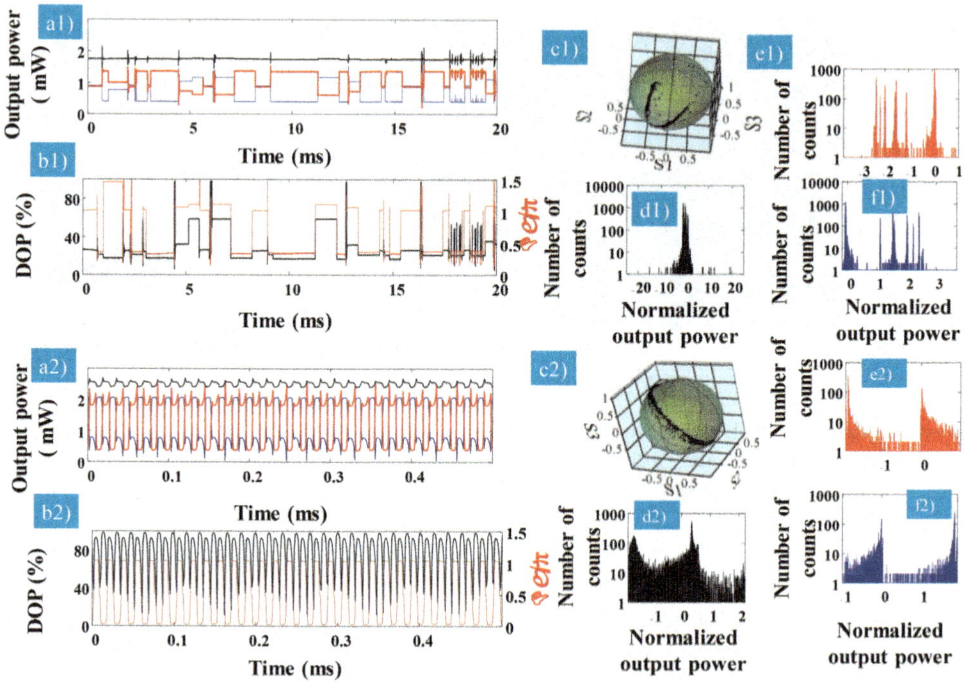

Figure 9.3. (a)–(c) Polarisation measurements (1 μs resolution, i.e. averaging over approximately 33 roundtrips, 20 slices with 1024 points per slice): (a1)–(a2) output powers versus time: I_x (red), I_y (blue) and $I = I_x + I_y$ (black); (b1)–(b2) phase differences (red) and DOP (black) versus time; (c1)–(c2) trajectories in normalised Poincaré sphere. (d)–(f) Probability distribution histograms for the total output power $I = I_x + I_y$ (d1, d2), I_x (e1, e2), and I_y (f1, f2). Each output power I_i (I, I_x, I_y) is normalised as shown in figure 9.2. Parameters: (a1)–(f1) J_p = 140 mA; (a2)–(f2) J_p = 170 mA.

SOPs. The PDHs in figures 9.3(e1) and (f1) demonstrate that there are no RW statistics for the output powers I_x and I_y. The 'bright' and 'dark' RW emergence is a result of interaction of the orthogonal SOP leading to the 'antiphase' dynamics of the output powers I_x and I_y (figure 9.3(b1)). In addition, we found that there is no RW in the case of regular phase difference oscillations between 0 and π (figures 9.3(d2)–(f2)). Thus, the experimental data for the fast (roundtrip scale) and slow (tens of thousands of roundtrips) timescales demonstrate: (i) soliton rain (fast timescale) and 'bright' and 'dark' RWs (slow timescale) for I_p = 140 mA, and (ii) soliton rain (fast timescale) and no RWs (slow timescale) for I_p = 170 mA.

9.2.2 Theoretical results

To justify that the RW emergence at the slow timescale is driven by polarisation instabilities, we use the vector model of an erbium-doped fibre laser mode-locked with CNT developed by Sergeyev and co-workers [16, 17, 29, 30, 35]. The model describes the evolution of the lasing field for two orthogonal SOPs ($u = \sqrt{I_x}\exp(i\phi_x)$, $v = \sqrt{I_x}\exp(i\phi_y)$, $\Delta\phi = \phi_x - \phi_y$) averaged over the pulse width and the population of the first excited level in the Er^{3+}-doped active medium [16, 17, 29, 30, 35]. We introduced a slow-time variable $t_s = z/(V_g t_R)$, where $t_R = L/V_g$ is the photon roundtrip time and L is the cavity length, and assumed an ansatz in the form $E_{x,y}(t, t_s) = u(t_s)\mathrm{sech}(t/T_p)$, $E_x(t, t_s) = v(t_s)\mathrm{sech}(t/T_p)$ (where T_p is the pulse width). After averaging over the time $T_p \ll t \ll t_R$, we obtained the following equations [16, 17, 29, 30, 35]:

$$
\begin{aligned}
\frac{d|u|}{dt} &= \left(\frac{\alpha_1(f_1 + f_2)}{1 + \Delta^2} - \alpha_2 - F(|u|, |v|)\right)|u| + \frac{\alpha_1 f_3}{1 + \Delta^2}|v|\cos(\Delta\varphi) \\
&\quad - \frac{\alpha_1 f_3 \Delta}{1 + \Delta^2}|v|\sin(\Delta\varphi) + \frac{2\gamma}{3}|v|^2|u|\sin(2\Delta\varphi), \\
\frac{d|v|}{dt} &= \left(\frac{\alpha_1(f_1 - f_2)}{1 + \Delta^2} - \alpha_2 - F(|u|, |v|)\right)|v| + \frac{\alpha_1 f_3}{1 + \Delta^2}|u|\cos(\Delta\varphi) \\
&\quad + \frac{\alpha_1 f_3 \Delta}{1 + \Delta^2}|u|\sin(\Delta\varphi) - \frac{2\gamma}{3}|u|^2|v|\sin(2\Delta\varphi), \\
\frac{d\Delta\varphi}{dt} &= \Delta\Omega + K_{NL}\sin^2(\Delta\varphi) + K_s\sin(\Delta\varphi) + K_{as}\cos(\Delta\varphi), \\
\frac{df_1}{dt} &= \varepsilon\left[\frac{(\chi - 1)I_p}{2} - 1 - \left(1 + \frac{I_p}{2} + d_1 S_0\right)f_1 - \left(d_1 S_1 + \frac{I_p}{2}\frac{(1 - \delta^2)}{(1 + \delta^2)}\right)f_2 - d_1 S_2 f_3\right], \\
\frac{df_2}{dt} &= \varepsilon\left[\frac{(1 - \delta^2)}{(1 + \delta^2)}\frac{I_p(\chi - 1)}{4} - \left(\frac{I_p}{2} + 1 + d_1 S_0\right)f_2 - \left(\frac{(1 - \delta^2)}{(1 + \delta^2)}\frac{I_p}{2} + d_1 S_1\right)\frac{f_1}{2}\right], \\
\frac{df_3}{dt} &= -\varepsilon\left[\frac{d_1 S_2 f_1}{2} + \left(\frac{I_p}{2} + 1 + d_1 S_0\right)f_3\right], \\
\Delta\Omega &= \frac{2\alpha_1 f_2 \Delta}{1 + \Delta^2} - 2\beta,\ K_{NL} = \frac{\gamma L I_{ss}}{6}(|v|^2 - |u|^2), \\
K_s &= -\frac{2(|v|^2 + |u|^2)}{|u||v|}\frac{\alpha_1 f_3}{(1 + \Delta^2)},\ K_{as} = \frac{2(|v|^2 - |u|^2)}{|u||v|}\frac{\alpha_1 f_3 \Delta}{(1 + \Delta^2)}, \\
F(|u|, |v|) &= \frac{\alpha_0}{\pi}\left(1 - \sqrt{\frac{\alpha_s(|u|^2 + |v_2|)}{1 + \alpha_s(|u|^2 + |v|^2)}}\arctan\left(\sqrt{\frac{\alpha_s(|u|^2 + |v|^2)}{1 + \alpha_s(|u|^2 + |v|^2)}}\right)\right).
\end{aligned}
\tag{9.3}
$$

Here, time is normalised to the roundtrip time; pump and lasing powers (I_p and $|u|^2$, $|v|^2$) are normalised to the corresponding saturation powers [16, 17, 29, 30, 35]. α_1 is the total absorption of erbium ions at the lasing wavelength; α_0 is the CNT absorption coefficient at the lasing wavelength; α_s is the ratio of saturation powers for CNT and erbium-doped fibre; α_2 is the total insertion losses in cavity; β is the birefringence strength ($2\beta = 2\pi L/L_b$, where L_b is the beat length); $\gamma = \gamma_k L I_{ss}$ (where γ_k is the Kerr coupling constant); δ is the ellipticity of the pump wave; $\varepsilon = t_R/\tau_{Er}$ is the ratio of the roundtrip time t_R to the lifetime of erbium ions at the first excited level τ_{Er}; $\chi = (\sigma_a^{(L)} + \sigma_e^{(L)})/\sigma_a^{(L)}$, where $\sigma_a^{(L)}$ and $\sigma_e^{(L)}$ are absorption and emission cross sections at the lasing wavelength; Δ is the detuning of the lasing wavelength with respect to the maximum of the gain spectrum (normalised to the gain spectral width); $d_1 = \chi/((1+\Delta^2))$; functions $f_i (i = 1,2,3)$ are related to the angular distribution of the dipole moments of the excited ions $n(\theta)$ expanded into a Fourier series [16, 17, 29, 30, 35]:

$$n(\theta) = \frac{n_0}{2} + \sum_{k=1}^{\infty} n_{1k}\cos(k\theta) + \sum_{k=1}^{\infty} n_{2k}\sin(k\theta), \quad f_1 = \left(\chi\frac{n_0}{2} - 1\right) + \chi\frac{n_{12}}{2}, \quad f_2 = \left(\chi\frac{n_0}{2} - 1\right) - \chi\frac{n_{12}}{2}, \quad f_3 = \chi\frac{n_{22}}{2}. \tag{9.4}$$

Equation (9.4) uses the assumption that the dipole moments of the absorption and emission transitions for erbium-doped silica are located in the plane orthogonal to the direction of the light propagation. This approximation results in the finite dimensional system presented by equation (9.3), where only the n_0, n_{12}, and n_{21} components contribute to the dynamics. In contrast, the more general assumption of the 3D orientation distribution of the dipole moments' orientation [31–34] results in an infinite-dimensional system with the more complex expansion of the angular distribution of the dipole moments into series of the Wigner's D^l_{mn} or spherical Y_{mn} functions [34].

We solved equation (9.3) numerically for 660 K roundtrips (20 ms) using parameters quite close to the experimental: $\alpha_1 = 21.53$, $\alpha_0 = 0.136$, $\alpha_s = 1.8{\cdot}10^{-5}$, $\alpha_2 = 2.533$, $\chi = 2.3$, $\Delta = 0.1$, $I_p = 75$, $\gamma = 2 \times 10^{-6}$, and $\varepsilon = 10^{-4}$. The results for $\delta = 0.85$ (elliptically polarised pump), $\beta = \pi \cdot 10^{-2}$, $I_p = 75$, and $\delta = 1$ (circularly polarised pump), $\beta = 0.8\pi \cdot 10^{-2}$, $I_p = 80$ shown in figure 9.4 demonstrate the interplay between in-cavity birefringence and the ellipticity of the pump wave in the context of the RW emergence [15]. As follows from figure 9.4(a1), the dynamics for the elliptically polarised pump take the form of the chaotic oscillations with the probability distribution histogram satisfying rogue wave criteria. Low-pass filtering of the theoretical data over 30 roundtrips corresponds to the experimental characterisation based on a polarimeter with 1 MHz resolution (figure 9.3).

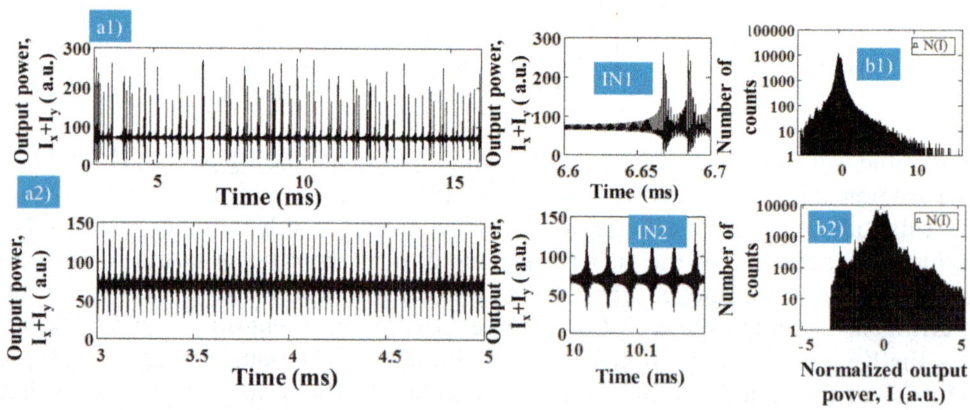

Figure 9.4. (a1)–(a2) Dynamics of the output power averaged over the pulse width; (b1)–(b2) probability distribution histograms (the output power I is normalised as shown in figure 9.2). (IN1–IN2) enlarged fragment of the output power dynamics at 100 and 150 μs intervals. Parameters: $I_p = 75$, $\alpha_1 = 21.53$, $\alpha_0 = 0.136$, $\alpha_s = 1.8{\cdot}10^{-5}$, $\alpha_2 = 2.533$, $\chi = 2.3$, $\Delta = 0.1$, $\gamma = 2 \times 10^{-6}$, $\varepsilon = 10^{-4}$; (a1), (b1) $\delta = 0.85$ (elliptically polarised pump), $\beta = \pi \cdot 10^{-2}$; (a2), (b2) $\delta = 1$ (circularly polarised pump), $\beta = 0.8\,\pi \cdot 10^{-2}$

The obtained results shown in figure 9.5 are in good correspondence with the experimental data (figure 9.3): the extreme spikes in the output power (figure 9.5(a1)) correspond to the spikes in DOP and phase difference jumps in π (figure 9.5(b1)) leading to the transitions between orthogonally polarised SOPs (figure 9.5(c1)). Asymmetrical probability distribution histograms for the powers I_x and I_y (figures 9.5(e1) and (f1)) and histogram for the total power $I = I_x + I_y$ (figure 9.4(d1)) correspond to the histograms shown for $J_p = 140$ mA in polarimetry experiments (figure 9.3) in the context of presence of 'dark' and 'bright' RWs. For the circularly polarised pump ($\delta = 1$) and $I_p = 80$, the dynamics takes the form of more regular oscillations as shown in figure 9.4(a2). Unlike the experimental data for $J_p = 170$ mA, the RWs are absent (figure 9.4(b2)) because the approximation of the average over the pulse output power neglects the contribution of the soliton rain in

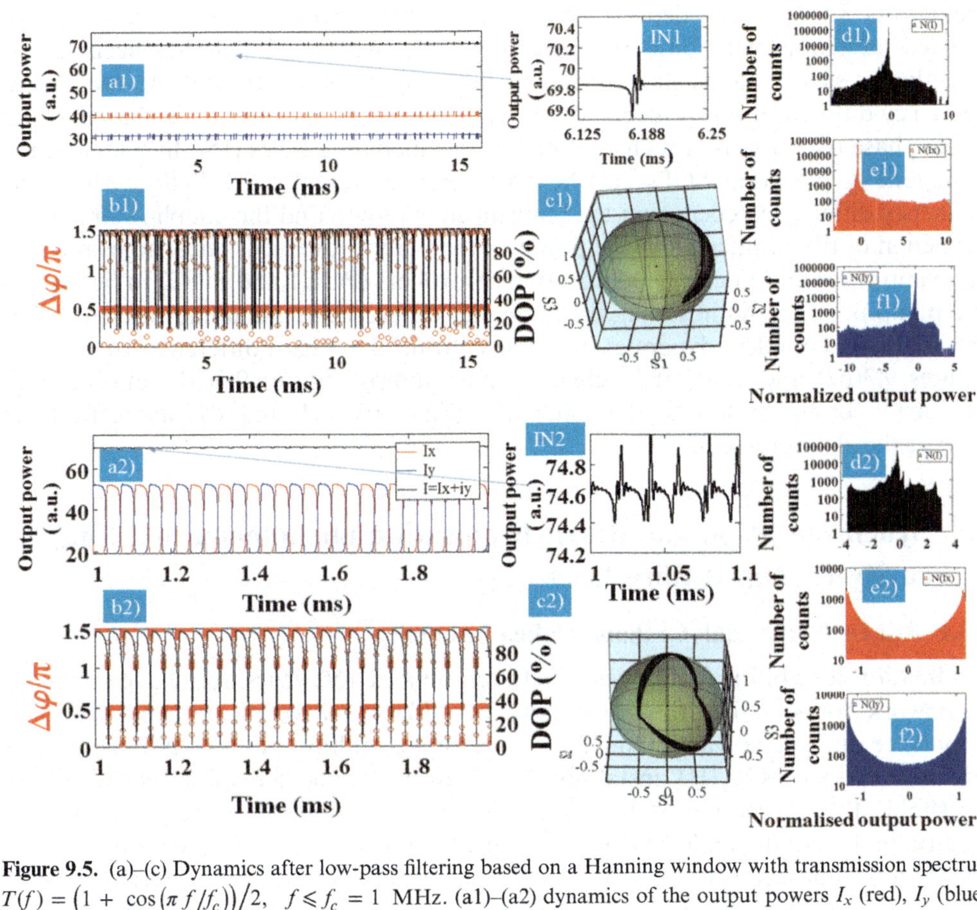

Figure 9.5. (a)–(c) Dynamics after low-pass filtering based on a Hanning window with transmission spectrum $T(f) = \left(1 + \cos\left(\pi f/f_c\right)\right)/2$, $f \leqslant f_c = 1$ MHz. (a1)–(a2) dynamics of the output powers I_x (red), I_y (blue), $I = I_x + I_y$ (black); (b1)–(b2) dynamics of the phase difference $\Delta\varphi$ (red) and DOP (black); (c1)–(c2) trajectories on the Poincaré sphere; (d)–(f) probability distribution histograms for the total output power $I = I_x + I_y$ (d1), I_x (d2), and I_y (d3). Each output power I_I (I, I_x, I_y) is normalised as shown in figure 9.2. Parameters: $I_p = 75$, $\alpha_1 = 21.53$, $\alpha_0 = 0.136$, $\alpha_s = 1.8{\cdot}10^{-5}$, $\alpha_2 = 2.533$, $\chi = 2.3$, $\Delta = 0.1$, $\gamma = 2 \times 10^{-6}$, $\varepsilon = 10^{-4}$; (a1)–(f1) $\delta = 0.85$ (elliptically polarised pump), $\beta = \pi \cdot 10^{-2}$; (a2)–(f2) $\delta = 1$ (circularly polarised pump), $\beta = 0.8\ \pi \cdot 10^{-2}$

the statistics of the output power. The results for low-pass filtering are shown in figure 9.4(a2). The antiphase dynamics of the output powers I_x and I_y results in the regular oscillations of the total output power, DOP, and the phase difference that is quite similar to the dynamics observed experimentally (figure 9.3). The difference in the theoretical and experimental data for output powers I_x and I_y at $J_p = 140$ mA can be explained by the different settings of the in-cavity polarisation controller that transforms the output SOP by different ways for $J_p = 140$ and 170 mA. The detailed explanation for this is found in annex I.

In the context of the synchronisation scenarios for the coupled nonlinear oscillators, the equation for the phase difference (equation 9.1) is a further generalisation of the Adler equation [26–28] accounting for asymmetry in the coupling of polarisation modes (K_{as} coefficient) and polarisation mode coupling based on the Kerr effect (K_{NL}). In addition, I_x and I_y are also time-dependent variables and coupling coefficients K_s, K_{as}, K_{NL} depend on I_x and I_y; this means we have a new class of coupled oscillators with time-dependent coupling strength [27]. It has been shown recently for such systems that by tuning the coupling strength and the oscillators' frequency difference, it is possible to tune the phase dynamics from the phase locking to the phase oscillations and chaotic phase slips including RWs [15]. In our case, the orthogonally polarised SOPs can be considered as coupled oscillators where the frequency difference depends on the birefringent strength and the coupling strength is a function of the ellipticity and the power of the pump wave [15]. Thus, adjustment of the pump wave and in-cavity polarisation controllers (POC1 and POC2 in figure 9.1) and pump power result in tuning the frequency difference and coupling strength. This provides the complex dynamics tunability from the phase-difference oscillations (figures 9.3(b2) and 9.5(b2)) to chaotic phase jumps (figures 9.3(b1) and 9.5(b1)), resulting in 'bright' and 'dark' RW emergence (figures 9.3(a1) and (d1) and 9.5(a1) and (d1)) at the slow timescale.

9.3 Synchronisation and desynchronisation phenomena in a long cavity Er-doped fibre laser

9.3.1 Risken–Nummedal–Graham–Haken instability

Modulation instabilities, discovered between 50 and 150 years ago, have created a framework for studying the complexity of different wave phenomena including turbulence and rogue waves [36–53]. For example, multimode Risken–Nummedal–Graham–Haken (RNGH) instability leads to the existence of a second lasing threshold, the exceeding of which results in excitation of a large number of longitudinal spatial modes and generation of a pulse train with a period equal to the photon roundtrip time [45–53]. The history of RNGH instability started in 1968, and since then the extensive theoretical and experimental study of Er-doped fibre lasers has revealed a decreased second lasing threshold to values slightly exceeding the value for the first threshold [48–53]. In this section, we review our recent experimental and theoretical results and demonstrate a new type of multimode instability, namely vector resonance multimode instability (VRMI), that is behind complex behaviours ranging from the turbulent, including different types of rogue waves, to stable pulse trains similar to the laser mode-locking regime [22, 23]. We have justified experimentally and theoretically that increased in-cavity birefringence strength causes spatial modulation of the SOP of the in-cavity lasing field (with a period related to the beat length) and thus leads through dispersion relation to the emergence of additional satellite frequencies with frequency splitting proportional to the birefringence strength. When the splitting approaches a fundamental frequency inversely proportional to the photon roundtrip time, parametric resonance results in longitudinal mode synchronisation and emergence of a pulse train with narrow pulse width. In contrast, in the case of weak birefringence and strong fibre twist adjusted with the help of an in-cavity polarisation controller, we demonstrate rogue wave emergence through transition from quasiperiodic motion (torus in the phase space) and further to chaotic behaviour satisfying RW criteria through the Ruelle–Takens–Newhouse scenario (route to chaos via two-dimensional torus destruction [54]).

9.3.2 Experimental data

The details of the schematic of the unidirectional cavity fibre laser are found in annex II [22]. The pump power for the first lasing threshold of the continuous wave operation was 16 mW, whereas the second threshold of the multimode instability was 18 mW. The paddles of the polarisation controllers (POC1 and POC2) were set with respect to the vertical position. The paddle of POC1 was oriented at $\theta_1 = -59°$ and the paddle of POC2 was adjusted at three different positions ($\theta_2 = -78°, -74°, -69°$). The radiofrequency (RF) spectrum comprises three peaks which were resolved better near the 1000th harmonic of the fundamental self-mode locking frequency, i.e. for 325.2 MHz (figure 9.6(a)): the central peak (1000th harmonic), two adjustable with the help of the POC2 satellites, and two satellites independent of POC2 orientation. As follows from the figures 9.6(a) and (b), the adjustment of the POC results in

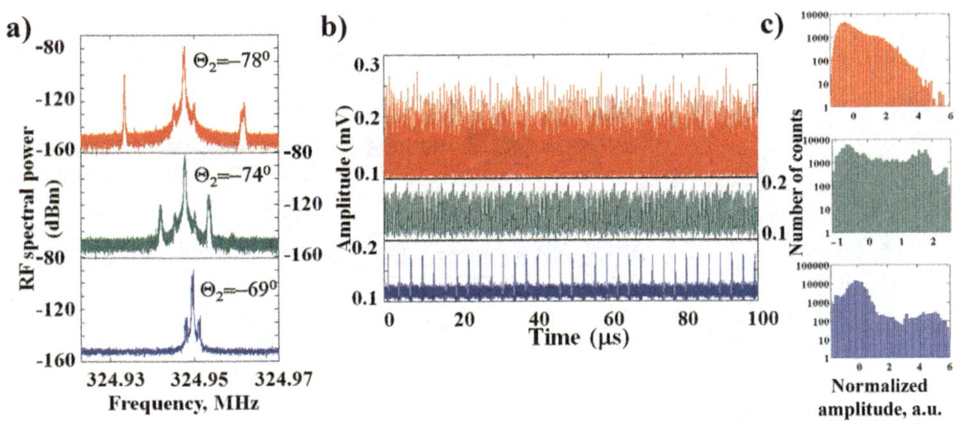

Figure 9.6. (a) RF spectrum, (b) oscillograms, and (c) probability distribution histograms (the output amplitude is normalised as shown in figure 9.2) for different setting of the POC2: $\theta_2 = -78°$, $\theta_2 = -74°$, $\theta_2 = -69°$.

approaching of the satellites from the neighbouring to 325.2 MHz lines to the main peak of 325.2 MHz. Such adjustment results in regime stabilisation similar to mode-locking for the case when satellites match the main line (figures 9.6(a) and (b)). Unlike multimode RNGH instability [48–53] demonstrating oscillations close to the harmonic with the period of photon roundtrip time, the pulse train in figure 9.6 has a pulse width of 40 ns. This period is much less than the roundtrip time of 3 μs that indicates the excitation of many phase-synchronised longitudinal modes. Analysis of the amplitude distribution shown in figure 9.6(c) reveals that the complex oscillatory dynamics shown in figure 9.6(b) do not satisfy RW criteria.

The results on polarisation dynamics are shown in figure 9.7. As follows from figure 9.7(a), the size of the trajectory on the Poincaré sphere indicates that N-fold beat length deviates from the cavity length. Thus, N-fold beat length matches the cavity length when the trajectory is shrinking to the dot. This mismatch is accompanied by oscillations of the output power S_0 and phase difference $\Delta\varphi$ (figures 9.7(a1) and (b1)). DOP of 60% indicates that the laser dynamics oscillate with a timescale smaller than 1 μs for the POC2 setting at $\theta_2 = -78°$ (figures 9.7(a1) and (b1)). Setting the POC2 at $\theta_2 = -74°$ demonstrates suppression of the fast dynamics. The constant outputs and high DOP (over 80%) in figures 9.7(a), (a3), and (b3) indicate stable mode locking (figure 9.6(b)) along with stable phase difference and SOP locking (figure 9.7(b3)). As follows from figure 9.7, all the aforementioned vector dynamics are localised nearby one SOP, and this may be a reason for the RW absence.

As follows from section 9.2, the random switching between orthogonal SOPs results in emergence of spikes with statistics satisfying RW criteria. This is why we readjust the POC1 and POC2 settings to enable switching between the orthogonal SOPs leading to RW emergence.

As follows from the obtained results (figure 9.8(a)), adjustment of POC1 and POC2 results in excitation complex multi-pulsing (middle figure) including soliton rain (bottom figure). The corresponding RF spectrum indicates excitation of the satellites with frequencies: $f_1 = 4.8 \cdot 10^{-3} f_0$, $f_2 = f_1/4$, $f_3 = f_2/4$ (figure 9.8(b)). As follows from the figures 9.8(a) and (c), though dynamics for all three cases

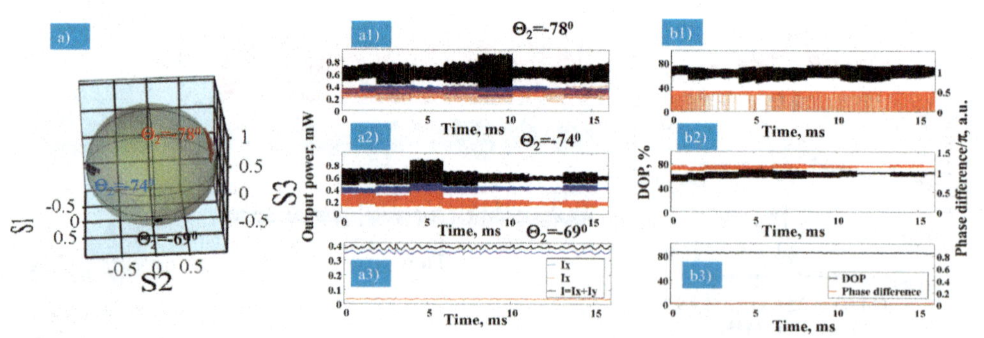

Figure 9.7. (a) trajectories in normalised Poincaré sphere: (a1)–(a3) output powers versus time: I_x (red), I_y (blue), and $I = I_x + I_y$ (black); (b1)–(b3) phase differences $\Delta\varphi/\pi$ (red) and DOP (black) versus time; POC2 settings: $\theta_2 = -78°, -74°, -69°$.

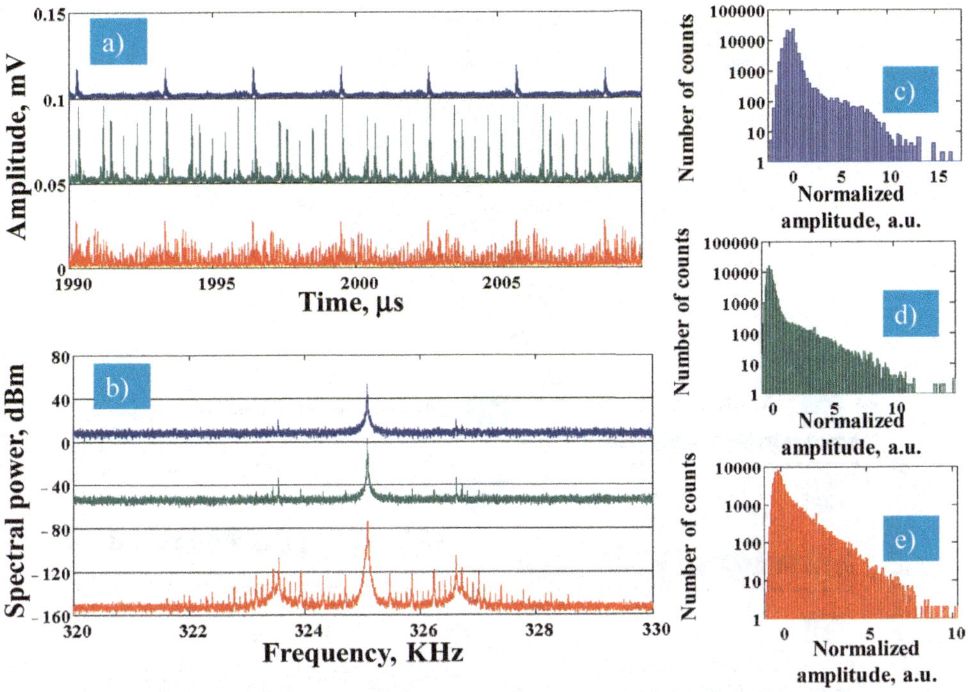

Figure 9.8. (a) Oscillograms, (b) RF spectrum, (c)–(e) probability distribution histograms (the output amplitude is normalised as shown in figure 9.2) for different setting of POC1 and POC2.

demonstrate the L-shape statistics typical for RWs, the case of mode-locking leads to predictable dynamics and so pulses cannot be counted as RWs.

The results for polarisation dynamics shown in figure 9.9 confirm the connection of the RW emergence with random switching between orthogonal SOPs. In addition, as follows from figures 9.7 and 9.9, complex dynamics including RWs exist if the output powers for the orthogonal SOPs are approximately equal, whereas the mode locking appears when the output power for one of the SOPs is close to zero, i.e. for linearly polarised SOPs. This means that, by controlling POC1 and POC2, it is possible to adjust anisotropy and in-cavity birefringence to enable strong orthogonal SOP coupling that, according to the theory of the coupled oscillators, leads to the synchronisation that in our case corresponds to the longitudinal mode locking. In contrast to the mode locking, weak coupling between the SOPs leads to phase-difference random slips and so to the emergence of spikes in the dynamics of the output power.

9.3.3 Theoretical results

The experimental results can be well understood based on the vector model of an Er-doped fibre laser. The following rate equations are derived from the model suggested by Sergeyev *et al* [22]:

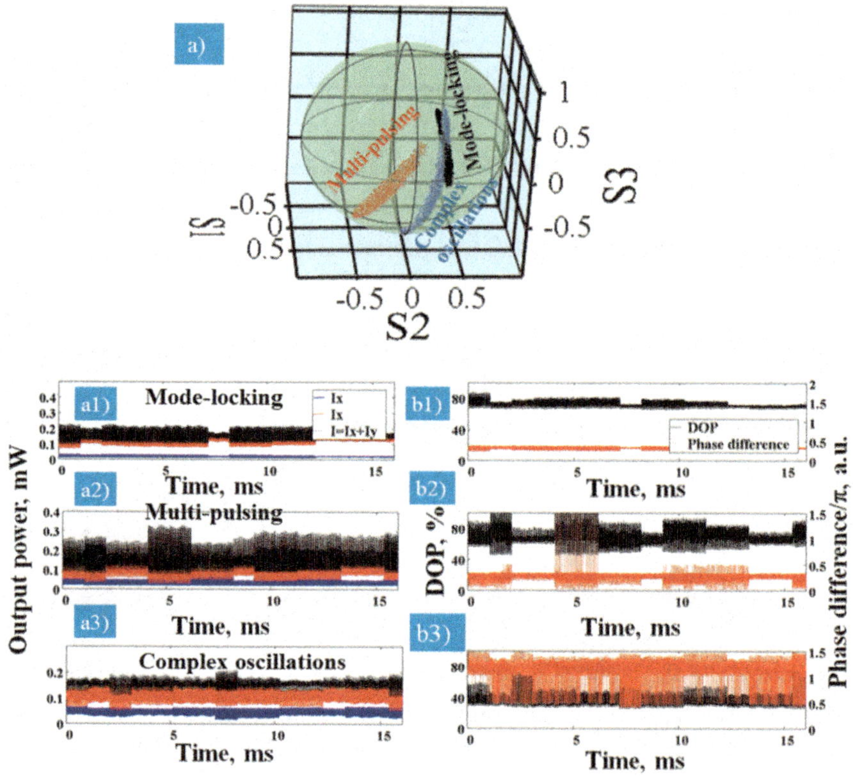

Figure 9.9. (a) Trajectories in normalised Poincaré sphere. (a1)–(a3) output powers versus time: I_x (red), I_y (blue), and $I = I_x + I_y$ (black); (b1)–(b3) phase differences $\Delta\varphi/\pi$ (red) and DOP (black) versus time.

$$
\begin{aligned}
&\frac{\partial S_0}{\partial z} + \frac{\partial S_0}{\partial t} = \left(\frac{2\alpha_1 f_1}{1+\Delta^2} - 2\alpha_2\right)S_0 + \frac{2\alpha_1 f_2}{1+\Delta^2}S_1 + \frac{2\alpha_1 f_3}{1+\Delta^2}S_2,\\
&\frac{\partial S_1}{\partial z} + \frac{\partial S_1}{\partial t} = \gamma S_2 S_3 + \left(\frac{2\alpha_1 f_1}{1+\Delta^2} - 2\alpha_2\right)S_1 + \frac{2\alpha_1 f_2}{1+\Delta^2}S_0 - \frac{2\alpha_1 f_3 \Delta}{1+\Delta^2}S_3 + \rho S_2,\\
&\frac{\partial S_2}{\partial z} + \frac{\partial S_2}{\partial t} = -\gamma S_1 S_3 + \frac{2\alpha_1 f_3}{1+\Delta^2}S_0 + \left(\frac{2\alpha_1 f_1}{1+\Delta^2} - 2\alpha_2\right)S_2 + \left(\frac{2\alpha_1 f_2 \Delta}{1+\Delta^2} - 2\beta\right)S_3 - \rho S_1,\\
&\frac{\partial S_3}{\partial z} + \frac{\partial S_3}{\partial t} = \frac{2\alpha_1 \Delta f_3}{1+\Delta^2}S_1 - \frac{2\alpha_1 \Delta f_2}{1+\Delta^2}S_2 + 2\beta S_2 + \left(\frac{2\alpha_1 f_1}{1+\Delta^2} - 2\alpha_2\right)S_3,\\
&\qquad \frac{df_1}{dt} = \varepsilon\left[\frac{(\chi_s - 1)I_p}{2} - 1 - \left(1 + \frac{I_p\chi_p}{2} + d_1 S_0\right)f_1 - \left(d_1 S_1 + \frac{I_p}{2}\xi\right)f_2 - d_1 S_2 f_3\right],\\
&\qquad \frac{df_2}{dt} = \varepsilon\left[\xi\frac{I_p(\chi_s - 1)}{4} - \left(\frac{I_p\chi_p}{2} + 1 + d_1 S_0\right)f_2 - \left(\frac{I_p\chi_p}{2}\xi + d_1 S_1\right)\frac{f_1}{2}\right],\\
&\qquad \frac{df_3}{dt} = -\varepsilon\left[\frac{d_1 S_2 f_1}{2} + \left(\frac{I_p\chi_p}{2} + 1 + d_1 S_0\right)f_3\right].
\end{aligned}
\tag{9.5}
$$

Equation (9.5) is the further generalisation of equation (9.3) in the context of accounting for the angle of the SOP rotation around the S_3-axis induced by POC2 adjustment (ρ), pump at 1480 nm ($\chi_{p,s} = (\sigma_a^{(s,p)} + \sigma_e^{(s,p)})/\sigma_a^{(s,p)}$, where $\sigma_a^{(s,p)}$ and $\sigma_e^{(s,p)}$ are absorption and emission cross sections at the lasing (s) and pump (p) wavelengths), and the absence of averaging over the pulse width.

As follows from the structure of equation (9.5), the in-cavity birefringence of the passive fibre combined with birefringence tunable by the in-cavity polarisation controller ($2\beta = 2\pi L/L_b$) and birefringence induced by polarisation hole burning in the active fibre $\Delta_\beta = 2\alpha_1 f_2 \Delta/(1 + \Delta^2)$. In the experiments, we set the pump power at 18 mW and so exclude mode locking based on nonlinear polarisation rotation that requires much larger power (approximately 800 mW) [8–12]. The pulse width in our experiments (figures 9.6 and 9.8) was estimated to be of 40 ns, and so the second-order dispersion can be neglected in the context of the qualitative description of the experiments. The dynamics of the medium polarisation were also ignored because the pulse width is also much longer than the time of transverse relaxation of 160 fs.

To specify conditions for the vector resonance mode locking, we linearise equation (9.5) in the vicinity of the steady-state solution $\boldsymbol{F}_0 = (S_{00}, \pm S_{00}, 0, 0, f_{10}, f_{20}, 0)^T$ and find eigenvalues as follows (details are found in annex II):

$$\begin{aligned}
&\text{(I)}\ \lambda_0 = iq + A_0(I_p, \xi),\\
&\text{(II)}\ \lambda_1 = A_1(q, I_p, \xi) + i\Omega_1(q, I_p, \xi),\\
&\lambda_2 = A_2(q, I_p, \xi) + i\Omega_2(q, I_p, \xi),\\
&\lambda_3 = A_3(q, I_p, \xi) + i\Omega_3(q, I_p, \xi),\\
&\text{(III)}\ \lambda_4 = A_4(q, I_p, \beta, \xi) + i\Omega_4(q, I_p, \beta, \zeta),\\
&\lambda_5 = A_5(q, I_p, \beta, \xi) + i\big(q + \Delta\Omega(q, I_p, \beta, \xi)\big),\\
&\lambda_6 = A_6(q, I_p, \beta, \xi) + i\big(q - \Delta\Omega(q, I_p, \beta, \xi)\big),\\
&A_0(I_p, \xi) > 0,\quad A_1(q, I_p, \xi) < 0,\\
&A_2(q, I_p, \xi) < 0,\quad A_3(q, I_p, \xi) < 0,\quad A_4(q, I_p, \xi) < 0,\\
&A_5(q, I_p, \xi) > 0,\quad A_6(q, I_p, \xi) > 0.
\end{aligned} \tag{9.6}$$

Here, $q = 0, \pm 1, \pm 2, \ldots \pm N$ is the wave number of the longitudinal mode. All eigenvalues are normalised to the fundamental frequency $\omega = 2\pi/\tau_R$. Here, we neglected the fibre twist, i.e. we use condition $\rho = 0$.

The obtained results are shown in figures 9.10(a)–(c) [22]. As follows from figure 9.10(a), the threshold of the multimode instability is the same for all longitudinal modes and coincides with the first lasing threshold for the circularly polarised pump and slightly exceeds the first one with increased pump anisotropy parameter ξ. In contrast, threshold pump values for RGHM are different for excitation of different longitudinal modes. The threshold for multimode instability (branch I) is the same as the threshold of birefringence-dependent RF satellite excitation (branch III). As follows from figure 9.10(b), the scalar branch II comprises additional birefringence-independent satellites

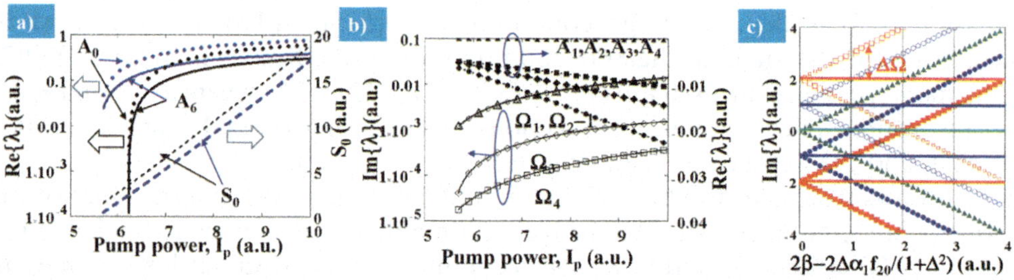

Figure 9.10. (a)–(c) Results of linear stability analysis of equation (9.5) in vicinity of steady solutions: (a1)–(c1) $\boldsymbol{F}_0 = (S_{00} \pm S_{00}\, 0\, 0\, f_{10} f_{20}\, 0)^T$ and (a2–b2) $\boldsymbol{F}_0 = (S_{00}\, 0\, 0 \pm S_{00} f_{10}\, 0\, 0)^T$. (a1) Vector multimode instability in terms of positive real parts of eigenvalues, i.e. A_0 (dots), A_5 (solid line) (A_6 is close to the A_5 and so is not shown here), and the output signal S_0 (dashed line) versus pump power I_p for $\xi = 0$ (blue lines), $\xi = 0.1$ (black line); (b1) frequencies of the scalar branch: Ω_1 (empty circles), $\Omega_2 - 1$(empty triangles), Ω_3 (empty squares), and vector branch Ω_4 (empty diamonds) along with the real parts of the scalar branch: A_1 (filled circles), A_2 (filled triangles), A_3 (filled squares) and the vector branch: A_4 (filled diamonds) versus pump power I_p for $\xi = 0.1$. (c1) Frequencies for the branch I, i.e. $\Omega_0 = \pm q$, $q = 0, 1, 2, 3$ (solid lines), and the vector branch (III), i.e. $\Omega_{5,6} = q \pm \Delta\Omega$ (squares, triangles, circles) versus the birefringence strength $2\beta - 2\Delta\alpha_1 f_{20}/(1+\Delta^2)$ for $\xi = 0.1$. The equality $\Omega_0 = \Omega_{5,6}$ for $2\beta - 2\Delta\alpha_1 f_{20}/(1+\Delta^2) = 1, 2, \ldots$ results in the resonance mode locking shown in figure 9.6. All frequencies are normalised to the fundamental frequency f. Parameters: $L = 615\,\text{m}, 2\alpha_1 = \ln(10)6.4, 2\alpha_2 = \ln(10)0.5, \chi_s = 3/2, \chi_p = 1/0.7, \Delta = 0.1, I_p = 10(c), \gamma = 2 \times 10^{-6}, \varepsilon = 10^{-3}, \beta = 1$ (a, b), $q = 1$ (b), $\xi = 0.1$ (b, c). Reprinted figure with permission from [22]. Copyright 2017.

which are located in the RF spectrum at $0.01f$ (f is the fundamental frequency) with respect to the longitudinal mode frequency q. The satellites can be excited through parametric phase-locking with the frequencies of branch I. If we consider branch III, we can find that increased birefringence strength can result in resonance conditions where satellites frequencies for the longitudinal mode q (branch III) match the frequency of the longitudinal mode $q + N$ (where N is integer) from branch I.

Thus, if we compare figures 9.6 and 9.10(a)–(c), we can conclude that the theoretical results are in a good correspondence with the experimental data, i.e. (i) the threshold of the multimode instability (the second threshold) slightly exceeds the first lasing threshold; (ii) if the birefringence-dependent satellites frequencies are off-resonance with the longitudinal mode frequencies, then the dynamics take the form of complex oscillations; (iii) resonance of the longitudinal modes' frequencies with the birefringence-dependent satellites results in their synchronisation and so to the stable mode locking is accompanied by stable SOP locking.

As follows from the experimental data shown in figures 9.8 and 9.9, the dynamics nearby the orthogonal SOPS (with $I_x = I_y$), including random switching between them, results in complex multi-pulse dynamics RW emergence. To prove this, we linearise equation (9.5) in the vicinity of the steady-state solution $\boldsymbol{F}_0 = (S_{00}\, 0\, 0 \pm S_{00} f_{10}\, 0\, 0)^T$ and find eigenvalues as follows [23]:

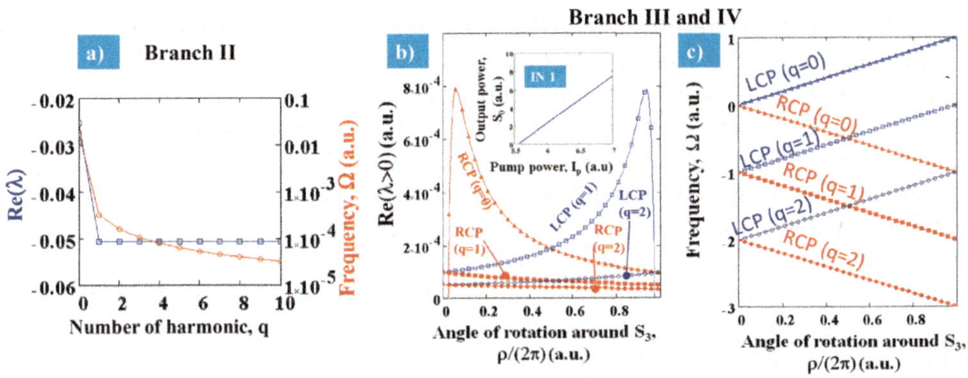

Figure 9.11. (a–c) Results of linear stability analysis of equation (9.5) in the vicinity of steady-state solution $F_0 = (S_{00}\ 0\ 0 \pm S_{00} f_{10}\ 0\ 0)^T$. (a) Branch II: Re($\lambda$) > 0 (squares) and corresponding frequencies $\Omega = |\mathrm{Im}(\lambda)|/(2\pi) - q$ (circles) as a function of the harmonic number q; (b) Branch III and IV: Re(λ) > 0 as a function of the angle of the SOP rotation around axis S_3 (INSET: the total output power S_0 as a function of I_p); (c) Branch III and IV: frequencies $\Omega = |\mathrm{Im}(\lambda)|/(2\pi)$ as a function of the angle of the SOP rotation around axis S_3. Parameters are the same as for figure 9.10 ($\beta = 0$). All frequencies are normalised to the fundamental frequency f.

$$\begin{aligned}
&\text{(I)}\quad \lambda_0 = i2\pi q,\\
&\text{(II)}\quad \lambda^2 + (c_2 + i2\pi q)\lambda + b_1c_1 + i2\pi c_2 q = 0,\\
&\text{(III)}\quad \lambda^2 + (c_2 + i2\pi q - i\rho)\lambda + \frac{b_1c_1}{2}(1 - ia\Delta) + ic_2(2\pi q - \rho) = 0,\\
&\text{(IV)}\quad \lambda^2 + (c_2 + i2\pi q + i\rho)\lambda + \frac{b_1c_1}{2}(1 + ia\Delta) + ic_2(2\pi q + \rho) = 0,\\
&\qquad b_1 = 2\alpha_1 S_{00},\quad c_1 = \varepsilon d_1 F_1,\quad c_2 = \varepsilon\left(1 + \frac{I_p\chi_p}{2} + d_1 S_{00}\right),\\
&\qquad S_{00} = \frac{\alpha_1}{\alpha_2 d_1}\left(\frac{\chi_s - 1}{2}I_p - 1\right) - \frac{1}{d_1}\left(1 + \frac{I_p\chi_2}{2}\right),\quad F_1 = \frac{\left(\frac{\chi_s - 1}{2}I_p - 1\right)}{\left(1 + \frac{I_p\chi_p}{2} + d_1 S_{00}\right)}.
\end{aligned}\tag{9.7}$$

Here, $a = 1$ for $S_{30} = S_{00}$ (right circularly polarised SOP) and $a = -1$ for $S_{30} = -S_{00}$ (left circularly polarised SOP). The results of the eigenvalue analysis are shown in figure 9.11(a)–(c).

As follows from figure 9.11(a), the position of satellites located nearby ($\sim 10^{-2} f$) of the main peak in the RF spectrum with $q = 0$ (branch II) are independent of the in-cavity polarisation controller tuning the same as shown in figure 9.8(b). As follows from figure 9.11(b) for $\rho = 0$, $\mathrm{Re}(\lambda_{q=1}) > \mathrm{Re}(\lambda_{q=2}) > \mathrm{Re}(\lambda_{q=3})$, i.e. unlike the previous case shown in figure 9.10, the vector MMI results in the excitation of few longitudinal modes. Tuning the POC2 leads to increased ρ and thus results in increasing the maximum of the $\mathrm{Re}(\lambda_{q=1})$, which is proportional to the amplitude of the oscillations (figure 9.11(b)). Though Re(λ)<0 for the branch II, the satellites can be excited through parametric phase locking (PPHL) with the frequencies of branch

III and IV. The PPHL results in increased spectral power in the RF spectrum, shown in figure 9.8(b), caused by the adjustment of the angle ρ leading to the mismatch between the frequencies from the branches I and III/IV proportional to the frequency from branch II (figure 9.11(c)). In view of the fact that the frequency's splitting is proportional to the angle ρ (figure 9.11(c)), harmonic oscillations appear as shown in figure 9.8(a). Coupling between left- and right-circularly polarised SOPs results in switching between SOPs having different eigenfrequencies, as shown in figure 9.11(c). By adjusting ρ, a rational relation between SOP frequencies can be realised, which results in quasiperiodic motion (torus in the phase space) and further to chaotic behaviour through the Ruelle–Takens–Newhouse scenario (route to chaos via two-dimensional torus destruction [54]).

9.4 Summary

To summarise, we reviewed our recent experimental and theoretical data, demonstrating a new type of vector rogue wave emerging at a slow timescale of tens of thousands of roundtrips and resulting from the interaction between the orthogonal SOPs in an Er-doped mode-locked fibre laser with a saturable absorber. By tuning the in-cavity and pump wave polarisation controllers, we adjust the coupling between SOPs and thus provide conditions for the appearance of polarisation instabilities driven by slow bright and dark rogue waves. Such RWs complement fast RWs in the form of soliton rain visible in the timescale of a roundtrip. For a long fibre laser without a generic mode-locking mechanism, we reviewed our recent experimental and theoretical results on vector resonance multimode instability, which can result in vector resonance mode locking and complex multi-pulsing behaviour satisfying rogue wave criteria.

Annex I

Experimental setup: The laser (figure 9.1) comprises 1.1 m long erbium-doped fibre (EDF) with a nominal absorption ratio of 80 dB m^{-1} at 1530 nm. The group velocity dispersion of the EDF is +59 ps^2 km^{-1}. A fibre pigtailed optical isolator (OISO) has been used to support a unidirectional operation of the laser. A 975 nm laser diode (LD) with maximum output power of 200 mW is used to pump the laser via a 980/1550 wavelength division multiplexer (WDM). A standard 70:30 fused coupler (OUTPUT C) has been installed to redirect 30% of the laser light out of the cavity. The rest of the laser cavity includes 1.22 m of OFS980 fibre and 4.4 m of SMF 28 fibre. The CNT mode-locker is used in the form of a piece of carbon nanotube polymer composite film placed between two standard fibre connectors supplemented by index matching for minimisation of insertion losses. The total cavity GVD is of −0.04 ps^2 nm^{-1}. Two polarisation controllers (POC1 and POC2) are installed to adjust the SOP of the pump wave and the in-cavity birefringence. A polarimeter (*Thorlabs IPM5300,* 1 µs resolution, 1024 samples) has been connected to the output. The laser signal is measured using a *UDP–15–IR–2FC* detector with a bandwidth of 17 GHz; the electric signal from the detector has been recorded by a *Tektronix DPO7254* oscilloscope (with 25 ps sample interval and 20 K in each

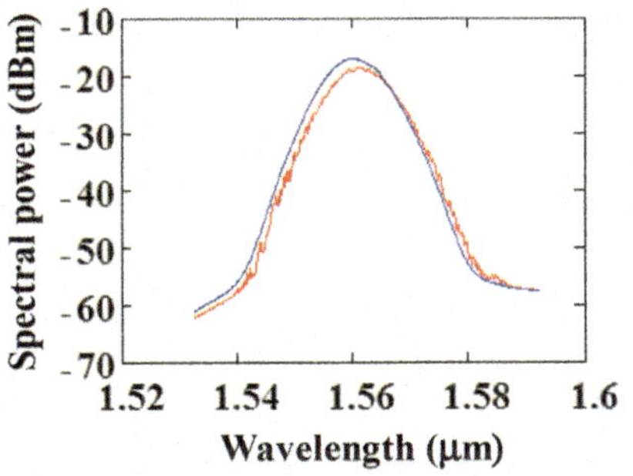

Figure A1. Optical spectrum of the laser emission: J_p = 140 mA (red) and 170 mA (blue).

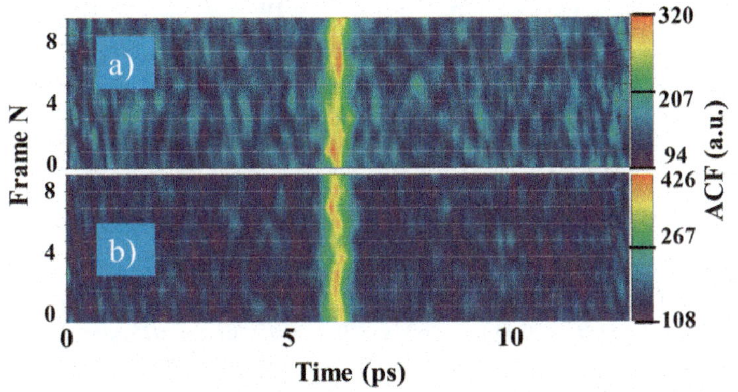

Figure A2. Autocorrelation traces for I_p = 140 mA (a); 170 mA (b).

oscillogram). POC1 and POC2 have been adjusted to find conditions for RW emergence. The optical spectrum has been measured with the optical spectrum analyser (ANDO AQ6317B) and the pulse width, with the help of the autocorrelator (Pulsecheck). The oscilloscope data comprises 11 oscillograms of the pulse dynamics with 20 K points in each oscillogram, the autocorrelation data consists of 8 traces, and polarimetry data includes 20 traces.

Optical spectrum: The dispersion map comprises anomalous dispersion of the EDF and normal dispersion of the SMF, and so the laser configuration is typical for stretched pulse laser as per the spectrum shown in figure A1 [55].

Autocorrelation traces: Detailed autocorrelation analysis shown in figure A2(a) reveals pulse width of about 700 fs and a fluctuating distance between the main pulse and satellites from 1 to 7 ps for J_p = 140 mA. The presence of soliton rain (figure A2(a), J_p = 140 mA), pedestal in the autocorrelation trace, and the smooth spectrum (figure A1) indicate the presence of loosely bound solitons moving in the range of delays from ps to ns with respect to the main pulse. With the increase of pump current to 170 mA (figure A2(b)), the satellites are located far from the main peak (> 5 ps), which means that loosely bound soliton interaction is weaker as compared to the previous case.

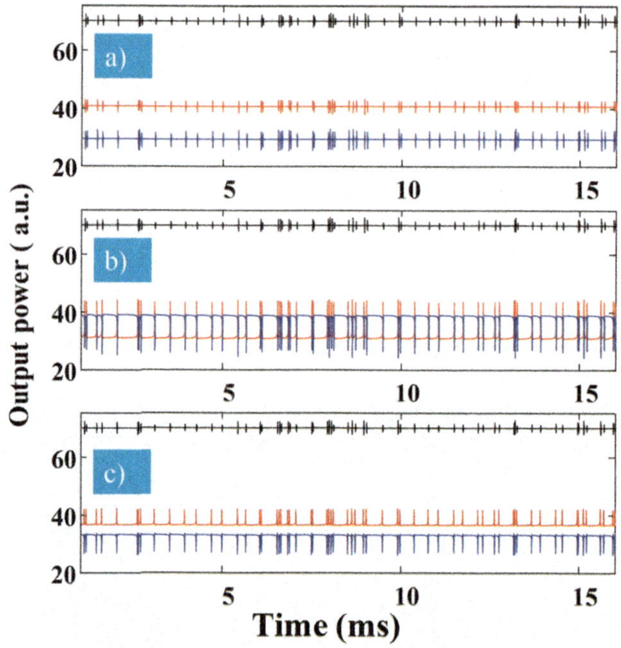

Figure A3. Dynamics of the output powers at orthogonally polarised SOPs: I_x (red), I_y (blue), and $I = I_x + I_y$ (black). Parameters: (a) $\alpha = \psi = \gamma = 0$; (b) $\alpha = \pi/3$, $\psi = 0$, $\gamma = \pi/5$; (c) $\alpha = \pi/25$, $\psi = \pi/15$, $\gamma = \pi/10$; $I_p = 75$; $\alpha_1 = 21.53$, $\alpha_0 = 0.136$, $\alpha_s = 1.8{\cdot}10^{-5}$, $\alpha_2 = 2.533$, $\chi = 2.3$, $\Delta = 0.1$, $I_p = 75$, $\gamma = 2{\cdot}10^{-6}$, $\varepsilon = 10^{-4}$; $\delta = 0.85$ (elliptically polarised pump), $\beta = \pi \cdot 10^{-2}$.

Numerical simulation (3D SOP rotation): In-cavity polarisation controller and a fibre patchcord connecting the laser output with the polarimeter provide 3D rotation (around axes S_1, S_2, S_3) of the initial SOP as follows [56]:

$$\begin{pmatrix} \tilde{s}_1 \\ \tilde{s}_2 \\ \tilde{s}_3 \\ \tilde{s}_0 \end{pmatrix} = \begin{bmatrix} a_{11} & a_{12} & a_{13} & 0 \\ a_{21} & a_{22} & a_{23} & 0 \\ a_{31} & a_{32} & a_{33} & 0 \\ 0 & 0 & 0 & 1 \end{bmatrix} \begin{pmatrix} s_1 \\ s_2 \\ s_3 \\ s_0 \end{pmatrix},$$

$$\begin{aligned} &a_{11} = \cos(\psi)\cos(\gamma), \quad a_{12} = \cos(\gamma)\sin(\alpha)\sin(\psi) - \cos(\alpha)\sin(\gamma), \\ &a_{13} = \cos(\alpha)\cos(\gamma)\sin(\psi) + \sin(\alpha)\sin(\gamma), \\ &a_{21} = \cos(\psi)\sin(\gamma), \quad a_{22} = \cos(\alpha)\cos(\gamma) + \sin(\alpha)\sin(\psi)\sin(\gamma), \\ &a_{23} = -\cos(\gamma)\sin(\alpha) + \sin(\psi)\sin(\gamma), \\ &a_{31} = -\sin(\gamma), \quad a_{32} = \cos(\psi)\sin(\alpha), \quad a_{33} = \cos(\alpha)\cos(\psi). \end{aligned} \tag{A1}$$

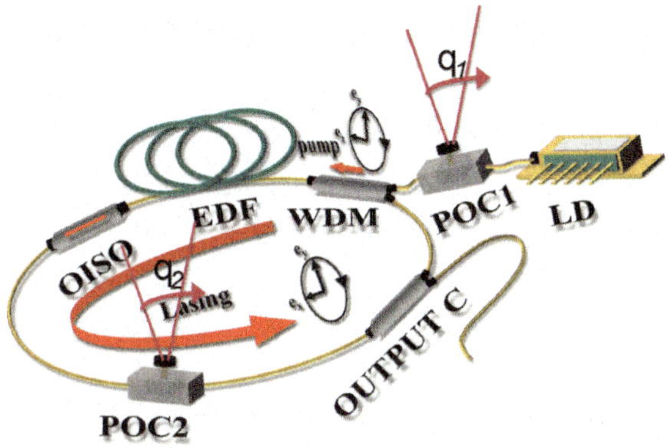

Figure A4. Erbium-doped fibre laser. EDF: erbium-doped fibre; LD: 1480 nm laser diode for pump; POC1 and POC2: polarisation controllers; OISO: optical isolator; WDM: wavelength division multiplexer (WDM); OUTPUT C: 80:20 output coupler. Reprinted figure with permission from [22]. Copyright 2017.

Here, α, ψ, and γ are Euler angles [56]. Transformation of the output power dynamics for two orthogonal SOPs (I_x and I_y) is shown in figure A3. As follows from figure A3(b), 3D rotations lead to a dynamic similar to the dynamics observed experimentally in figure 9.3(a1).

Annex II

Experimental setup: The schematic of the unidirectional cavity fibre laser is shown in figure A4. The cavity is comprised of 1 m of Er-doped fibre (Liekki Er80-8/125) and 614 m of single-mode fibre SMF-28. An optical isolator (OISO) with 51 dB back losses has been used to provide unidirectional lasing. The 80/20 fibre coupler was used to redirect the part of the signal outside the cavity. The cavity was pumped via a 1480/1550 WDM, using a 1480 nm laser diode (FOL14xx series) with an in-built isolator. The pump power was measured after the pump polarisation controller (POC1) and wavelength division multiplexing (WDM). The signal was detected using a photodetector with a bandwidth of 17 GHz (InGaAsUDP-15-IR-2 FC) connected to a 2.5 GHz sampling oscilloscope (Tektronix DPO7254) and electrical spectrum analyzer (ROHDE and SCHWARZ FSV Signal Analyzer, 10 Hz/13.6 GHz). An in-line polarimeter (Thorlabs IPM5300) was used to record the state and degree of polarisation (SOP and DOP), respectively.

Linear stability analysis: To find eigenvalues, we substitute the following ansatz into equation (9.3):

$$\mathbf{F}(t, z) \equiv \left[S_0 \, S_1 \, S_2 \, S_3 f_1 f_2 f_3\right]^T = \mathbf{F}_0 + [x_0 \, x_1 x_2 x_3 x_4 x_5 x_6 x_7]^T \exp{(\lambda t + qz)}, \qquad \text{(A2)}$$

where $\boldsymbol{F}_0 = (S_{00} \; \pm S_{00} \; 0 \; 0 \; f_{10} \, f_{20} \; 0)^T$.

As a result, we find the following equation for eigenvalues:

$$\det\begin{bmatrix} a_1 - iq - \lambda & 0 & 0 & 0 & a_2 & a_2 & 0 \\ -a_1 & a_1 - iq - \lambda & 0 & 0 & a_2 & a_2 & 0 \\ 0 & 0 & a_1 - iq - \lambda & a_3 + a_4 & 0 & 0 & a_2 \\ 0 & 0 & -a_3 & a_1 - iq - \lambda & 0 & 0 & \Delta a_2 \\ b_2 & b_4 & 0 & 0 & b_1 - \lambda & b_3 & 0 \\ b_4 & \frac{b_2}{2} & 0 & 0 & \frac{b_3}{2} & b_1 - \lambda & 0 \\ 0 & 0 & \frac{b_2}{2} & 0 & 0 & 0 & b_1 - \lambda \end{bmatrix} = 0, \quad \text{(A3)}$$

where

$$a_1 = \frac{2\alpha_1}{1+\Delta^2}f_{10} - \alpha_2,\ a_2 = \frac{2\alpha_1}{1+\Delta^2}S_{00},\ a_3 = \frac{2\alpha_1\Delta}{1+\Delta^2}f_{20} - 2\beta,\ a_4 = -\gamma S_{00},$$

$$b_1 = -\varepsilon\left(1 + \frac{I_p\chi_p}{2} + d_1 S_{00}\right),\ b_2 = -\varepsilon d_1 f_{10},\ b_3 = -\varepsilon\left(d_1 S_0 + \frac{I_p\chi_p\xi}{2}\right),\ b_4 = -\varepsilon d_1 f_{20},$$

$$S_{00} = \frac{-Q_2 - \sqrt{Q_2^{\,2} - 4Q_1Q_3}}{2Q_1},\quad Q_1 = -\frac{\alpha_2 d_1^{\,2}}{2},$$

$$Q_2 = -\alpha_2\left(2d_1\left(1 + I_p\chi_p\right) - \frac{d_1 I_p\xi}{2}\right) + \left(\frac{\chi_s - 1}{2}I_p - 1\right).$$

As a result, equation (A3) has three branches of eigenvalues:

$$\lambda - 2a_1 + 2iq = 0, \quad \text{(I)}$$

$$\lambda^3 + [-2b_1 + iq]\lambda^2 + \left(b_1^{\,2} - \frac{b_3^{\,2}}{2} - \frac{3a_2b_2}{2} - 2a_2b_4 - 2iqb_1\right)\lambda + \frac{3a_2b_1b_2}{2} + 2a_2b_1b_4$$
$$- 2a_2b_1b_4 - a_2b_2b_3 - \frac{3a_2b_3b_4}{2} - \frac{ib_3^{\,3}q}{2} + ib_1^{\,2}q = 0, \quad \text{(II)}$$

$$\lambda^3 + (-2a_1 - b_1 + 2iq)\lambda^2 + \left(a_1^{\,2} + 2b_1a_1 + a_3^2 + a_4a_3 - q^2 - \frac{a_2b_2}{2} - 2ib_1q - 2ia_1q\right)$$
$$+ b_1q^2 - a_3^2b_1 - a_1^{\,2}b_1 + \frac{a_1a_2b_2}{2} - a_3a_4b_1 - \frac{a_2a_3b_2\Delta}{2} - \frac{a_2a_4b_2\Delta}{2} - \frac{ia_2b_2q}{2}$$
$$+ 2ia_1b_1q = 0. \quad \text{(III)}$$

Acknowledgements

SVS, HK and SAK acknowledge support from the Leverhulme Trust (Grant ref: RPG-2014-304); CM acknowledges support from National Science Foundation of China 61605107, the Young Eastern Scholar program (QD2015027) at Shanghai Institutions of Higher Learning, and the 'Young 1000 Talent Plan' program of

China; AM acknowledges support from the H2020 Marie-Sklodowska-Curie IF scheme; and VK acknowledges support from the Austrian Science Fund (FWF project P24916-N27).

References

[1] Kharif C, Pelinovsky E and Slunyaev A 2009 *Rogue Waves in the Ocean* (Heidelberg: Springer)

[2] Zhen-Ya Y 2010 Financial rogue waves *Commun. Theor. Phys.* **54** 947

[3] Onorato M, Residori S, Bortolozzo U, Montina A and Arecchi F T 2013 Rogue waves and their generating mechanisms in different physical contexts *Phys. Rep* **528** 47

[4] Solli D R, Ropers C, Koonath P and Jalali B 2007 Optical rogue waves *Nature* **450** 1054

[5] Liu Z, Zhang S and Wise F W 2015 Rogue waves in a normal-dispersion fiber laser *Opt. Lett.* **40** 1366

[6] Gao L, Zhu T, Wabnitz S, Liu M and Huang W 2016 Optical polarization rogue waves in fiber laser *IEEE Conf. Lasers and Electro-Optics (CLEO)* 1–2

[7] Runge A F J, Aguergaray C, Broderick N G R and Erkintalo 2014 Raman rogue waves in a partially mode-locked fiber laser *Opt. Lett.* **39** 319

[8] Akhmediev N *et al* 2016 Roadmap on optical rogue waves and extreme events *J. Opt.* **18** 063001

[9] Dudley J M, Dias M, Erkintalo M and Genty G 2014 Instabilities, breathers and rogue waves in optics *Nat. Photon.* **8** 755–64

[10] Lecaplain C, Grelu Ph, Soto-Crespo J M and Akhmediev N 2013 Dissipative rogue wave generation in multiple-pulsing mode-locked fiber laser *J. Opt.* **15** 064005

[11] Lecaplain C, Grelu Ph, Soto-Crespo J M and Akhmediev N 2012 Dissipative rogue waves generated by chaotic pulse bunching in a mode-locked laser *Phys. Rev. Lett.* **108** 233901

[12] Soto-Crespo J M, Grelu Ph and Akhmediev N 2011 Dissipative rogue waves: extreme pulses generated by passively mode-locked lasers *Phys. Rev.* E **84** 016604

[13] Baronio F, Degasperis A, Conforti M and Wabnitz S 2012 Solutions of the vector nonlinear Schrödinger equations: evidence for deterministic rogue waves *Phys. Rev. Lett.* **109** 044102

[14] Chen S H, Grelu Ph and Crespo J M 2014 Dark-and bright-rogue-wave solutions for media with long-wave–short-wave resonance *Phys. Rev.* E **89** 011201

[15] Chang W, Soto-Crespo J M, Vouzas P and Akhmediev N 2018 Extreme pulse dynamics in mode-locked lasers *Recent Trends in Applied Nonlinear Mechanics and Physics* (Berlin: Springer) 171–89

[16] Sergeyev S V, Mou C, Kolpakov S and Kalashnikov V 2016 Bright and dark vector rogue waves *18th IEEE Int. Conf. Transparent Optical Networks (ICTON)* 1–4

[17] Sergeyev S V, Kolpakov S A, Mou C, Jacobsen G, Popov S and Kalashnikov V 2016 Slow deterministic vector rogue waves *Proc. SPIE* **9732** 97320K

[18] Kbashi H, Kolpakov S A and Sergeyev S V 2016 Temporal scaling of optical rogue waves in unidirectional ring fiber laser *18th IEEE Int. Conf. Transparent Optical Networks (ICTON)* 1–4

[19] Kolpakov S A, Kbashi H and Sergeyev S V 2016 Dynamics of vector rogue waves in a fiber laser with a ring cavity *Optica* **3** 870–5

[20] Kolpakov S, Kbashi H and Sergeyev S 2016 Slow optical rogue waves in a unidirectional fiber laser *CLEO: QELS Fundamental Science JW2A–56*

[21] Kolpakov S A, Kbashi H and Sergeyev S 2017 Rogue waves driven by polarization instabilities in a long ring fiber oscillator *Proc. SPIE* **10228** 102280Q

[22] Sergeyev S V, Kbashi H, Tarasov N, Loiko Yu and Kolpakov SA 2017 Vector-resonance-multimode instability *Phys. Rev. Lett.* **118** 033904

[23] Chang S, Baronio F, Soto-Crespo JM, Grelu P and Mihalache D 2017 Versatile rogue waves in scalar, vector, and multidimensional nonlinear systems *J. Phys. A: Math. Theor.* **50** 463001

[24] Gordon J P 1983 Interaction forces among solitons in optical fibers *Opt. Lett.* **8** 596–8

[25] Mitschke F M and Mollenauer L F 1987 Experimental observation of interaction forces between solitons in optical fibers *Opt. Lett.* **12** 355

[26] Pikovsky A, Rosenblum M and Kurths J 2001 *Synchronisation: A Universal Concept in Nonlinear Science* (Cambridge: Cambridge University Press)

[27] Arenas A, Díaz-Guilera A, Kurths J, Moreno Y and Zhou Ch 2008 Synchronization in complex networks *Phys. Rep* **469** 93

[28] Ansmann G, Karnatak R, Lehnertz K and Feudel U 2013 Extreme events in excitable systems and mechanisms of their generation *Phys. Rev.* E **8** 052911

[29] Sergeyev S V, Mou C, Turitsyna E G, Rozhin A, Turitsyn S K and Blow K 2014 Spiral attractor created by vector solitons *Light Sci. Appl* **3** e131

[30] Sergeyev S V 2014 Fast and slowly evolving vector solitons in mode-locked fibre lasers *Phil. Trans. R. Soc. A* **372** 20140006

[31] Zeghlache H and Boulnois A 1995 Polarization instability in lasers. I. Model and steady states of neodymium-doped fiber lasers *Phys. Rev.* A **52** 4229

[32] Leners R and Stéphan G 1995 Rate equation analysis of a multimode bipolarization Nd3+ doped fibre laser *Quantum Semiclass. Opt.* **7** 757

[33] Sergeyev S V 1999 Spontaneous light-polarization symmetry breaking for an anisotropic ring-cavity dye laser *Phys. Rev.* A **59** 3909

[34] Gaisenok V A, Krylov G G and Sergeyev S V 1994 Dynamic operations of a dye laser with a saturable absorber: pumping anisotropy and orientational relaxation *Physica D* **72** 161

[35] Sergeyev S V 2016 Vector self-pulsing in erbium-doped fiber lasers *Opt. Lett.* **41** 4700–3

[36] Tlidi M *et al* 2014 Localized structures in dissipative media: from optics to plant ecology *Phil. Trans. R. Soc. A* **372** 20140006

[37] Agrawal G P 2013 *Nonlinear Fiber Optics* (San Diego, CA: Academic)

[38] Zakharov V E and Ostrovsky L A 2009 Modulation instability: the beginning *Physica D* **238** 540

[39] Benjamin T B and Feir J E 1967 The disintegration of wave trains on deep water Part 1. Theory *J. Fluid Mech.* **27** 417

[40] Faraday M 1831 On a peculiar class of acoustical figures; and on certain forms assumed by groups of particles upon vibrating elastic surfaces *Phil. Trans. R. Soc. London* **121** 299

[41] Szwaj Ch, Bielawski S, Derozier D and Erneux T 1998 Faraday instability in a multimode laser *Phys. Rev. Lett.* **80** 3968

[42] Turitsyna E G, Smirnov S V, Sugavanam S, Tarasov N, Shu X, Babin S A, Podivilov E V, Churkin D V, Falkovich G and Turitsyn S K 2013 The laminar-turbulent transition in a fibre laser *Nat. Photon.* **7** 783–6

[43] Onorato M *et al* 2009 Statistical properties of directional ocean waves: the role of the modulational instability in the formation of extreme events *Phys. Rev. Lett.* **102** 114502

[44] Perego A M, Tarasov N, Churkin D V, Turitsyn S K and Staliunas K 2016 Pattern generation by dissipative parametric instability *Phys. Rev. Lett.* **116** 028701

[45] Risken H and Nummedal 1968 Instability of off resonance modes in lasers *Phys. Lett.* A **26** 275

[46] Risken H and Nummedal 1968 Self-pulsing in lasers *J. Appl. Phys.* **39** 4662

[47] Graham R and Haken H 1968 Quantum theory of light propagation in a fluctuating laser-active medium *Z. Phys.* **213** 420

[48] Fontana F, Begotti M, Pessina E M and Lugiato L A 1995 Maxwell–Bloch modelocking instabilities in erbium-doped fiber lasers *Opt. Commun.* **114** 89

[49] Pessina E M, Bonfrate G, Fontana F and Lugiato L A 1997 Experimental observation of the Risken–Nummedal–Graham–Haken multimode laser instability *Phys. Rev.* A **56** 4086

[50] Pessina E M, Prati F, Redondo J, Roldán E and de Valcárcel G J 1999 Multimode instability in ring fiber lasers *Phys. Rev.* A **60** 2517

[51] Roldán E, de Valcárcel G J, Prati F, Mitschke F and Voigt T 2004 Multilongitudinal mode emission in ring cavity class B lasers arXiv:physics/0412071

[52] Voigt T, Lenz M, Mitschke F, Roldán E and de Valcárcel G J 2004 Experimental investigation of Risken–Nummedal–Graham–Haken laser instability in fiber ring lasers *Appl. Phys.* B **79** 175

[53] Lugiato L, Prati F and Brambila M 2015 *Nonlinear Optical Systems* (Cambridge: Cambridge University Press)

[54] Anishchenko V S, Astakhov V, Neiman A, Vadivasova T and Schmansky-Geier L 2007 *Nonlinear Dynamics of Chaotic and Stochastic Systems* (Berlin, Heidelberg, New York: Springer) 121–39

[55] Gumenyuk R and Okhotnikov O G 2013 Impact of gain medium dispersion on stability of soliton bound states in fiber laser *IEEE Photon. Technol. Lett* **25** 133–135

[56] Varshalovich D A, Moskalev A N and Khersonskii V K 1988 *Quantum Theory of Angular Momentum* (Singapore: World Scientific) 21–3

IOP Publishing

Nonlinear Guided Wave Optics
A testbed for extreme waves
Stefan Wabnitz

Chapter 10

Fundamental rogue waves and their superpositions in nonlinear integrable systems

Nail Akhmediev, Adrian Ankiewicz and José M Soto-Crespo

10.1 Introduction

High-amplitude waves must be considered as nonlinear. They are normally described by nonlinear partial differential equations. A paradigmatic example is the nonlinear Schrödinger equation [1] and its extensions [2]. The integrability of this equation plays a significant role in our ability to derive exact solutions for physical processes of interest. Until recently, most research activity has focused on solitons and their interactions [1, 2]. The discovery of rogue waves played a major role in shifting this interest towards breathers [3] and rational solutions [4]. This latter category represents doubly localized solutions that naturally describe rogue waves [4]. The fact of the existence of such solutions and their higher-order counterparts [5] introduced a new vision for these formations. This chapter summarizes the theoretical knowledge accumulated until now in this area of research. Importantly enough, most of the theoretical predictions have been confirmed in experiments [6–9], thus solidifying the theoretical accomplishments.

Solitons are solutions that are localized only along the transverse variable (retarded time in fiber optics), but their shape remains constant in the longitudinal direction (distance along the fiber). In contrast, rogue waves are localized both in time and in space [4]. As the energy is conserved, a complete localization with zero field at infinity is impossible. This type of localized solution usually appears on a constant background that serves as the source of its energy. When a rogue wave grows, energy is concentrated towards its center and is absorbed afterwards by the background when the rogue wave disappears. This fact is in agreement with the observations of rogue waves in the ocean. As the NLSE describes the wave envelope, the background represents the average wave pattern of the ocean surface. When conditions allow, a rogue wave may grow out of this background, gaining energy from its surrounding waves and exceeding their amplitude several times. The central

doi:10.1088/978-0-7503-1460-2ch10 © IOP Publishing Ltd 2017

bump of the envelope may contain more than a single wave. This case corresponds to the so-called 'three sisters,' where several waves of high amplitude can appear, one after another.

In recent years, the rogue wave concept has been expanded significantly. Now, there are not only extreme waves in the oceans, but there are also pulses with extreme amplitudes in fiber optics [10], formations with high amplitudes in wide-aperture optical cavities [11, 12], extreme pulses in laser outputs [13, 14], and areas of high concentration in Bose–Einstein condensates (BEC) [15]. Spreading the ideas to other areas of physics plays a positive role in enriching the rogue wave concept [16]. Finding common features between objects of different scale and nature allows us to use the laboratory observations and apply them to ocean waves that are normally difficult to study due to the dangerous conditions they involve and the enormity of the oceans. Collective efforts of experts from various disciplines are providing rapid progress in this research field [17]. Laboratory physics allows a faster accumulation of data from smaller scale processes than a direct study of the evolution of ocean waves. Optics in particular is revealing itself as the most useful field in leading this progress [18].

One of the factors that has contributed to the expansion of the rogue-wave concept is the fact that wave dynamics in these seemingly different fields is described by similar evolution equations. At the basic level, when ignoring higher-order effects, this evolution equation is the nonlinear Schrödinger equation (NLSE). This equation accurately describes nonlinear waves in optical fibers [19], the dynamics of ocean waves [20], and the physics of BEC. Despite being known for several decades, and even with the development of elaborate mathematical techniques for solving it, the NLSE still remains one of the most studied equations of nonlinear dynamics. Its popularity has increased since the rogue wave concept has been transferred to mathematics. New types of NLSE solutions have been found to describe rogue waves. This chapter presents the new developments in this area during the last few years.

10.2 NLSE rogue waves

The nonlinear Schrödinger equation is one of the first nonlinear partial differential evolution equations that was found to be completely integrable [1]. In a normalized form, it can be written as:

$$i\frac{\partial\psi}{\partial x} + \frac{1}{2}\frac{\partial^2\psi}{\partial t^2} + |\psi|^2\psi = 0. \tag{10.1}$$

Here, we shall adhere to one of the two alternative notations. Namely, we consider x as the propagation distance and t as the retarded time in a reference frame moving with the group velocity. This notation was originally adopted in nonlinear fiber optics [19] and recently in the theory of ocean waves [21]. Zakharov and Shabat developed a technique for solving the initial value problem for this equation [1]. For localized initial conditions, the general solution of equation (10.1) consists of a finite number of solitons and small amplitude radiation waves. However, when the initial conditions contain a background that is not zero at infinity, the solutions turn out to be more complicated, and the problem cannot be described by the original theory [1]. The initial value problem then becomes excessively complicated and additional assumptions are needed. Solitons and radiation waves are no longer the only possible nonlinear excitations in the medium, and it is necessary to take into account the existence of other modes possessed by the NLSE.

A more complete set of first-order solutions of the NLSE was found in [22]. In addition to ordinary solitons, this set contains solitons on a background (Kuznetsov–Ma solitons), cnoidal waves, doubly periodic solutions, Akhmediev breathers, and rogue waves (*viz.* the Peregrine solution). All of them play essential roles in nonlinear wave evolution when the initial conditions are not localized. Before solving a complicated initial value problem, each of these fundamental solutions should be understood in detail.

The closest analog of a rogue wave in mathematics is the rational Peregrine solution [4, 23]:

$$\psi(x, t) = \left[4\frac{1 + 2\,ix}{1 + 4x^2 + 4t^2} - 1\right]e^{ix}. \tag{10.2}$$

This is also known as the Peregrine soliton [6] or Peregrine breather [24]. The reason for this is that this solution is the limiting case of Akhmediev breathers or the Kuznetsov–Ma solitons when the period of either solution becomes infinite. Then, only one of the peaks of the solution remains and the solution becomes localized both in time and space.

Finding this limit is far from trivial. Breather or soliton solutions are usually expressed in terms of trigonometric and/or hyperbolic functions. These solutions become undefined at the exact limit. Special techniques are required to circumvent this problem [25]. The resulting solutions become rational, i.e. they are expressed as a ratio of polynomials. Rational solutions of the NLSE and other integrable equations have been known and analyzed in detail by mathematicians. However, they mostly dealt with singular solutions [26, 27]. The latter cannot be used in physics, since nature avoids singularities.

Peregrine derived this solution in 1983 as the limit of the Kuznetsov–Ma soliton [28]. The amplitude profile of the Peregrine solution is shown in figure 10.1(a). It has

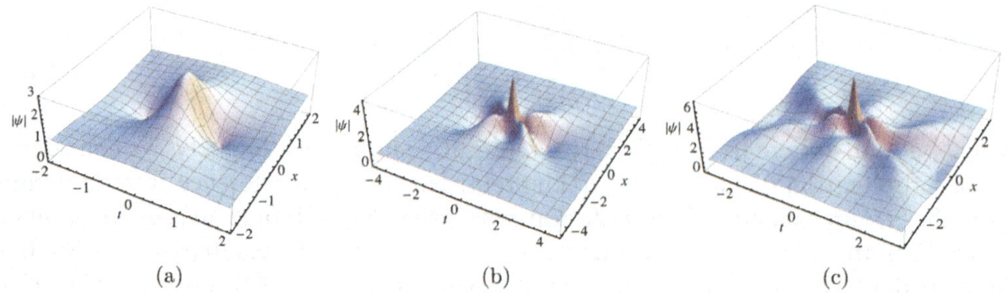

Figure 10.1. (a) Peregrine solution. (b) Akhmediev–Peregrine solution of second order. (c) Akhmediev–Peregrine solution of third order.

a maximum amplitude of 3 which is positioned at the origin and a background magnitude of 1. In addition to the peak at the center, the solution has two depressions with zero amplitude along the t-axis, offset from the maximum. These zeros are the distinctive feature of the Peregrine solution that allows us to distinguish it from other excitations in a chaotic wave field. This has been observed in experiments, both in water waves [7] and in optical fibers [6]. Recent studies have shown that this solution appears in field evolution dynamics even in the case of localized initial conditions. That is, the Peregrine solution is a universal structure emerging during the intensity localization of high-power pulses in the semi-classical limit of the NLSE [29].

The NLSE is an envelope equation. Thus, the profile shown in figure 10.1(a) is the wave envelope, both in optics and in the water wave case, that shows the maxima of the waves within this envelope. There could be more than one maximum within the central elevated part of the envelope. In this case, more than one rogue wave can be experienced. Moreover, the origin may correspond to a depression (trough) of the water wave rather than to an elevation (crest). In this case, this trough could be called a rogue wave hole [30].

The phase factor e^{ix} in equation (10.2) is a correction to the total wave number of these waves. It influences the ratio of the group and phase velocities of the waves. Group and phase velocities of the waves normally differ. In the water wave case, they differ by a factor of two. As a consequence, the elevated rogue wave may be transformed into a rogue wave hole during wave propagation. This may happen within the limited time existence of the rogue wave. In optics, we measure the intensity of waves, i.e. we deal with the absolute value of the optical field. Therefore, rogue wave holes cannot be observed. This difference must be taken into account when building analogies between optical and oceanic rogue waves. The above considerations are only valid for the NLSE case. Multi-component solutions, like those in [31] or in the works of Baronio [32–34] describe different physical processes, and conclusions may differ depending on the component that we are measuring. Still, in optics, even negative values of the envelope correspond to positive intensities.

The Peregrine solution is not the only one which is localized both in time and in space. The NLSE admits higher-order rational solutions that have the same property. One such solution has been presented in [35] and its relation to rogue waves is revealed in [4]. This is the second-order solution, and is shown in figure 10.1(b). Several methods have been developed for constructing higher-order solutions of integrable equations. One of them is known as the Darboux transformation [36]. Alternative methods also exist [37–39]. All these techniques show that there is a hierarchy of rogue wave solutions with progressively increasing central amplitudes. They are known as Akhmediev–Peregrine breathers [40–43]. Their general form is given by

$$\psi_N(x,\,t) = \left[(-1)^N + \frac{G_N(x,\,t) + ixH_N(x,\,t)}{D_N(x,\,t)}\right]e^{ix}, \tag{10.3}$$

where G, H, and D are polynomials in the two variables x and t, and N is the order of the solution. In the simplest case of the Peregrine solution, where $N = 1$, we have [2, 28]

$$G_1(x, t) = 4, \quad H_1(x, t) = 8, \quad D_1(x, t) = 1 + 4x^2 + 4t^2. \tag{10.4}$$

For the second-order solution, we have

$$\begin{aligned} G_2 &= \frac{3}{8} - 3t^2 - 2t^4 - 9x^2 - 10x^4 - 12t^2x^2, \\ H_2 &= \frac{15}{4} + 6t^2 - 4t^4 - 2x^2 - 4x^4 - 8t^2x^2, \\ D_2 &= \frac{1}{8}\left[\frac{3}{4} + 9t^2 + 4t^4 + \frac{16}{3}t^6 + 33x^2 + 36x^4 + \frac{16}{3}x^6 - 24t^2x^2 \right. \\ &\quad \left. + 16t^4x^2 + 16t^2x^4\right]. \end{aligned} \tag{10.5}$$

The third-order solution is given below in section 10.3. It is shown in figure 10.1(c) for the case $b = 0$. The analytical expressions for even higher-order solutions are too cumbersome to be presented here. They can be found in [39].

The maximal amplitude of the solutions of this hierarchy is $(2N + 1)$ times the background, where N is the order of the solution. Thus, for the first-order solution shown in figure 10.1(a), the amplitude is 3 times the background, while for the second-order solution shown in figure 10.1(b), the amplitude is 5 times the background. The solution shown in figure 10.1(c) has an amplitude of 7 times the background. This remarkable result has been first presented in [5] (see table 1 of this work). Though simple to write down, it is far from trivial. The first mathematical proof of this expression for the maximal amplitude was given in [44]. Namely, the recurrence relations (7) and (8) of this work lead directly to the explicit formula $(2N + 1)$ for the maximal amplitude. Gaillard presented an alternative proof of the same expression in [40]. Additional mathematical evidence was given in [45]. The maximal amplitude of a rogue wave is one of its main characteristics. Therefore, interest regarding this matter has been very high. This issue has been also addressed in [46].

A convenient way of expressing the rogue wave solutions is to represent their intensity minus the background. Namely, the Peregrine solution can be written as

$$|\psi_1(x, t)|^2 - 1 = 8\frac{(1 - 4t^2 + 4x^2)}{(4t^2 + 4x^2 + 1)^2}. \tag{10.6}$$

We can refer to it as the elementary rogue wave solution (ERW). Equivalently, we can consider it as a rogue wave quantum. An interesting fact about this representation is that, for any order N, it can be expressed in terms of just the denominator of the whole solution [47]:

$$|\psi_N(x, t)|^2 - 1 = [\log(D_N(x, t))]_{tt}. \tag{10.7}$$

This shows that the denominator D_n contains all the information concerning the intensity pattern of the rogue wave solutions.

10.3 Splitting of higher-order rogue waves

General rogue wave patterns are defined by the order of the solution, N, and may depend on additional free parameters. The free parameters split the solution into separate parts. They specify mutual separations and relative positions of the individual components on the (x, t) plane. Generally, a solution of Nth order contains $Q_N = N(N + 1)/2$ individual components that, when well-separated, can be identified as Peregrine solutions [48]. A detailed classification of the rogue wave patterns has been given earlier in [49].

Independent of separation, the Nth order solution pattern always contains $N(N + 1)/2$ elementary particles. In other words, solutions of orders $N = 1, 2, 3, 4, 5, \ldots$ consist of 1, 3, 6, 10, 15, … ERWs, respectively. The latter, known as 'triangular numbers,' are illustrated in figure 10.2. These are the total number of points uniformly distributed within a triangle with N points along its edges.

Let us consider the simplest example. The second-order solution with free parameters has been given earlier in [37, 50]. As explained previously, the solution intensity $|\psi|^2$ is completely defined by the denominator:

$$\begin{aligned} D_2(x, t) = {} & \beta^2 + \gamma^2 + 64t^6 + 48t^4(4x^2 + 1) - 16\beta t^3 \\ & + 12t^2(16x^4 - 24x^2 + 4\gamma x + 9) + 12t\beta(1 + 4x^2) \\ & + 64x^6 + 432x^4 - 16\gamma x^3 + 396x^2 - 36\gamma x + 9, \end{aligned}$$

where β and γ are free parameters. Let us suppose that the three components are well separated, i.e. at least one of the parameters β or γ is large. It was shown [50] that the components of the resulting rogue wave triplet are then located at the vertices of an equilateral triangle. An example of this solution is shown in figure 10.3(a). Despite being of second order, the solution contains three elementary rogue waves. This is a distinctive new feature of rogue wave solutions, distinguishing them from multi-soliton and multi-breather solutions. In the two latter cases, the number of individual components is equal to the order of the solution. Namely, the Nth order multi-soliton solution contains exactly N solitons. The physical reason for this difference is still unclear.

To see the idea more clearly, let us consider the third-order circular rogue wave solution [51]. The solution is illustrated in figure 10.3(b). This is a particular case of a split third-order solution. Various forms of the denominator, $D_3(x, t)$, have been presented in [47, 50, 53, 55] and [54]. The circular pattern shown in figure 10.3(b) is

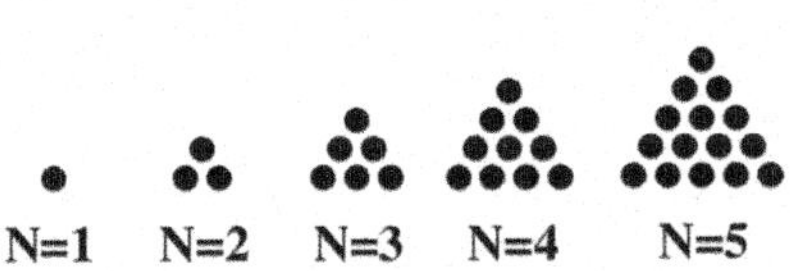

Figure 10.2. The first five triangular numbers; they are represented by the number of solid dots within each triangle.

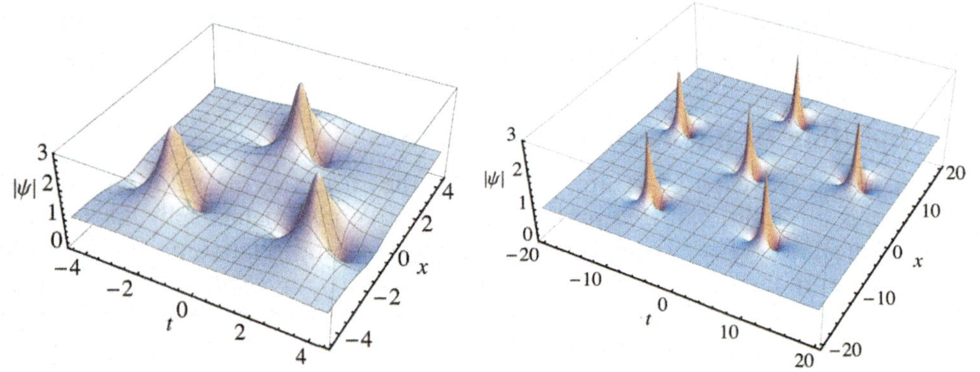

Figure 10.3. (a) Rogue wave triplet with parameters $\gamma = 200$ and $\beta = 0$ [50]. (b) Third-order rogue wave solution with a circular pattern; it is given by equations (10.7) and (10.8) with $b = 2 \times 10^7$.

described with a single parameter, b. Let us write the denominator following the notation used in [55]:

$$D_3(x, t) = \sum_{j=0}^{12} d_j(T)X^j, \tag{10.8}$$

where $X = 2x$ and $T = 2t$, while the T-polynomials $d_j(T)$ are:

$$\begin{aligned}
d_0(T) &= b^2(T^2 + 1) + T^{12} + 6T^{10} + 135T^8 + 2340T^6 + 3375T^4 \\
&\qquad + 12\,150T^2 + 2025, \\
d_1(T) &= -10b(T^6 - 9T^4 - 45T^2 + 45), \\
d_2(T) &= b^2 + 6(T^{10} - 15T^8 + 90T^6 + 2250T^4 - 6075T^2 + 15\,525), \\
d_3(T) &= 10b(T^4 + 18T^2 + 9), \\
d_4(T) &= 15(T^8 - 12T^6 - 90T^4 + 5220T^2 + 9585), \\
d_5(T) &= 6b(3T^2 - 17), \\
d_6(T) &= 20(T^6 + 3T^4 + 675T^2 + 765), \\
d_7(T) &= -2b, \\
d_8(T) &= 15(T^4 + 18T^2 + 249), \\
d_{10}(T) &= 6(T^2 + 21), \\
d_{12}(T) &= 1,
\end{aligned}$$

with $d_9(T) = d_{11}(T) = 0$. For b large enough, the central part of the solution can be described by

$$D_3(x, t) = b^2(1 + T^2 + X^2) = b^2(1 + 4t^2 + 4x^2).$$

As expected, there is a single ERW at the origin. When $b = 0$, D_3 reduces to the special form given in appendix E of [55] (with numerators and denominators multiplied by 16 200). Since D_3 has order 12, at large b, we can approximately factorize it into a product of six polynomials, $\prod_{j=0}^{5} F_j$, representing six separated ERWs. For a circular pattern [51], one of them is centered at the origin and the remaining 5 form a pentagon around it, as shown in figure 10.3(b). The total number of ERWs is again represented by a triangular number. Remarkably, there are no exact solutions with the number of ERWs being 2, 4, 5, 7, ... etc.

This is valid for all higher-order solutions. For simplicity, let us consider only circular solutions [51], and find the radii r of these circles. We only need to minimize D_N for large values of (x,t). In other words, we need to find the value of r which minimizes D_N. This step involves only a few higher-order terms in the polynomial. As before, we set $X = 2x$. For $N = 3$, the highest-order terms in X are

$$D_3 = X^{12} - 2bX^7 + b^2X^2 + \cdots.$$

Setting $b = (2u_3r)^5$ with $x = r$, $t = 0$ for example, shows that $(u_3^5 - 1)^2 r^{12} = 0$, so $u_3 = 1$ and the radius is $r \approx \frac{1}{2}b^{1/5}$. Figure 10.3(b) uses $b = 2 \times 10^7$, and it can be seen that the predicted radius of 14.4 is quite accurate. For $N = 4$, the highest-order terms in X are

$$D_4 = X^{20} - 2bX^{13} + b^2X^6 + \cdots.$$

Setting $b = (2u_4r)^7$ with $x = r$ and $t = 0$, for example, shows that $(u_4^7 - 1)^2 r^{20} = 0$, so $u_4 = 1$ and the radius is $r \approx \frac{1}{2}b^{1/7}$. Near the center, i.e. for small x and t, for large b we obtain:

$$\begin{aligned} D_3/b^2 \approx\ & 9 + 27T^2 + 3T^4 + T^6 + 99X^2 + 27X^4 \\ & + X^6 - 18T^2X^2 + 3T^4X^2 + 3T^2X^4, \end{aligned}$$

where $T = 2t$, i.e. it has the form of the second-order basic rogue wave given in [4, 35, 55]. This is an algebraic verification of the inherent quanta referred to earlier. Here, the central second order rogue wave contains three quanta while each of the seven in the surrounding ring contains one, making the expected total of 10.

For $N = 5$, the highest-order terms are $D_5 = X^{30} + 2bX^{21} + b^2X^{12} + \cdots$. Setting $b = (2u_5r)^9$ with $x = -r$ and $t = 0$, for example, shows that $(u_5^9 - 1)^2 r^{30} = 0$. Therefore, $u_5 = 1$ and the radius is $r \approx \frac{1}{2}b^{1/9}$.

For $N = 6$, the highest-order terms are $D_6 = X^{42} - 2bX^{31} + b^2X^{20} + \cdots$. Setting $b = (2u_6r)^{11}$ with $x = r$ and $t = 0$ shows that $(u_6^{11} - 1)^2 r^{42} = 0$. Then $u_6 = 1$ and the radius is $r \approx \frac{1}{2}b^{1/11}$. For an arbitrary value of N, the radius is well approximated by $r \approx \frac{1}{2}b^{1/(2N-1)}$.

The preceding radius estimates are approximate. However, from this study, it clearly appears that the number of ERWs does not depend on the free parameters. Their relative locations can change, and the position of each individual ERW may move, but the total number of them remains fixed and is given by a triangular number. Moreover, all ERWs can be built into a structure which does not allow us to see each part separately. However, even such a complicated structure consists of a triangular number of ERWs.

The number of components, Q_N, in a higher-order rogue wave can be calculated as the integral of the squared deviation from the background level across the space–time plane. Namely,

$$\begin{aligned} Q_N &= \frac{1}{8\pi}\int_{-\infty}^{\infty}\int_{-\infty}^{\infty}[|\psi_N(x,\, t)|^2 - 1]^2 dtdx \\ &= \frac{1}{2}N(N+1). \end{aligned} \tag{10.9}$$

In other words, Q_N, whether obviously apparent or not, is

$$Q_N = \frac{1}{8\pi}\int_{-\infty}^{\infty}\int_{-\infty}^{\infty}\{[\log(D_N(x,\, t))]_{tt}\}^2 dtdx. \tag{10.10}$$

As follows from (10.9), for the multi-rogue wave solution $\psi_N(x,t)$ of order N, the integral will be equal to a triangular number of order N, *viz.* Q_N. This integral does not depend on the free parameters included in the solution $\psi_N(x,t)$. Numerical evidence shows that the above integral is always equal to the number of ERWs in the solution, no matter what kind of pattern it has. For a single Peregrine solution, $\psi_1(x,t)$, we obtain analytically from equation (10.6):

$$\int_{-\infty}^{\infty}[|\psi_1(x,\, t)|^2 - 1]^2\, dt = \frac{8\pi}{(1+4x^2)^{3/2}}, \tag{10.11}$$

and then

$$\int_{-\infty}^{\infty}(1+4x^2)^{-3/2}dx = 1, \tag{10.12}$$

so that the number of components $Q_1 = 1$, as conjectured. For higher-order solutions, if the rogue wave components are 'well-separated' into a certain structure with $\frac{1}{2}N(N+1)$ peaks (that is, the free parameters of the solution are large), then Q_N will be just the sum of the integrals over each individual peak. As each integral equals unity, therefore $Q_N = \frac{1}{2}N(N+1)$.

When the separation is not complete (i.e. the free parameters are small), then we can expand the integrand to first-order in a small parameter. Suppose, for the second-order rogue wave, the integrand in equation (10.9) is $g(x,t,\beta,\gamma)$ with two free parameters. For $\gamma = 0$ and β small, this can be expanded as $g(x,t,0,0) + \beta\,\frac{\partial g}{\partial \beta}$. The surface integral of the first part is 3. The second function is an odd function of t for each x and hence the integral of it is zero. Similarly, for γ small and $\beta = 0$, we have $g(x,t,\beta = 0,\gamma) = g(x,t,0,0) + \gamma\,\frac{\partial g}{\partial \gamma} + \cdots$. The second term is an odd function of x

for any t, so the integral of it is also zero. Then, Q_2 remains equal to 3 as separation commences. The same effect occurs for the third order rogue wave: Q_3 remains equal to 6 as b starts to increase.

The value of Q_N is closely related to the central maximum of the integrand that can be found from the amplitude $\psi_N(0, 0) = (2N + 1)$ [44, 55]. When the free parameters are set to zero, the intensity at the origin can be found using the above expression; it is:

$$|\psi_N(0, 0)|^2 - 1 = c_N = 4N(N + 1).$$

Suppose that the integrand is approximated by a smooth function having its maximum at its center and decaying to zero at infinity. To be specific, let us take a Gaussian function with amplitude c_N^2, i.e.

$$S_N = c_N^2 \exp(-c_N r^2),$$

where $r^2 = x^2 + t^2$. Then,

$$\frac{1}{8\pi}(2\pi) \int_0^\infty S_N \, r \, dr = \frac{c_N}{8} = \frac{1}{2}N(N + 1) = Q_N. \tag{10.13}$$

This is exactly the same result as in equation (10.9).

This is a fundamental result: the number of ERWs in a multi-rogue-wave solution of the NLSE is given by a triangular number. No other number of ERWs can be present in a general multi-rogue-wave solution. This number does not depend on whether the higher-order rogue wave is well separated or partially separated into elementary components. In the latter case, the intensity pattern can be like the ones shown in figure 10.1(b) or figure 10.1(c), and the number of ERWs is not visually obvious. The integral approach can be used to determine the number of ERWs in any unknown rogue wave pattern.

This approach can be applied only to solutions localized both in time and in space. Otherwise the integral may become infinite. Clearly, no such integral can be proposed for soliton solutions. This simple fact shows that crucial differences exist between the classes of soliton solutions and rogue wave solutions.

Illustrative examples of higher-order solutions are shown in figure 10.4. This figure is a table of calculated rogue wave patterns that increase in order when going from top to bottom. It starts with the simplest Peregrine soliton in the top cell. Higher-order Akhmediev–Peregrine (AP) solutions are located in the left column. The figure is limited to solutions of sixth order, although it can be continued indefinitely. The second column represents triangular structures. The third column represents pentagram patterns. The fourth column provides examples of heptagram structures, i.e. patterns with seven ERWs on one or several circles of different radii. Further columns have 9, 11, … ERWs in the outer and inner shells of the structure.

'Atomic-like' circular structures are located along the diagonal line. Starting from order 3, they consist of a central nucleus containing $N - 2$ elementary rogue waves in the form of an AP solution and $2N - 1$ ERWs playing the role of the shell of 'electrons'. Patterns in inner cells of the table can be considered as 'atomic

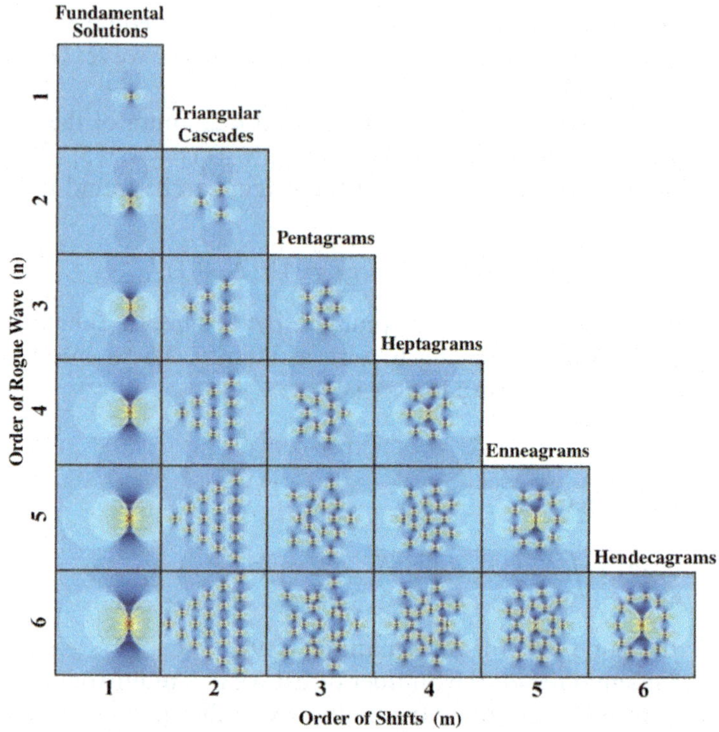

Figure 10.4. Illustration of the large variety of higher-order rogue wave solutions up to sixth order. There is only one first-order solution, the Peregrine soliton (first row), while there are two types of second-order solutions: the AP solution and the triplet (second row). The higher the order of the solution, the more patterns that are allowed due to the increasing number of free parameters. Circular clusters along the diagonal line display a ring of $(2N - 1)$ Peregrine solitons around a central rogue wave of order $N - 2$ (for $N > 2$). These can be viewed as atomic structures with a nucleus and a shell of 'electrons'.

structures' with several shells of 'electrons'. The larger variety of patterns for higher-order solutions is caused by the larger number of free parameters controlling the solution.

These complex NLSE solutions might presently look like a mathematical curiosity. Can they be observed experimentally? The Peregrine soliton has been observed both in optics [6] and in water waves [7]. The rogue wave triplet has been observed in a water wave tank [8]. Moreover, AP solutions up to fifth order have also been observed [9]. We have no doubt that even more complicated structures will soon be observed in one setting or another. As the NLSE has such a wide range of applications, these solutions may appear in situations that we cannot even predict right now. Can this approach be used for a description of real atomic structures? Only the future will give us an answer.

10.4 Extended equation

The NLSE is an equation that serves to approximately model nonlinear propagation phenomena. It includes the lowest orders of dispersion and nonlinearity that allow us to study the basic features of nonlinear waves. If more accurate modeling is required, this equation has to be extended to include higher-order terms. In optics, they have special names: the self-steepening term, $\sim\frac{\partial}{\partial t}(|\psi|^2\psi)$, a term related to the self-frequency shift, $\sim\psi\,\frac{\partial}{\partial t}(|\psi|^2)$, and the third order dispersion term, $\sim\psi_{ttt}$ [56–58]. Other forms of third-order contributions can be expressed as combinations of these three terms. By adding them to the NLSE, we obtain the following general equation of third order:

$$i\frac{\partial\psi}{\partial x} + \frac{1}{2}\frac{\partial^2\psi}{\partial t^2} + |\psi|^2\psi + i\,s\left[a_f\,\psi\,\frac{\partial}{\partial t}(|\psi|^2) + s_a\,\frac{\partial}{\partial t}(|\psi|^2\psi) - \gamma_3\,\psi_{ttt}\right] = 0, \quad (10.14)$$

where the coefficients s_a, a_f, and γ_3 are three independent parameters controlling the relative contribution of each higher-order effect, while s is generally a small factor that emphasizes that they are, in general, perturbative terms. When $s = 0$, this equation reduces to the NLSE. All of these new three terms introduce asymmetry along the t-variable and represent higher-order corrections to the group velocity dispersion term.

The modified NLSE is integrable only for some particular combinations of the three parameters [59]. When these conditions are not met, exact solutions for rogue waves may not exist. However, they can be found approximately for small values of s [60]. Namely, instead of the Peregrine solution (10.2), the corresponding approximate solution for equation (10.14) has the form:

$$\psi_p = \left[\frac{4}{D}(1 + 2i\,x) - 1 + \frac{i\,s}{D^2}[f(x,\,t) - i\,k(x,\,t)]\right]e^{ix}, \quad (10.15)$$

where $D = 1 + 4t^2 + 4x^2$ and the polynomials $f(x, t)$ and $k(x, t)$ are given by:

$$f(x,\,t) = 8t\big(4(a_f + 6\gamma_3 + 2s_a)t^2 + 12(a_f + 2(\gamma_3 + s_a))x^2 - 3a_f - 6\gamma_3 - 4s_a\big),$$

$$k(x,\,t) = 32\big(3a_f + 6\gamma_3 + 5s_a\big)tx.$$

Equation (10.15) is the solution of equation (10.14) to order s, while $f(x, t)$ and $k(x, t)$ are rational functions. Therefore, the full solution still belongs to the class of rational functions. Moreover, the highest power of these functions is lower than the highest power in the denominator polynomial, D^2. Thus, this solution remains localized, just like the original rogue wave solution. This perturbative solution is illustrated in figure 10.5. It becomes asymmetric relative to its mirror images along axes $t = 0$ and $x = 0$, but it still has inversion symmetry relative to the origin. Remarkably, the central part now shows two peaks. However, the main feature of a rogue wave solution, *viz.* its full localization, remains intact.

10.5 Integrable extensions

The general extension (10.14) can only be approximately solved when s is small. In this case, we can obtain perturbed solutions of the NLSE. For larger values of s, we have to

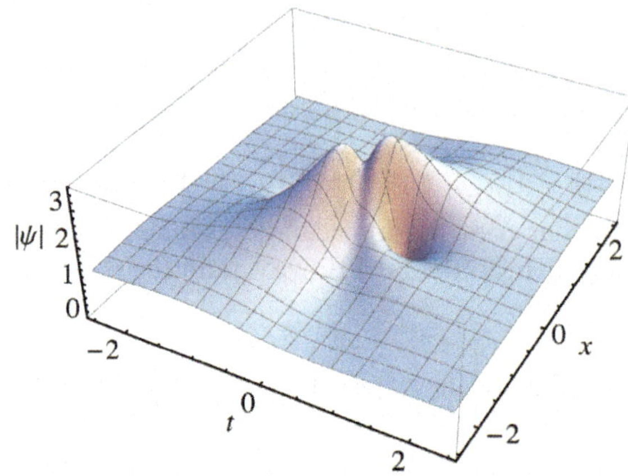

Figure 10.5. Peregrine rogue wave approximate solution of equation (10.14). The coefficients of the higher-order terms are $\gamma_3 = s_f = a_s = 1$, with the small parameter s being $s = 0.15$.

operate with special sets of coefficients for the extended equation to be integrable. A few such cases are considered below. The exact solutions obtained for them confirm the general conclusions derived for rogue wave solutions of equation (10.14).

10.5.1 Rogue waves of the Hirota equation

One of the integrable NLSE extensions is the so-called Hirota equation [61]:

$$i\frac{\partial\psi}{\partial x} + \frac{1}{2}\frac{\partial^2\psi}{\partial t^2} + |\psi|^2\psi - i\,\alpha_3\left[\frac{\partial^3\psi}{\partial t^3} + 6\,|\psi|^2\frac{\partial\psi}{\partial t}\right] = 0. \tag{10.16}$$

It can be obtained from equation (10.14) when the following conditions are met: $s\gamma_3 = \alpha_3$, $a_f = -s_a$, and $ss_a = -6\alpha_3$. Equation (10.16) is completely integrable for any arbitrary real value of the parameter α_3. The fundamental rogue wave solution of this equation is [61]:

$$\psi_1 = -\left[1 - 4\frac{1 + 2i\,x}{1 + 4(t + 6\alpha_3\,x)^2 + 4x^2}\right]e^{ix}. \tag{10.17}$$

It is similar to the Peregrine solution (10.2) apart from the term $6\alpha_3\,x$, which indicates a 'velocity' of $6\alpha_3$. The solution is shown in figure 10.6(a). The skew angle of the solution is the same feature as for the perturbed rogue wave shown in the previous section, although the sign of the effect is different because $ss_a = -6\alpha_3$. When $\alpha_3 = 0$, the skew is zero and the solution reduces to the NLSE one.

Just as in the NLSE case, the Hirota equation has an infinite hierarchy of rogue wave solutions with increasing amplitudes. In particular, the exact second-order rational solution has the form:

$$\psi_2(x,\,t) = \left(1 + \frac{G_2 + i\,x\,H_2}{D_2}\right)e^{ix}, \tag{10.18}$$

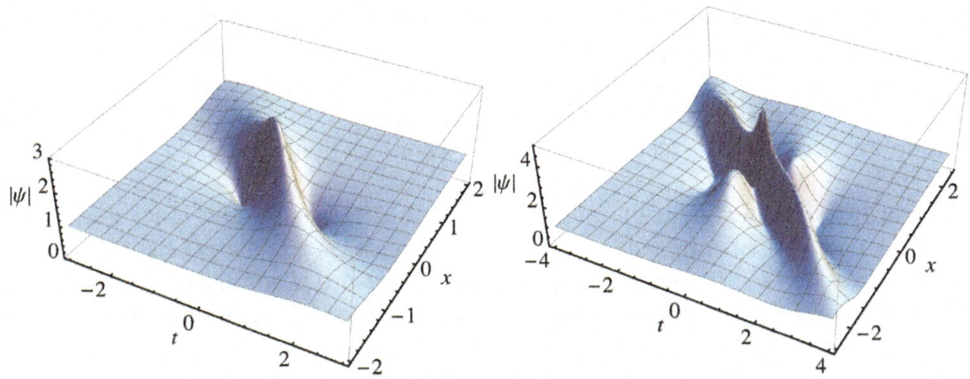

Figure 10.6. (a) First-order rational solution of the Hirota equation (equation (10.17)) for $\alpha_3 = 0.2$. (b) Second-order rational soliton of the Hirota equation for $\alpha_3 = 0.17$. The arbitrary real parameter α_3 is responsible for the skew angle of the 'ridge' of these solutions relative to the axes.

where

$$\begin{aligned} G_2/12 = &- 16t^4 - 384\alpha_3 x t^3 - 24(4(36\alpha_3^2 + 1)x^2 + 1)t^2 \\ &- 96\alpha_3 x(12(12\alpha_3^2 + 1)x^2 + 7)t - 16(1296\alpha_3^4 \\ &+ 216\alpha_3^2 + 5)x^4 - 72(44\alpha_3^2 + 1)x^2 + 3, \end{aligned} \tag{10.19}$$

$$\begin{aligned} H_2/24 = &- 16t^4 - 384\alpha_3 x t^3 - 8((432\alpha_3^2 + 4)x^2 - 3)t^2 \\ &- 96\alpha_3 x\, t(4(36\alpha_3^2 + 1)x^2 + 1) \\ &-16(36\alpha_3^2 + 1)^2 x^4 - 8(180\alpha_3^2 + 1)x^2 + 15, \end{aligned} \tag{10.20}$$

and

$$\begin{aligned} D_2 = &\, 64t^6 + 2304\alpha_3 x t^5 - 432(624\alpha_3^4 - 40\alpha_3^2 - 1)x^4 \\ &+ 36(556\alpha_3^2 + 11)x^2 + 9 + 64(36\alpha_3^2 + 1)^3 x^6 \\ &+ 384\alpha_3 t^3(12(60\alpha_3^2 + 1)x^2 - 1)x + 48t^4((720\alpha_3^2 + 4)x^2 + 1) \\ &+ 12(16(6480\alpha_3^4 + 216\alpha_3^2 + 1)x^4 - 24(60\alpha_3^2 + 1)x^2 + 9)t^2 \\ &+ 144\alpha_3 x(16(36\alpha_3^2 + 1)^2 x^4 + (8 - 864\alpha_3^2)x^2 + 17)t. \end{aligned} \tag{10.21}$$

The amplitude profile of this solution is shown in figure 10.6(b). It has a skewness similar to the one of the first-order solution. On the central line $x = 0$, where the solution reaches its maximum, we have

$$G_2(0,\, t) = 12(3 - 24t^2 - 16t^4)$$

and

$$D_2(0,\, t) = 9 + 108t^2 + 48t^4 + 64t^6.$$

These expressions show that the wave profile on this line is independent of α_3. It is therefore no surprise that the function $\psi_2(0,t)$ is the same as for the solution of the NLSE. Consequently, the absolute maximum value of the rogue wave profile occurs at $(x, t) = (0, 0)$ and is equal to 5 for any value of α_3. Obviously, when $\alpha_3 = 0$, this solution reduces to the second-order AP solution, defined by equations (10.3) and (10.5).

Similarly, we could write the higher-order AP solutions for the Hirota equation. The sequence of their amplitudes would be the same as for the NLSE rogue wave solutions, i.e. 3, 5, 7, 9, etc. More generally, higher-order solutions can be written with additional free parameters. Then, for instance, the second-order solution can be transformed into a triplet. Each component of the triplet will have a skewness which is the same for every component. A classification similar to the one shown in figure 10.4 would remain valid for the whole hierarchy of the higher-order solutions of the Hirota equation. Again, each component of these solutions will have an identical skewness.

10.5.2 Rogue waves of the Sasa–Satsuma equation

The Sasa–Satsuma equation (SSE) is another integrable extension of the NLSE. According to the original work of Sasa and Satsuma [62], the equation can be written in the form:

$$i\frac{\partial\psi}{\partial x} + \frac{1}{2}\frac{\partial^2\psi}{\partial t^2} + |\psi|^2\psi = i\varepsilon\left[i\frac{\partial^3\psi}{\partial t^3} + 3\frac{\partial|\psi|^2}{\partial t}\psi + 6\,|\psi|^2\frac{\partial\psi}{\partial t}\right]. \tag{10.22}$$

Here, the arbitrary real parameter ε scales the integrable perturbations of the NLSE. The correspondence with the coefficients on the right-hand side of equation (10.14) is the following: $ss_a = 6\varepsilon$, $s\gamma = -\varepsilon$, and $s(a_f + s_a) = 3\varepsilon$. When $\varepsilon = 0$, equation (10.22) reduces to the standard NLSE.

The exact solution describing a rogue wave of the SSE (10.22) was presented for the first time in the work [63]. It is given by:

$$\psi(x, t) = -\frac{1}{2\varepsilon}\left(1 - \frac{\zeta - \zeta^*}{c}F_q\right)\exp\left[i\left(\frac{k}{2\varepsilon}t + \frac{\omega}{8\varepsilon^2}x\right)\right], \tag{10.23}$$

with

$$\omega = 2c^2 - k^2 + (6kc^2 - k^3), \tag{10.24}$$

and

$$F_q = \frac{|u|^2\,\mathrm{Re}[\zeta](\zeta^*ug^*+\zeta u^*h) + (\zeta^*|g|^2 + \zeta|h|^2)(\zeta ug^*+\zeta^*u^*h)}{|\zeta|^2(|u|^2 + |g|^2 + |h|^2)^2 - |u^2 + 2hg|^2\,\mathrm{Im}[\zeta]2}, \tag{10.25}$$

where

$$\zeta = \pm \frac{i\sqrt{9c^2(9c^2 + 10K^2) + 3c(9c^2 - 4K^2)^{3/2} - 2K^4}}{3\sqrt{2}K},$$
$$u = \left(\frac{\nu_{21}}{2}\tau - 2\varepsilon x\right),$$
$$h = 3c\left(\frac{u}{M_1} + i\frac{12\varepsilon^2}{M_1^2}\right),$$
$$g = 3c\left(\frac{u}{M_2} - i\frac{12\varepsilon^2}{M_2^2}\right),$$
$$M_1 = K + d - \zeta,$$
$$M_2 = K - d + \zeta,$$
$$d = \left(\frac{b}{2} + \frac{2(K^2 + 18c^2 + 3\zeta^2)}{3b}\right),$$
$$b = (-1 + i\sqrt{3})[(K^2 - 9c^2 - \zeta^2)\zeta]^{1/3},$$
$$\nu_{21} = \frac{9(a - 6c^2)\zeta^4 + 3a(a - 1 - 18c^2)\zeta^2 + a^3}{3(2\zeta^2 + d\zeta + a)^2},$$
$$a = \frac{(K^2 - 3 - 36c^2)}{3}.$$

The solution depends on the variables t and x, as well as on the three real parameters ε, $K = 1 + 3k$, and c, which is proportional to the background plane wave amplitude $c/(2\varepsilon)$.

A comparison of this rogue wave solution with the corresponding one of the NLSE or of the Hirota equation shows that the SSE rogue wave has a significantly more complicated structure. In particular, while the two former equations involve polynomials of second order, the SSE involves polynomials of fourth order. This can be seen from the structure of the expression (10.25) with the numerator and denominator being of fourth power in u. These complications are related to the fact that the spectral problem for the SSE involves 3×3 matrices rather than 2×2 matrices, as for the two other cases. Consequently, solitons, rogue waves, and any other similar solutions take significantly more complex forms.

Rogue waves (10.23) do exist provided that $9c^2 - 4K^2 < 0$. This follows from the requirement that the eigenvalue ζ of the spectral problem must have a nonzero real part, and this implies that $|1 + 3k| > 3c/2$. Explicitly, the condition is either

$$k > \frac{c}{2} - \frac{1}{3} \quad \text{or} \quad k < -\frac{c}{2} - \frac{1}{3}. \tag{10.26}$$

Thus, the wavenumber k can be zero only when $c < 2/3$. Otherwise, the plane wave propagation direction has to be skewed for the rogue wave to exist.

The solution given by equation (10.23) is illustrated in figure 10.7(a) for the following values of the parameters: $c = 1$, $\varepsilon = 0.5$, and $k = 0.8$. It exhibits a double peak and has a maximum amplitude of around 2.5. The possibility of having a rogue

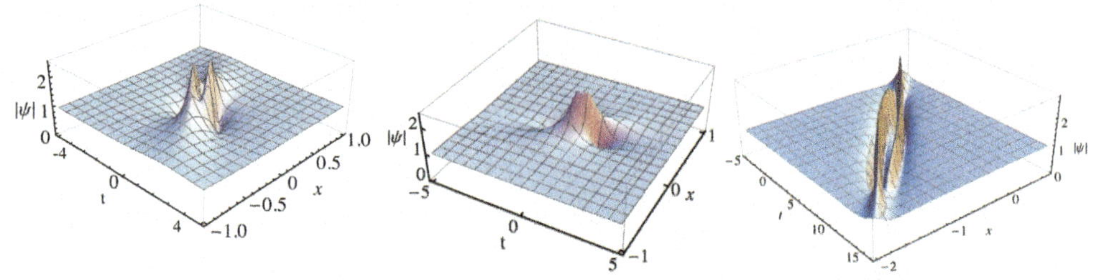

(a) (b) (c)

Figure 10.7. (a) Rogue waves of the Sasa–Satsuma equation for $c = 1$ and $\varepsilon = 0.5$. The parameter k is: (a) $k = 0.8$, (b) $k = 2$, and (c) $k = 0.25$.

wave with a double peak structure in the case of an extended NLSE was already noticed in section 10.4. The background amplitude is $c/(2\varepsilon)$, which is equal to 1 here. For $c = 1$ and any ε, the wavenumber k has to be larger than 1/6. Thus, the plane wave propagates at a certain angle with respect to the t-axis. The solution itself is also tilted and has a double-peak structure for all values of k in the interval $1/6 < k \lesssim 2$. However, at larger values of k, the two maxima merge and the solution has a single peak. An example is shown in figure 10.7(b) where $k = 2$. For very small k-values, the solution retains its double-peak structure but becomes elongated [64]. A representative case for $k = 0.25$ is shown in figure 10.7(c). In all three cases, we have kept parameters c and ε fixed in such a way that the background remains constant, equal to 1.

Despite the seemingly singular structure of the solution, with ε being in the denominator, the amplitude of the background for the rogue wave is finite and equal to $\psi_0 = c/(2\varepsilon)$. This follows from the fact that the plane wave solution of the SSE has the form:

$$\psi_0(x, t) = -\frac{c}{2\varepsilon} \exp\left[i\left(\frac{k}{2\varepsilon} t - \frac{\omega}{8\varepsilon^2} x \right) \right], \tag{10.27}$$

where c, the wave-number k, and the frequency ω are coupled through the dispersion relation, equation (10.24). We can keep a constant background by choosing the ratio c/ε to be a constant. When taking the limit $\varepsilon \to 0$, we should simultaneously take the limit $c \to 0$. Then k and ω should also be considered in the same zero limit. This should be done carefully, to keep k inside the allowed limits (10.26).

When reducing the parameter ε to zero while keeping the plane wave wave-number $k = 0$, the limit is rather complicated, but the numerator and denominator can be factorized. After cancellation of the same polynomial expression $4^8(2x^2 + 2t(t + 1) + 1)$ in both, the degree of polynomials is reduced and the solution becomes the rogue wave of the NLSE:

$$\psi = -\left[1 + \frac{4i(x + it)(ix + t + 1)}{2x^2 + 2t(t + 1) + 1} \right] e^{ix}. \tag{10.28}$$

After the simple translation $t \rightarrow t - 1/2$ along the t-axis, we obtain the standard expression for the Peregrine solution given by equation (10.2). In normalized form, it does not have any free parameters.

From a physical point of view, the two additional terms in the Hirota equation (10.16), responsible for third-order dispersion and self-frequency shift, are perfectly balanced. As a result, the rogue wave solution (10.17) shows only a trivial velocity shift. The presence of the additional third term in the SSE ruins this delicate balance, thus resulting in a more complicated structure of the rogue wave. Revealing the influence of these additional terms on rogue waves, within and beyond integrable cases, deserves further study.

Higher-order rogue wave solutions for the Sasa–Satsuma equation must exist, but have not been found yet. Finding them is a challenge that requires developing a technique involving 3×3 matrix operations and solving cubic polynomial equations at each step. While technically this may be difficult to implement, the basic principles of rogue wave construction are the same as for the NLSE. Thus, we can expect rogue wave triplets for the second-order solutions and a whole classification pattern similar to the one shown in figure 10.4, where the elementary rogue waves would be given by equation (10.23).

The presence of free parameters in the equation may influence the rogue waves and significantly distort the solutions. Examples of such influence have been considered for the Hirota equation in [65, 66]. In particular, the shape of the rogue waves may become elongated and resemble a soliton [65]. These are special cases which have to be kept in mind when considering patterns of elementary rogue waves and their interactions.

10.6 Infinitely long NLSE extensions

The two special NLSE extensions presented in the previous section are not the only ones that are integrable. We can add more higher-order terms to the right-hand side and find the conditions for which the whole equation becomes integrable. The technique has been developed in [67, 68]. In what follows, we present only the final results. Namely, the infinite NLSE hierarchy of integrable equations can be written in the form:

$$\begin{aligned} F[\psi(x, t)] = i\psi_x &+ \alpha_2 K_2[\psi(x, t)] - i\alpha_3 K_3[\psi(x, t)] \\ &+ \alpha_4 K_4[\psi(x, t)] - i\alpha_5 K_5[\psi(x, t)] \\ &+ \alpha_6 K_6[\psi(x, t)] - i\alpha_7 K_7[\psi(x, t)] \\ &+ \alpha_8 K_8[\psi(x, t)] - i\, \alpha_9 K_9[\psi(x, t)] \\ &+ \cdots = 0, \end{aligned} \tag{10.29}$$

where the combined operator $F[\psi(x, t)]$ represents the whole hierarchy of integrable equations, while each individual functional K_j represents a particular operator of order j. Coefficients α_j are arbitrary real parameters. They do not have to be small. This allows us to go well beyond the simple extension of the NLSE with corrective perturbative terms.

In the lowest, second order, we obtain the fundamental nonlinear Schrödinger equation:

$$i\psi_x + \alpha_2 K_2 = i\psi_x + \alpha_2(\psi_{tt} + 2\psi|\psi|^2) = 0. \tag{10.30}$$

Taking $\alpha_2 = \frac{1}{2}$ or rescaling the x-variable, we get the NLSE (10.1) in standard form. By adding the third-order operator, K_3, we obtain the Hirota equation [61, 72]:

$$i\psi_x + \alpha_2(\psi_{tt} + 2\psi|\psi|^2) - i\alpha_3[\psi_{ttt} + 6|\psi|^2\psi_t] = 0, \tag{10.31}$$

which was already discussed in the previous section. In fact, when $\alpha_2 = \frac{1}{2}$, this equation is identical to equation (10.16). Now, as a particular case of equation (10.31), we can take $\alpha_2 = 0$. The resulting equation

$$\psi_x - \alpha_3[\psi_{ttt} + 6|\psi|^2\psi_t] = 0, \tag{10.32}$$

is known as the 'basic' Hirota equation or as the 'complexified' modified Korteweg–de Vries (mKdV) equation [73]. Again, taking $\alpha_3 = 1$ leads to its standard form. Remarkably, the common factor i is canceled when transforming (10.31) into equation (10.32). This cancellation can be done for all equations (10.29) when the coefficients with even indices are zero, i.e. $\alpha_{2n} = 0$. This can be seen clearly from equation (10.29). However, solutions of these equations can be sought in either complex or real forms For example, when the function ψ is real (and $\alpha_3 = 1$), equation (10.32) becomes the real mKdV:

$$\psi_x - \psi_{ttt} + 6\psi^2\psi_t = 0. \tag{10.33}$$

Solutions of the real mKdV equation are related to the solutions of the Korteweg–de Vries (KdV) equation through the Miura transformation [74]. equation (10.33) has rational solutions [75] that also have a single high peak that can be viewed as a rogue wave at the center. However, these elevated peaks are not completely localized, but are positioned on soliton-like structures [75].

With the fourth-order operator, K_4, the equation is known as the Lakshmanan–Porsezian–Daniel (LPD) [69–71] equation. In this case, the equation describes a Heisenberg spin chain. As we can see, particular cases in the hierarchy (10.29) have physical relevance. K_4 then contains a fourth-order derivative term:

$$K_4[\psi(x, t)] = \psi_{tttt} + 8|\psi|^2\psi_{tt} + 6\psi|\psi|^4 + 4\psi\,|\psi_t|^2 + 6\psi_t^2\psi^* + 2\psi^2\psi^*_{tt}. \tag{10.34}$$

The fifth-order operator, K_5 is given by:

$$K_5[\psi(x, t)] = \psi_{ttttt} + 10|\psi|^2\psi_{ttt} + 10(\psi\,|\psi_t|^2)_t + 20\psi^*\psi_t\psi_{tt} + 30|\psi|^4\psi_t. \tag{10.35}$$

It was considered earlier in the work [76]. Even from this brief analysis, we can see the wide range of possibilities that the general equation (10.29) provides. It includes many particular cases and it allows us to combine them into a unified model.

Moreover, the original NLSE does not have to be part of it (we can have $\alpha_2 = 0$), but it can serve to suggest the form of some solutions.

Further, the sixth-order operator, K_6, has the following explicit form:

$$\begin{aligned} K_6[\psi(x, t)] = {} & \psi_{tttttt} + [60\psi^*|\psi_t|^2 + 50(\psi^*)^2\psi_{tt} + 2\psi^*_{tttt}]\psi^2 \\ & + \psi\left[12\psi^*\psi_{tttt} + 8\psi_t\psi^*_{ttt} + 22\,|\psi_{tt}|^2 + 18\psi_{ttt}\psi^*_t + 70(\psi^*)^2\psi_t^2\right] \\ & + 20(\psi_t)^2\psi^*_{tt} + 10\psi_t[5\psi_{tt}\psi^*_t + 3\psi^*\psi_{ttt}] + 20\psi^*\psi_{tt}^2 \\ & + 10\,\psi^3\left[\left(\psi_*^{t}\right)^2 + 2\psi^*\psi^*_{tt}\right] + 20\psi\,|\psi|^6, \end{aligned} \tag{10.36}$$

while the seventh-order operator, K_7, is:

$$\begin{aligned} K_7[\psi] = {} & \psi_{ttttttt} + 70\psi_{tt}^2\psi^*_t + 112\psi_t\,|\psi_{tt}|^2 + 98\,|\psi_t|^2\psi_{ttt} \\ & + 70\psi^2\left[\psi_t[(\psi^*_t)^2 + 2\psi^*\psi^*_{tt}] + \psi^*(2\psi_{tt}\psi^*_t + \psi_{ttt}\psi^*)\right] + 28\psi_t^2\psi^*_{ttt} \\ & + 14\psi[\psi^*(20|\psi_t|^2\psi_t + \psi_{ttttt}) + 3\psi_{ttt}\psi^*_{tt} + 2\psi_{tt}\psi^*_{ttt} + 2\psi_{tttt}\psi^*_t \\ & + \psi_t\psi^*_{tttt} + 20\psi_t\psi_{tt}(\psi^*)^2] + 140\,|\psi|^6\psi_t + 70\psi_t^3(\psi^*)^2 \\ & + 14(5\psi_{tt}\psi_{ttt} + 3\psi_t\psi_{tttt})\psi^*. \end{aligned} \tag{10.37}$$

The process can be continued up to infinity. The complexity of the operators grows with the order j. The general iterative rules for obtaining these operators are given in [67, 68]. Explicit expression for the operator of the eighth order is given in [68]. The equation when the two coefficients α_3 and α_4 are arbitrary has been considered earlier in [77, 78]. In particular, soliton solutions of this equation were given in [77], while rogue wave solutions were presented in [78].

The NLSE hierarchy of integrable equations (10.29) has been derived for the case of 2×2 matrix Lax pairs. Therefore, the Sasa–Satsuma equation, which requires the Lax pairs to be based on 3×3 matrices, is not part of this hierarchy. As a result, complications related to this type of equation can be avoided and solutions can be presented in relatively simple forms.

The power of this approach can be demonstrated by the fact that we can seek solutions of equation (10.29) in a general form, taking into account the whole infinite set of operators involved in the equation. For example, the first-order rogue wave solutions for equation (10.29) can be written in explicit form [78]:

$$\psi(x, t) = c\left[4\frac{1 + 2i\,B_r\,x}{D(x, t)} - 1\right]e^{i\phi_r x}, \tag{10.38}$$

where c is an arbitrary background, while

$$D(x, t) = 1 + 4B_r^2x^2 + 4(ct + v_r\,x)^2, \tag{10.39}$$

$$\begin{aligned} B_r &= \sum_{n=1}^{\infty}\frac{n(2n)!}{(n!)^2}\alpha_{2n}c^{2n} \\ &= 2c^2(\alpha_2 + 6c^2\alpha_4 + 30c^4\,\alpha_6 + 140c^6\,\alpha_8 + 630c^8\alpha_{10} + \cdots), \end{aligned} \tag{10.40}$$

and

$$\begin{aligned} v_r &= \sum_{n=1}^{\infty} \frac{(2n+1)!}{(n!)^2} \alpha_{2n+1} c^{2n+1} \\ &= 2c^3(3\alpha_3 + 15c^2\,\alpha_5 + 70c^4\,\alpha_7 + 315c^6\alpha_9 + 1386c^8\,\alpha_{11} + \cdots). \end{aligned} \tag{10.41}$$

The coefficient ϕ_r in the exponential factor of (10.38) is given by:

$$\begin{aligned} \phi_r &= c^2 \sum_{n=1}^{\infty} \frac{(2n)!}{(n!)^2} \alpha_{2n} c^{2n-2} \\ &= 2c^2(\alpha_2 + 3c^2\alpha_4 + 10c^4\,\alpha_6 + 35c^6\,\alpha_8 + 126c^8\alpha_{10} + \cdots). \end{aligned} \tag{10.42}$$

The tilt factor v_r (or 'velocity') depends only on the coefficients of the operators of odd-order, α_{2n+1}, while the exponential factor, ϕ_r, and the stretching factor, B_r, depend only on the coefficients of the even-order operators α_{2n}. These observations allow us to make some general conclusions about the rogue wave profiles. Two illustrative examples are given in figure 10.8. We can see from this figure that the tilt appears only when α_5 is non-zero.

In this chapter, due to limited space, we restrict ourselves to these illustrations only. The work in this direction continues [79, 80] and it may bring many new unexpected results. NLSE extensions are usually considered to be improved models for a more accurate description of nonlinear wave propagation in the ocean [81–83] and in optical fibers [84, 85]. Normally, these extensions are approximate as they use small coefficients when dealing with higher-order terms. The exactly integrable cases described above may expand the range of applicability of these models. Moreover, linear dispersion can be modeled accurately up to an infinite number of terms in the expansion. Although nonlinear terms become fixed in this case, the deviations from realistic situations may be small. Perturbation techniques developed based on these models may be a better solution [86] than choosing the NLSE as a 'zero order' approximation. Further progress in this direction would certainly be desirable.

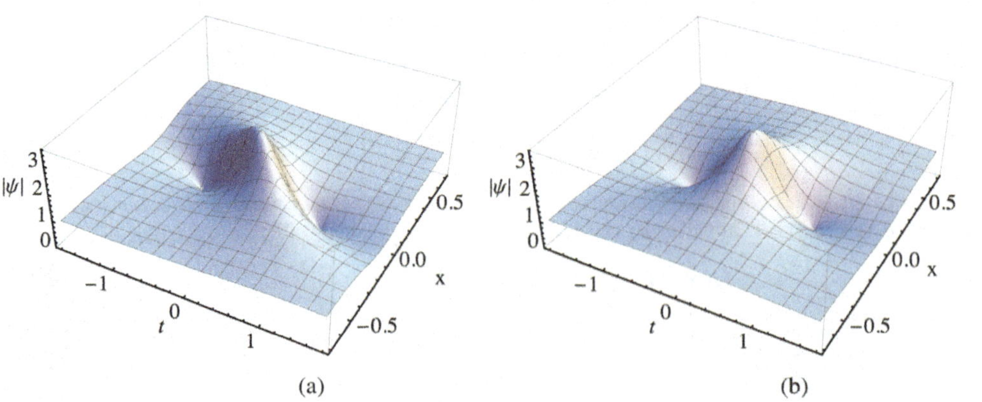

Figure 10.8. Rogue waves of equation (10.38) when $c = 1$, (a) $\alpha_4 = \dfrac{1}{4}$ and (b) $\alpha_4 = \dfrac{1}{4}$, $\alpha_5 = \dfrac{1}{16}$. All other α_j's in both cases are zero.

10.7 Conclusions

Integrable evolution equations have been studied for a few decades now. Most attention has been paid to soliton solutions of the KdV, NLSE, sine-Gordon and other similar equations. Rogue waves have been left aside, although their significance could be as vital as the soliton ones. Clearly, they differ from solitons in several ways. Firstly, they appear on a background which has infinite energy, so that rogue waves can be considered as formations that concentrate part of this energy into a small spot. Secondly, rogue waves last only a short time, in addition to being localized in space. These features make them unique solutions that may allow us to provide explanations to oceanic rogue waves. On the other hand, these solutions appear in various circumstances. One of them is a chaotic wave field [87] which contains all types of excitations including solitons, breathers, and rogue waves [88, 89]. This is in contrast to early views, when only solitons and radiation waves were taken into account [1]. This field of research is presently known as 'integrable turbulence' [90, 91]. Rogue waves also appear in a semi-classical soliton context when the initial condition contains many solitons [92]. The universal mechanism of the appearance of wave trains of Peregrine solutions has been found in [93]. Collisions of breathers also take the shape of rogue waves [94], thus extending this universality.

An average background energy is a common concept in physics. For example, vacuum energy exists in space throughout the entire Universe [95]. The processes of creation and annihilation of particle–antiparticle pairs is a possible mechanism for it. Quantum fluctuations are fundamental occurrences in quantum mechanics. This means that energy does not vanish anywhere in space. Thus, the zero field at infinity, commonly used in mathematics as a boundary condition, is an idealization that never occurs. Therefore, rogue waves that require infinite background energy for existence are natural formations that can be more realistic than any other particles in physics. We do not exclude the possibility of the appearance of rogue waves in the vast spaces of the Universe.

Rogue waves as solutions of evolution equations were first found for the NLSE [4]. This was a natural step, as the NLSE describes waves in deep water [23]. These findings have been generalized to optical fibers [10], BECs [15], plasmas [96], and the atmosphere [97] since the NLSE is a generic model for many branches of wave physics which feature weak nonlinearities. This universality of the rogue wave concept allows us to take further steps. We have shown that rogue waves are not only solutions of the NLSE. Similar solutions are found for other integrable equations considered in our review. Moreover, the existence of rogue waves has been demonstrated for non-integrable extensions of the NLSE [60]. This means that rogue waves are even more widespread than previously imagined. Revealing this universality should be the aim of future works: mathematical, theoretical and experimental. Our review presents only the initial steps of this progress. More work still has to be done.

Acknowledgments

The authors acknowledge the support of the Volkswagen Foundation. The work of JMSC was supported by MINECO under contract TEC2015-71127-C2-1-R, and by

CAM under contract S2013/MIT-2790. NA and AA acknowledge the support of the Australian Research Council (Discovery Project numbers DP140100265 and DP150102057).

References

[1] Zakharov V E and Shabat A B 1972 Exact theory of two-dimensional self-focusing and one-dimensional self-modulation of waves in nonlinear media *J. Exp. Theor. Phys.* **34** 62–9

[2] Akhmediev N and Ankiewicz A 1997 *Solitons, nonlinear pulses and beams* (London: Chapman and Hall)

[3] Akhmediev N and Korneev V I 1986 Modulation instability and periodic solutions of nonlinear Schrödinger equation *Theor. Math. Phys.* **69** 1089–93 (in Russian)

[4] Akhmediev N, Ankiewicz A and Taki M 2009 Waves that appear from nowhere and disappear without a trace *Phys. Lett.* A **373** 675–8

[5] Akhmediev N, Ankiewicz A and Soto-Crespo J M 2009 Rogue waves and rational solutions of the nonlinear Schrödinger equation *Phys. Rev.* A **80** 026601

[6] Kibler B, Fatome J, Finot C, Millot G, Dias F, Genty G, Akhmediev N and Dudley J M 2010 The Peregrine soliton in nonlinear fibre optics *Nat. Phys.* **6** 790–5

[7] Chabchoub A, Hoffmann N P and Akhmediev N 2011 Rogue wave observation in a water wave tank *Phys. Rev. Lett.* **106** 204502

[8] Chabchoub A and Akhmediev N 2013 Observation of rogue wave triplets in water waves *Phys. Lett.* A **377** 2590–93

[9] Chabchoub A, Hoffmann N, Onorato M, Slunyaev A, Sergeeva A, Pelinovsky E and Akhmediev N 2012 Observation of a hierarchy of up to fifth-order rogue waves in a water tank *Phys. Rev.* E **86** 056601

[10] Solli D R, Ropers C, Koonath P and Jalali B 2007 Optical rogue waves *Nature* **450** 1054–7

[11] Residori S, Bortolozzo U, Montina A, Lenzini F and Arecchi F T 2012 Rogue waves in spatially extended optical systems *Fluct. Noise Lett.* **11** 1240014

[12] Montina A, Bortolozzo U, Residori S and Arecchi F T 2009 Non-Gaussian statistics and extreme waves in a nonlinear optical cavity *Phys. Rev. Lett.* **103** 173901

[13] Zaviyalov A, Egorov O, Iliew R and Lederer F 2012 Rogue waves in mode-locked fibre lasers *Phys. Rev.* A **85** 013828

[14] Soto-Crespo J M, Grelu Ph and Akhmediev N 2011 Dissipative rogue waves: extreme pulses generated by passively mode-locked lasers *Phys. Rev.* E **84** 016604

[15] Bludov Yu V, Konotop V V and Akhmediev N 2009 Matter rogue waves *Phys. Rev.* A **80** 033610

[16] Akhmediev N and Pelinovsky E 2010 Editorial—introductory remarks on 'discussion & debate: rogue waves—towards a unifying concept?' *Eur. Phys. J. Special Topics* **185** 1–4

[17] Akhmediev N *et al* 2016 Roadmap on optical rogue waves and extreme events *J. Opt.* **18** 063001

[18] Akhmediev N, Dudley J M, Solli D R and Turitsyn S K 2013 Recent progress in investigating optical rogue waves *J. Opt.* **15** 060201

[19] Hasegawa A and Tappert F 1973 Transmission of stationary nonlinear optical pulses in dispersive dielectric fibers. I. Anomalous dispersion *Appl. Phys. Lett.* **23** 142–4

[20] Zakharov V E 1968 Stability of periodic waves of finite amplitude on a surface of deep fluid *J. Appl. Mech. Tech. Phys.* **9** 190–4

[21] Onorato M, Residori S, Bortolozzo U, Montina A and Arecchi F T 2013 Rogue waves and their generating mechanisms in different physical contexts *Phys. Rep.* **528** 47–89

[22] Akhmediev N, Eleonskii V M and Kulagin N E 1987 Exact first-order solutions of the nonlinear Schrödinger equation *Theor. Math. Phys.* **72** 809 (in Russian)

[23] Shrira V I and Geogjaev V V 2009 What makes the Peregrine soliton so special as a prototype of freak waves *J. Eng. Math.* **67** 11–22

[24] Gaillard P and Gastineau M 2015 The Peregrine breather of order nine and its deformations with sixteen parameters solutions to the NLS equation *Phys. Lett.* A **379** 1309–13

[25] Kedziora D J, Ankiewicz A and Akhmediev N 2012 Second-order nonlinear Schrödinger equation breather solutions in the degenerate and rogue wave limits *Phys. Rev.* A **85** 066601

[26] Clarkson P A 2006 Special polynomials associated with rational solutions of the defocusing nonlinear Schrödinger equation and the fourth Painlevé equation *Eur. J. Appl. Math.* **17** 293–322

[27] Hone A N W 1997 Crum transformation and rational solutions of the non-focusing nonlinear Schrödinger equation *J. Phys. A: Math. Gen.* **30** 7473–83

[28] Peregrine D H 1983 Water waves, nonlinear Schrödinger equations and their solutions *J. Aust. Math. Soc.* B **25** 16

[29] Bertola M and Tovbis A 2012 Universality for the focusing nonlinear Schrödinger equation at the gradient catastrophe point: Rational breathers and poles of the Tritronquéé solution to Painlevé I *Commun. Pure Appl. Math.* **66** 678

[30] Chabchoub A, Hoffmann N P and Akhmediev N 2012 Observation of rogue wave holes in a water wave tank *J. Geophys. Res.* **117** C00J02

[31] Chen S, Ph Grelu and Soto-Crespo J M 2014 Dark- and bright-rogue-wave solutions for media with long-wave short-wave resonance *Phys. Rev.* E **89** 011201

[32] Baronio F, Degasperis A, Conforti M and Wabnitz S 2012 Solutions vector nonlinear Schrödinger equations: evidence for deterministic rogue waves *Phys. Rev. Lett.* **109** 044102

[33] Baronio F 2017 Akhmediev breathers and Peregrine solitary waves in a quadratic medium *Opt. Lett.* **42** 1756

[34] Baronio F, Conforti M, Degasperis A and Lombardo S 2013 Rogue waves emerging from the resonant interaction of three waves *Phys. Rev. Lett.* **111** 114101

[35] Akhmediev N, Eleonskii V M and Kulagin N E 1985 Generation of periodic trains of picosecond pulses in an optical fiber: exact solutions *Sov. Phys. JETP* **62** 894–9

[36] Matveev V B and Salle M A 1991 *Darboux Transformations and Solitons* (Berlin: Springer)

[37] Dubard P, Gaillard P, Klein C and Matveev V B 2010 On multi-rogue wave solutions of the NLS equation and positon solutions of the KdV equation *Eur. Phys. J. Special Topics* **185** 247–58

[38] Ohta Y and Yang J 2012 General high-order rogue waves and their dynamics in the nonlinear Schrödinger equation *Proc. R. Soc.* A **468** 1716–40

[39] Gaillard P 2013 Degenerate determinant representation of solutions of the nonlinear Schrödinger equation, higher order Peregrine breathers and multi-rogue waves *J. Math. Phys.* **54** 013504

[40] Gaillard P *et al* 2015 Other $2N - 2$ parameters solutions of the NLS equation and $2N + 1$ highest amplitude of the modulus of the Nth order AP breather *J. Phys. A; Math. Theor.* **48** 145203

[41] Dontsop P Y G, Essama B G O, Doing J M, Dedzo M M, Atangana J, Yemele D and Kofane T C 2016 Akhmediev-Peregrine rogue waves generation in a composite right/left-handed transmission line *Opt. Quant. Electron.* **48** 59

[42] Chabchoub A and Fink M 2014 Time-reversal generation of rogue waves *Phys. Rev. Lett.* **112** 124101

[43] Vitanov N K, Chabchoub A and Hoffmann N 2013 Deep-water waves: on the nonlinear Schrödinger equation and its solutions *J. Theor. Appl. Mech* **43** 43–54

[44] Ankiewicz A, Clarkson P and Akhmediev N 2010 Rogue waves, rational solutions, the patterns of their zeros and integral relations *J. Phys. A: Math. Theor.* **43** 122002

[45] Wang L, Yang C, Wang J and He J 2017 The height of an *n*th order fundamental rogue wave for the nonlinear Schrödinger equation *Phys. Lett.* A **381** 1714–8

[46] Chin S A, Ashour O A, Nikolić S N and Belić M R 2016 Maximal intensity higher-order Akhmediev breathers of the nonlinear Schrödinger equation and their systematic generation *Phys. Lett.* A **380** 3625–9

[47] Ling L and Zhao L-C 2013 Simple determinant representation for rogue waves of the nonlinear Schrödinger equation *Phys. Rev.* E **88** 043201

[48] Ankiewicz A and Akhmediev N 2017 Multi-rogue waves and triangular numbers *Rom. Rep. Phys.* **69** 104

[49] Kedziora D J, Ankiewicz A and Akhmediev N 2013 Classifying the hierarchy of nonlinear Schrödinger equation rogue wave solutions *Phys. Rev.* E **88** 013207

[50] Ankiewicz A, Kedziora D J and Akhmediev N 2011 Rogue wave triplets *Phys. Lett.* A **375** 2782–5

[51] Kedziora D J, Ankiewicz A and Akhmediev N 2011 Circular rogue wave clusters *Phys. Rev.* E **84** 056611

[52] Kedziora D J, Ankiewicz A and Akhmediev N 2012 Triangular rogue wave cascades *Phys. Rev.* E **86** 056602

[53] Gaillard P 2013 Deformations of third-order Peregrine breather solutions of the nonlinear Schrödinger equation with four parameters *Phys. Rev.* E **88** 042903

[54] Gaillard P 2013 Degenerate determinant representation of solutions of the nonlinear Schrödinger equation, higher order Peregrine breathers and multi-rogue waves *J. Math. Phys.* **54** 013504

[55] Akhmediev N, Ankiewicz A and Soto-Crespo J M 2009 Rogue waves and rational solutions of the nonlinear Schrödinger equation *Phys. Rev.* E **80** 026601

[56] Agrawal G P 2007 *Nonlinear Fiber Optics* 4th edn (New York: Academic)

[57] Trippenbach M and Band Y B 1998 Effects of self-steepening and self-frequency shifting on short-pulse splitting in dispersive nonlinear media *Phys. Rev.* E **57** 4791–803

[58] Anderson D and Lisak M 1983 Nonlinear asymmetric self-phase modulation and self-steepening of pulses in long optical waveguides *Phys. Rev.* A **27** 1393

[59] Brugarino T and Sciacca M 2006 Singularity analysis and integrability for a HNLS equation governing pulse propagation in a generic fiber optics *Opt. Commun.* **262** 250–6

[60] Ankiewicz A, Devine N and Akhmediev N 2009 Are rogue waves robust against perturbations? *Phys. Lett.* A **373** 3997–4000

[61] Ankiewicz A, Soto-Crespo J M and Akhmediev N 2010 Rogue waves and rational solutions of the Hirota equation *Phys. Rev.* E **81** 046602

[62] Sasa N and Satsuma J 1991 New type of soliton solutions for a higher-order nonlinear Schrödinger equation *J. Phys. Soc. Japan* **60** 409–17

[63] Bandelow U and Akhmediev N 2012 Persistence of rogue waves in extended nonlinear Schrödinger equations: Integrable Sasa-Satsuma case *Phys. Lett.* A **376** 1558–61
[64] Akhmediev N, Soto-Crespo J M, Devine N and Hoffmann N P 2015 Rogue wave spectra of the Sasa-Satsuma equation *Physica* D **294** 37–42
[65] Liu C, Yang Z-Y, Zhao L-C and Yan W-L 2015 State transition induced by higher-order effects and background frequency *Phys. Rev.* E **91** 022904
[66] Chowdury A, Ankiewicz A and Akhmediev N 2015 Moving breathers and breather-to-soliton conversions for the Hirota equation *Proc. R. Soc.* A **471** 20150130
[67] Kedziora D J, Ankiewicz A, Chowdury A and Akhmediev N 2015 Integrable equations of the infinite nonlinear Schrödinger equation hierarchy with time variable coefficients *Chaos* **25** 103114
[68] Ankiewicz A, Kedziora D J, Chowdury A, Bandelow U and Akhmediev N 2016 Infinite hierarchy of nonlinear Schrödinger equations and their solutions *Phys. Rev.* E **93** 012206
[69] Lakshmanan M, Porsezian K and Daniel M 1988 Effect of discreteness on the continuum limit of the Heisenberg spin chain *Phys. Lett.* A **133** 483–8
[70] Porsezian K, Daniel M and Lakshmanan M 1992 On the integrability aspects of the one–dimensional classical continuum isotropic biquadratic Heisenberg spin chain *J. Math. Phys.* **33** 1807–16
[71] Porsezian K 1997 Completely Integrable nonlinear Schrödinger type equations on moving space curves *Phys. Rev.* E **55** 3785–9
[72] Hirota R 1973 Exact envelope-soliton solutions of a nonlinear wave equation *J. Math. Phys.* **14** 805
[73] Ono H 1976 Algebraic soliton of the modified Korteweg–de Vries equation *J. Soc. Japan* **41** 1817–8
[74] Miura R M 1968 Korteweg–de Vries equation and generalizations. I. A remarkable explicit nonlinear transformation *J. Math. Phys.* **9** 1202–4
[75] Chowdury A, Ankiewicz A and Akhmediev N 2016 Periodic and rational solutions of modified Korteweg–de Vries equation *Eur. Phys. J.* D *70 104*
[76] Hoseini S M and Marchant T R 2006 Solitary wave interaction and evolution for a higher-order Hirota equation *Wave Motion* **44** 92–10
[77] Ankiewicz A and Akhmediev N 2014 Higher-order integrable evolution equation and its soliton solutions *Phys. Lett.* A **378** 358–61
[78] Ankiewicz A, Wang Y, Wabnitz S and Akhmediev N 2014 Extended nonlinear Schrödinger equation with higher-order odd and even terms and its rogue wave solutions *Phys. Rev.* E **89** 012907
[79] Chowdury A, Kedziora D J, Ankiewicz A and Akhmediev N 2015 Breather-to-soliton conversions described by the quintic equation of the nonlinear Schrödinger hierarchy *Phys. Rev.* E **91** 032928
[80] Chowdury A, Kedziora D J, Ankiewicz A and Akhmediev N 2015 Breather solutions of the integrable quintic nonlinear Schrödinger equation and their interactions *Phys. Rev.* E **91** 022919
[81] Dysthe K B and Trulsen K 1999 Note on breather type solutions of the NLS as models for freak-waves *Phys. Scr.* **T82** 48–52
[82] Sedletskii Y V 2003 The fourth-order nonlinear Schrödinger equation for the envelope of Stokes waves on the surface of a finite-depth fluid *J. Exp. Theor. Phys.* **97** 180–93

[83] Slunyaev A V 2005 A high-order nonlinear envelope equation for gravity waves in finite-depth water *J. Exp. Theor. Phys.* **101** 926–41

[84] Potasek M J 1991 Exact solutions for an extended nonlinear Schrödinger equation *Phys. Lett.* A **60** 449–52

[85] Cavalcanti S B, Cressoni J C, da Cruz H R and Gouveia-Neto A S 1991 Modulation instability in the region of minimum group-velocity dispersion of single-mode optical fibers via an extended nonlinear Schrödinger equation *Phys. Rev.* A **43** 6162–5

[86] Ankiewicz A, Soto-Crespo J M, Chowdhury M A and Akhmediev N 2013 Rogue waves in optical fibers in presence of third-order dispersion, self-steepening, and self-frequency shift *J. Opt. Soc. Am.* B ***30*** *87–94*

[87] Akhmediev N, Soto-Crespo J M and Ankiewicz A 2009 Extreme waves that appear from nowhere: On the nature of rogue waves *Phys. Lett.* A **373** 2137–45

[88] Soto-Crespo J M, Devine N and Akhmediev N 2016 Integrable Turbulence and Rogue Waves: Breathers or Solitons *Phys. Rev. Lett.* **116** 103901

[89] Akhmediev N, Soto-Crespo J M and Devine N 2016 Breather turbulence versus soliton turbulence: rogue waves, probability density functions, and spectral features *Phys. Rev.* E **94** 022212

[90] Agafontsev D S and Zakharov V E 2015 Integrable turbulence and formation of rogue waves *Nonlinearity* **28** 2791–821

[91] Walczak P, Randoux S and Suret P 2015 Optical rogue waves in integrable turbulence *Phys. Rev. Lett.* **114** 143903

[92] Tikan A *et al* 2017 Universal Peregrine soliton structure in nonlinear pulse compression in optical fiber *Phys. Rev. Lett.*

[93] Grimshaw R H J and Tovbis A 2013 Rogue waves: analytical predictions *Proc. R. Soc.* A **469** 20130094

[94] Akhmediev N, Soto-Crespo J M and Ankiewicz A 2009 How to excite a rogue wave *Phys. Rev.* A **80** 043818

[95] Saunders S and Brown H R 1991 *The Philosophy of Vacuum* (Oxford: Clarendon)

[96] Moslem W M, Shukla P K and Eliasson B 2011 Surface plasma rogue waves *Eur. Phys. Lett* **96** 25002

[97] Stenflo L and Marklund M 2010 Rogue waves in the atmosphere *J. Plasma Phys.* **76** 293

Chapter 11

Are rogue waves really rogue?

Günter Steinmeyer and Ayhan Demircan

11.1 Introduction

The appearance of unexpectedly high waves on the ocean has intrigued laymen, seafarers, as well as scientists in the field. It may be our obvious failure to predict rogue waves that has generated a feeling of awe of nature's forces, but then there may also be some aspect of sensationalism that appears warranted to explain the popular interest in rogue waves. Despite years of discussion, it seems that we are still rather far away from a generally accepted explanation of this fascinating yet also highly destructive scenario [1, 2]. The adjective used in 'rogue wave' suggests the large, destructive, and unpredictable nature of such waves. Considering that the word 'rogue' has been used in Shakespeare's times, it even seems plausible that malicious intent of a higher force was originally suggested. Other languages of famous seafarer nations further corroborate evil will behind these extreme waves, e.g. the Spanish 'onda del granuja' or the French 'vague scélérate'. At this stage, it is interesting to note that implications of evil will or erratic behavior are not universal. The commonly used Italian word is 'ondo anomala', which just implies statistical anomaly. In a similar sense, the somewhat mysterious German word for 'rogue wave' is 'Kaventsmann', which is most likely an ancient word for a wealthy, influential, yet at the same time also heavily overweight merchant in a Hanseatic city. Based on these observations, it seems that the overall reception of rogue waves is rather split around the languages: it is either the statistical behavior, sheer size, or general unpredictability of the individual event which serves to define a rogue wave. The purpose of this chapter is a critical evaluation of these characteristics of rogue waves in different physical systems.

11.2 Definition of rogue waves: predictability

So far, events have been considered rogue whenever Gaussian statistics do not hold and when a heavy tail of extreme events forms. Using the common definition of rogue waves in oceanography, a wave is considered extreme when it exceeds the

doi:10.1088/978-0-7503-1460-2ch11 © IOP Publishing Ltd 2017

significant wave height $H_{1/3}$ by a factor of at least 2, with $H_{1/3}$ being the average of the upper third of all measured waves. For the ocean, long-term measurements indicate that about 10^{-5} of all waves exceed this threshold. Assuming a wave period of 10 s, e.g. one therefore expects to see one anomalously high wave within 2 weeks. It is the very rareness of these waves that makes it very difficult and time-consuming to collect experimental data directly on the ocean. Optical systems which show a heavy tail in the statistical distribution of events allow for much faster collection of significant amounts of data [3, 4]. Fiber-optical rogue waves, for example, can be recorded at the 100 MHz repetition rates of mode-locked lasers. Assuming the same 10^{-5} probability, one would expect to observe a rogue wave within a millisecond, but fiber-optical rogue waves are in fact much more frequent than their ocean counterparts. This comparison already indicates a possible problem in the study of analogous systems: if rogue waves are actually frequent events that appear with rather high probability, then they are not really surprising anymore. They may qualify as extreme events, but one cannot attest to 'rogueness' in a statistical sense anymore. However, such an interpretation of the word 'rogue' falls short of the predictability of rogue events. Even when every tenth wave exceeds the significant wave height by a factor of 2, this does not necessarily mean that one can predict extreme events in the system.

Previous studies on ocean rogue waves have mostly addressed statistical aspects of this phenomenon. Knowing that 1 in 100 000 waves is possibly rogue enables us to design ships that are strong enough to survive any likely force on their hull for hundreds or thousands of years, statistically. Such statistics can be solely derived from measurements, and no understanding of the underlying physical mechanisms seems necessary for this kind of prediction. A much more difficult and currently unsolved question is the prediction of an individual extreme event in the various rogue-wave supporting systems. At this stage, it is not even clear whether such a prediction is possible after all. In particular, one needs to be very careful with the meaning of predictability, as different systems may rely on different types of nondeterministic processes, i.e. either driven by classical stochastic dynamics or fundamental quantum mechanical indeterminism. We will address this point below in more detail, but we start here with predictability in stochastic systems. Restricting us to the notion of unpredictability, we therefore provocatively ask: are rogue waves really rogue?

One can approach the question of predictability of rogue waves from different practical angles. One idea may be that rogue waves require some time to build up to a dangerous height and then eventually decay, but nobody can safely say, currently, how long a rogue wave survives once it has formed. Rogue wave prediction could then be possible by monitoring the ocean surface within a certain radius around the ship or oil rig, i.e. we effectively reduce the prediction to observation succeeded by the estimation of the likely course of the emerging rogue wave. Transferring such an observation-based prediction to the fiber context would require information at early stages of the propagation dynamics within the fiber to predict formation of a rogue wave. While such an approach seems difficult from an experimental point of view, one can certainly easily explore this concept in numerical simulations [5–7]. Given that very similar partial differential equations rule the propagation of the waves in both cases, both systems, i.e. nonlinear fiber propagation as well as the propagation on the ocean surface appear, in principle, predictable. However, one could come up with a different definition of predictability in both contexts. Imagine, for example, that a long series of single-shot spectra has been recorded at the output of the nonlinear fiber or that a long time series of ocean surface elevation is available for a single location on the ocean. Would this information suffice to predict the future emergence of a rogue wave? One could certainly widen the amount of information in the ocean context: would complete knowledge of the history of the ocean surface enable prediction of rogue formation before it even reaches a critical height?

One problem in answering these questions seems obvious: in the ocean context, there simply exist very few measurements of rogue waves, which were all exclusively measured at one fixed location, e.g. an oil rig. The rather small number of available measurements makes a statistical analysis currently impossible. Optical rogue waves, in contrast, may appear at a very high rate, enabling us to record a statistically relevant number of such events within a minute or less. In the following, we use measured data in two independent optical systems to provide an answer to the predictability of rogue waves. One of these systems is the well-known nonlinear fiber scenario, i.e. a one-dimensional scenario, and the other is the less-explored multi-filament scenario [8, 9], which we will discuss below.

11.3 Rogue waves in the multi-filament scenario

A filament forms when the critical power for small-scale self-focusing is exceeded [10]. In a simple way of speaking, self-focusing narrows down the beam diameter, making the beam even more intense, which then further increases the self-focusing action. This positive feedback causes a rapid collapse of the beam radius, which one would expect to only be arrested when the beam radius is of the order of the wavelength. However, measurements of collapsed beam profiles clearly indicate that the beam collapses are already arrested at values at least 10–100 times larger in solid-state materials and gases, respectively. This effect is usually explained by the onset of multi-photon ionization processes. Close to the critical power, a single plasma string forms that self-guides the laser light. The length of this self-guided propagation zone may significantly exceed the Rayleigh range, i.e. they are sometimes referred to as subdiffractive. Further increasing the power, single filamentation will eventually also become unstable, and two or more separated plasma strings form. This transition from single filamentation to multiple filamentation is illustrated in figure 11.1(a).

Compared to single filamentation, which is typically observed to lead to relatively stable beam profiles, beam profiles during multiple filamentation are sometimes observed to be highly dynamic, with completely different subsequent beam profiles [11, 12]. It is interesting to note that this highly dynamic scenario only appears under certain experimental conditions. Beam profiles of multi-filaments of low-repetition-rate lasers are often remarkably stable and reproduce quite accurately from shot to shot. This behavior may even not be overly surprising as it shows that the individual filament strings are obviously seeded by hot spots in the laser beam profile which

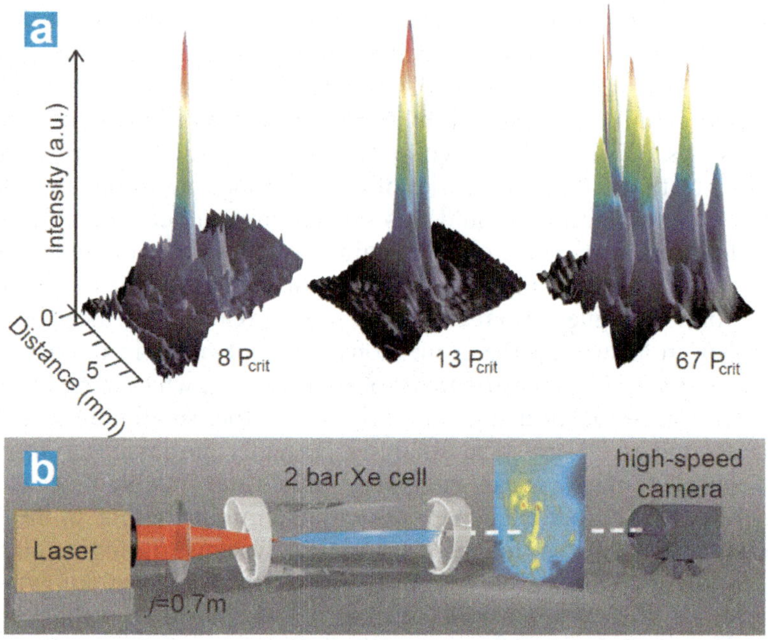

Figure 11.1. Single-shot beam profiles recorded for three different peak powers ranging from 8 through 67 critical powers for plasma formation. Beam patterns change from single to multiple filamentation. Intensities are plotted as elevation and are additionally color-coded. (b) Experimental setup employed in the experiments. The amplified femtosecond laser is focused into a 2-bar xenon cell, and the resulting dynamic beam patterns are recorded by high-speed cameras after suitable attenuation.

often stem from imperfections of the optical components in the beam path, in particular intracavity components inside the laser. Going to kHz repetition rates, it is actually not straightforward to observe the highly dynamic scenario that gives rise to rogue waves. If multi-filament rogue waves behave similarly to their one-dimensional scenario in the fiber context, one would naturally expect their formation to be prevalent when only a medium with sufficiently high nonlinearity is used. Following this idea, we started our experimental investigation of rogue waves in multiple filamentation with a water cell. With all its additional rotational and vibrational degrees of freedom, water has a relatively high optical nonlinearity ($n_2 = 4 \times 10^{-16}$ cm^2 W^{-1}), and the additional advantage of being self-healing, i.e. a suddenly appearing hot spot (or rogue wave) in the beam profile would not cause lasting catastrophic damage to the medium. However, despite several weeks of experimentation, we could not observe any indication for rogue wave formation in such a cell, using a 3 kHz driver laser with 0.7 mJ pulse energy and 30 fs pulse duration (cf figure 11.1). Pulses from this laser display a peak power of about 20 GW, which exceeds the critical power for filamentation of about 2.5 MW by nearly 4 orders of magnitude. While the beam profiles emerging from the water cell could certainly not be considered static, a detailed statistical analysis clearly provided no indication for a heavy tail, as was otherwise always considered a necessary observation for rogue wave dynamics.

After the initial experiments with the water cell, we switched to a more powerful laser (1 kHz, 5 mJ, 200 GW peak power) and continued our study with a gas cell (figure 11.1) [9]. Using this slightly adapted experimental setup, the statistical character of the beam profile changed, and we observed several hundred rogue events within a minute of observation, i.e. a total of 60 000 laser shots. A statistical evaluation of these measurements is shown in figure 11.2 in comparison with the single-filament scenario and the statistics of the laser all by itself. The latter two cases, which served as a control experiment, clearly show a near-Gaussian distribution, which gives rise to a parabolic distribution of events in the semi-log representation of figure 11.2. In order to quantify deviations from a pure Gaussian scenario, we fitted two-parameter Weibull distributions

$$\mathrm{PDF}_W(\xi||\alpha, \beta) = \beta\alpha^{-\beta}\xi^{\beta-1}\exp\left(-\frac{\xi}{\alpha}\right)^{\beta} \tag{11.1}$$

to the measured intensity histograms, where ξ is either the measured fluence F or the electric field strength equivalent E. Here, α and β are the scale parameter and shape parameter, respectively. Large values of β indicate a near-normal distribution of events, whereas values of $\beta < 3$ typically indicate a heavy tail in the probability density function [13]. Fitting a Weibull distribution to ocean wave data results in shape parameters in the range from 1.7 to 1.8, which is fairly close to the value of 2 that is theoretically expected for the Rayleigh distribution resulting from two-dimensional random linear interference of waves. For our experimental situation, we find significantly smaller values of $\beta < 0.5$ in the fluence or intensity histograms of multi-filament dynamics in gaseous media, see figure 11.2(c). Reducing the input fluence below the threshold of multiple filamentation, a near-Gaussian distribution of events is observed (figure 11.2(b)). Differences to pulse energy fluctuations of the laser itself (figure 11.2(a)) are negligible. In fact, single filamentation seems to suppress rather than amplify pulse energy fluctuation. The fairly obvious difference between the shape parameters of ocean rogue waves and multiple filamentation may certainly indicate a different physical origin of the observed

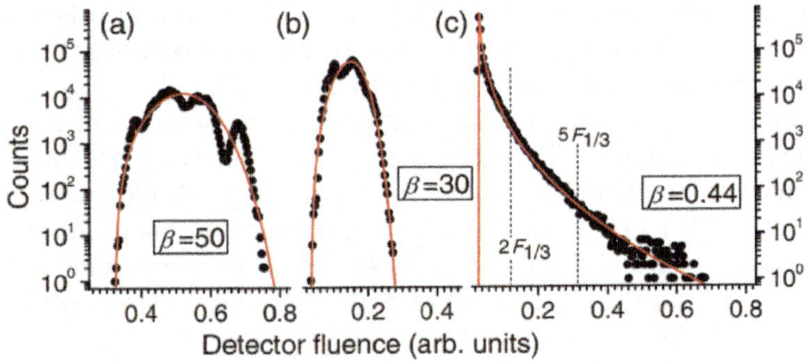

Figure 11.2. Histogram analysis of 1 min line scan camera recordings (60 000 laser shots). Symbols represent the histogram of the accumulated 20 neighboring pixels in the most intense part of the beam profile. Line: Weibull fit. Detector fluence calibrated relative to the maximum limiting value. (a) Laser measured directly, $P \ll P_{\mathrm{cr}}$. (b) Single filamentation, $P \approx 8P_{\mathrm{cr}}$. (c) Multiple filamentation, $P \approx 19P_{\mathrm{cr}}$. Dashed lines indicate multiples of the significant fluence $F_{1/3}$. Data were taken out of the small marked part of the beam profile in figure 11.4(c). Reprinted figure with permission from [9] Copyright 2013 by the American Physical Society.

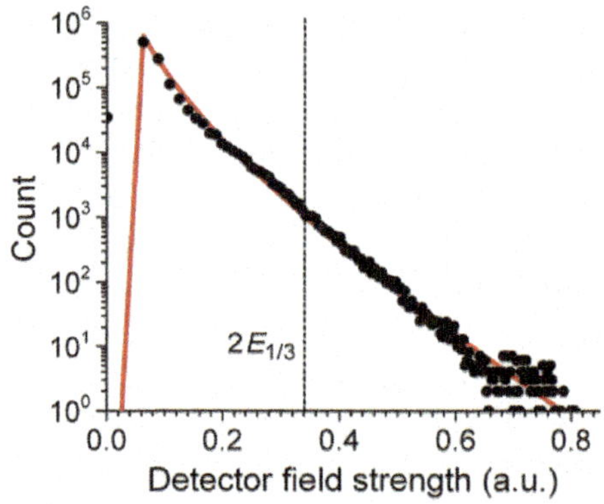

Figure 11.3. Field histogram obtained by computing the square root of the abscissa data in figure 11.2(c) after background subtraction. This gives rise to a near-exponential probability density function and enables a fairer comparison with ocean wave height histograms. Note that regardless of representation, multi-filament probability densities exhibit a much stronger heavy tail than ocean dynamics. Reprinted with permission from [9]. Copyright 2013 by the American Physical Society.

phenomena. In contrast to the ocean system, multiple filamentation is a highly nonlinear optical scenario. On the other hand, one may argue that the comparison is not quite fair, as the electric field strengths are the true optical equivalent of wave heights in the ocean scenario. We therefore rescaled our histograms to field strength equivalents by computing the square root of the measured intensities (see figure 11.3). This analysis yields a near-exponential probability density function ($\beta \approx 1$), but rogue waves still appear with a 1% probability, i.e. they are much more abundant in the multiple-filamentation scenario than on the ocean.

It seems interesting to further explore the phase-transition-like emergence of rogue waves that seems strongly linked to multiple filamentation. To this end, we made a spatially resolved statistical analysis, computing significant wave height and the maximum and average fluence for all pixels of a line-scan camera (see figure 11.4). These measurements clearly reveal that the emergence of a heavy tail is not confined to the most intense part of the beam profile, but takes over the dynamics in its entirety, similar to a phase transition from solid to liquid.

A second interesting question is the spatio-temporal isolation of rogue events in the multi-filament scenario. To address this question, we performed additional measurements with an area-scan camera (see figure 11.5). While this camera cannot keep up with the full 1 kHz repetition rate of the laser, we synchronized it to 1/10th of this frequency. Figure 11.5 shows five subsequent frames measured by the camera and clearly reveals that the rogue event appeared without apparent precursor within a time frame of 10 ms or less.

11.4 Comparison of the three different rogue wave supporting systems

Unfortunately, the area scan camera proved to be too slow to allow for higher temporal resolution of the emergence of a rogue wave. The relevance of the duration

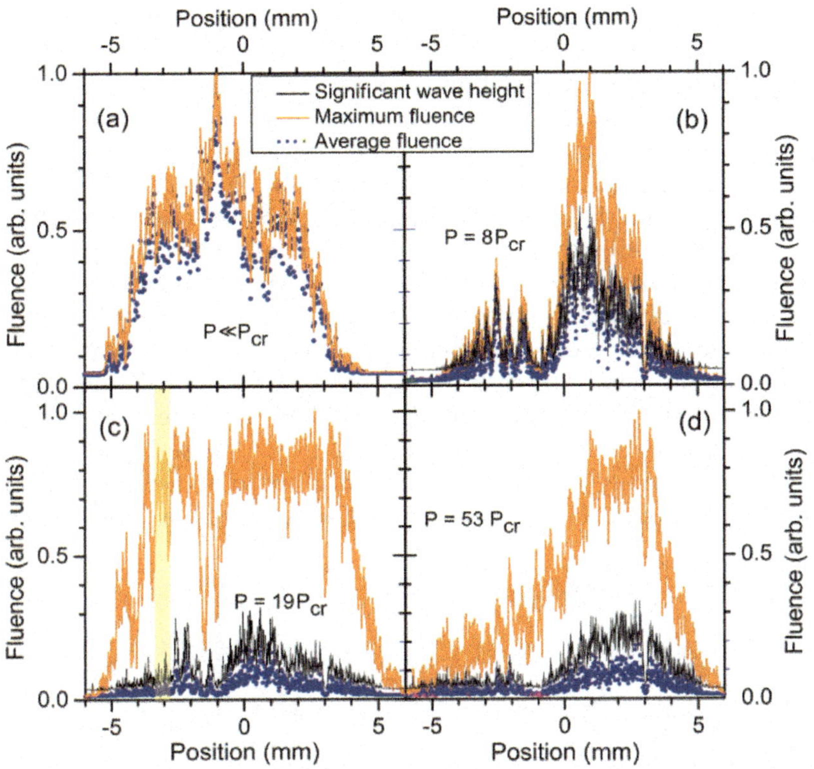

Figure 11.4. Statistical analysis of 1 min line scan camera recording that includes 60 000 individual scans. Solid orange lines: maximum fluence $F_{\max}$. Blue dots: average fluence F. Solid black lines: significant fluence $F_{1/3}$. (a) Input profile ($P \ll P_{\mathrm{cr}}$). (b) Profile after single filamentation ($P \approx 8P_{\mathrm{cr}}$). (c)–(d) Profile after multiple filamentation ($P \approx 19$ and $53P_{\mathrm{cr}}$). The region examined in the statistical analysis of figure 11.3(c) has been marked by a yellow semitransparent rectangle in (c). The sensitivity dip at 3 mm is an artifact of the camera. Reprinted figure with permission from [9]. Copyright 2013 by the American Physical Society.

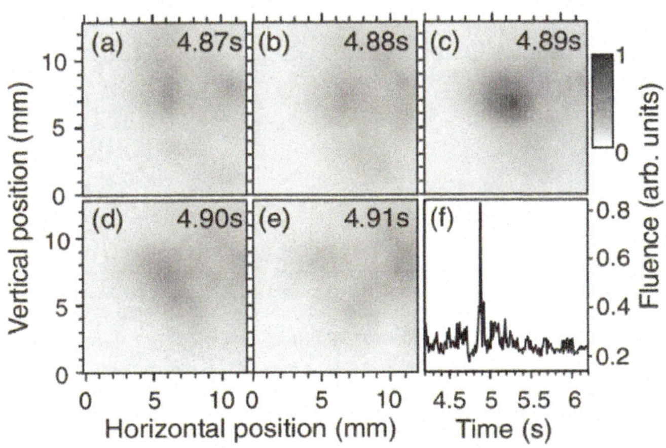

Figure 11.5. (a)–(e) Sequence of five multi-filament fluence profiles recorded with high-speed two-dimensional camera, with an extreme event appearing 4.89 s after the start of recording. (f) Fluence observed at position (6 mm, 7 mm). Reprinted figure with permission from [9]. Copyright 2013 by the American Physical Society.

of a rogue event becomes clear when comparing the dynamics of three different rogue wave supporting systems in figure 11.6. Ocean rogue waves are often accompanied by one preceding and one trailing deep trough, as shown exemplarily for the Draupner wave data in figure 11.6(b). An autocorrelation analysis of the Draupner data (inset of figure 11.6(b)) results in a strongly damped oscillatory function, with a characteristic damping time constant that corresponds to about one period of the oscillation. Here, it is most interesting to note that the Draupner wave has a nearly identical shape as the linear autocorrelation function of the entire 40 min long wave height record. This indicates a formal similarity to laser mode-locking, i.e. when all spectral components are incidentally phase-locked then this results in the emergence of one extremely large wave. The exact ratio of correlation time and oscillation period depends on the spectral width of the underlying wave dynamics, with fewer broadband scenarios enabling the formation of rogue wave groups, which are known, e.g. as the 'three sisters.' These highly characteristic autocorrelation functions have to be compared to those obtained in the analogous optical scenarios. Figure 11.6(c) shows a fiber-optical rogue wave that has been

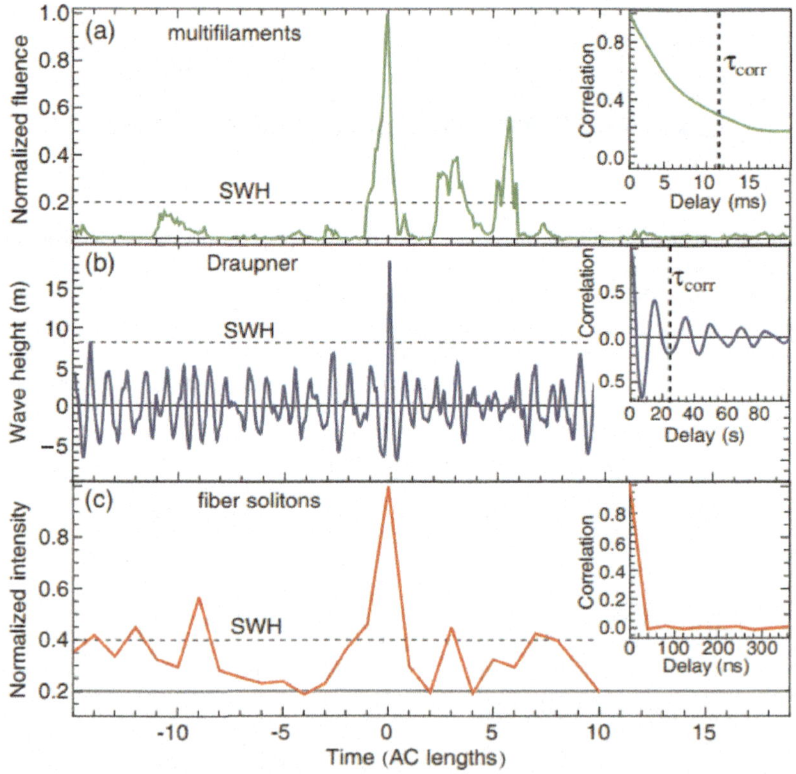

Figure 11.6. Examples of rogue events in (a) multi-filament dynamics [9], (b) the oceanographic context (Draupner data set), and (c) soliton dynamics in a nonlinear fiber [14]. Abscissas in (a) and (b) have been rescaled to equal units in terms of autocorrelation width; see insets. Solid lines: long-term averages. Dashed horizontal lines: significant wave height (SWH). Whenever applicable, the linear $1/e$ autocorrelation width is shown as dashed vertical lines in the insets. Reprinted figure with permission from [22]. Copyright 2015 by the American Physical Society.

sampled at the multi-megahertz repetition rate of a mode-locked laser [14]. More specifically, optical spectra have been recorded at a 20 MHz sampling rate, and a small spectral region in the infrared tail has been filtered out. Figure 11.6(c) shows the dynamics of a single rogue wave in this narrow spectral band. The specific large wave shown exceeds the significant wave height by a factor of ~2.5, i.e. it clearly qualifies as a rogue wave, but the resulting autocorrelation functions are qualitatively totally different from those observed in ocean wave dynamics. The correlation time constant in the nonlinear fiber scenario is apparently faster than the sampling time interval of 50 ns, i.e. it turns out to be beyond the scope of this measurement approach. Now the interesting question is how the correlation of multi-filaments behave (see figure 11.6(a)). In fact, the autocorrelation function of multi-filaments shows features of both previously discussed examples. They show a monotonic decrease like the fiber soliton scenario, but the decay of the correlation is actually longer than the inverse repetition rate of the laser, i.e. about 10 ms.

Figure 11.6 therefore actually appears to be a key to understanding the differences of the physics ruling the three rogue wave supporting systems. While it appears impossible to pinpoint a linear or nonlinear origin, ocean rogue waves form when different spectral components are phase-locked or synchronized. The characteristic hallmark of fiber rogue waves is the complete absence of correlation (or coherence) between two successive pulses. In fact, this loss of pulse-to-pulse coherence is well understood for fiber-based supercontinua [15–19]. The high sensitivity of the supercontinuum generation process towards spontaneous emission noise on the driver laser results in pulse splitting or a soliton fission process, leading to output pulses with strong shot-to-shot variations of spectral amplitudes and phases. This renders such supercontinua useless for pulse-compression purposes. Even though the spectra of these supercontinua cover more than an optical octave, which should allow their compression to the duration of a single optical cycle, such attempts have proven to be rather futile, as static dispersion compensation could only compress a single pulse in a train with megahertz repetition rate but none of the others. Along the same line of explanation, an optical rogue soliton is rarely generated given the strong shot-to-shot variations in the fission scenario. As these dynamics are ultimately ruled by quantum fluctuations on the driver laser, there is in particular no correlation to be expected between successive pulses. In this analog system, we therefore have real unpredictability given by the fundamental quantum mechanical indeterminism.

Given these considerations, a lack of correlation in the two optical supercontinuum systems appears quite plausible. Being well acquainted with the fiber supercontinuum scenario, one could argue that similar quantum fluctuations must also exist for the driver laser generating multiple filamentation. Therefore, it seems highly puzzling that pulse-to-pulse correlations actually seem to exist in the multi-filament scenario. Nevertheless, pump lasers that can generate millijoule pulses with kilohertz repetition rates are typically dominated by technical noise sources. In particular, this class of lasers shows a marked three-phase power supply ripple (at 300 Hz), and we find this signature also for all our kilohertz pump lasers (see figure 11.7). Doing a similar Fourier analysis for the dynamics in figure 11.6(a), however, shows practical absence of

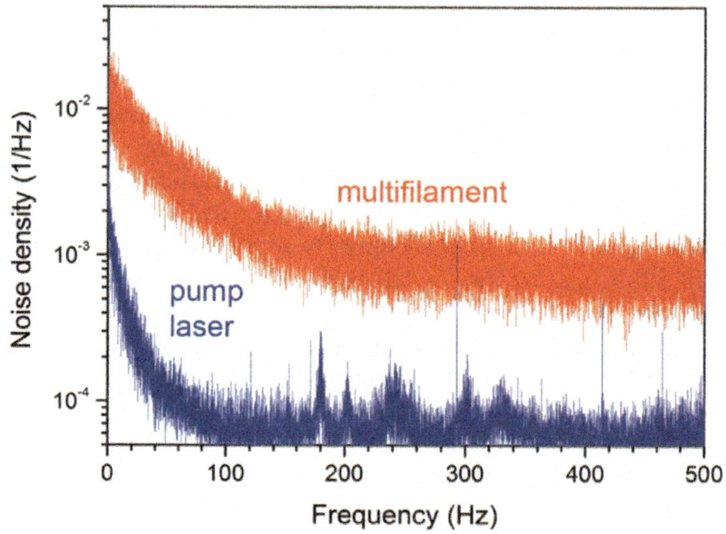

Figure 11.7. Noise spectra of the multi-filament dynamics in figure 11.6(a) compared to a spectral analysis of the pump laser. While the pump laser shows a number of spectrally narrow artifacts that are most likely being caused by harmonics of the three-phase power supply, the multi-filament data is virtually free of any narrowband artifacts and only shows a broadband noise signature (pink noise). Reprinted figure with permission from [9]. Copyright 2013 by the American Physical Society.

these artifacts in the rogue wave dynamics. In fact, the Fourier spectra of the time series rather resemble a pink noise signature, so it seems unlikely that the observed correlation time has anything to do with noise-generating mechanisms in the pump laser, which gives rise to the question of the physical mechanism that can actually store information in the gas cell for a duration of a few milliseconds. As there actually is plasma generated inside the cell, a straightforward assumption may be that a very small part of the ionization may survive for one millisecond or longer. However, recombination times reported for xenon ions consistently amount to much less than a microsecond. Instead, the much longer millisecond correlation times seem to fit much better to the phenomenon of 'pressure holes' [20], so it appears that the heated plasma strings generate pressure wave around them which then locally modify the refractive index in the xenon cell, causing a deflection of light beams in successive laser shots. In fact, the rather narrow dimensions of the cell and the relatively poor heat conductivity of xenon are expected to result in convective cooling, i.e. a transition from laminar to turbulent heat transport inside the cell. This transition is usually expected for local Grashof number in the range from 2×10^8 to 2×10^9 that we estimate for our gas cell. Moreover, the formation of turbulent roles inside filamentation cells was also confirmed later by Schlieren experiments [21]. Quite interestingly, the thermodynamic origin of the rogue wave dynamics is further corroborated by their absence in the first experiments with a water cell. As water has a much higher viscosity and thermal conductivity, one can quickly estimate that convective cooling would only set in for an about three orders of magnitude larger heating power, i.e. a laser with multiple kW pump power. Alternatively, we also convinced ourselves that the multi-filament rogue waves disappear when we reduce the thermal intake of the xenon cell by a factor of 10, suitably reducing our laser repetition rate to 100 Hz with a synchronized chopper. These test experiments clearly show that multi-filament rogue waves are actually a thermodynamic phenomenon in their origin, whereas the fiber-optical rogue wave scenario is related to quantum noise.

11.5 Filament rogue waves

To understand the multi-filament scenario exhibiting rogue events, we further ran a series of numerical simulations that are based on a variant of the nonlinear Schrödinger equation (NLSE) [10]:

$$\partial_z E = \frac{i}{2k_0}\nabla_\perp^2 E + \frac{i\omega_0}{c} n_2 \mid E \mid^2 E - \frac{i\omega_0}{2n_0 c\rho_c}\rho(I)E \tag{11.2}$$

Here, k_0 and ω_0 are wavevector and angular frequency of the drive laser, respectively, c is the speed of light, and n_0 and n_2 are the linear and nonlinear index of refraction, respectively. Compared to other commonly used variants of the NLSE, equation (11.2) restricts itself to modeling in the transverse plane but does not account for the detailed temporal shape of the pulses. Consequently, the first term on the right-hand side (rhs) describes diffraction, the second self-focusing, and the third the action of the plasma. To this end, we compute the electron density

$$\rho(I) = \rho_{nt}\left(1 - \exp\left(-\int_{-\infty}^{t} dt' \; W[I(t')]\right)\right) \tag{11.3}$$

under the assumption of a constant temporal intensity profile $I(t)$ as given by the pump laser. Here, ρ_{nt} is the density of the neutral xenon atoms at 2 bars pressure. We seed the simulation with a constant intensity to which we add 1% random noise on the initial spatial profile. A sample calculation is shown in figure 11.8. Depending on the random seed, a number of individual filament strings form, which initially travel in parallel. Our rather homogeneous initial intensity profile prevents two filaments from forming initially in immediate transverse vicinity. However, after about 20 cm of propagation, characteristic mergers appear between individual filament strings of similar phase. In figure 11.8, phases are color coded and the most prominent merger event is highlighted by a dashed box in figure 11.8(b). Snapshots of the beam profiles in this boxed area are additionally shown in figure 11.8(c). One can see that the merger, at about 22 cm, is rather short lived and decays after another 5 cm of propagation. If we separate two filament strings from the surrounding photon bath and run the simulation again, we end up with a stable merger that shows a breather-like behavior. This merger contains double the fluence of an individual filament string, but is obviously unstable in the presence of perturbations induced by the photon bath and other surrounding filament strings of different phase. Running a large number of numerical simulations with different noise seeds and analyzing the fluence histograms, finally we can clearly identify merger events as giving rise to the most extreme fluences (or pulse energies) in the statistics (see figure 11.8(d)). These simulations therefore corroborate that, despite the highly nonlinear character of

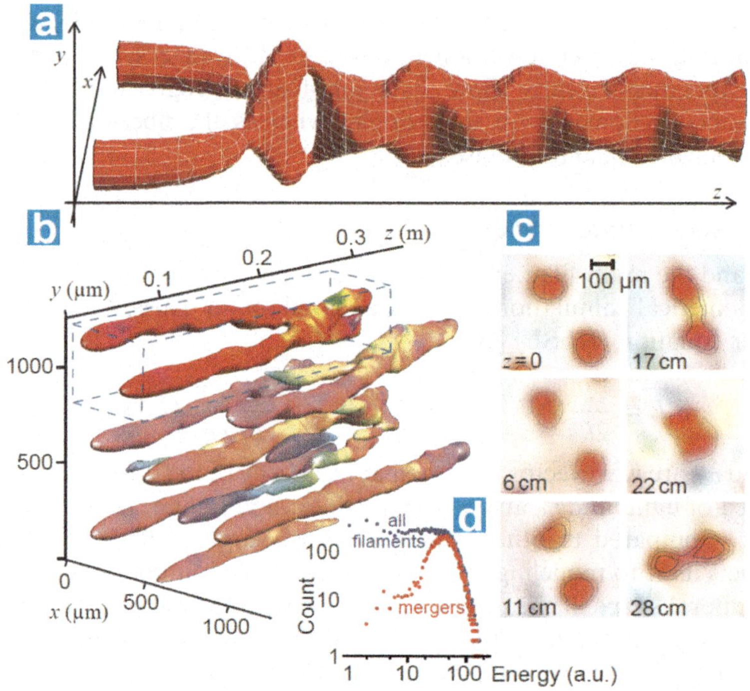

Figure 11.8. Numerical simulations of the evolution of fluence distributions along the propagation coordinate z during multiple filamentation. (a) Isofluence surfaces, with color coding showing relative phases. (b) Fluence statistics of filament strings, with separated merger events. Reprinted figure with permission from [9]. Copyright 2013 by the American Physical Society.

filament propagation, the microscopic mechanism behind rogue wave formation is actually a collision process of individual filament strings, i.e. rather similar to soliton interaction models in the one-dimensional optical fiber analog system, as well as to wave interference in the ocean scenario.

11.6 Predictability of rogue waves

After understanding the microscopic origin of rogue waves in the multi-filament context, and after considering that elementary linear interference and random phase synchronization are at the heart of all three systems, i.e. the ocean context, multiple filaments, and the optical fiber scenario, let us return to the original open question: are rogue waves actually predictable, or do they really appear without a warning? So are rogue waves really rogue, following the meaning of the English word 'rogue'? As this question cannot be answered for ocean rogue waves due to the scarcity of available data, the vast data of the multi-filament system offers enough substance to allow sound statistical conclusions. In the following, we therefore use this available data to evaluate the predictability of rogue waves.

In order to be predictable, a physical system must be ruled by deterministic dynamics. The simplest example may be a simple mechanical system, e.g. a harmonic oscillator. Knowing the equation of motion of this system and knowing the initial state of the system, we can perfectly predict the future of this system. This holds also for complex systems such as Brownian motion if all initial values are exactly taken into account. One realization of a particle trajectory occurs in a classical deterministic way. Limitations of prediction appear when a stochastic description is used, where probabilistic estimates are introduced rather than the required huge number of exact initial values. The system then exhibits a certain amount of stochastic indeterminism, making predictions difficult but still possible up to a certain degree. This behavior is contrasted by a quantum mechanical system, e.g. an atom. If this atom is in an excited state, it will eventually emit a photon and return to the ground state. However, prediction is limited here by the fundamental uncertainty principle and quantum indeterminism. In our investigations, we have to consider both kinds of indeterminism.

When a system is deterministic, there has to be a certain amount of correlation between behavior in the past and future development of the system. Restricting ourselves to the case of having only time series $\{x_1, x_2, x_3, \ldots, x_N\}$ of one physical quantity available, we therefore expect that particular sequences of values $\{y_1, y_2, y_3, \ldots, y_m\}$ with embedding dimension $m \ll N$ appear more frequently in the time series than others. Going back to the example of a harmonic oscillator, e.g. one observes sinusoidal sequences with one particular oscillation frequency but not with others; nor does one expect to see other functional behavior in the time series. In the quantum mechanical example cited above, observation of one fluorescent photon from one atom does not allow us to predict the fluorescence of a second independent atom, i.e. there are no preferential sequences in the time series. In order to evaluate the amount of determinism in the data, one therefore computes a histogram

$$C_m(r) = \frac{2\left|\left\{r_{ijm}\colon (1-\varepsilon)r < r_{ijm} \leqslant r\right\}\right|}{(N-m)(N-m-1)} \tag{11.4}$$

of metric distances

$$r_{ijm} = \left\| \mathbf{y}_{i,m} - \mathbf{y}_{j,m} \right\| = \sqrt{\sum_{k=i}^{i+m} \left| x_k - x_{k+j-i} \right|^2}, \tag{11.5}$$

where ε defines the bin size. Expanding equation (11.3), one can immediately see the connection to Pearson's correlation coefficient in the product term $x_k x_{k+j-i}$ that results from expanding the rhs of equation (11.4), i.e. similar subsequences that are highly correlated to each other result in small values of r. Statistical anticorrelation, in contrast, maximizes r. Example histograms are shown in figure 11.9 for the three systems under discussion, and they typically show a characteristic near-parabolic behavior in the log–log representation chosen. As indicated by the parabola, strong correlation or anticorrelation are relatively rare in all three systems. For a judgment on the predictability of the system, there has to be an anomalously high number of counts at low r, i.e. there are particular sequences of events that happen much more often than one would expect in a time series without temporal correlation. The obvious problem is that one does not have an uncorrelated time series available that

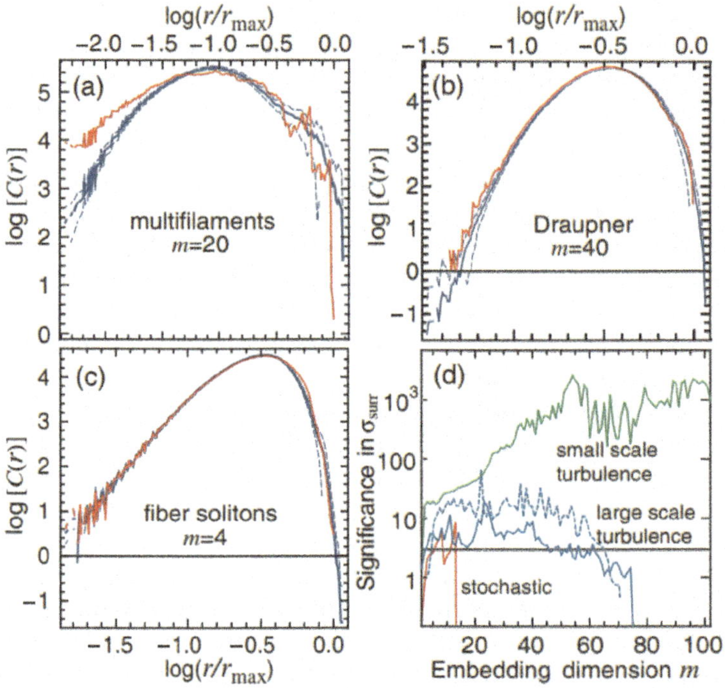

Figure 11.9. Nonlinear correlation analysis of rogue wave systems. Shown is the histogram count C as a function of Euclidian distance r. Red line: histogram of original data set. Blue line: average of surrogates. Dashed blue lines: surrogate standard deviation spread. Embedding dimensions m have been adapted to the correlation widths in (a) the multi-filament system and (b) an analysis of the Draupner event. (c) Analysis for solitons in a nonlinear fiber; an embedding of 4 has been chosen to avoid trivial behavior. (d) Maximum detected statistical significance of deterministic behavior as a function of embedding dimension. Significances < 3 can safely be ignored as they indicate stochastic behavior within the expected credibility limits. Green curve: multi-filament. Dark blue curve: Draupner1 data set. Dashed blue curve: Draupner2 data set. Red curve: fiber solitons [14]. Reprinted figure with permission from [22]. Copyright 2015 by the American Physical Society.

one could compare with. However, one can generate such an uncorrelated time series by randomly shuffling the x_i. This procedure is known as the generation of surrogate time series, which also conserves the Fourier transform and the linear autocorrelation function of the time series as best as possible. Therefore, the original time series and its surrogates are undistinguishable by all means of linear descriptive statistics. Moreover, it has proven important to use 'true random numbers' for the random shuffling process, as pseudo-random sequences bear a certain amount of temporal correlation since they have been generated by a deterministic algorithm on a computer. True random numbers, in contrast, are always based on measurements of a stochastic physical process, e.g. radioactive decay, or Johnson or shot noise. The random numbers used in our study have been deduced from the noise of a radio receiver, i.e. they are related to Johnson noise.

Applying the surrogate method to the time series and reanalyzing the data with equation (11.3) gives rise to the blue curves in figure 11.9. We generated a rather large number of 100 surrogates for each physical system, which then allowed us to compute mean and variance of the surrogate histograms. The difference between surrogate and original data set is most clearly illustrated for the case of multi-filaments in figure 11.9(a). Looking at the counts at lowest r at about 1% of the maximum, one immediately sees that very similar events are at least one order of magnitude more frequent in the original than in the surrogates, in which all temporal correlation has been removed. This clearly proves that the multi-filament scenario is deterministic and not driven by quantum noise. The nonlinear fiber propagation scenario, in contrast, does not show any statistically significant deviation between original and surrogates (see figure 11.9(c)). This clearly underlines the previous explanation of these two different rogue wave supporting systems. The rogue waves that appear after nonlinear fiber propagation have their origin in quantum-noise-induced pulse energy fluctuations of the driver laser. Due to the modulation instability, a high sensitivity to the initial energy fluctuation leads to different scenarios of random splitting of the pulse into completely different distributions of pulses. These pulses then propagate through the fiber in a deterministic fashion, which is described by the nonlinear Schrödinger equation. No further random driven or fluctuating processes in the fiber have to be taken into account. However, the break-up at the beginning is driven by quantum mechanical uncertainty, resulting in a complete destruction of coherence (or correlation) between successive pulses. Similar to the example of atomic fluorescence, it is impossible to predict from the history of pulse energies when conditions for rogue wave generation inside the fiber are met and when a 'rogue soliton' is observed at the end of the fiber. If a similar noise amplification process were also to rule the multi-filament dynamics, one would also expect to see no difference between original and surrogates in figure 11.9(a), but this is obviously not the case. Therefore, despite dissimilar underlying nonlinear propagation equations and modulation instability, there must be a different generation mechanism behind the rogue waves in this system, and this conclusion clearly matches the previously found explanation that turbulence inside the gas cell is the driving mechanism behind multi-filament rogue waves. As it is ultimately a deterministic mechanical process, the small-scale turbulence in the cell provides the correlation mechanism between successive shots of the kHz pump laser.

After understanding the two prototypical types of behavior, let us return to the original question: are ocean rogue waves predictable or not? As shown in figure 11.9(b), the difference between original and surrogates is actually quite small in this case. We have therefore computed the statistical significance of determinism as a function of embedding dimension in figure 11.9(d). For the two ocean time series analyzed, the significance amounts to values of about 10 standard deviations, which confirms that the ocean is a deterministic system and therefore bears some amount of predictability. However, as the ocean is obviously a much larger system than a gas cell with 5 cm diameter, ocean dynamics are admittedly much more complex and prediction of a large wave is therefore expected to be more difficult than for the small-scale turbulence in our cell.

As the multi-filament data contains several hundred rogue events in a one-minute record length, this feat offers the unique possibility to ask the question much more specifically: even if the system shows overall deterministic behavior, are the rogue events heralded by particular sequences that might allow us to predict them? To answer this question, we modified our surrogate analysis. First we specifically isolate small segments of the time series that are immediately preceding a rogue event in the multi-filament data. Then we compute a histogram of the latter (see figure 11.10(a)). For comparison, we select an equal number of random segments from the same trace and compute a surrogate histogram accordingly. Again, the surrogate process is repeated to build meaningful statistics, and the resulting mean and standard deviation are shown as blue lines in figure 11.10. This analysis clearly shows a shift of the observed maximum count between surrogates and original. In particular, there is also a marked increase of counts at low r in the original data set. We therefore conclude that rather characteristic patterns precede rogue events in the multi-filament scenario. We also redid this analysis using time sequences to follow up on individual rogue events, and see basically an identical behavior. However, if we increase the delay between rogue event and preceding sequence, we start to lose the statistical anomaly. This loss of correlation already appears on the 10 ms scale, and the correlation is completely lost for delays of 100 ms (or roughly 10 times the length of a typical rogue event).

From our correlation analysis, we can therefore conclude that rogue wave prediction on the ocean will be a difficult task. Transferring the results of the multi-filament scenario to the ocean system, we expect to be able to predict a rogue wave at best a few oscillation cycles before impact of the wave. In practical terms, this means a few tens of seconds for an average storm on the North Sea. It should be clear that this estimate is probably too optimistic, as ocean dynamics are clearly more complex than the small-scale turbulence in our cell. In the end, one may only be able to predict a rogue wave less than ten seconds before impact, and this short

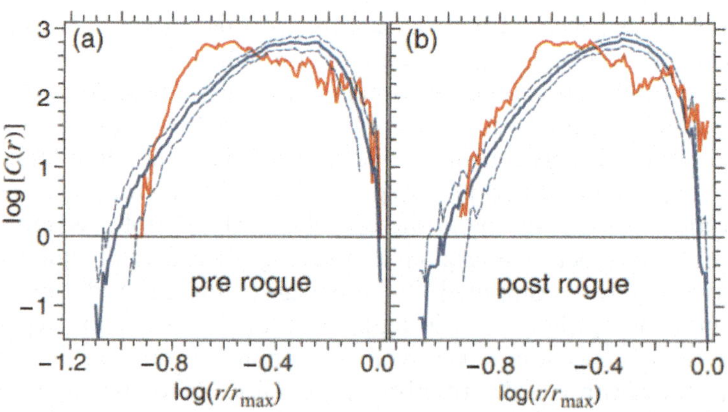

Figure 11.10. Histogram of Grassberger–Procaccia algorithm analysis of time series segments in immediate vicinity of a rogue event in the multi-filament data (red curve), in comparison to arbitrarily chosen segments of equal length (solid blue curve). (a) Pre-rogue statistics. Reference time series immediately preceding 289 isolated rogue events. (b) Post-rogue statistics. The statistical spread for arbitrarily chosen data (100 data sets) is indicated by the dashed blue curves. Reprinted figure with permission from [22]. Copyright 2015 by the American Physical Society.

timespan would make a prediction completely worthless. One may certainly argue that observation at many points surrounding a maritime structure may provide additional data for a more meaningful predictions, but in the end this would probably qualify more as observation rather than prediction. Once a rogue wave is seen at the perimeter of the detector array surrounding a ship, one can certainly predict the trajectory of the wave, but this has probably already been done for hundreds of years and has clearly proven insufficient to avoid deadly encounters with extreme waves.

11.7 Conclusion

Returning to our initial question of the predictability of ocean rogue waves, our discussion clearly corroborates that the ocean is a deterministic system and therefore, at least in principle, predictable. However, in practical terms, warning times seem to be too short or predictions too unreliable to be anywhere helpful. Comparing rogue waves in our three different systems, it also becomes clear that the appearance of a heavy-tail distribution is completely independent of the predictability of the system. Fiber-optical rogue waves are positively unpredictable, while those in the multi-filament context clearly show a certain amount of predictability. Rogueness, i.e. the surprising or unpredictable character of an individual event, is therefore a rather weak criterion and extreme events are not automatically rogue. Going back to the initial discussion on different languages, it seems that the Italian 'ondo anomala' is probably the most precise term, whereas the much more popular English term 'rogue wave' is rather inaccurate. Nevertheless, even though scientifically much less precise, the German 'Kaventsmann' appears to be the most pictorial word for extreme ocean waves among the languages that we are aware of.

Acknowledgments

We gratefully acknowledge Simon Birkholz for his experimental work on multi-filament rogue waves at MBI, and Carsten Brée for conducting pertinent numerical simulations at the Weierstrass Institute. We further thank Daniel Solli (UCLA, Los Angeles, California) for providing original rogue wave data for the fiber-optical system and Statoil ASA for their permission to use the data recorded on the Draupner oil platform in January 1995. Mads Haahr is acknowledged for giving us generous access to true random number series at random.org. Moreover, we thank Michel Olagnon (IFREMER, Brest, France), Sverre Haver (Norwegian University of Science and Technology, Trondheim), and Anne Karin Magnusson (Norwegian Meteorological Institute, Bergen) for their support. We further acknowledge Goëry Genty (Tampere University of Technology), John Dudley (Université de Franche-Comté), Fedor Mitschke (University of Rostock), and Claus Ropers (University of Göttingen) for helpful discussions on various aspects of rogue waves. Finally, we thank Luc Bergé (Commissariat à l'Energie Atomique et aux Energies Alternative) and Giuseppe Sansone (University of Freiburg) for their insights into the exact meaning of the French and Italian translations, respectively, of 'rogue wave'.

References

[1] Onorato M, Residori S, Bortolozzo U, Montinad A and Arecchi F T 2013 Rogue waves and their generating mechanisms in different physical contexts *Phys. Rep.* **528** 47

[2] Akhmediev N and Pelinovsky E 2010 Editorial – Introductory remarks on 'discussion & debate: rogue waves - towards a unifying concept?' *Eur. Phys. J. Spec. Top.* **185** 1

[3] Solli D R, Ropers C, Koonath P and Jalali B 2007 Optical rogue waves *Nature* **450** 1054

[4] Akhmediev N, Kibler B and Baronio F *et al* 2016 Roadmap on optical rogue waves and extreme events *J. Opt.* **18** 063001

[5] Dudley J M, Genty G and Eggleton B J 2008 Harnessing and control of optical rogue waves in supercontinuum generation *Opt. Express* **16** 3644–51

[6] Genty G, de Sterke C M, Bang O, Dias F, Akhmediev N and Dudley J M 2010 Collisions and turbulence in optical rogue wave formation *Phys. Lett.* A **374** 989–96

[7] Dudley J M, Dias F, Erkintalo M and Genty G 2014 Instabilities, breathers and rogue waves in optics *Nat. Photon.* **8** 755

[8] Kasparian J, Béjot P, Wolf J-P and Dudley J M 2009 Optical rogue wave statistics in laser filamentation *Opt. Express* **17** 12070

[9] Birkholz S, Nibbering E T J, Brée C, Skupin S, Demircan A, Genty G and Steinmeyer G 2013 Spatiotemporal rogue events in optical multiple filamentation *Phys. Rev. Lett.* **111** 243903

[10] Bergé L, Skupin S, Nuter R, Kasparian J and Wolf J-P 2007 Ultrashort filaments of light in weakly ionized, optically transparent media *Rep. Prog. Phys.* **70** 1633

[11] Hosseini S A, Luo Q, Ferland B, Liu W, Chin S, Kosareva O, Panov N, Aközbek N and Kandidov V 2004 Competition of multiple filaments during the propagation of intense femtosecond laser pulses *Phys. Rev.* A **70** 033802

[12] Bergé L *et al* 2004 Multiple filamentation of terawatt laser pulses in air *Phys. Rev. Lett.* **92** 225002

[13] Weibull W 1951 A statistical distribution function of wide applicability *J. Appl. Mech. Trans. ASME* **18** 293

[14] Solli D R, Herink G, Jalali G and Ropers C 2012 Fluctuations and correlations in modulation instability *Nat. Photon.* **6** 463

[15] Dudley J M, Genty G and Coen S 2006 Supercontinuum generation in photonic crystal fiber *Rev. Mod. Phys.* **78** 1135

[16] Newbury N R, Washburn B R, Corwin K L and Windeler R S 2003 Noise amplification during supercontinuum generation in microstructure fiber *Opt. Lett.* **28** 944

[17] Corwin K L, Newbury N R, Dudley J M, Coen S, Diddams S A, Weber K and Windeler R S 2003 Fundamental noise limitations to supercontinuum generation in microstructure fiber *Phys. Rev. Lett.* **90** 113904

[18] Demircan A and Bandelow U 2007 Analysis of the interplay between soliton fission and modulation instability in supercontinuum generation *Appl. Phys.* B **86** 31–9

[19] Genty G, Surakka M, Turunen J and Friberg A T 2011 Complete characterization of supercontinuum coherence *J. Opt. Soc. Am.* B **28** 2301

[20] Cheng Y-H, Wahlstrand J K, Jhajj N and Milchberg H M 2013 The effect of long timescale gas dynamics on femtosecond filamentation *Opt. Express* **21** 4740

[21] Ryabtsev A, Pouya S, Koochesfahani M and Dantus M 2014 Vortices in the wake of a femtosecond laser filament *Opt. Express* **22** 26098

[22] Birkholz S, Brée C, Demircan A and Steinmeyer G 2015 Predictability of Rogue Events *Phys. Rev. Lett.* **114** 213901

IOP Publishing

Nonlinear Guided Wave Optics
A testbed for extreme waves
Stefan Wabnitz

Chapter 12

Rogue waves in integrable turbulence: semi-classical theory and fast measurements

Pierre Suret, Gennady El, Miguel Onorato, Stéphane Randoux

12.1 Introduction

Understanding the mechanisms of formation of rogue waves is nowadays considered a challenge in different fields of physics. Originally observed on the surface of the ocean, rogue waves have been investigated in many optical [1–3] and hydrodynamical [4] laboratories, and numerical simulations of nonlinear dispersive partial differential equations. While the measurements of rogue waves in a typical water wave tank experiment are more or less straightforward, efforts in the field of optics have been devoted in order to develop measurement devices suitable for recording in time the envelope of a rogue wave [5, 6].

In the mathematical community, the term 'rogue wave' has been associated with exact rational or breather solutions of partial differential equations, the prominent examples being the Peregrine soliton (PS), Akhmediev breather (AB), and Kuznetsov–Ma breather (KM) solutions of the focusing nonlinear Schrödinger equation (NLSE) [7–13]. Starting from the works in [14–16] on the NLSE, techniques have been developed in order to find rational solutions of many different systems of equation such as the three wave or the vector NLS systems [17]. The success in finding these interesting solutions that are considered the prototypes of rogue waves is largely due to the fact that the mentioned partial differential equations share the property of being integrable in terms of the inverse scattering transform (IST). Besides the possibility of finding various complex nonlinear structures, starting from the pioneering work of A R Osborne (see [18] and references therein), the IST has been used also as a tool for analyzing experimental or numerical data (see e.g. [19–21]), just as the linear Fourier analysis is used for linear systems. In particular, an IST-based method developed in [22] allows one to recognize locally a nonlinear structure; for example, a breather embedded in a random field. The IST-based methodology so far has been applied to data assuming

doi:10.1088/978-0-7503-1460-2ch12 © IOP Publishing Ltd 2017

that the dynamics are ruled by the NLSE, which acts as a bridge connecting different communities of nonlinear physics [23, 24]. In particular, it is well known that to the leading order in nonlinearity and dispersion, deep water surface gravity waves can be described with some degree of accuracy by such an equation. Similarly, the dynamics in an optical Kerr fiber characterized by an anomalous dispersion (focusing) regime are also very well described by the NLSE.

The above mentioned 'standard' rogue wave solutions found for the NLSE live on a coherent background, i.e. are characterized by a carrier wave modulated by long perturbations; a more demanding task is the understanding of rogue waves in a partially coherent system: such conditions are typical of wind waves and are known in the field of oceanography as random sea states. In such conditions, the coexistence of random waves with coherent structures makes the research extremely interesting and suitable for a statistical description [25–28]. In an integrable system, such structures outlive interactions and are preserved during the evolution; for non-integrable systems (as all physical systems), they have a finite life time which depends on how close the system is to an integrable one.

In general, the nonlinearity of ocean waves, measured by the wave steepness, is small (except in very shallow water): even though the wave capping observed in ocean waves is a strongly nonlinear process, it has the end effect of keeping the waves sufficiently small (for a given wavelength). The outcome of such a phenomenon is that a weakly nonlinear statistical approach is sufficient for extracting the main features of the water wave dynamics. However, in an optical fiber, much stronger nonlinear conditions can be achieved and in such cases any standard weakly nonlinear approach would fail dramatically, and new tools should be developed.

It should be mentioned that, while in an optical fiber and in a long water wave tank experiment one can propagate unidirectional waves, ocean waves generated by the wind, as in the ocean, are characterized by a directional spectrum, i.e. wave energy is distributed as a function of frequency and direction (in polar coordinates). This fact limits the usability of the one-dimensional NLSE and sets new challenges in the field of rogue waves on a ground where integrability is not a feature of the governing equations.

In the next sections, a more detailed analysis of some experimental work carried out in optics and hydrodynamics will be outlined; particular attention will be given to the theoretical framework under which such experimental results can be interpreted. The lack of a statistical theory that describes a strongly nonlinear regime appears to be the most intriguing facet of the field.

12.1.1 Deviations from Gaussian statistics and the appearance of heavy tails

Given a time series of the surface elevation recorded in the ocean, it is hardly possible to recognize correlations between Fourier phases; therefore, the general description of ocean waves, also for engineering applications, consists in assuming that the Fourier phases are distributed randomly on the $[0, 2\pi)$ interval and the Fourier amplitudes are described by some spectral shape, usually the JONSWAP spectrum [29].

Under these assumptions, according to the central limit theorem, the surface elevation is a stochastic variable described by a normal (Gaussian) distribution; this implies that the wave envelope is distributed according to the Rayleigh distribution and the intensity (the square of the envelope) distribution is exponential. Thus rogue waves, defined as those waves larger than a certain threshold, can always emerge in linear systems [30]: it is only a question of collecting enough data in a simulation or in an experiment. However, it is nonlinearity that can make these waves to appear more (or less) often than predicted by Gaussian statistics.

Any keen observer of the ocean can notice that the ocean waves are not symmetric in the sense that crests are narrower that the troughs: this is a first nonlinear effect due mostly to nonresonant (mismatched) three-wave interactions (quadratic nonlinearity). Overall, such effect leaves a signature on the probability density function (PDF) of the surface elevation by making it skewed; marginal effects, if any, can be observed in the PDF of the envelope or intensity.

Drastic changes of the PDF of the envelope or intensity can be detected if higher-order nonlinearities are taken into account: in particular, four-wave interactions seem to play a crucial role in determining the deviation from the Gaussian statistics. Indeed, experiments performed in a long wave tank [31–33] (unidirectional propagation) with initial conditions prescribed by the JONSWAP spectrum and random phases have confirmed the existence of large deviations (heavy tails) of the PDF from the Rayleigh distribution of the wave envelope. The spontaneous emergence of coherent structures from a random initial condition can be observed directly from the experimental data (see figure 12.1 in [32]): the most energetic wave packets go through a modulational process that leads to a growing of a single wave at the expense of the energy in the packet. It has been observed that such mechanism is more probable when the ratio between the mean steepness of the waves to the relative spectral bandwidth of the initial condition is large. Such ratio is known as the Benjamin–Feir index [34, 35] in the oceanographic community of rogue waves, and it can be easily related to the ratio between the nonlinear to linear Hamiltonian of the NLS equation or to the ratio between a linear and nonlinear time scale of the NLS equation. The PDF of the wave envelope is strongly influenced by those structures, and their end effect is to populate the tails of the PDF of wave envelope. Similar behavior in the PDFs has been observed in experiments in optical fibers [36] and in numerical simulations of the NLS equation [37] (see the next sections).

12.1.2 Weakly nonlinear theory

The standard tool for studying a system of a large number of interacting (nonlinear) dispersive waves is the so-called weak wave turbulence theory [25, 38]. The main outcome of the theory is the wave kinetic equation, which represents an evolution equation *à la* Boltzmann for the wave spectral density; the collision integral is characterized by the presence of two Dirac-δ functions over four wave numbers and four frequencies that assure the conservation of momentum and linear energy, besides that of the number of waves. The energy transfer between modes is therefore associated with a resonant process described by the above conservation laws, all

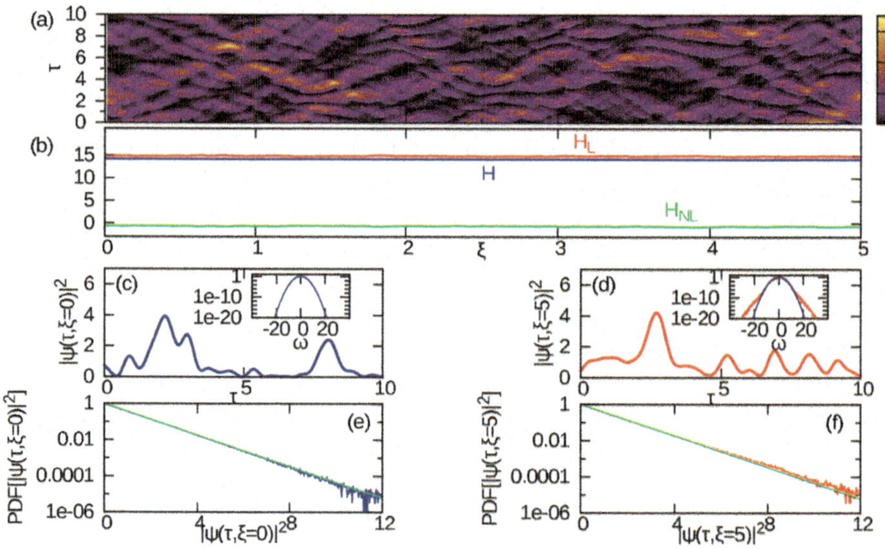

Figure 12.1. Integrable turbulence in the weakly nonlinear regime. Numerical simulations of equation (12.1) between $\xi = 0$ and $\xi = 5$, with $\varepsilon = 3$ and a random field having Gaussian statistics at initial stage ($\xi = 0$). (a) Spatiotemporal evolution of $|\psi(\tau,\xi)|^2$. (b) Evolution of the linear kinetic energy H_L (red line), of the nonlinear energy H_{NL} (green line) and of the Hamiltonian of the wave system $H = H_L + H_{NL}$ (blue line). (c) Time evolution of the power $|\psi(\tau,\xi = 0)|^2$ of the initial condition and corresponding power spectrum $|\hat{\psi}(\omega,\xi = 0)|^2$ in inset. (d) Same as (c) but at $\xi = 5$. (e) PDF of the power fluctuations of the initial condition (blue line) and (f) PDF of the power fluctuations at $\xi = 5$ (red line). The green lines in (e) and (f) represent the exponential distribution.

weighted by an interaction coefficient. In a closed nonintegrable system, the theory predicts for large times the appearance of a thermalized spectrum known as the Rayleigh–Jeans spectrum [25, 28, 38, 39]. One of the main assumptions of the theory is the randomness of the phases which allow for a Wick decomposition of the moments higher than the second. Such theoretical framework does not allow one to gain any information on higher-order moments; in particular, no information can be obtained on the fourth-order one, i.e. the first that weights the relevance of the tails of the PDF and gives information on rogue waves. Moreover, it should be mentioned that in the case of an integrable system, the theory does not provide any prediction because, for such systems either there are no resonances or the interaction coefficient is zero on the resonant manifold [40].

Such circumstances have been overcome in [35], see also [41, 42], by considering nonresonant interactions, i.e. interactions for which linear energy is not conserved. An explicit calculation has revealed that the kurtosis, i.e. the normalized centered fourth-order moment of the surface elevation, is proportional to the square of the Benjamin–Feir Index. We stress here once more that the key point, essential to get nontrivial statistics, is to allow for interactions that do not preserve the linear energy; in the NLS framework, this is the only possibility for letting the nonlinear part of the energy (directly related to the fourth-order moment) to change in time and assume large values. In particular, an exact relation between the statistics and the spectral width can be easily demonstrated from the conservation of the Hamiltonian; further discussion on this interesting issue can be found in [43].

The predictions of the theory on the evolution of the spectral shape and of the kurtosis work fine in the weakly nonlinear regime; however, in integrable systems, as the nonlinearity becomes strong, other tools need to be considered and this is the topic of the next section.

12.1.3 Integrable turbulence

The theoretical framework combining a statistical approach to random waves together with the property of integrability of the 1D-NLSE is known as *integrable turbulence*. This emerging field of research, recently introduced by V Zakharov, relies on the analysis of complex phenomena found in nonlinear random wave systems described by an integrable equation [36, 42, 44–52]. One of the key features of integrable turbulence is the establishment, at long evolution time, of a state in which the statistical properties of the wave system remain stationary. Due to integrable nature of the system, the long-time statistics depend on the statistics of the input random process (see [22, 36, 47, 48, 50, 51]). So far, there has been no satisfactory theoretical framework developed for the description of statistical features of integrable turbulence due to high complexity of nonlinear wave interactions occurring over the course of its development. In this section, we briefly review the main dynamical, spectral, and statistical features typifying integrable turbulence both in strongly and in weakly nonlinear regimes.

We consider the focusing integrable 1D-NLSE in dimensionless form:

$$i\varepsilon\frac{\partial\psi}{\partial\xi} + \frac{\varepsilon^2}{2}\frac{\partial^2\psi}{\partial\tau^2} + |\psi|^2\psi = 0. \tag{12.1}$$

In an optical fiber experiment, $\psi(\tau, \xi)$ is the slowly varying envelope of the electric field A that is normalized to the square root of the mean optical power $\overline{\rho_0}$ of the partially coherent field propagating inside the fiber ($\psi = A/\sqrt{\overline{\rho_0}}$). It is usual in nonlinear fiber optics to introduce a nonlinear length $L_{NL} = 1/(\gamma\overline{\rho_0})$ and a linear dispersion length $L_D = 2/(\beta_2[\Delta\nu_0]^2)$. γ and β_2 are the Kerr and the second-order dispersion coefficients of the optical fiber, respectively. $\Delta\nu_0$ represents the width of the spectrum of power fluctuations of the wave injected inside the optical fiber. With our notations, the propagation distance z along the fiber is normalized as $\xi = z/\sqrt{L_{NL}L_D}$, the physical time t is normalized as $\tau = t/T_0$ with $T_0 = 1/\Delta\nu_0$, and $\varepsilon = \sqrt{L_{NL}/L_D}$ is the dispersion parameter. However, since ξ is an evolution variable in (12.1), we sometimes shall be referring to ξ and τ as to 'time' and 'space,' respectively when describing the mathematical developments. We shall also allow τ to assume negative values, implying the necessary physical time shift.

Equation (12.1) conserves the Hamiltonian $H = H_L + H_{NL}$ that has a nonlinear contribution $H_{NL} = -(1/2)\int|\psi(\tau, \xi)|^4 d\tau$ and a linear (kinetic) contribution $H_L = (\varepsilon^2/2)\int\omega^2|\hat{\psi}(\omega, \xi)|^2 d\omega$, the Fourier transform being defined as

$\hat{\psi}(\omega, \xi) = 1/\sqrt{2\pi} \int \psi(\tau, \xi)e^{-i\omega\tau}d\tau$. Equation (12.1) also conserves the number of particules (or power) $N = \int |\psi(\tau, \xi)|^2 d\tau$ and the momentum $P = \int \omega|\hat{\psi}(\omega, \xi)|^2 d\omega$.

The field $\psi(\tau, 0)$ taken as initial condition in our numerical simulations is made from the superposition of many independent Fourier modes with random phases. As discussed in detail in [37], the statistics of this initial field is Gaussian, which implies that the PDF of the initial power fluctuations $P_0(\tau) = |\psi(\tau,0)|^2$ is given by the exponential distribution, i.e. $\mathrm{PDF}[P_0/<P_0>] = \exp(-P_0/<P_0>)$, where the brackets denote the operation of the ensemble averaging over thousands of realizations of the random process.

Figure 12.1 shows dynamical, spectral, and statistical features typifying integrable turbulence in the weakly nonlinear regime where $\varepsilon = 3$ and $<P_0> =1$. In this regime, the linear kinetic energy is much greater than the nonlinear energy ($H_L \gg |H_{NL}|$), see figure 12.1(b). The spatiotemporal evolution plotted in figure 12.1(a) reveals that power fluctuations present at the initial stage slowly evolve with ξ without any significant increase in their peak amplitude. This is confirmed by Fourier analysis that shows that the spectra of the input ($\xi = 0$) and of the output ($\xi = 5$) power fluctuations are very similar, see insets in figures 12.1(c) and (d). In the weakly nonlinear interaction regime, the PDF of power fluctuations exhibits very small heavy-tailed deviations from the initial exponential distribution, compare figure 12.1(e) and figure 12.1(f).

The spectral and statistical features evidenced in the weakly nonlinear regime of figure 12.1 have been described by implementing a modified treatment of the conventional wave turbulence (WT) theory. This approach has been initially proposed by Janssen [35], and it has been relatively recently applied to optics [28, 41, 42]. The conventional WT theory is commonly used to treat random wave systems that are ruled by *non*-integrable equations and in which the long-time evolution is dominated by resonant interactions among the Fourier components composing the random wave. However, it has been shown that the short-term evolution of WT can be influenced by non-resonant terms [53]. In the revisited treatment of WT theory, 'quasi-kinetic' equations that describe the evolution of the wave spectrum are derived by keeping the contribution of non-resonant terms. This WT treatment describes adequately the transient evolutions and the stationary state of the power spectrum, and also of the PDF of field fluctuations that are observed in the weakly nonlinear regime [35, 41, 42].

Figure 12.2 shows dynamical, spectral, and statistical features typifying integrable turbulence in the strongly nonlinear regime, where $\varepsilon = 0.3$ and $<P_0> =1$. In this regime, the nonlinear energy is much greater than the linear kinetic energy ($|H_{NL}| \gg H_L$) at the initial stage ($\xi \ll 1$), see figure 12.2(b). However, the wave system undergoes some rapid evolution and a state where $|H_{NL}| = 2H_L$ is quickly reached for $\xi > 0.25$. This evolution is accompanied by the emergence of coherent structures that have peak powers significantly greater that the peak power reached by the fluctuations of the initial condition, see figures 12.2(a), (c), and (d). This spatiotemporal evolution is correlated with a significant broadening of the Fourier spectrum of the random field that takes a triangular shape in vertical logarithmic

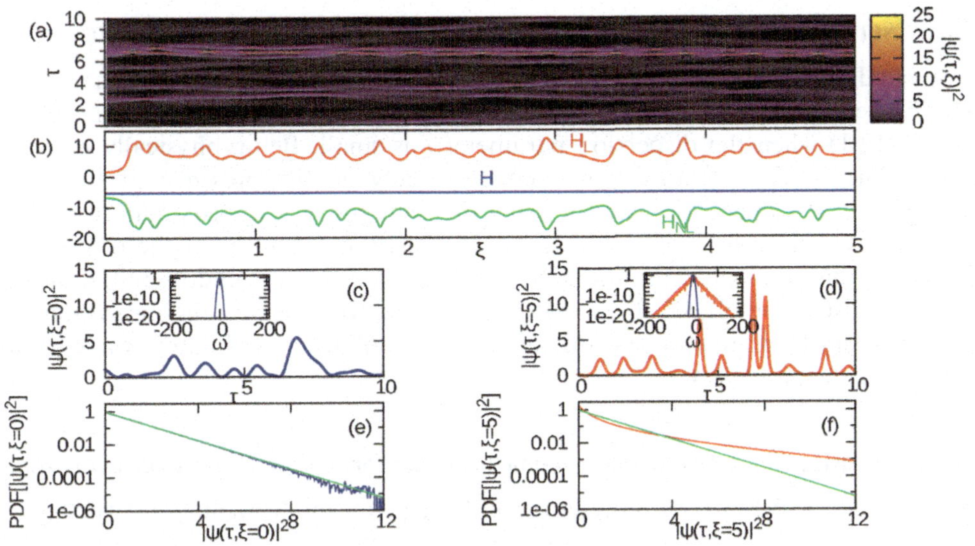

Figure 12.2. Integrable turbulence in the strongly nonlinear regime. Same numerical simulations as in figure 12.1 but with $\varepsilon = 0.3$.

scale at $\xi = 5$, see insets in figures 12.2(c) and (d). In the strongly nonlinear interaction regime, the PDF of power fluctuations exhibits pronounced heavy-tailed deviations from the initial exponential distribution, see figures 12.2(e) and (f). Those heavy-tailed deviations from the Gaussian statistics have been evidenced in experiments reported in [5, 36].

As previously mentioned, there is no satisfactory theoretical framework developed for the description of statistical features of integrable turbulence. However recent developments on the analysis of integrable turbulence in the *defocusing* NLSE reported in [52] have shown that the evolution in the strongly nonlinear regime of figure 12.2 can be decomposed into two distinct stages. In the initial, so-called pre-breaking stage of integrable turbulence, nonlinear effects dominate linear dispersion, and the wave fronts of the random initial field experience some gradual steepening leading to the formation of gradient catastrophes that are of the wave-breaking type. As discussed in [52], the gradient catastrophes in the defocusing case are regularized through the generation of dispersive shock waves (DSWs) representing nonlinear unsteady, rapidly oscillating structures [54], which are analytically described by a semi-classical theory.

In the focusing case considered in this chapter, the NLSE evolution is very different compared to the defocusing case but the two main qualitative features characterizing the strongly nonlinear regime remain the same: the initial evolution leads to the formation of gradient catastrophe, which is then regularized by weak dispersion via the generation of ε-scaled oscillations. However, the gradient catastrophe occurring in the focusing case is of a different type [55], and its regularization occurs not through the generation of a DSW but through the universal emergence of a Peregrine soliton followed by the formation of a quasi-periodic 'breather lattice,' as discussed more in detail in section 12.2.2.

12.2 Semi-classical limit of focusing 1D-NLSE and rogue waves

Motivated by the experiments involving the strongly nonlinear regime of the development of integrable turbulence, we shall be interested in solutions of the NLSE (12.1) when the dispersion parameter ε is small. Being physically relevant, this regime is also advantageous mathematically as the presence of the small parameter $\varepsilon \ll 1$ opens the way for using the semi-classical approximation.

In this section, we describe two basic *dynamical* mechanisms of the rogue wave formation that are realized in the process of the evolution of random initial data in the small-dispersion NLSE. These mechanisms are: (i) dispersive regularization of a gradient catastrophe, and (ii) interaction of nonlinear modulated wave trains (dispersive shock waves).

12.2.1 Focusing dispersive hydrodynamics and the formation of gradient catastrophe

We introduce a WKB-type representation for $\psi(x,t)$ (the Madelung transform)

$$\psi = \sqrt{\rho}\, e^{i\frac{\phi}{\varepsilon}}, \qquad \phi_\tau = u \tag{12.2}$$

to convert NLSE (12.1) into a system:

$$\begin{aligned} \rho_\xi + (\rho u)_\tau &= 0, \\ u_\xi + u u_\tau - \rho_\tau - \varepsilon^2 \left(\frac{\rho_\tau^2}{8\rho^2} - \frac{\rho_{\tau\tau}}{4\rho} \right)_\tau &= 0. \end{aligned} \tag{12.3}$$

System (12.3) is a 'focusing' analog of Eulerian dispersive hydrodynamics [54, 56]. It describes the flow of a 'photon fluid' with density $\rho \geqslant 0$, velocity u, and the negative hydrodynamic pressure $p = -\rho^2/2$. The optical significance of $\rho = |\psi|^2$ is the instantaneous power, and u is the instantaneous frequency (the chirp). The key feature of dispersive hydrodynamics is the scale separation due to the presence of the small parameter $\varepsilon \ll 1$ specifying the typical 'dispersion time' in the system. A natural mathematical framework in which to study the dynamics described by (12.3) with $\varepsilon \ll 1$ is the semi-classical approximation.

If functions $\rho(\tau,\xi)$ and $u(\tau,\xi)$ vary sufficiently slowly so that their derivatives are $O(1)$, one can neglect the dispersive $O(\varepsilon^2)$ term in (12.3) and describe the evolution in ξ by the leading-order approximation system (focusing nonlinear geometric optics):

$$\rho_\xi + (\rho u)_\tau = 0, \qquad u_\xi + u u_\tau - \rho_\tau = 0. \tag{12.4}$$

The dispersionless system (12.4) has complex eigenvalues $u \pm i\sqrt{\rho}$ and so is of an elliptic type, implying ill-posedness of initial-value problems for all but analytic initial data $\rho(\tau,0) = \rho_0(\tau)$, $u(\tau,0) = u_0(\tau)$. Even for analytic initial data, the solution of (12.4) has finite lifespan due to the formation of a *gradient catastrophe* characterized by infinite τ- and ξ-derivatives of ρ and u at some point τ_c, when ξ approaches a certain critical value ξ_c [55].

Let us first assume that $u_0(\tau) = 0$ (zero initial chirp), and $\rho_0(x)$ is some analytic bell-shaped function with the maximum at $\tau = 0$, say $\mathrm{sech}^2(\tau)$. Then the initial evolution of ρ and u with ξ can be found by looking for an asymptotic solution of (12.4) in the form of 'short-time' ($\xi \ll 1$) expansions for $\rho(\tau,\xi)$ and $u(\tau,\xi)$ which readily yields:

$$\rho(\tau,\, \xi) \simeq \rho_0(\tau) - \frac{1}{4}\left[\rho_0^2{,}(\tau)\right]''\xi^2, \qquad u(\tau,\, \xi) \simeq \rho'_0(\tau)\xi. \tag{12.5}$$

At the point $\tau = 0$ of the local maximum of $\rho_0(\tau)$, we have $[\rho_0^2(0)]'' = 2\rho_0(0)[\rho_0(0)]''$ <0, so the solution (12.5) for ρ predicts the growth of the instantaneous power at the point of the maximum, which, together with conservation of $\int \rho d\tau$, implies the self-compression of the initial pulse. On the other hand, the second expression in (12.5) shows that the chirp grows with ξ outside the point of maximum of $\rho_0(\tau)$ and changes the sign across this point. Both outlined tendencies are clearly seen in figure 12.3, showing the numerical solution of system (12.4) with initial conditions $\rho_0(\tau) = \mathrm{sech}^2\,\tau,\ u_0(\tau) = 0$.

An explicit analytic solution of (12.4) for $0 < \xi < \xi_c$ is only available for several specific initial profiles (see [57, 58]). Particularly, for $\rho_0(\tau) = \mathrm{sech}^2(\tau)$, $u_0(\tau) = 0$, the gradient catastrophe point (τ_c, ξ_c, ρ_c) was found to be (0,0.5,2), see figure 12.3. More recently, this result was extended in [59] to the chirped initial data $\rho_0(\tau) = \mathrm{sech}^2(\tau)$, $u_0(\tau) = -\mu \tanh \tau$, where $\mu \geqslant 0$. In that case, the gradient catastrophe was shown to occur at $\tau_c = 0$, $\xi_c = 1/(\mu + 2)$, $\rho_c = \mu + 2$. Note that, for $\mu = 0$, $\varepsilon = 1/N$, where $N = 1, 2, \ldots$ the initial function $\psi_0(\tau)$ represents an exact N-soliton solution of the NLSE (12.1) [60]; on the other hand, when $\mu \geqslant 2$, the initial data $\psi_0(\tau)$ is solitonless, i.e. its IST spectrum is purely continuous.

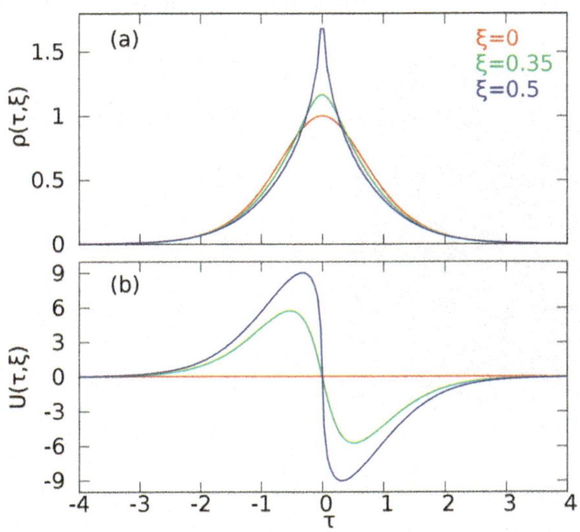

Figure 12.3. Gradient catastrophe formation in the dispersionless evolution (12.4) of the pulse $\rho(\tau,0) = \mathrm{sech}^2(\tau)$, $u(\tau,0) = 0$. The catastrophe occurs at $\xi_c = 0.5$, $\tau_c = 0$, $|\psi_c| = \sqrt{\rho(0,\, \xi_c)} = \sqrt{2}$.

12.2.2 Regularization of gradient catastrophe and the universal Peregrine soliton dynamics

In the vicinity of the gradient catastrophe point, the contribution of the dispersive term in (12.3) becomes significant so that the emerging regularized dynamics start to exhibit rapid, ε-scaled oscillations of finite amplitude signifying the singular nature of the semi-classical $\varepsilon \to 0$ limit for $\xi > \xi_c$. The small-dispersion evolution of pulsed initial data in the NLSE has been the subject of very active mathematical research in the last two decades (see [55, 59, 61, 62] and references therein). This research has revealed a number of fundamental properties of the evolution exhibiting the formation of highly coherent multi-scale spatiotemporal structures, see figure 12.4, in contrast with the traditional notion of modulational instability of plane wave typically associated with the generation of globally incoherent oscillations (see e.g. [47]).

One of the most remarkable features of the asymptotic behavior of the small-dispersion NLSE solutions is the universal emergence of the Peregrine soliton near the point of gradient catastrophe discovered by Bertola and Tovbis in [62]. Specifically, it was shown in [62] that the solution at the point of maximum temporal localization $\xi_{\rm m} = \xi_c + O(\varepsilon^{4/5})$ assumes the form $|\psi(\tau, \xi_{\rm m})| = a_0[1 - 4/(1 + 4a_0^2(\tau/\varepsilon)^2)](1 + O(\varepsilon^{1/5}))$, where the background amplitude $a_0 = \sqrt{\rho(0, \xi_c)} + O(\varepsilon^{1/5})$, i.e. is determined at leading order by the value of instantaneous power at the gradient catastrophe point. The maximum value of $|\psi|$ in the local PS is then $3a_0$ and is determined up to $O(\varepsilon^{1/5})$.

Importantly, the effect of the emergence of the PS right beyond the gradient catastrophe point is *local* and does not depend on the global spectral content of the initial data defined by the reflection coefficient in the IST formalism [64]. In other words, the initial condition for (12.3) can have high soliton content or be completely solitonless—in both cases the PS appears as a nonlinear coherent structure locally

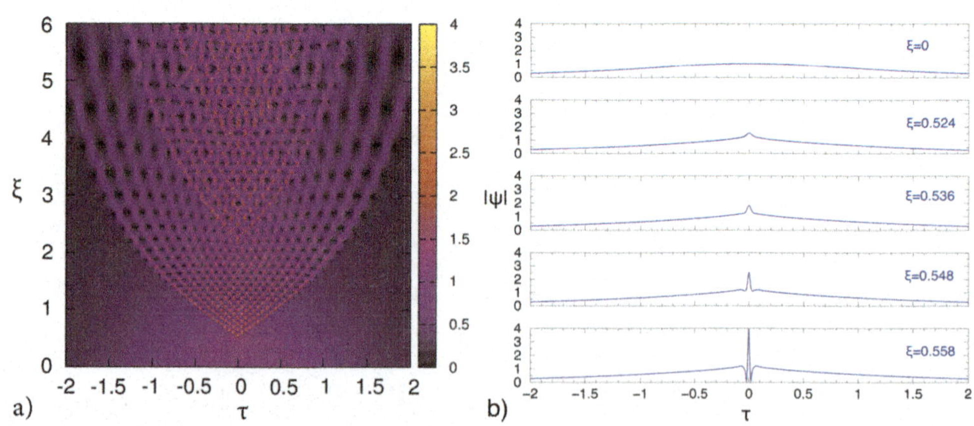

Figure 12.4. Dispersive regularization of gradient catastrophe via the generation of ordered oscillations of finite amplitude. (a) Contour plot of $\sqrt{\rho(\tau,\xi)}$ of the solution of the NLSE (12.1) with $\varepsilon = 1/33$ and the initial data $\psi = \operatorname{sech} x$. Reproduced from [63]. (b) Local emergence of the Peregrine soliton near the gradient catastrophe point $\tau_c = 0$, $\xi_c = 1/2$, $|\psi|_c = \sqrt{\rho(0, \xi_c)} = \sqrt{2}$.

regularizing the gradient catastrophe. The specific form of the initial condition affects only the propagating distance ξ_m of the PS development.

We note that it has been widely considered that the PS was uniquely associated with the process of plane wave modulation instability. The results outlined above show that it actually appears more generally as a universal coherent structure during intensity localization of high-power pulses. The persistence of the PS as the universal local coherent structure regularizing gradient catastrophe suggests its fundamental role in problems involving the propagation of coherent and *partially coherent* optical signals whose typical temporal scale is significantly larger than the coherence time of the fiber. Some evidence of this universality is presented in numerical simulations shown in figure 12.5, where several PS-like structures emerge from a random field having Gaussian statistics at the initial stage ($\xi = 0$). This scenario has been confirmed by experiments presented in [5]. Further careful experiments have revealed both intensity and phase signatures of the PS during the initial nonlinear evolution stage of pulses propagating inside optical fibers [65], see section 4 for the description of the relevant experimental techniques.

The semi-classical dynamics of pulsed signals beyond the formation of the PS are detailed in [62] and are characterized by the generation of multi-peaked 'breather lattices' (see figure 12.4(a)). These breather lattices are asymptotically described by modulated multiphase (finite-gap) NLSE solutions [18, 66]. A breather lattice is characterized by a *genus* g which is related to the number N of nonlinear oscillatory modes in the finite-gap theory formalism by a simple relation $N = g + 1$ (see [18, 37, 63] and section 12.3). In particular, the 'standard' breathers (PS, AB, and KM) are

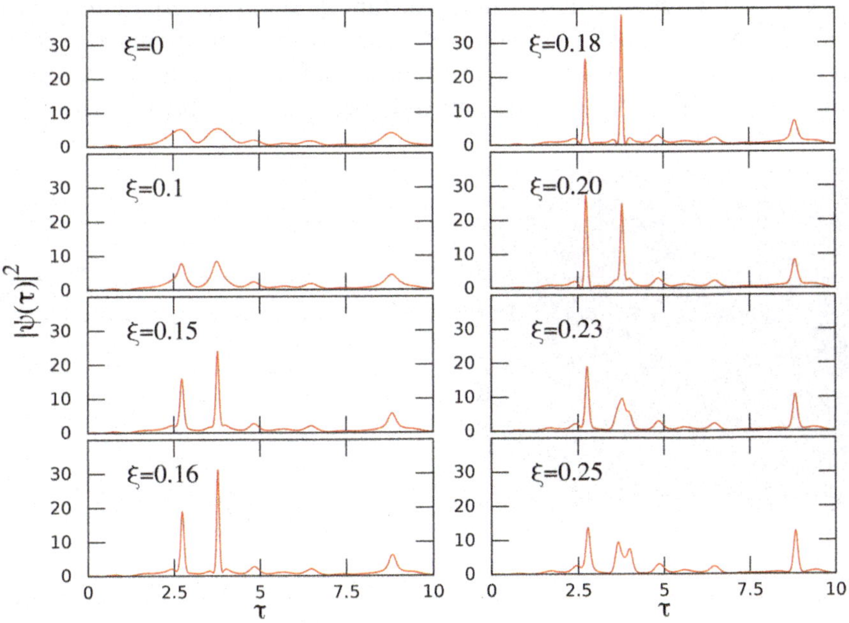

Figure 12.5. Numerical simulations of equation (12.1) with $\varepsilon = 0.3$, showing the spontaneous formation of Peregrine-like solitons from an initial random field.

all examples of degenerate genus two NLSE solutions (see e.g. [37]). The initial evolution of analytic profiles before the formation of the gradient catastrophe shown in figure 12.3 is characterized by genus zero. The genus of the solution changes in the process of the evolution, $g = 0, 2, 4, 6, \ldots$ as the solution undergoes a chain of bifurcations, see figure 12.4(a) and [63].

12.2.3 Rogue wave formation via the interaction of modulated wavetrains

Consider the focusing dispersive hydrodynamics (12.3) with the initial condition in the firm of rectangular 'box' of length $2L$ and height q:

$$\rho(\tau, 0) = \begin{cases} q^2 > 0 & \text{for} \quad |\tau| < L, \\ 0 & \text{for} \quad |\tau| > L, \end{cases} \qquad u(\tau, 0) = 0. \tag{12.6}$$

As shown in [63], the initial evolution of the box data for the small-dispersion focusing NLS equation can be viewed as a combination of two 'dam break' problems, which give rise to the instantaneous formation of nonlinear wave trains described by modulated genus one solutions regularizing the discontinuities at the opposite edges of the initial profile (see figure 12.6). These wave trains have the structure similar to DSWs [54]: they exhibit a solitary wave (a stationary fundamental soliton in the NLSE case) at one edge and gradually degenerate, via a modulated nonlinear wave, into a linear wavepacket propagating with group velocity at the opposite edge.

The 'focusing' DSWs regularizing the box data (12.6) expand inside the interval $-L < x < L$ and, after a certain moment of time $\xi_0 = L/(2\sqrt{2}q)$, defined as a point of collision of the small-amplitude edges, start to interact, giving rise to a rapid formation of a breather with the maximum amplitude close to 3 (see the bottom panel in figure 12.6(b)). While the amplitude profile of this breather resembles the

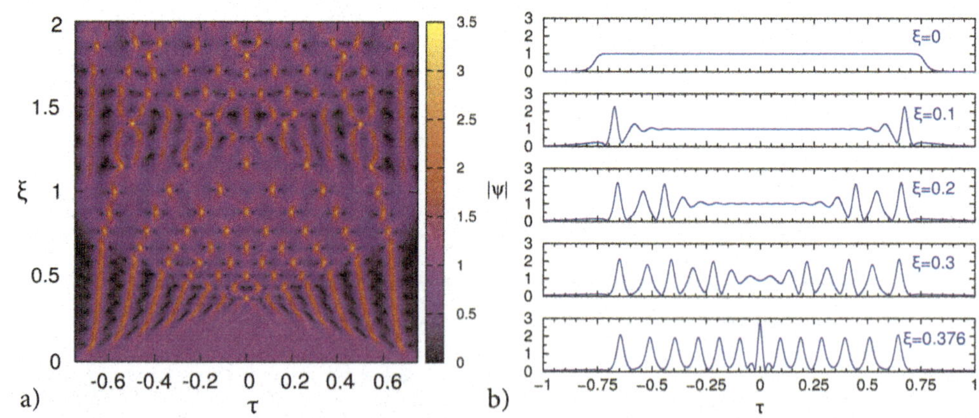

Figure 12.6. Collision of two counter-propagating focusing dispersive dam break flows results in the formation of large-amplitude breather lattices. Numerical simulation of the (smoothed) NLS box problem (12.3), (12.6) with $\varepsilon = 1/33$, $q = 1$, $L = 25/33$. The large-amplitude breather is formed at about $\xi = 0.376$. (a) Contour plot for $|\psi(\tau, \xi)|$; (b) amplitude profiles for different ξ. Reproduced from [63].

PS, an IST-based analysis of [63] and [37] shows that this wave is part of a more general, modulated genus 2 *breather lattice* which represents a multiphase quasi-periodic structure developing with ξ and filling a diamond-shaped region in the τ–ξ plane shown in figure 12.6(a). It is shown in [63] that this breather lattice gradually evolves towards the PS-like coherent structure realized at $\tau = 0$, $\xi \approx 1.166$ (see figure 12.6(a)). The subsequent evolution of the 'box', similar to the 'bell' evolution scenario described in section 12.2.2, occurs through the chain of bifurcations; however, the genus of the solution now assumes odd values: $g = 1, 3, 5, \ldots$ everywhere for $|\tau| < L$ except $\tau = 0$, see [63].

Concluding, we have presented two fundamental dynamical scenarios of the rogue wave formation realized in physical systems described by the small-dispersion NLSE (12.1). Importantly, in both scenarios, the evolution of the NLSE field in the semi-classical limit is asymptotically described in terms of modulated multiphase NLSE solutions. These solutions enter in the framework of the so-called finite gap theory (FGT), which represents an extension of the celebrated inverse scattering transform (IST) [67] to the case of periodic boundary conditions (see [18] and references therein).

12.3 Integrable turbulence and the inverse scattering transform method

In this section, we review some applications of the traditional IST method of the analysis of random wave fields and then outline the application of the FGT-based approach [22] to the characterization of rogue wave structures in integrable turbulence.

12.3.1 A brief review of the application of the IST method to the analysis of random wave fields

The IST method is a well-established technique for solving nonlinear partial differential equations of integrable nature such as the 1D NLSE or the Korteweg–de-Vries equation [60, 67]. The method of IST, developed for NLSE by Zakharov and Shabat in [64], is based on a transformation between the space–time domain and a spectral domain, in which the description of the nonlinear evolution of some given input field reduces to solving a linear problem. The first step of the IST method is the so-called direct scattering transform (DST), which gives the scattering data of a given initial field considered as a potential in the Zakharov–Shabat operator of the NLSE Lax pair. The scattering data consists of the spectrum itself and a set of 'scattering parameters' (norming constants, reflection coefficient). Crucially, the spectrum is preserved during the nonlinear evolution of the wave system while the evolution of the remaining scattering data is found to be particularly simple. The evolved scattering data are then transformed back to the physical space–time domain by solving a (linear) inverse problem. The whole procedure can be viewed as a nonlinear counterpart of the classical Fourier transform method for solving linear partial differential equations. Thus, the DST can be regarded as a 'nonlinear Fourier transform' that provides a spectral signature of the nonlinear modes composing the field under consideration (see [68–70] for some recent applications of this concept in the nonlinear optics context).

If the input field is a sufficiently rapidly decaying function of τ, i.e. $|\psi(\tau, \xi = 0)| \to 0$ as $|\tau| \to \infty$, see equation (12.1), the scattering data generally consist of two components: a discrete spectrum component related to the soliton content of the input field and a continuous spectrum component associated with dispersive radiation. However, if the input field is not decaying but periodic in space and/or in time [71], the traditional IST method does not apply and its modification, called the finite gap theory (FGT), represents the appropriate framework for analyzing the content of the field in terms of nonlinear modes [18, 72]. These nonlinear modes are characterized by Zakharov–Shabat spectra made of finite numbers of bands representing segments in the complex spectral plane.

As was mentioned in section 12.2.2, within the FGT, the multi-phase NLSE solutions are characterized by a *genus*, calculated as $N - 1$, where N is the number of spectral bands. Physically, the genus characterizes the number of degrees of freedom (i.e. the number of fundamental oscillatory modes, or phases) within the nonlinear periodic or quasi-periodic solution for the envelope of the plane wave [18]. Mathematically, the genus of a finite-band NLSE solution represents the genus of the hyperelliptic Riemann surface, on which this solution is defined in terms of theta-functions [18]. From the viewpoint of the FGT, the plane wave solution $\psi = qe^{iq^2t/\varepsilon}$ of the NLSE (12.1) is classified as a regular genus 0 solution while the fundamental soliton represents a *degenerate* genus 1 solution so that its FGT spectrum consists of two complex conjugate double spectral points (degenerate bands), the counterparts of the discrete spectrum in the Zakharov–Shabat scattering problem [64] with decaying potentials. The standard solitons on finite background (AB, KM soliton, and PS) all are the degenerate *genus 2* solutions so that their spectral portraits are made of a single vertical band in the spectral plane, complemented by two complex conjugate double points (that can be viewed as degenerate bands). The spectral portraits of the AB, KM soliton, and PS determined from the resolution of the Zakharov–Shabat problem are shown in figures 12.1 and 12.2 of [22].

From a practical point of view, the IST method has been shown to represent a promising approach to overcoming transmission limitations in fiber communications channels by encoding information in the nonlinear IST spectrum [70, 73–76]. From another perspective, the FGT and the IST methods have been introduced as tools for nonlinear Fourier analysis of hydrodynamical random wave trains [77–80]. These tools have been successfully implemented in several circumstances to determine the content of random wave trains in terms of nonlinear oscillating modes [18]. Recently, FGT has been used to evidence the presence of soliton turbulence in shallow water ocean surface waves [81].

Classical IST analysis has been used to determine the soliton content in freak (rogue) wave time series. In the approach reported in [21], the author uses a time window that is sliding along the experimental time series. The signal outside this time window is artificially set to zero so that the corresponding IST spectrum computed from numerical resolution of the Zakharov–Shabat scattering problem is pure discrete. This procedure enables one to interpret the local dynamics in the randomly fluctuating wavetrain in terms of solitons. Recently, the process of RW formation has also been studied by computing *global* IST spectra characterizing multiple random fluctuations that are found inside a box having a large size [50, 51].

Regarding the specific question of the prediction of RWs, previous numerical works have shown from computations of IST spectra that the development of RWs can be statistically correlated with the proximity to homoclinic solutions of the 1D NLSE [19, 20]. Taking typical JONSWAP sea-state initial data, the authors of [20] have also used numerical simulations of the 1D NLSE together with Floquet analysis to show that the developing sea state is well approximated by the first few dominant modes of the nonlinear spectrum, and can be described in terms of 2- or 3-phase periodic NLSE solutions.

In [22], another FGT-based approach has been developed to identify and characterize nonlinear modes that emerge *locally* in space and time from random wave fields. The approach consists first in isolating some specific coherent structure that emerges locally in space and time, from a globally random wave train. The structure under consideration is subsequently analyzed by a numerical IST procedure relying on a periodization of the considered structure. This procedure, motivated by the outcomes of the semi-classical analysis (see section 12.2), amounts to performing a local finite-band approximation of the wave field which includes only the most essential nonlinear modes. This approach provides the effective IST spectra giving an accurate signature of the nature of the pulse extracted from the random wavetrain. Importantly, the procedure accounts for the interaction of the specific coherent structure with its surrounding so the local IST spectrum can change in time (in contrast with the global spectrum, which is stationary). The local IST analysis of [22] showed the necessity of extending the existing classifications of the prototypes of RWs from standard breathers and their collisions to more general NLSE solutions characterized by their non-degenerate finite-band nonlinear spectra, supporting the general idea that has been put forward by Osborne [18].

Note that soliton radiation beat analysis is another technique that has been introduced in [82, 83] to determine the soliton content of arbitrarily shaped light pulses. The technique basically tracks the nonlinear phase rotation that is the signature of solitons and it can be applied, with some limitations, to wave systems described by non-integrable equations.

12.3.2 Regularization of gradient catastrophes in the focusing regime: local IST analysis

In section 12.2.2, we described the universal mechanism of regularization of gradient catastrophe in the focusing NLSE through the generation of PS, discovered in [62]. This mathematical result, however, is asymptotic and relies on the smallness of the dispersion parameter ε in (12.1). It is therefore, important to establish relevance of this important result to realistic physical and numerical settings when ε is not necessarily very small.

With this aim in view, in this subsection, we apply the local IST analysis of [22] to the process of dispersive regularization of a gradient catastrophe described in section 12.2.2. We shall analyze the IST spectrum of numerical simulations of equation (12.1) by taking a sech profile as the initial condition, i.e. $\psi(\xi = 0, \tau) = \text{sech}(\tau)$, and analyzing the solution in the neighborhood of the gradient catastrophe point at $\tau = 0$, $\xi = 0.5$. These numerical simulations essentially represent a zoom-in of the evolution shown in figure 12.4(a) near the gradient catastrophe point and are also very similar to those presented in [65].

Here, we use the tool of the local IST analysis introduced in [22] to compute spectral portraits of the coherent structure that emerge in the neighborhood of the nonlinear focusing point.

Figure 12.7 shows the gradual emergence of a PS-like structure and the corresponding local IST spectra for $\varepsilon = 0.02$. At $\xi = 0.42$, the process of nonlinear self-focusing is not very pronounced: the intensity profile (red line) is nearly flat and the phase profile (green line) of the solution is only a bit curved. The solution can be locally considered as being a slowly modulated plane wave, which is confirmed by IST analysis that reveals that the solution is locally a genus 0 NLSE solution. Self-focusing becomes significantly more pronounced at $\xi = 0.53$, where a small hump emerges from the center of the initial sech profile. The local IST analysis [22] reveals that the spectrum now contains three bands, which means that the solution has become locally a genus 2 NLSE solution. The local IST spectrum slightly evolves

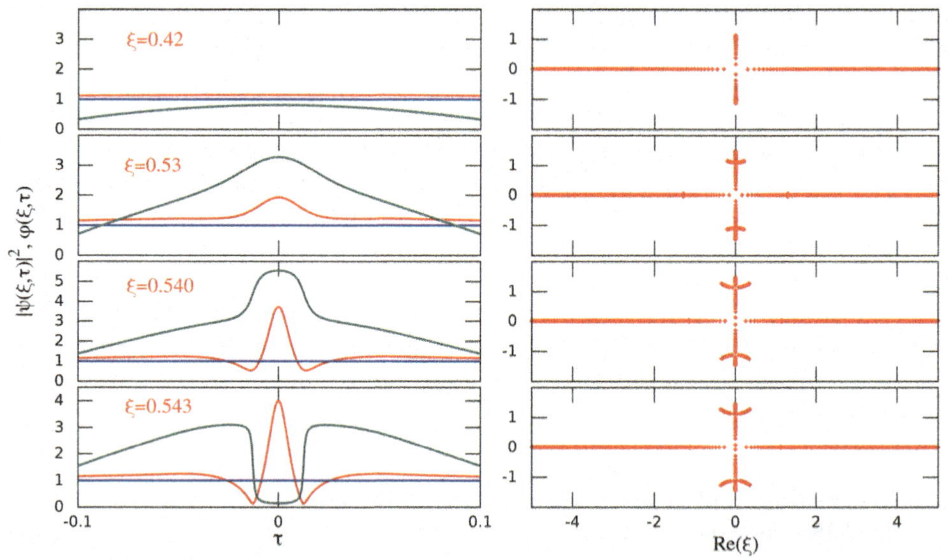

Figure 12.7. Left column: numerical simulation of equation (12.1) with $\varepsilon = 0.02$ for the initial condition given by $\psi(\tau,\xi = 0) = \text{sech}(\tau)$ plotted in blue lines. Red lines represent the power profile $|\psi(\tau,\xi)|^2$ and green lines represent the phase profile $\phi(\tau,\xi)$, $\psi(\xi,\tau) = |\psi(\xi = 0,\tau)|\exp[i\phi(\tau,\xi)]$ of the solution. Regularization of the gradient catastrophe occurs via the emergence of a coherent structure that is very close to the Peregrine soliton. Right column: local IST spectra of coherent structures shown in the left column at $\xi = 0.42$, $\xi = 0.53$, $\xi = 0.540$, $\xi = 0.543$ (from top to bottom). For $\xi < 0.5$ (upper row), the local IST spectrum contains only one band, showing that the solution is locally a plane wave. For $\xi > 0.5$, the local IST spectra computed from the procedure described in [22] contains three bands, thus showing that the coherent structure is locally a genus 2 solution of equation (12.1). IST spectra obtained for $\xi > 0.5$ are close to IST spectra of the pure Peregrine soliton, see figures 12.1 and 12.2 of [22].

between $\xi = 0.53$ and $\xi = 0.540$ but is kept composed of three bands, demonstrating that the solution remains locally a genus 2 NLSE solution.

It should be emphasized that IST spectra computed slightly before ($\xi = 0.540$) and after ($\xi = 0.543$) the maximum compression point of the PS are not strictly identical to the IST spectrum of the pure PS, as defined by its rational mathematical expression. As discussed in detail in [22], the exact PS can be classified as a degenerate genus 2 NLSE solution having a spectrum composed of three identical bands that are merged on the vertical imaginary axis in the complex plane. The resulting spectrum then consists of a single band with two added complex conjugate double points coinciding with the main band's simple spectrum endpoints. The slight discrepancies between the computed spectra and the degenerate spectrum of the exact PS arise from the fact that the phase and intensity profiles of the solutions that are analyzed at $\xi = 0.540$ and at $\xi = 0.543$ are not exactly identical to the phase and intensity profiles of the pure PS. In particular, the pure 'mathematical' PS has an infinite temporal extension together with flat phase and intensity profiles outside the region where the PS reaches its peak intensity. As shown in figure 12.7, the structure that emerges at $\varepsilon = 0.02$ does not display flat phase and intensity profiles outside the peaked region; see also figure 12.2 of [65] for a direct comparison between the exact PS solution and the asymptotic PS emerging from the sech profile. Greater proximity between the calculated IST spectrum and the spectrum of the exact PS is expected to come from a decrease of the parameter ε below 0.02. However, the process of modulational instability driven by the small noise coming from the numerical integration of the 1D NLSE starts to play a significant role for $\varepsilon < 0.02$, and this significantly modifies the scenario shown in figure 12.7.

In conclusion, we note that the rogue waves generated due to the interaction of modulated wavetrains in the dam break problem (see section 12.2.3) have also been analyzed using the local IST analysis in [22].

12.4 Experiments in optical fibers

Optical fiber experiments are considered as favorable tabletop laboratories for the investigation of hydrodynamics-like phenomena [1, 84–89] and of integrable turbulence [5, 36]. Optical fiber experiments open the investigation of strongly nonlinear interaction regimes, which is not feasible in hydrodynamical experiments that are limited by the occurrence of the wave-breaking phenomenon. A lot of optical fiber experiments described by the 1D NLSE (12.1) can be made at a value $\varepsilon \sim 1$ with optical fields reaching peak powers of several tens of watts over time scales of ~1 ps. Transient dynamics in nonlinear regimes (that correspond to the limit $\varepsilon \ll 1$ used in the semi-classical NLSE) are easily achieved in optical fiber experiments by using an initial optical field having either high power ($\gg$10 W) or slow time scales ($\gg$1 ps).

In both cases (strongly and weakly nonlinear regimes), the typical timescale of the structures generated in optical fibers is picoseconds or less. As a matter of fact, the typical timescales underlying the dynamics have prevented—until recent studies—the real-time observation of structures and rogue waves emerging in optical fiber turbulence. Very recently, two major advances have been demonstrated in the context of RWs emerging from the propagation of continuous waves (cw) or partially coherent waves in optical fiber:

(i) it has been shown that it is possible to use asynchronous optical sampling in order to measure precisely the statistical distribution of the optical power exhibiting sub-picosecond fluctuations [36];
(ii) the precise single-shot recordings of RWs have been demonstrated using two different setups based on the so-called 'time lens' technique [5, 6].

In this section, we review these different results related to the propagation of nonlinear random waves.

12.4.1 Formulation of the problem

The observation of fast dynamics is one of the most challenging and important question in optics. The time scales involved in an optical field can be estimated from the optical spectrum or from the autocorrelation of the electric field. The optical spectrum and the autocorrelation are easily measured in optics but do not provide the precise shape of the structures [90].

Based on photoelectric effect, photodetectors allow direct measurements of optical intensity. In particular, photodiodes can be used to observe the dynamics of the optical power varying on time scales slower than 10 ps. For the detection of pulses having 100 fs time scales, a streak camera that transforms the time variations of a light pulse into a spatial profile can be used.

Since the demonstration of mode-locked lasers emitting picosecond or femtosecond pulses in the 60s, a huge number of publications have been devoted to the ultrafast measurement of optical pulses [90–92]. Since the 90s, techniques based on nonlinear optics have been developed to measure the temporal profile of femtosecond and picosecond pulses. In particular, frequency-resolved optical gating (FROG) and spectral phase interferometry for direct electric-field reconstruction (SPIDER) allow the measurement of both the intensity and phase evolution. The temporal resolution of these techniques can reach several tens of femtoseconds.

Optical techniques such as SPIDER or FROG were initially developed to measure the intensity profile of *isolated pulses* and not the evolution of random 'cw' light. Most of the time, the pulses under investigation are emitted by periodic (mode-locked) lasers, and the measurement is performed on many pulses. Single-shot versions of these techniques are available [90] but they are characterized by time-observation windows that are strongly limited by the geometrical constraints. These single-shot techniques are generally limited to short (typically ps) temporal windows of observation.

The requirements and the balance between the resolution, temporal span of observation, and the dynamical range (i.e. the signal/noise ratio) guide the choice among the different optical measurement techniques described above. In optical fiber experiments, the typical time scale that corresponds to the duration of solitons is the picosecond. As a matter of fact, contrary to the experiments performed in the spatial domain [93, 94], the experimental study of optical turbulence and RWs in optical fibers requires ultrafast techniques that combine a large temporal window of observation and sub-picosecond resolution.

Surprisingly, despite the numerous experimental works devoted to ultrafast optical measurement and optical RWs [1, 13, 36, 84, 85, 88, 95], the real-time observation of RWs in the context of continuous or *random waves* has been demonstrated only very recently [5, 6] by using devices based on time lens. The pioneering works hence naturally provided only *indirect evidence* of RWs using e.g. spectral filtering [1, 85, 95, 96].

In 2015, it was shown that the statistical distribution of optical power can be precisely measured using asynchronous optical sampling techniques [36]. The main idea is inspired by the time-resolved fluorescence upconversion experiments [97] and by the optical sampling oscilloscope [98].

The principle of asynchronous optical sampling is to measure *samples* of the instantaneous power of the partially coherent waves under study. In principle, these samples can be generated by various nonlinear optical effects. In [36], sum frequency generation is achieved in a $\chi^{(2)}$ crystal from the interaction of the random signal and periodic femtosecond pulses. More precisely, the pulses are generated from a mode-locked Ti:Sa laser and have a temporal width ~140 fs. The final uncertainty on the measurements is estimated to be 250 fs. This techniques allows the measurement of the PDF of optical power of partially coherent waves i.e. random light having finite spectral width (<4 THz).

In the experiments reported in [36], the initial power fluctuations exhibit exponential distribution that corresponds to the Gaussian statistics of the field. On the contrary, at the output of a nonlinear fiber, the statistics of the power are strongly heavy-tailed. Note that the experiments reported in [36] are extremely well reproduced by numerical simulations of the *focusing* 1D NLSE. As reviewed in the Introduction, this phenomenon is one of the key points of integrable turbulence: an initial condition having Gaussian statistics evolves toward strongly non-Gaussian statistics.

These results published in 2015 on the statistics have clearly demonstrated that the partially coherent waves propagating in an optical fiber could become strongly non-Gaussian[1]. However, the dynamics and precise shapes of the coherent structures underlying these statistics have been observed only very recently using techniques based on time lens [5, 6]. The next section describes briefly these techniques and the consequences on the experimental study of optical RWs.

12.4.2 Observation of ultrafast random dynamics: time lens and time microscope

Demonstrated almost 30 years ago, the *time lens* is a technique that is based on a one-to-one correspondence between the spatial evolution of an optical beam in

[1] Note that the L-shaped statistics reported in the context of RWs in optical fibers are often observed using spectral filtering and/or in pulsed experiments (the initial conditions are periodic pulses). In [36], the statistics are measured without any filtering.

optical instrument such as lens or microscope and the temporal evolution of an optical field, for example in dispersive media [99–101]. In the last 20 years, various works have been devoted to the development and to the applications of time lens [99–105]. The main advantage of time lens techniques is to provide single-shot observation of optical power on tunable temporal windows that can be easily of several tens of picoseconds.

In imaging optical devices such as single lenses or a microscope, one has to distinguish between (i) the free propagation (between the object and the lens and between the lens and the image for example, see figure 12.8), and (ii) the lenses.

(i) **Correspondence between free space propagation and group velocity dispersion**

At leading order, the (spatial) propagation of a beam in the free space is formally equivalent and the (temporal) evolution of a field in a dispersive media. The paraxial approximation in the spatial propagation is equivalent to the second-order dispersion in the temporal domain. Under the paraxial approximation, the one-dimensional spatial evolution of monochromatic light is described by [106]:

$$\frac{\partial A(x, z)}{\partial z} = \frac{j}{2k}\frac{\partial^2}{\partial x^2}A(x, z) \tag{12.7}$$

where $k = 2\pi/\lambda$ is the modulus of the wavevector and where A is the slow varying amplitude of the electric field propagating in the z direction. In the Fourier space, one immediately gets

$$\begin{aligned} \tilde{A}(z, k_x) &= \tilde{A}(0, k_x)\exp\left[-jzk_x^2/(2k)\right] \quad \text{where} \\ \tilde{A}(z, k_x) &= \int_{-\infty}^{+\infty} A(z, x)e^{-ik_x x}dx. \end{aligned} \tag{12.8}$$

On the other hand, the propagation of the slowly varying envelope $A(z, t)$ of a non-monochromatic field in a dispersive media having a group velocity dispersion coefficient $\beta_2 = \frac{d^2k}{d\omega^2}$ is described by [107]:

$$\frac{\partial A(z, t)}{\partial z} = -j\frac{\beta_2}{2}\frac{\partial^2}{\partial t^2}A(z, t). \tag{12.9}$$

In the Fourier space, one immediately gets

$$\begin{aligned} \tilde{A}(z, \omega) &= \tilde{A}(0, \omega)\exp\left[j\beta\omega^2/2\right] \quad \text{where} \\ \tilde{A}(z, \omega) &= \int_{-\infty}^{+\infty} A(z, t)e^{-i\omega t}dt. \end{aligned} \tag{12.10}$$

In practice, the dispersion is achieved equivalently by using the propagation in a dispersive media such as ($\beta = \beta_2 z$) or by using a dispersive device such as the Treacy compressor ($\beta = \frac{d^2\phi}{d\omega^2}$ where $\phi(\omega)$ is the phase experienced by the wave in the device) [108].

Equations (12.10) and (12.8) show that there is an exact analogy between the free-space propagation and the second-order dispersion.

(ii) **Correspondence between 'spatial lens' and 'time lens'**

Assuming paraxial approximation, optical waves propagating through a thin lens experience a phase having a quadratic dependence in the transverse coordinate x. Considering the one-dimensional case of cylindric lens, this means that the field at the output of a thin lens $A_L'(x)$ just reads

$$A'_L(x) = A_L(x)\exp\left[-j\frac{\pi}{\lambda f'}x^2\right], \tag{12.11}$$

where f' is the focal length of the lens.

The key point of the time lens device is thus to add a quadratic phase on the temporal profile of the field:

$$A'(t) = A(t)\exp\,[jCt^2]. \tag{12.12}$$

In a practical device, the quadratic temporal phase is achieved by using second-order or third-order nonlinear interaction between a chirped pulse and the signal. A chirped pulse is easily generated using the properties of the group velocity dispersion [107]: for example, a Fourier limited pulse emitted by a mode-locked laser can be either launched into a dispersive media or into a dispersive device such as a Treacy compressor.

Finally, realizing the focus of the time lens device means to adjust the quadratic temporal phase C and the spectral quadratic β in order to satisfy the usual conjugation relations needed in imaging devices. Two devices can be distinguished: (i) the single time lens and (ii) the time microscope.

(i) The single time lens device involves two dispersive elements before and after the time lens in order to satisfy the famous conjugation relation $1/\overline{OA'} - 1/\overline{OA} = 1/f'$ of the thin lens (see figure 12.8(a)). These two different dispersive elements are generally single-mode optical fibers [6, 104]. In this case, the time lens is used as a time magnifier: the ultrafast temporal

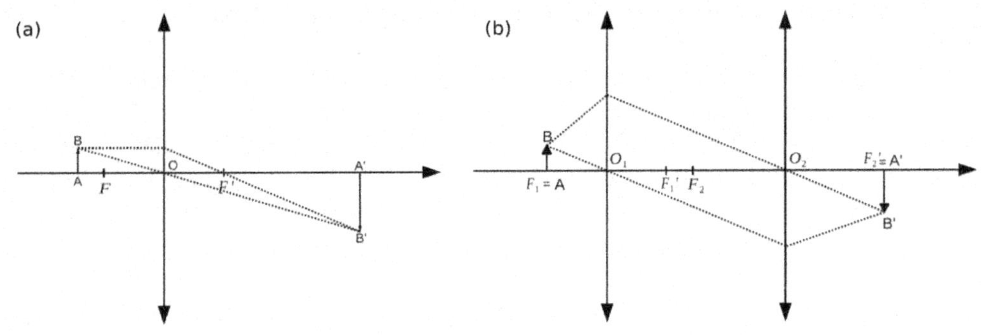

Figure 12.8. Geometrical optics principles of imaging with simple lens (a) and microscope (b). By using a simple lens, the distance $\overline{OA}$ and $\overline{OA'}$ are not equal to the focal length. On the contrary, in a microscope involving a camera, $\overline{O_1A} = -f'_1$ and $\overline{O_2A'} = +f'_2$.

evolution (<ps) of the optical power is imaged onto a slower dynamics (~10 ps) that can be observed using a fast photodiode.

(ii) A microscope is made of two lenses: the objective and the tube lens (see figure 12.8(b)). In the common modern configuration, the object is fixed in the object focal plane of the objective and a camera records the image in the focal plane of the tube lens. It is well known that the observation in the focal plane of the tube lens gives the 'far field,' i.e. the (spatial) spectrum of the field at the output of the first lens of the microscope. It is possible to use this property to design a time microscope [5, 101].

In this case, the first lens of the microscope is a time lens (see above) and the quadratic coefficient C is adjusted in order to put the temporal object exactly at the 'temporal object focal plane' of the lens. The second part of the time microscope is simply a single-shot spectrometer that can be achieved using a simple grating and a camera.

The advantage of the single time lens device is a relatively high repetition rate of the recording (typically 10^8 snapshots per second), whereas the use of a camera limits up to now the time microscope to low repetition rate (typically 10^3 snapshots per second in [5]). The advantage of the time microscope is the use of an extremely low noise detector such as sCMOS camera. In the case of a single time lens device, the signal is detected with standard photodiode and has to be filtered (see [6]). In the case of time microscope, without any filtering, the dynamics of measurement are typically 40 dB (see [5]).

Today, these two techniques exhibit similar temporal resolution (200–500 fs) and temporal window of observation (20–100 ps). In 2016, these two complementary techniques were used to record for the first time the coherent structures emerging from the nonlinear propagation of partially coherent waves [5] and from the nonlinear propagation of continuous waves with additional noise [6].

In [5] the direct single-shot recordings of optical RWs using a specially designed *upconversion time microscope* ultrafast acquisition system are reported. The objective is a time lens operating from sum-frequency generation (SFG) between the 1560 nm signal and a chirped pump pulse (at 800 nm). The temporal resolution of ~250 fs and the temporal window ~20–40 ps of the time microscope allows the investigation of the fast dynamics arising from the nonlinear propagation of partial waves in an optical fiber.

In these experiments (see [5] for details), the input random optical field is emitted by an amplified spontaneous emission (ASE) light source at $\lambda \sim 1560$nm. Using a programmable optical filter, the optical spectrum of the partially coherent light is filtered and designed to exhibit typical spectral width $\Delta\nu \sim 0.1$ THz. The observations made with the time microscope confirm that the initial random fluctuations of the optical power have typical time scale around 5–10 ps (see figure 12.1(a) of [5]).

The partially coherent waves are launched into a 500 m-long single-mode polarization-maintaining fiber at a wavelength falling into the anomalous (focusing) regime of dispersion. Observations performed with the time microscope at the output of the fiber immediately reveal the emergence of intense peaks, with powers frequently exceeding the average power $\langle P \rangle$ by factors of 10–50 (see figures 12.1(b)–(d) in [5]).

Those extreme peaks are extremely narrow, with time scales of the order of several hundreds of femtoseconds. In figure 12.9, original data recorded with this setup are presented.

As mentioned above, the time microscope strategy and the use of a 16 bit sCMOS camera (RMS dark noise of ≈2 electrons and a saturation value of 30 000 electrons) enables us to easily reach extremely high dynamical ranges (up to 40 dB). As a consequence, the signal/noise ratio observed in figure 12.9 is very good.

As already observed in the optical sampling experiments [36], the experiments performed with the time microscope confirm the emergence of strongly non-Gaussian statistics. As expected, the PDF of the optical power emitted by the ASE source is systematically very close to the exponential distribution (see figure 12.9(a)). The exponential distribution of the power $|\psi|^2$ corresponds to a Rayleigh distribution of $|\psi|$ and to Gaussian statistics for the field ψ under the assumption of uniformly distributed phases [36, 47, 109, 110]. Accordingly to the central limit theorem, the Gaussian statistics of the field used as an initial condition arise from the linear superposition of numerous independent waves (i.e. independent spectral components

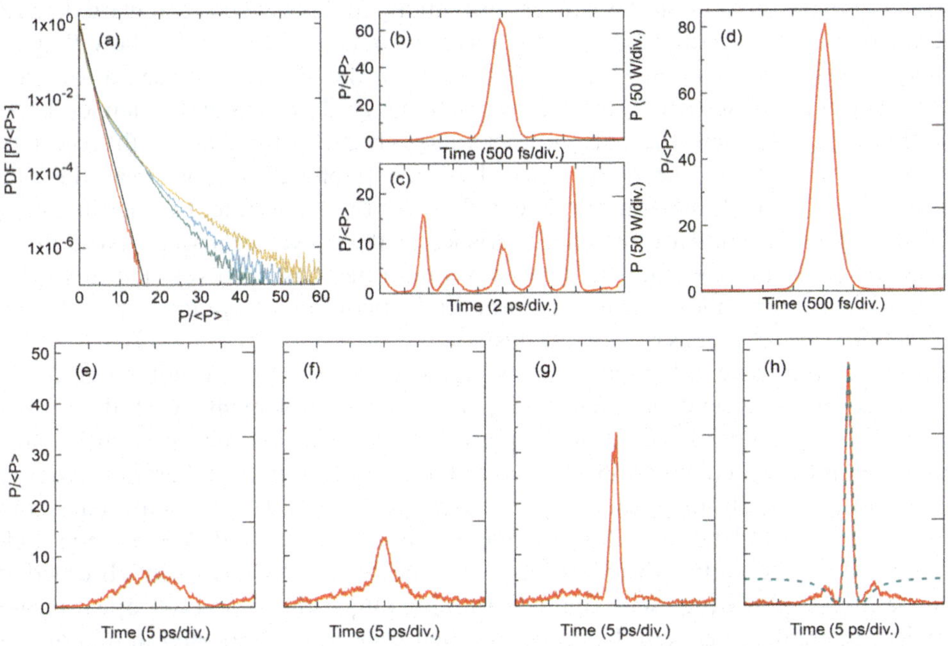

Figure 12.9. Time Microscope ultrafast measurement of Rogue waves (Experiments performed with a 500 m-long Polarization Maintaing Fiber) (a) Probability density function (PDF) of the initial condition (red line) and of the output power at 0.5 W (green line), 2.6 W (blue line), and 4 W (yellow line). The black line represents the normalized exponential distribution: $\text{PDF}(P/\langle P\rangle) = \exp(-P/\langle P\rangle)$. (b)–(d) Typical random signals recorded by the TM at the output of the nonlinear PM fiber. (b) PS-like structure ($\langle P\rangle = 2.6$ W, $\Delta\nu = 0.1$ THz). (c) Series of multiple peaks ($\langle P\rangle = 2.6$ W, $\Delta\nu = 0.1$ THz). (d) Giant optical RW reaching a peak power ~80 times greater than the mean power of the random wave ($\langle P\rangle = 4$ W, $\Delta\nu = 0.1$ THz). (e)–(h) Typical snapshots showing several stages leading to the emergence of a breather-like structure ($\langle P\rangle = 300$mW, $\Delta\nu = 0.05$ THz). The green dashed line in (h) represents the best fit to the analytical expression of a Peregrine soliton. Reprinted by permission Macmillan Publishers Ltd: [5] (© 2016).

of the ASE source). The experiments are well described by the 1D NLSE, and the PDF of light power measured at the output of the nonlinear fiber is found to exhibit heavy-tailed deviations from the exponential distribution, thus confirming the generation of RWs (see figure 12.9(a)). Moreover the PDFs demonstrate that the number of structures having high peak power increases while the mean power $\langle P \rangle$ of random optical waves (i.e. the strength of nonlinearity) increases.

Simultaneously with these results obtained using time microscope [5], other experiments based on time lens were published in 2016 [6]. In this article, the authors investigate the breakup of a continuous wave field induced by the so-called modulation instability in optical fiber. The time lens is based on four-wave mixing in a silicon waveguide, and the dispersion is achieved in optical fiber. The resolution ~300 fs and the temporal window of observation ~50 ps allow us to study the emergence of breather pulses expected from the modulation instability of the plane wave. As in [5], the experiments are well described by the 1D NLSE.

We believe that these two works [5, 6] published in 2016 are of extremely high importance in the context of nonlinear statistical optics and optical turbulence. In several aspects, these two separate studies are complementary and pave the way toward detailed experimental study of nonlinear random waves in optics. These works show that the techniques based on time lens, first demonstrated in the 90s, can be used to measure the statistics of ultrafast partially coherent waves having sub-picosecond dynamics. From the technical point of view, the 'pure' time lens allows high repetition rate of the recording but requires filtering of the noise induced by the photodetector, whereas the time microscope allows high dynamical range with lower repetition rate. Further investigations are needed to compare the characteristics of these two techniques. Note that a heterodyne version of the time microscope has been demonstrated very recently [111]. This new device now allows phase and amplitude recording of random waves having ~100 fs timescales over tens of picoseconds.

From the fundamental point of view, these two works [5, 6] enter within the framework of *integrable turbulence* [36, 47]. The complementary study of two different kinds of initial conditions experimentally confirms the apparently paradoxical theoretical prediction [36, 47, 50]. In the case of partially coherent waves (i.e. with random fluctuations of the optical power reaching 100%), the stationary state of integrable turbulence exhibits strongly non-Gaussian statistics of the field (strongly non-exponential distribution of the power). This has been observed in the experiments performed with optical sampling [36] and time microscope [5] using partially coherent waves. On the contrary, the observations reported in [6] confirm that in the long propagation distances, the complex dynamics emerging from the modulation instability of continuous waves (with small additional noise) leads to exponential statistics. The transition between these two limiting cases has been theoretically investigated in [50, 51] and deserves further theoretical and experimental study. We think the recent experimental results in optical fibers [5, 6, 36] show that integrable turbulence is a hot topic of fundamental but also practical importance in the field of nonlinear random waves. Moreover, the recent heterodyne version of the time microscope will allow further investigation of both the phase and amplitude dynamics of nonlinear random waves [111].

12.4.3 Universal regularization of gradient catastrophe: the generation of Peregrine solitons

In section 12.2, we presented the fundamental result by Bertola and Tovbis obtained in the semi-classical limit of the 1D focusing NLSE [62]. This mathematical study shows that the Peregrine soliton locally universally emerges as the dispersive regularization of the gradient catastrophe occurring in the dispersionless limit of the 1D NLSE (see section 12.2.1). Experimentally, reasonably small values of ε are achieved with large time scale of the initial pulse and/or large values of the initial peak power. This corresponds to the *nonlinearity dominated regime* of the propagation. In section 12.3.2 we demonstrated, using the local IST analysis, that the coherent structure regularizing gradient catastrophe is indeed close to PS for $\varepsilon = 0.02$.

Very recent experiments performed in optical fibers have confirmed that the theoretical prediction of the universal nature of the PS as a coherent structure regularizing gradient catastrophe is relevant to a very broad range of values of ε [65], far beyond the formal applicability of the semi-classical analysis. As seen in figure 12.10, the exact Peregrine soliton solution of 1D NLSE fits extremely well the shape of an isolated pulse after some propagation in an optical fiber. Remarkably, the value of $\varepsilon \simeq 0.45$ corresponding to the experimental conditions is not particularly small. In the experiments reproduced here, the temporal evolution of the optical power is measured using a (synchronized) optical sampling technique. As the initial field is made of isolated (periodic) pulses, a complete characterization of the amplitude and phase of the field dynamics can be also achieved using the FROG technique. The phase and amplitude measurements reported in [65] unmistakably confirm that a structure having *locally* the phase and amplitude of the Peregrine soliton shape indeed emerges from the propagation of pulses in optical fibers.

In the context of integrable turbulence arising in the propagation of partially coherent waves, the humps observed in the initial fluctuations might be seen as randomly distributed pulses. The experiments performed with the time microscope

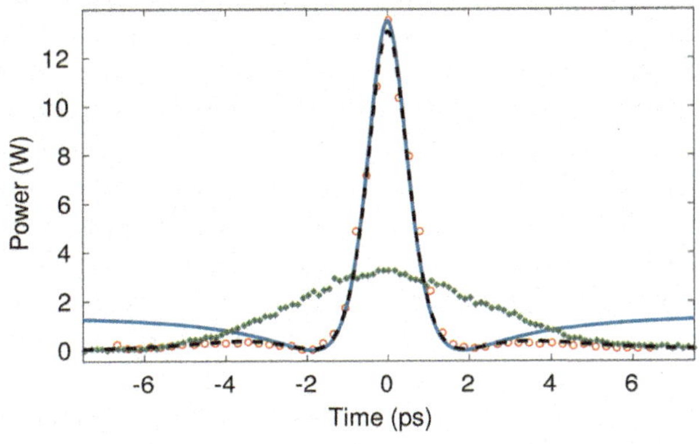

Figure 12.10. Reproduced from [65]. Temporal evolution of a single pulse (experiments and numerical simulations). Input pulse (green points) corresponding to $\varepsilon \simeq 0.45$. Output of the fiber (red circles). Numerical simulations of NLSE (black line) and theoretical Peregrine soliton (blue line).

and presented above are performed in nonlinear regimes characterized by a dispersion length L_D much larger than the nonlinear length L_{NL} in the initial stage [5]. Typically, for a time scale $T_0 \sim 10$ ps corresponding to the initial spectral width $\Delta\nu = 0.1$ THz, $L_D = T_0^2/|\beta_2| \sim 5$ km. Considering a hump with an maximum power $P = 5 \times \langle P \rangle$ and $\langle P \rangle = 0.5W$, the nonlinear length is $L_{NL} = 1/(\gamma P) \sim 0.2$ km. The experiments are thus compatible with the scenario described above in the semi-classical (small-dispersion) regime [62]. Remarkably, the *power* profile of the exact analytical Peregrine soliton coincides with some of the structures experimentally observed (see figure 12.9).

We believe that these observations strongly support the idea that the semi-classical approach of 1D NLSE provides a powerful theoretical framework to investigate the emergence of coherent structures in integrable turbulence.

12.5 Conclusion

Optical fibers are known to provide a favorable tabletop laboratory to investigate 1D NLS-like phenomena [107]. In particular, in the last years, using specifically designed *coherent* initial conditions, various solitons on finite background described by exact solutions of the focusing 1D NLSE have been generated in optical fiber experiments [2, 86, 87]. *Integrable turbulence* is the name of the complex spatio-temporal dynamics emerging from the propagation of partially coherent waves in systems described by integrable equations, e.g. by 1D NLSE [36, 42, 44–48, 50–52], and it is thus natural to study integrable turbulence in optical fibers. However, the study of partially coherent waves in optical fibers is extremely challenging from the measurement point of view because of their randomness along with their sub-picosecond time scale.

Recently, it has been demonstrated that asynchronous optical sampling can been used to precisely measure the probability density function of the randomly fluctuating optical power [36]. The important result is that, starting from initial Gaussian statistics of the field (i.e. the exponential distribution for the power), the statistics measured at the output of the fiber are strongly non-Gaussian. As reviewed in this chapter, the implementation of *time lens* techniques has allowed the measurement of the single shots of the structures themselves (and not only their statistics) [5, 6]. We believe that these experiments will pave the way to a deep experimental investigation of the mechanisms underlying integrable turbulence. The experimental statistics and observed structures are well reproduced by numerical simulations of the 1D NLSE. These results show that rogue waves, as extreme events appearing more frequently than predicted by the normal law, do emerge in *integrable turbulence* when initial conditions are strongly fluctuating. Note that the very recent demonstration of a heterodyne time microscope will now allow the study of phase dynamics of nonlinear random waves with short timescales (~100 fs) [111].

Up to now, there is no satisfactory theory of integrable turbulence. The theoretical approach based on quasi-kinetic equations that take into account non-resonant interactions have been developed *for the weakly nonlinear regime* several years ago [35, 41, 42]. In this chapter, we have shown that the semi-classical approach to the 1D NLSE represents a very promising framework in which to analyze the mechanisms underlying integrable turbulence [22, 62, 63]. The semi-classical regime of wave propagation corresponds to the small-dispersion limit and thus, in the context of nonlinear optics, to the strongly nonlinear regime.

As an example, in the framework of semi-classical limit of the 1D NLSE, it has been mathematically proven that the Peregrine soliton emerges as the regularization of the gradient catastrophe when a single (large) hump is propagated [62]. Remarkably, this phenomenon has been very recently observed in optical fiber experiments for a very broad range of parameters [65]. Moreover, this mechanism has been identified in experiments involving partially coherent waves as one of the possible scenarios of the emergence of rogue waves in optical fibers [5].

Acknowledgments

This work has been partially supported by the Ministry of Higher Education and Research, Nord-Pas de Calais Regional Council and European Regional Development Fund (ERDF) through the Contrat de Projets Etat-Région (CPER Photonics for Society P4S), as well as by the Agence Nationale de la Recherche through the LABEX CEMPI project (ANR-11-LABX-0007) and the OPTIROC project (ANR-12-BS04-0011 OPTIROC). MO has been funded by Progetto di Ricerca d'Ateneo CSTO160004. GE thanks the London Mathematical Society (grant number 41368) and EPSRC (grant EP/R00515X/1). The authors acknowledge Tamara Grava, Goery Genty, John Dudley, Eduardo Khamis, and Alexander Tovbis for fruitful discussions and collaborations. The authors are grateful to Emmanuel Courtade, Francois Anquez, Arnaud Mussot, Rémi habert, the Biophysics of Cellular Stress Response group and the photonics of the PhLAM, for their important help in the experiments. Finally, the authors would like to thank Serge Bielawski, Christophe Szwaj, Clément Evain, Pierre Walczak, Rebecca El Koussaifi, Francois Gustave, and Alexey Tikan for their crucial contributions to the results presented in this chapter.

References

[1] Solli D R, Ropers C, Koonath P and Jalali B 2007 Optical rogue waves *Nature* **450** 1054–1057

[2] Kibler B, Fatome J, Finot C and Millot G 2016 Experiments on breathers in nonlinear fibre optics *Rogue and Shock Waves in Nonlinear Dispersive Media* (Berlin: Springer) pp 89–115

[3] Akhmediev N, Kibler B, Baronio F, Belić M, Zhong W-P, Zhang Y, Chang W, Soto-Crespo J M, Vouzas P and Grelu P *et al* 2016 Roadmap on optical rogue waves and extreme events *J. Opt.* **18** 063001

[4] Chabchoub A, Onorato M and Akhmediev N 2016 Hydrodynamic envelope solitons and breathers *Rogue and Shock Waves in Nonlinear Dispersive Media* (Berlin: Springer) 55–87 pp

[5] Suret P, El Koussaifi R, Tikan A, Evain C, Randoux S, Szwaj C and Bielawski S 2016 Single-shot observation of optical rogue waves in integrable turbulence using time microscopy *Nat. Commun.* **7** 13136

[6] Närhi M, Wetzel B, Billet C, Toenger S, Sylvestre T, Merolla J-M, Morandotti R, Dias F, Genty G and Dudley J M 2016 Real-time measurements of spontaneous breathers and rogue wave events in optical fibre modulation instability *Nat. Commun.* **7** 13675

[7] Akhmediev N, Soto-Crespo J and Ankiewicz A 2009 Extreme waves that appear from nowhere: on the nature of rogue waves *Phys. Lett.* A **373** 2137–2145

[8] Akhmediev N, Ankiewicz A and Taki M 2009 Waves that appear from nowhere and disappear without a trace *Phys. Lett.* A **373** 675–678

[9] Dudley J M, Genty G, Dias F, Kibler B and Akhmediev N 2009 Modulation instability, Akhmediev breathers and continuous wave supercontinuum generation *Opt. Express* **17** 21497–21508

[10] Dubard P, Gaillard P, Klein C and Matveev V 2010 On multi-rogue wave solutions of the NLS equation and positon solutions of the KdV equation *Eur. Phys. J. Spec. Top.* **185** 247–258

[11] Ankiewicz A, Kedziora D J and Akhmediev N 2011 Rogue wave triplets *Phys. Lett.* A **375** 2782–2785

[12] Kedziora D J, Ankiewicz A and Akhmediev N 2013 Classifying the hierarchy of nonlinear-Schrödinger-equation rogue-wave solutions *Phys. Rev.* E **88** 013207

[13] Dudley J M, Dias F, Erkintalo M and Genty G 2014 Instabilities, breathers and rogue waves in optics *Nat. Photonics* **8** 755

[14] Kuznetsov E 1977 Solitons in a parametrically unstable plasma *Akademiia Nauk SSSR Doklady* **236** 575–577

[15] Peregrine D 1983 Water waves, nonlinear Schrödinger equations and their solutions *J. Austral. Math. Soc. Ser.* B **25** 16–43

[16] Akhmediev N, Eleonskii V and Kulagin N 1985 Generation of periodic trains of picosecond pulses in an optical fiber: exact solutions *Sov. Phys. JETP* **62** 894–899

[17] Degasperis A and Lombardo S 2016 Integrability in action: solitons, instability and rogue waves *Rogue and Shock Waves in Nonlinear Dispersive Media* (Berlin: Springer) 23–53 pp

[18] Osborne A 2010 *Nonlinear Ocean Waves & the Inverse Scattering Transform* vol 97 (New York: Academic)

[19] Islas A L and Schober C M 2005 Predicting rogue waves in random oceanic sea states *Phys. Fluids* **17** 031701

[20] Calini A and Schober C M 2017 Characterizing JONSWAP rogue waves and their statistics via inverse spectral data *Wave Motion* **71** 5–17

[21] Slunyaev A 2006 Nonlinear analysis and simulations of measured freak wave time series *Eur J. Mechanics B Fluids* **25** 621–635

[22] Randoux S, Suret P and El G 2016 Inverse scattering transform analysis of rogue waves using local periodization procedure *Sci. Rep.* **6** 29238

[23] Onorato M, Residori S and Baronio F 2016 *Rogue and Shock Waves in Nonlinear Dispersive Media* (Berlin: Springer)

[24] Chabchoub A, Kibler B, Finot C, Millot G, Onorato M, Dudley J and Babanin A 2015 The nonlinear Schrödinger equation and the propagation of weakly nonlinear waves in optical fibers and on the water surface *Ann. Phys* **361** 490–500

[25] Nazarenko S 2011 Wave turbulence *Lecture Notes in Physics* (Berlin: Springer)

[26] Barviau B, Kibler B and Picozzi A 2009 Wave-turbulence approach of supercontinuum generation: Influence of self-steepening and higher-order dispersion *Phys. Rev.* A **79** 063840

[27] Hammani K, Kibler B, Finot C and Picozzi A 2010 Emergence of rogue waves from optical turbulence *Phys. Lett.* A **374** 3585–3589

[28] Picozzi A, Garnier J, Hansson T, Suret P, Randoux S, Millot G and Christodoulides D 2014 Optical wave turbulence: towards a unified nonequilibrium thermodynamic formulation of statistical nonlinear optics *Phys. Rep.* **542** 1–132
[29] Hasselmann K, Barnett T, Bouws E, Carlson H, Cartwright D, Enke K, Ewing J, Gienapp H, Hasselmann D and Kruseman P *et al* 1973 Measurements of wind-wave growth and swell decay during the Joint North Sea Wave Project (JONSWAP) *Technical Report* Deutches Hydrographisches Institut
[30] Fedele F, Brennan J, De León S P, Dudley J and Dias F 2016 Real world ocean rogue waves explained without the modulational instability *Sci. Rep.* **6** 2016
[31] Onorato M, Osborne A R, Serio M, Cavaleri L, Brandini C and Stansberg C T 2004 Observation of strongly non-Gaussian statistics for random sea surface gravity waves in wave flume experiments *Phys. Rev.* E **70** 067302 Dec 2004
[32] Onorato M, Osborne A, Serio M and Cavaleri L 2005 Modulational instability and non-Gaussian statistics in experimental random water-wave trains *Phys. Fluids* **17** 078101
[33] Onorato M, Osborne A, Serio M, Cavaleri L, Brandini C and Stansberg C 2006 Extreme waves, modulational instability and second order theory: wave flume experiments on irregular waves *Eur. J. Mech. B Fluids* **25** 586–601
[34] Onorato M, Osborne A R, Serio M and Bertone S 2001 Freak waves in random oceanic sea states *Phys. Rev. Lett.* **86** 5831–5834
[35] Janssen A E M 2003 Nonlinear four-wave interactions and freak waves *J. Phys. Oceanogr.* **33** 863
[36] Walczak P, Randoux S and Suret P 2015 Optical rogue waves in integrable turbulence *Phys. Rev. Lett.* **114** 143903
[37] Randoux S, Walczak P, Onorato M and Suret P 2016 Nonlinear random optical waves: integrable turbulence, rogue waves and intermittency *Physica* D **333** 323–335
[38] Zakharov V, L'vov B and Falkovich G 1992 *Kolmogorov Spectra of Turbulence I* (Berlin: Springer)
[39] Picozzi A 2007 Towards a nonequilibrium thermodynamic description of incoherent nonlinear optics *Opt. Express* **15** 9063
[40] Zakharov V E and Schulman E 1988 On additional motion invariants of classical Hamiltonian wave systems *Physica D: Nonlinear Phenomena* **29** 283–320
[41] Soh D B S, Koplow J P, Moore S W, Schroder K L and Hsu W L 2010 The effect of dispersion on spectral broadening of incoherent continuous-wave light in optical fibers *Opt. Express* **18** 22393–22405
[42] Suret P, Picozzi A and Randoux S 2011 Wave turbulence in integrable systems: nonlinear propagation of incoherent optical waves in single-mode fibers *Opt. Express* **19** 17852–17863
[43] Onorato M, Proment D, El G, Randoux S and Suret P 2016 On the origin of heavy-tail statistics in equations of the nonlinear Schrödinger type *Phys. Lett.* A **380** 3173–3177
[44] Zakharov V E 2009 Turbulence in integrable systems *Studies Appl. Math* **122** 219–234
[45] Zakharov V E and Gelash A A 2013 Nonlinear stage of modulation instability *Phys. Rev. Lett.* **111** 054101
[46] Pelinovsky E, Shurgalina E, Sergeeva A, Talipova T, El G and Grimshaw R 2013 Two-soliton interaction as an elementary act of soliton turbulence in integrable systems *Phys. Lett.* A **377** 272–275
[47] Agafontsev D and Zakharov V E 2015 Integrable turbulence and formation of rogue waves *Nonlinearity* **28** 2791

[48] Randoux S, Walczak P, Onorato M and Suret P 2014 Intermittency in integrable turbulence *Phys. Rev. Lett.* **113** 113902

[49] Barviau B, Randoux S and Suret P 2006 Spectral broadening of a multimode continuous-wave optical field propagating in the normal dispersion regime of a fiber *Opt. Lett.* **31** 1696

[50] Soto-Crespo J M, Devine N and Akhmediev N 2016 Integrable turbulence and rogue waves: breathers or solitons? *Phys. Rev. Lett.* **116** 103901

[51] Akhmediev N, Soto-Crespo J M and Devine N 2016 Breather turbulence versus soliton turbulence: Rogue waves, probability density functions, and spectral features *Phys. Rev.* E **94** 022212

[52] Randoux S, Gustave F, Suret P and El G 2017 Optical random Riemann waves in integrable turbulence *Phys. Rev. Lett.* **118** 233901

[53] Annenkov S Y and Shrira V I 2009 'Fast' nonlinear evolution in wave turbulence *Phys. Rev. Lett.* **102** 024502

[54] El G and Hoefer M 2016 Dispersive shock waves and modulation theory *Physica* D **333** 11–65

[55] Dubrovin B, Grava T and Klein C 2009 On universality of critical behavior in the focusing nonlinear Schrödinger equation, elliptic umbilic catastrophe and the *Tritronquée* solution to the Painlevé-I equation *J. Nonlinear Sci.* **19** 57–94

[56] Biondini G, El G, Hoefer M and Miller P 2016 Dispersive hydrodynamics: preface *Physica* D **333** 1–5

[57] Akhmanov S, Sukhorukov A and Khokhlov R 1966 Self-focusing and self-trapping of intense light beams in a nonlinear medium *Sov. Phys. JETP* **23** 1025–1033

[58] Gurevich A and Shvartsburg A 1970 Exact solutions of the equations of nonlinear geometric optics *Sov. Phys. JETP* **31** 1084–1089

[59] Tovbis A, Venakides S and Zhou X 2004 On semiclassical (zero dispersion limit) solutions of the focusing nonlinear Schrödinger equation *Comm. Pure Appl. Math* **57** 877–985

[60] Yang J 2010 *Nonlinear Waves in Integrable and Nonintegrable Systems* (Philadelphia, PA: SIAM)

[61] Kamvissis S, McLaughlin K D T R and Miller P D 2003 *Soliton Ensembles for the Focusing Nonlinear Schrödinger Equation* (Princeton, NJ: Princeton University)

[62] Bertola M and Tovbis A 2013 Universality for the focusing nonlinear Schrödinger equation at the gradient catastrophe point: rational breathers and poles of the *Tritronquée* solution to Painlevé I *Commun. Pure Appl. Math.* **66** 678–752

[63] El G A, Khamis E G and Tovbis A 2016 Dam break problem for the focusing nonlinear Schrödinger equation and the generation of rogue waves *Nonlinearity* **29** 2798–2836

[64] Zakharov V E and Shabat P B 1972 Exact theory of two-dimensional self-focusing and one-dimensional self-modulation of waves in nonlinear media *Sov. Phys. JETP* **34** 62–69

[65] Tikan A *et al* 2017 Universality of the peregrine soliton in the focusing dynamics of the cubic nonlinear Schrödinger equation *Phys. Rev. Lett.* **119** 033901

[66] Bertola M, El G A and Tovbis A 2016 Rogue waves in multiphase solutions of the focusing nonlinear Schrödinger equation *Proc. R. Soc. A: Mathematical, Phys. Eng. Sci* **472** 20160340

[67] Ablowitz M J and Segur H 1981 *Solitons and Inverse Scattering Transform* (Philadelphia, PA: SIAM)

[68] Wahls S and Poor H V 2013 Introducing the fast nonlinear fourier transform 2013 *IEEE Int. Conf. on Acoustics, Speech and Signal Processing* 5780–5784

[69] Wahls S and Poor H V 2015 Fast numerical nonlinear Fourier transforms *IEEE Trans. Inf. Theor* **61** 6957–6974

[70] Turitsyn S K, Prilepsky J E, Le S T, Wahls S, Frumin L L, Kamalian M and Derevyanko S A 2017 Nonlinear fourier transform for optical data processing and transmission: advances and perspectives *Optica* **4** 307–322

[71] Ma Y-C and Ablowitz M J 1981 The periodic cubic Schrödinger equation *Stud. Appl. Math.* **65** 113–158

[72] Tracy E and Chen H 1988 Nonlinear self-modulation: an exactly solvable model *Phys. Rev.* A **37** 815

[73] Hasegawa A and Nyu T 1993 Eigenvalue communication *J. Lightwave Technol.* **11** 395–399

[74] Prilepsky J E, Derevyanko S A, Blow K J, Gabitov I and Turitsyn S K 2014 Nonlinear inverse synthesis and eigenvalue division multiplexing in optical fiber channels *Phys. Rev. Lett.* **113** 013901

[75] Le S T, Prilepsky J E and Turitsyn S K 2015 Nonlinear inverse synthesis technique for optical links with lumped amplification *Opt. Express* **23** 8317–8328

[76] Frumin L L, Gelash A A and Turitsyn S K 2017 New approaches to coding information using inverse scattering transform *Phys. Rev. Lett.* **118** 223901

[77] Boffetta G and Osborne A 1992 Computation of the direct scattering transform for the nonlinear Schrödinger equation *J. Comp. Phys* **102** 252–264

[78] Osborne A R, Segre E, Boffetta G and Cavaleri L 1991 Soliton basis states in shallow-water ocean surface waves *Phys. Rev. Lett.* **67** 592–595

[79] Osborne A R 1993 Behavior of solitons in random-function solutions of the periodic Korteweg21de Vries equation *Phys. Rev. Lett.* **71** 3115–3118

[80] Osborne A R and Petti M 1994 Laboratory generated, shallow water surface waves: Analysis using the periodic, inverse scattering transform *Phys. Fluids* **6** 1727–1744

[81] Costa A, Osborne A R, Resio D T, Alessio S, Chriv E, Saggese E, Bellomo K and Long C E 2014 Soliton turbulence in shallow water ocean surface waves *Phys. Rev. Lett.* **113** 108501

[82] Böhm M and Mitschke F 2006 Soliton-radiation beat analysis *Phys. Rev.* E **73** 066615

[83] Böhm M and Mitschke F 2007 Soliton content of arbitrarily shaped light pulses in fibers analysed using a soliton-radiation beat pattern *Appl. Phys.* B **86** 407–411

[84] Hammani K, Finot C, Dudley C and Millot G J M 2008 Optical rogue-wave-like extreme value fluctuations in fiber Raman amplifiers *Opt. Express* **16** 16467–16474

[85] Erkintalo M, Genty G and Dudley J M 2009 Rogue-wave-like characteristics in femto-second supercontinuum generation *Opt. Lett.* **34** 2468–2470

[86] Kibler B, Fatome J, Finot C, Millot G, Dias F, Genty G, Akhmediev N and Dudley J M 2010 The peregrine soliton in nonlinear fibre optics *Nat. Phys.* **6** 790–795

[87] Kibler B, Fatome J, Finot C, Millot G, Genty G, Wetzel B, Akhmediev N, Dias F and Dudley J M 2012 Observation of Kuznetsov–Ma soliton dynamics in optical fibre *Sci. Rep.* **2** 2012

[88] Onorato M, Residori S, Bortolozzo U, Montina A and Arecchi F 2013 Rogue waves and their generating mechanisms in different physical contexts *Phys. Rep.* **528** 47–89

[89] Turitsyna E G, Smirnov S V, Sugavanam S, Tarasov N, Shu X, Podivilov S B E, Churkin D, Falkovich G and Turitsyn S 2013 The laminar-turbulent transition in a fibre laser *Nat. Photon.* **7** 783–786

[90] Trebino R 2012 *Frequency-Resolved Optical Gating: The Measurement of Ultrashort Laser Pulses* (New York: Springer)

[91] Krausz F, Fermann M E, Brabec T, Curley P F, Hofer M, Ober M H, Spielmann C, Wintner E and Schmidt A J 1992 Femtosecond solid-state lasers *IEEE J. Quantum Electron.* **28** 2097–2122

[92] Weiner A 2011 *Ultrafast Optics* (New York: Wiley)

[93] Bromberg Y, Lahini U, Small E and Silberberg Y 2010 Hanbury Brown and Twiss interferometry with interacting photons *Nat. Photon.* **4** 721–726

[94] Pierangeli D, Di Mei F, Conti C, Agranat A J and DelRe E 2015 Spatial rogue waves in photorefractive ferroelectrics *Phys. Rev. Lett.* **115** 093901

[95] Mussot A, Kudlinski A, Kolobov M, Louvergneaux E, Douay M and Taki M 2009 Observation of extreme temporal events in CW-pumped supercontinuum *Opt. Express* **17** 17010–17015

[96] Walczak P, Randoux S and Suret P 2015 Statistics of a turbulent raman fiber laser *Opt. Lett.* **40** 3101

[97] Mahr H and Hirsch M D 1975 An optical up-conversion light gate with picosecond resolution *Opt. Commun.* **13** 96–99

[98] Duguay M A and Hansen J W 1968 Optical sampling of subnanosecond light pulses *Appl. Phys. Lett.* **13** 178–180

[99] Kolner B H and Nazarathy M 1989 Temporal imaging with a time lens *Opt. Lett.* **14** 630–632

[100] Kolner B H 1994 Space–time duality and the theory of temporal imaging *IEEE J. Quantum Electron.* **30** 1951–1963

[101] Bennett C and Kolner B 1999 Upconversion time microscope demonstrating 103 magnification of femtosecond waveforms *Opt. Lett.* **24** 783–785

[102] Kauffman M, Godil A, Auld B, Banyai W and Bloom D 1993 Applications of time lens optical systems *Electron. Lett.* **29** 268–269

[103] Foster M A, Salem R, Geraghty D F, Turner-Foster A C, Lipson M and Gaeta A L 2008 Silicon-chip-based ultrafast optical oscilloscope *Nature* **456** 81–84

[104] Salem R, Foster M A, Turner A C, Geraghty D F, Lipson M and Gaeta A L 2009 Optical time lens based on four-wave mixing on a silicon chip *Opt. Lett.* **33** 1047–1049

[105] Okawachi Y, Salem R, Johnson A R, Saha K, Levy J S, Lipson M and Gaeta A L 2012 Asynchronous single-shot characterization of high-repetition-rate ultrafast waveforms using a time-lens-based temporal magnifier *Opt. Lett.* **37** 4892–4894

[106] Goodman J W 2005 *Introduction to Fourier Optics* (Greenwood Village, CO: Roberts and Company)

[107] Agrawal G P 2001 *Nonlinear Fiber Optics (Optics and Photonics)* 3rd edn (New York: Academic)

[108] Treacy E 1968 Compression of picosecond light pulses *Phys. Lett.* A **28** 34–35

[109] Goodman J W 1985 *Statistical Optics* (New York: Wiley)

[110] Kharif C, Pelinovsky E and Slunyaev A 2009 *Rogue Waves in the Ocean* (Berlin: Springer)

[111] Alexey Tikan, Serge Bielawski, Christophe Szwaj, Stephane Randoux and Pierre Suret 2017 Phase and amplitude single-shot measurement by using heterodyne time-lens and ultrafast digital time-holography arXiv:1707.07567

Chapter 13

Rogue wave formation in highly birefringent fiber

L Drouzi, S Coulibaly, C G L Tiofack, M Taki and K Laabidi

13.1 Introduction

Rogue wave (RW) is a nonlinear wave phenomenon in which temporally and spatially localized waves with abnormally high amplitudes are generated from a low-level average background. In fact, these phenomena first emerged in the ocean and then were introduced to optics by Solli *et al* [1] with the terminology *optical rogue waves*. Theoretically, the behavior of a RW can be well described by the nonlinear Schrödinger equation (NLSE) [2]. Experimentally, this phenomenon has been observed in nonlinear fiber optics [1, 3] and water wave tanks [4]. Nowadays, studies on RWs have been extended to a variety of complex systems, such as Bose–Einstein condensates [5], plasma physics [6], and superfluid helium [7], among others.

Significant efforts have been made to explain the RW excitation through a nonlinear process. These waves may arise from the instability of a certain class of initial conditions that tends to grow exponentially and thus has the possibility of increasing up to very high amplitudes due to modulation instability (MI). Conversely, not every kind of MI leads to the formation of an optical RW. The generation of high-intensity optical RWs in a more controlled way, based on induced MI, has been recently studied in [8]. Soliton collisions have also been reported as one of the major sources of RWs during supercontinuum generation in optical fibers [9, 10]. This happens when more complicated physical effects are acting, such as third-order dispersion (TOD) [11] and Raman effect [12]. Recent studies have confirmed that the presence of TOD in optical fibers turns the system convectively unstable and generates extraordinary optical intensities due to multiple soliton collisions. This process involves cascades of soliton interactions, with energy exchange between them, before the most energetic solitons emerge at the fiber output [13]. The appearance of RWs in supercontinuum generation through the collision of solitons has been validated by using the energy landscape of the most simplified

doi:10.1088/978-0-7503-1460-2ch13 © IOP Publishing Ltd 2017

NLSE model [14]. Moreover, the collision between optical solitons and dispersive waves at event horizons has also been proposed as a possible mechanism for the generation of RW events [15].

In view of the nonlinear phase change resulting from the cross-phase modulation (XPM) in the birefringent or multi-mode fibers, one must consider interactions of several field components at different frequencies or polarizations, since complex systems usually involve more than one component [16, 17]. The linear stability analysis of the coupled system indicates that the XPM term can vary the instability regime characters [18, 19]. In this case, the co-propagating orthogonally polarized fields are found to exhibit MI both in cases of normal and anomalous group velocity dispersion (GVD) [20–23], in contrast to scalar fibers where the MI takes place only in anomalous GVD regions. Alternatively, the presence of a group velocity mismatch (walk-off) might produce the same effect. The study of RW in multi-component wave systems is another hot topic where several advances have been recently reported [24, 25]. It has been demonstrated that vector RWs exhibit a richer structure than scalar ones, such as dark RWs [25, 26], four-petaled RWs [27], and interactions between RWs and other nonlinear waves [28, 29]. These studies indicate that nonlinear waves in coupled systems are much more diverse than the ones in uncoupled systems.

Very recently, the occurrence of RWs in non-integrable birefringent optical fibers has been analyzed in [30], and the birefringence angle for which the maximum number of rogue events appears has been identified. However, in the context of fiber optics, symmetry-breaking terms have been shown to introduce a rich variety of phenomena [31–35]. Therefore, the extended study of the generation of optical RWs, taking into account the symmetry-breaking terms in the birefringent fibers, must be considered. The particular goal of this work is to consider the effects of symmetry-breaking terms introduced by the walk-off parameters and TOD on the generation of extreme events in the birefringent coupled NLSE. Although the TOD generates asymmetry such that coherent structures can acquire nonzero relative velocity and undergo collisions, we show here that the interplay between walk-off and TOD plays a crucial role in the generation of extreme events.

The paper is organized as follows. In section 13.2, we give the governing equations describing the evolution of a highly birefringent optical fiber including nonlinear birefringent coupling, TOD, and the group velocity mismatch between the two polarized states of light propagating along the fiber. Standard modulational stability is recalled with a new emphasis on the role of the nonlinear birefringent coupling term in the characteristics and thresholds of instabilities. Section 13.3 is dedicated to an introduction of the main statistical properties of RWs, and the description of the criteria used to characterize extreme events. These include the deviation from the Gaussian statistics of the probability density function (PDF) of the peak intensities, the calculation of the *shape* parameter extracted from the generalized Gamma distribution, and the evaluation of the *factor of abnormality* of the peaks. Section 13.4 provides the results of numerical simulations in the normal dispersion regime, and section 13.5 deals with the case of the anomalous dispersion regime. Concluding remarks are summarized in the last section.

13.2 Model and linear stability analysis

Let's start with the coupled generalized NLSE satisfied by the orthogonally polarized components of a pulse in a lossless and highly birefringent optical fiber [20–23]:

$$\frac{\partial A_x}{\partial Z} = -i\frac{\beta_2}{2}\frac{\partial^2 A_x}{\partial t^2} + \frac{\beta_3}{6}\frac{\partial^3 A_x}{\partial t^3} + i\gamma\left(|A_x|^2 + B\left|A_y\right|^2\right)A_x, \tag{13.1a}$$

$$\frac{\partial A_y}{\partial Z} = -2b_0\frac{\partial A_y}{\partial t} - i\frac{\beta_2}{2}\frac{\partial^2 A_y}{\partial t^2} + \frac{\beta_3}{6}\frac{\partial^3 A_y}{\partial t^3} + i\gamma\left(\left|A_y\right|^2 + B\,|A_x|^2\right)A_y, \tag{13.1b}$$

where $A_x(Z,t)$ and $A_y(Z,t)$ stand for the two orthogonal components of an arbitrary polarized optical field $\mathbf{E}(z,t)$ at the carrier frequency ω_0, i.e.

$$\mathbf{E}(z, t) = \frac{1}{2}\left[A_x(z, t)\hat{x} + A_y(z, t)\hat{y}\right]e^{i(\beta_0 z - \omega_0 t)} + \text{c.c.},$$

where $\beta_0 = (\beta_{0x} + \beta_{0y})/2$ is the average propagation constant. β_2, β_3, B, and γ are the GVD, TOD, nonlinear birefringent coupling (or cross phase modulation), and Kerr nonlinearity parameters, respectively. $b_0 = \left(\beta_{1y} - \beta_{1x}\right)/2$ accounts for the group velocity mismatch between the two polarization states. Due to the short distances of only a few kilometers in a supercontinuum experiment, the effects of shock term and delayed Raman response can be neglected [12]. The nonlinear coupled Schödinger equations (1) have a continuous wave (CW) solution which reads: $A_j = A_{j0}\exp\left(i\phi_j z\right)$ where $\phi_j = \gamma\left(\left|A_{j0}\right|^2 + B\left|A_{j'0}\right|^2\right)$ with $(j, j') = (x, y)$ and (y, x) for the x-axis and y-axis, respectively. The standard stability analysis of this solution against *monochromatic* perturbations of the form $e^{i(\omega t - kz)}$ leads to the following dispersion relation:

$$\mathcal{D}(k, \omega) = [(k - 2b(\omega))^2 - H(\omega)][k^2 - H(\omega)] - C(\omega)^2 = 0, \tag{13.2}$$

with k being the wavenumber and ω the pulsation. In this relation, $H(\omega) = \beta_2\omega^2\,(\beta_2\omega^2/4 + \gamma P)$, $C(\omega) = B\beta_2\gamma P\omega^2$ and $b(\omega) = b_0\omega$, where we have set $|A_{x0}|^2 = \left|A_{y0}\right|^2 = P$; the pump power on each polarization state. Next, linear stability analysis allows us to determine when the wavenumber k possesses an imaginary part for some real values of the pulsation ω. Hence, the CW is said to be unstable (stable) if $k_i = \Im(k)$ is positive (negative). From linear stability analysis, a cumbersome but straightforward calculation leads to the instability frequency bands that are delimited by the following four frequencies [36]:

$$\omega_1^{\pm} = \pm 2\sqrt{(B - 1)\beta_2\gamma P + b_0^2/\beta_2}, \tag{13.3a}$$

$$\omega_2^{\pm} = \pm 2\sqrt{-(1 + B)\beta_2\gamma P + b_0^2/\beta_2}. \tag{13.3b}$$

We focus here on the normal dispersion regime and we investigate the impact of the walk-off parameter and nonlinear birefringence on the MI gain, which is given by $g = 2k_i$. In the absence of the walk-off ($b_0 = 0$), the MI spectra consists of symmetric instability bands on either side of the zero-propagation constant (with respect to the pump). Further, MI exhibits a threshold behavior, as it only occurs for the nonlinear birefringence $B > 1$ (figure 13.1(a)). For the lower values of the walk-off parameter, the threshold of the nonlinear birefringence is now extended into the region where $0 < B < B_c$, and $B_c < 1$ corresponds to the critical value at the threshold (figure 13.1(b)). The instability bands are limited by the frequencies $\omega_1^{\pm}$. By increasing the value of the walk-off, a transition is reached at $b_{0c} = \sqrt{(1 + B)\beta_2\gamma P}$, where two new frequencies $\omega_2^{\pm}$ are generated, giving rise to two separated lobes of instability. In this case, MI always occurs for arbitrary value of the nonlinear birefringence, which can be seen from figure 13.1(c). Therefore, MI can be controlled by adjusting both the values of the walk-off parameter b_0 and the nonlinear birefringence B.

13.3 Statistical analysis of the RW in the highly birefringent fiber

Rogue events in optics have been identified in a number of different ways. One approach uses the idea from PDF theory that associates rogue events with particular extreme-value probability distributions [37]. The most salient feature of RW statistics is that the PDF of the wave amplitude deviates from the Gaussian probability expected from the central limit theorem when a large number of uncorrelated events are recorded. In this work, we have chosen to model the PDF by a generalized Gamma (GG) distribution [38]. The GG PDF is a well-known distribution in the statistical literature for describing extreme events; it reads as [38, 39]

$$P(x; a, \beta, m) = \frac{a}{\beta\Gamma(m)}\left(\frac{x}{\beta}\right)^{am-1} e^{-(x/\beta)^a}, \tag{13.4}$$

where $a > 0$ and $m > 0$ refer to the *shape* parameters, and $\beta > 0$ is the *scale* parameter. The variable x is positive. In our case, the random variable x corresponds to the peak intensity $|A_x|^2$ ($|A_y|^2$) on the fast (slow) axis of the birefringent fiber. This distribution is very flexible and includes as special cases the Gamma ($a = 1$),

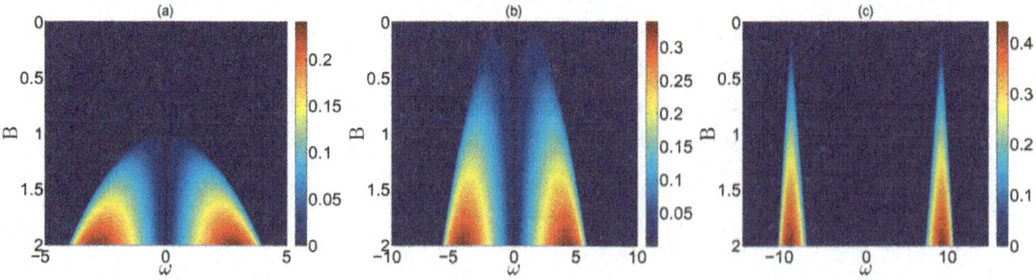

Figure 13.1. Modulational instability gain spectrum with different values of the walk-off parameter b_0. (a) $b_0 = 0$, (b) $b_0 = 0.13$ ps m^{-1}, (c) $b_0 = 0.3$ ps m^{-1}. The other parameters are $\beta_2 = 0.06$ ps^2 m^{-1}, $\gamma = 0.015$ W^{-1} m^{-1}, $P = 16$ W.

exponential ($a = 1$, $m = 1$), Rayleigh ($a = 2$, $m = 1$), the Weibull ($m = 1$), and Log-normal ($m \to \infty$) distributions. By using the GG distribution, any localized structures with *shape* parameter $a < 1$ are considered as an extreme event. This parameter will be used in the next section to characterize the appearance of a RW in nonlinear birefringent fibers.

Another approach used to characterize the RW phenomenon is based on the *abnormality index* A_I, which is the ratio between the height of the wave and the average wave height among one-third of the highest waves in a time series [40]. Every event whose abnormality index is larger than 2 is considered as a rogue wave. In optics, however, the accessible data is not the field amplitude but rather the intensity. In this way, we adapt the criterion commonly used in hydrodynamics to define RWs by introducing a RW intensity threshold $I_{\mathrm{RW}} = 4I_s$, where I_s is the mean intensity of the highest third of events. Note that in the literature the RW threshold can also be taken as $I_{\mathrm{RW}} = 2I_s$. Here we adopt the factor 4, which is more restrictive but surely ensures the occurrence of rogue waves.

In the following section, we will check whether the interplay between the walk-off and the TOD contribute to generate RWs. To this end, we performed an extensive simulation of MI in the coupled NLSE with a noise-perturbed plane wave initial condition, based on a standard split-step scheme [41]. In order to quantify the statistical properties of the simulations, we performed 100 numerical shots corresponding to 100 different initial noisy conditions. We propagate the whole signal in a fiber long enough and we record the time signals. We take the maximum values of the intensity, and then we measure the *abnormality index* coefficient A_I and the *shape* parameter a extracted from the GG distribution. Finally, we represent the linear and semi-logarithmic PDFs, and evaluate the rogue intensity threshold I_{RW}, which defines the intensity above which events can be classified as RW.

13.4 Results in the normal dispersion regime

We start by checking the possibility of generating RWs in highly birefringent fiber in normal dispersion regime ($\beta_2 > 0$). As we already know, in this dispersion regime, a standard fiber cannot even be subject to MI. However, the nonlinear wave coupling between two orthogonal polarization modes may still lead to the MI of an intense CW field, which induces the formation of a supercontinuum in a strong nonlinear regime. Moreover, the walk-off and TOD introduce a symmetry breaking which strongly influences the dynamic of the nonlinear systems.

13.4.1 Impact of the walk-off parameter

In the absence of the walk-off ($b_0 = 0$), we observe in table 13.1 that the *abnormality index* coefficient A_I and the *shape* parameter a have the same value on the two axes of the birefringent fiber. In addition, the criteria for generating extreme events ($A_I > 2$, $a < 1$) are filled on the two axes. Figure 13.2 represents the corresponding linear and non-linear PDFs as a function of the peak intensity along the two axes. The RW intensity threshold $I_{\mathrm{RW}} = 1092.5$ W is indicated by the vertical line, and we observe that very few peaks have intensities exceeding this value. It is obvious that the

Table 13.1. Fitted of the abnormality index A_I and the shape parameter a obtained in the normal GVD regime of the birefringent fiber without TOD. $B = 2$, $\beta_3 = 0$, $\beta_2 = 0.06$ ps^2 m^{-1}, $L = 15$ m, $\gamma = 6$ W^{-1} km^{-1}, $P = 100$ W.

	b_0	0	0.4	0.8	1.2	1.4	1.6
A_I	Fast axis	2.033	1.907	1.892	2.006	2.070	2.095
	Slow axis	2.033	1.895	1.944	1.925	1.965	1.954
a	Fast axis	0.9026	1.4242	1.1086	0.9857	0.9003	0.9645
	Slow axis	0.9026	1.4694	1.0461	1.0550	0.9950	0.9398

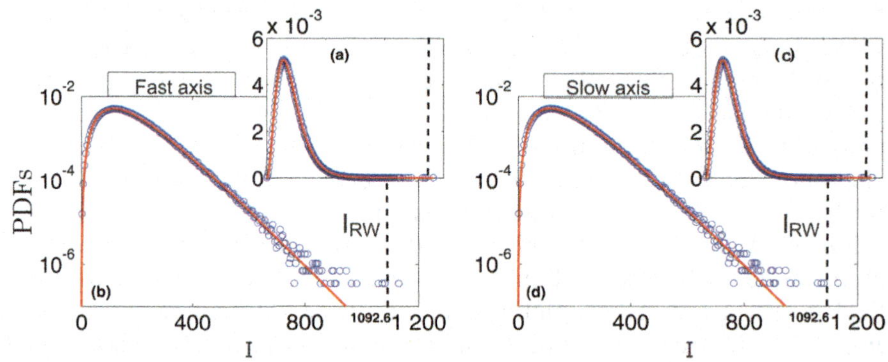

Figure 13.2. PDFs of the peak intensity in semi-logarithmic and linear scales (inset plot) in normal dispersion without walk-off and TOD (a) and (b) on fast axis and (c) and (d) on slow axis. The red line presents the fit by the GG distribution. Parameters of the simulations are: $B = 2$, $\beta_2 = 0.06$ ps^2 m^{-1}, $\beta_3 = 0$, $b_0 = 0$, $L = 15$ m, $\gamma = 6$ W^{-1} km^{-1}, $P = 100$ W.

PDF fits very well to a GG distribution. When the walk-off is taken into account, one can straightforwardly observe from table 13.1 that the values of the *abnormality index* coefficient A_I and *shape* parameter a are now different on the two axes. This is due to the fact that the walk-off parameter b_0 introduces a symmetry breaking which leads to a different trajectory, and hence to eventual different interaction processes of the pulses between the fast and slow axes of the birefringent fiber. Nevertheless, the walk-off can suppress the occurrence of extreme events on both axes simultaneously, as can be seen in table 13.1 for $b_0 = 0.8$ ps m^{-1}. It can suppress them only on one axis, and keep them on the other axis ($b_0 = 1.4$ ps m^{-1}). This indicates that there is a nontrivial dependence between the walk-off and the occurrence of rogue events. Therefore, the walk-off can be used to control the generation of extreme events in nonlinear birefringent fiber. The corresponding PDF for $b_0 = 1.4$ ps m^{-1} is displayed in figure 13.3. One can see that the intensity threshold I_{RW} grows and the high peaks gradually lose their RW character on the slow axis, as compared to the case of figure 13.2.

13.4.2 Impact of the third-order dispersion

Let us now focus on the simulation results obtained when the TOD β_3 is taken into account. It is well known that the TOD term strongly modifies the dynamics of

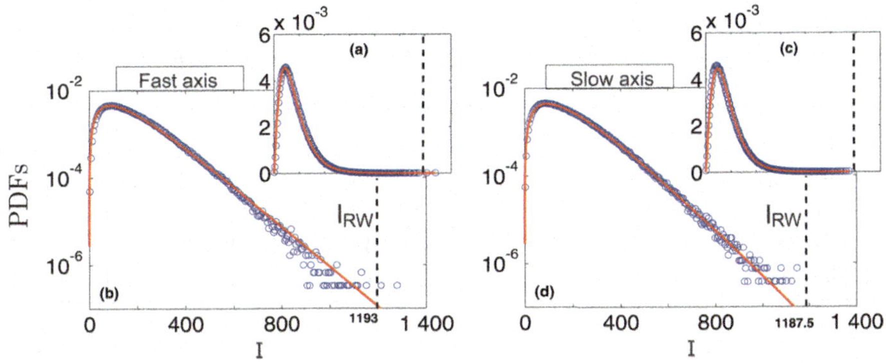

Figure 13.3. PDFs of the peak intensity in normal dispersion regime corresponding to the case where the walk-off acts alone: (a) and (b) on fast axis, (c) and (d) on slow axis. The red line presents the fit by the GG distribution. The inset plot uses the linear scale axis. Parameters of the simulations are: $B = 2$, $\beta_2 = 0.06$ ps^2 m^{-1}, $\beta_3 = 0$, $b_0 = 1.4$ ps m^{-1}, $L = 15$ m, $\gamma = 6$ W^{-1} km^{-1}, $P = 100$ W.

nonlinear systems by making them convectively unstable [11]. In optical fibers, it leads to the nonlinear symmetry-breaking of parametric processes [31–35], and to the generation of Cherenkov radiations by solitons [42], which are one of the key elements of supercontinuum generation.

When we consider only the TOD without involving the walk-off b_0, we find on the two axes that the fiber is at the limit of obtaining extreme events, both with the criterion of the *abnormality index* A_I and with the *shape* parameter a (see table 13.2 for $b_0 = 0$). The PDF of figure 13.4 indicates that the intensity threshold is relatively higher, and the number of rogue events is negligible. It can also be noted that the PDFs are well fitted by the GG distribution. The shape of our numerical PDFs depicts a bell-shape profile, in linear scale, on the two axes. One may conclude that for the set of parameters used, the presence of the third-order dispersion did not introduce any significant impact on the RW generation. Indeed, the high value of the nonlinear birefringent parameter ($B = 2$) still dominates the nonlinear interactions and dynamics, leading to a reduction of the importance of the asymmetry introduced by the third-order dispersion term.

13.4.3 Interplay between the walk-off parameter and the third-order dispersion

In order to have an overview of the interplay of the walk-off b_0 and the TOD β_3, we report in figure 13.5 the PDF for $b_0 = 1.4$ ps m^{-1}. We can clearly see that more intense RWs begin to appear and particularly, in a very pronounced way, on the fast axis. As already mentioned in the Introduction, multiple collisions between nonlinear waves have been suggested as a possible driver mechanism behind the RW formation. At the same time, the interaction between the dispersive and nonlinear waves is also an important mechanism for the RW generation. The latter result can then be explained by the fact that the TOD reinforces the difference in velocities of nonlinear waves between the fast and slow axes of the birefringent fiber. Multiple collisions between solitons or solitons and dispersive waves occur more frequently on the fast axis, leading to more powerful temporal events on this axis.

Table 13.2. Fitted of the abnormality index A_I and the shape parameter a obtained in the normal GVD regime of the birefringent fiber with TOD. $B = 2$, $\beta_3 = 9.10^{-3}$ ps^3 m^{-1}, $L = 15$ m, $\beta_2 = 0.06$ ps^2 m^{-1}, $\gamma = 6$ W^{-1} km^{-1}, $P = 100$ W.

	b_0	0	0.4	0.8	1.2	1.4	1.6
A_I	Fast axis	2.006	2.124	2.945	3.977	5.068	4.806
	Slow axis	2.006	1.820	1.950	2.129	2.315	2.028
a	Fast axis	0.9267	0.86	0.5461	0.21	0.38	0.6783
	Slow axis	0.9267	1.3259	1.01	0.98	0.91	0.9819

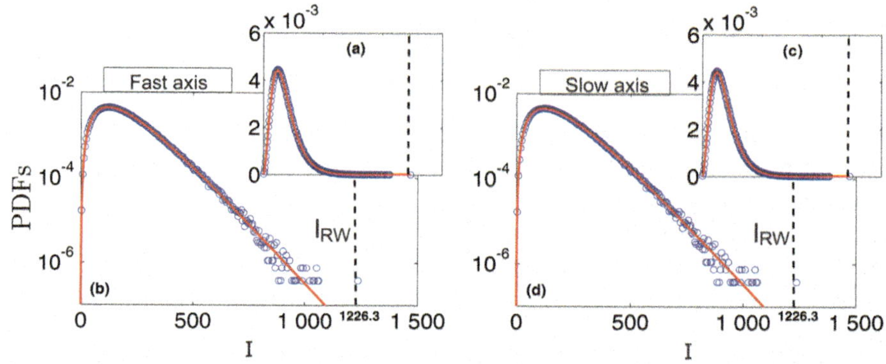

Figure 13.4. PDFs of the peak intensity in normal dispersion regime corresponding to the case where β_3 acts alone, (a) and (b) on fast axis, (c) and (d) on slow axis. The red line presents the fit by the GG distribution. The inset plot uses the linear scale axis. Parameters of the simulations are: $B = 2$, $\beta_2 = 0.06$ ps^2 m^{-1}, $\beta_3 = 9.10^{-3}$ ps^3 m^{-1}, $b_0 = 0$, $L = 15$ m, $\gamma = 6$ W^{-1} km^{-1}, $P = 100$ W.

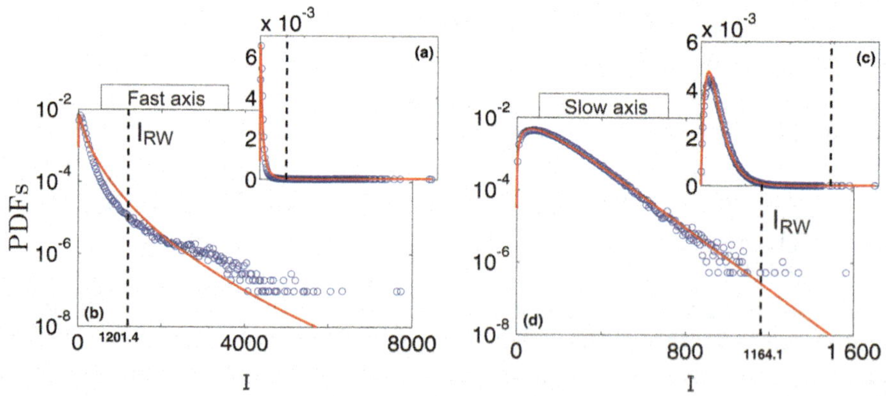

Figure 13.5. PDFs of the peak intensity in normal dispersion regime in presence of walk-off and TOD: (a) and (b) on fast axis, (c) and (d) on slow axis. The red line presents the fit by the GG distribution. The inset plot uses the linear scale axis. Parameters of the simulations are: $B = 2$, $\beta_2 = 0.06$ ps^2 m^{-1}, $\beta_3 = 9.10^{-3}$ ps^3 m^{-1}, $b_0 = 1.4$ ps m^{-1}, $L = 15$ m, $\gamma = 6$ W^{-1} km^{-1}, $P = 100$ W.

We can also see that the numerical PDF, in linear scale, depicts the L-shaped long-tailed characteristic of extreme events on the fast axis and the bell-shape profile on the slow axis.

13.5 Results in a normal dispersion

This section deals with the practicable case of pulse propagation, where β_2 takes negative values and thereby produces MI in the absence of the linear b_0 and/or the nonlinear birefringence B. We proceed along similar lines as discussed earlier. In this part, we will check whether a nonlinear birefringence fiber $B = 2$ can impact the RW generation.

13.5.1 Impact of the walk-off parameter

We first neglected the TOD in order to highlight the impact of the walk-off on the probability properties of extreme events. From table 13.3 one can see that, when the walk-off is neglected ($b_0 = 0$), the *abnormality index* A_I and the *shape* parameter a have the same values on the two axes of the birefringent fiber and verify the existence condition of RWs ($A_I > 2$, $a < 1$). With the increase of the walk-off parameter, the above criteria are always fulfilled on the two axes (see table 13.3). This is in contrast to the case of normal GVD, where we have observed that only the fast axis produces extreme events. Figure 13.6 represents the corresponding PDF for the particular value of the walk-off $b_0 = 1.2$ ps m^{-1}. The intensity threshold for RW formation $I_{RW} \approx 105$ W is indicated by the vertical line, and we observe that the percentage of rogue events is significantly greater than in the case obtained in normal GVD. The large number of events identified as RWs can be explained by the fact that, by comparison to the normal dispersion, the anomalous dispersion regime is more favorable to the development of MI and therefore to extreme events.

13.5.2 Impact of the third-order dispersion

When we consider only the TOD β_3 without the walk-off b_0, the criteria of the emergence of RW is fulfilled (see table 13.4 with $b_0 = 0$). The PDF of figure 13.7 shows that both the two axes contribute significantly to the formation of RW. The probability of RW emergence increases both in number and severity, as compared

Table 13.3. Fitted of the abnormality index A_I and the shape parameter a obtained in the anomalous GVD regime of the birefringent fiber without TOD. $B = 2$, $\beta_2 = -6.210^{-4}$ ps^2 m^{-1}, $\beta_3 = 0$, $\gamma = 10$ W^{-1} km^{-1}, $L = 200$ m, $P = 10$ W.

	b_0	0	0.4	1.2	1.4
A_I	Fast axis	3.213	2.403	2.400	2.497
	Slow axis	3.213	2.408	2.492	2.276
a	Fast axis	0.4829	0.8306	0.8373	0.3444
	Slow axis	0.4829	0.8315	0.8167	0.3239

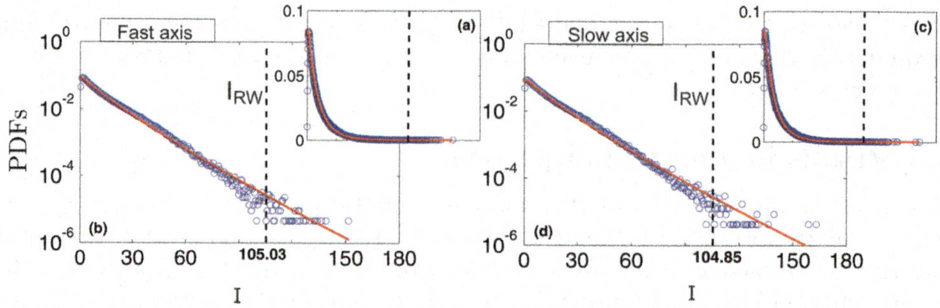

Figure 13.6. PDFs of the peak intensity in semi-logarithmic and linear scales (inset plot) in the anomalous dispersion regime corresponding to the case where the walk-off acts alone: (a) and (b) on fast axis, (c) and (d) on slow axis. The red line presents the fit by the GG distribution. The parameters of the simulations are: $B = 2$, $\beta_3 = 0$, $\beta_2 = -6.210^{-4}$ ps^2 m^{-1}, $\gamma = 10$ W^{-1} km^{-1}, $b_0 = 1.2$ ps m^{-1}, $L = 200$ m, $P = 10$ W.

Table 13.4. Fitted of the abnormality index A_I and the shape parameter a obtained in the anomalous GVD regime of the birefringent fiber with TOD. $B = 2$, $\beta_2 = -6.210^{-4}$ ps^2 m^{-1}, $\beta_3 = 3.510^{-5}$ ps^3 m^{-1}, $\gamma = 10$ W^{-1} km^{-1}, $L = 200$ m, $P = 10$ W.

	b_0	0	0.4	1.2	1.4
A_I	Fast axis	2.6425	2.9415	3.1712	2.9855
	low axis	2.6425	3.1279	3.1591	3.0268
a	Fast axis	0.4679	0.3225	0.3116	0.4686
	Slow axis	0.4679	0.3151	0.3137	0.4680

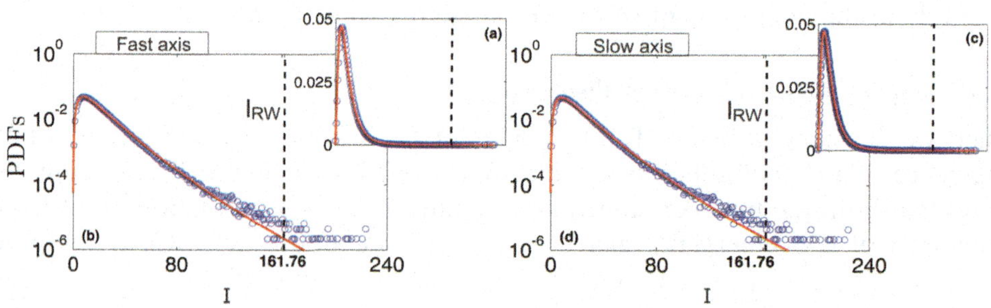

Figure 13.7. PDFs of the peak intensity in the anomalous dispersion regime corresponding to the case where β_3 act alone: (a) and (b) on fast axis, (c) and (d) on slow axis. The red line presents the fit by the GG distribution. The inset plot uses the linear scale axis. the parameters of the simulations are: $B = 2$, $\beta_3 = 3.510^{-5}$ ps^3 m^{-1}, $\beta_2 = -6.210^{-4}$ ps^2 m^{-1}, $\gamma = 10$ W^{-1} km^{-1}, $b_0 = 0$, $L = 200$ m, $P = 10$ W.

with the case obtained in figure 13.6, where the walk-off acts alone. Extreme events obtained here can be explained by the fact that under the action of the TOD, the noise-induced MI leads to soliton fission, which breaks up into a number of solitons that interact and transfer energy between each other during multiple collisions. The strong solitons have a tendency to accumulate energy from the weaker ones, thus

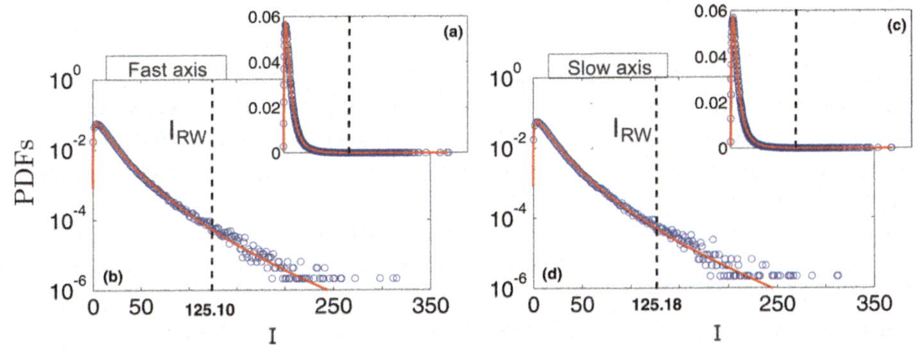

Figure 13.8. PDFs of the peak intensity in the anomalous dispersion regime in presence of walk-off and TOD: (a) and (b) on fast axis, (c) and (d) on slow axis. The red line presents the fit by the GG distribution. The inset plot uses the linear scale axis. $B = 2$, $\beta_3 = 3.510^{-5}$ ps^3 m^{-1}, $\beta_2 = -6.210^{-4}$ ps^2 m^{-1}, $\gamma = 10$ W^{-1} km^{-1}, $b_0 = 1.2$ ps m^{-1}, $L = 200$ m, $P = 10$ W.

exiting the fiber as RWs. Moreover, the numerical PDF is also well fitted by the GG function.

13.5.3 Interplay between the walk-off and the third-order dispersion

When we consider the association of the walk-off and TOD, we observe in the PDF of figure 13.8 that there is a significant increase in both the number and severity of events. By contrast, we find that the RW has a lower intensity threshold compared with figure 13.7, where the TOD acts alone. As mentioned above in the case of normal GVD, the difference in velocities of each individual soliton is more important when the walk-off and TOD act together. The RW can then be ejected by absorbing enough energy because of frequent soliton collisions. The numerical PDF depicts an L-shaped, long-tailed statistical profile of extreme events in linear scale, and the GG function perfectly models the PDF curve.

13.6 Conclusion

In conclusion, we have studied the process of RW formation in a birefringent fiber governed by the coupled NLSE with TOD. Linear stability analysis of this system has been performed in normal dispersion regime, with an emphasis on investigating the effect of the nonlinear birefringence and the walk-off parameters. It was found that these parameters can be used to control the MI generation. We have also performed a significant amount of statistical analysis using PDF of the peak intensities. The results of our investigations show that RWs occur in both anomalous and normal dispersion regimes, albeit more prominently in the anomalous ones. Moreover, the association of the third-order dispersion and the walk-off parameters promotes the formation of intense waves. We find that the statistics of these peak intensities are described very well by the GG distribution. These results shall stimulate experimental investigations aimed at observing such extreme wave events in birefringent fibers.

References

[1] Solli D R, Ropers C, Koonath P and Jalali B 2007 *Nature* **450** 1054

[2] Ankiewicz A, Soto-Crespo J M and Akhmediev N 2010 *Phys. Rev.* E **81** 046602
Akhmediev N, Ankiewicz A and Soto-Crespo J M 2009 *Phys. Rev.* E **80** 026601

[3] Kibler B, Fatome J, Finot C, Millot G, Dias F, Genty G, Akhmediev N and Dudley J M 2010 *Nat. Phys.* **6** 790

[4] Chabchoub A, Hoffmann N P and Akhmediev N 2011 *Phys. Rev. Lett.* **106** 204502

[5] Bludov Y V, Konotop V V and Akhmediev N 2010 *Eur. Phys. J. Special Topics* **185** 033610

[6] Moslem W M, Shukla P K and Eliasson B 2011 *Eur. Phys. Lett.* **96** 25002

[7] Efimov V B, Ganshin A N, Kolmakov G N, Mc-Clintock P V E and Mezhov-Deglin L P 2010 *Eur. Phys. J. Special Topics* **185** 181

[8] Zhao S, Yang H, Chen N and Zhao C 2017 *Sci. Rep.* **7** 39926

[9] Dudley J M, Genty G and Coen S 2006 *Rev. Mod. Phys.* **78** 1135

[10] Mussot A, Kudlinski A, Kolobov M, Louvergneaux E, Douay M and Taki M 2009 *Opt. Express* **17** 17010

[11] Taki M, Mussot A, Kudlinski A, Louvergneaux E, Kolobov M I and Douay M 2010 *Phys. Lett. A* **374** 691

[12] Dudley J M, Dias F, Erkintalo M and Genty G 2014 *Nat. Photon.* **8** 755

[13] Hammani K, Kibler B, Fatome J, Boscolo S, Genty G, Dudley J M, Millot G and Finot C 2012 *Opt. Fiber Technol.* **18** 248

[14] Armaroli A, Conti C and Biancalana F 2015 *Optica* **2** 497

[15] Demircan A, Amiranashvili S, Brée C, Mahnke C, Mitschke F and Steinmeyer G 2012 *Sci. Rep.* **2** 00850

[16] Hamner C, Chang J J, Engels P and Hoefer M A 2011 *Phys. Rev. Lett.* **106** 065302

[17] Tang D Y, Zhang H, Zhao L M and Wu X 2008 *Phys. Rev. Lett.* **101** 153904

[18] Wright O C and Forest M G 2000 *Physica* D **141** 104
Forest M G and Wright O C 2003 *Physica* D **178** 173

[19] Chow K W, Wong K K Y and Lam K 2008 *Phys. Lett.* A **372** 4596

[20] Wabnitz S 1988 *Phys. Rev.* A **38** 2018

[21] Rothenberg J E 1990 *Phys. Rev.* A **42** 682

[22] Millot G, Seve E, Wabnitz S and Haelterman M 1998 *J. Opt. Soc. Am.* B **15** 1266

[23] Agrawal G P 2008 *Nonlinear Fiber Optics* 3rd edn (Boston: Academic)

[24] Baronio F, Conforti M, Degasperis A, Lombardo S, Onorato M and Wabnitz S 2014 *Phys. Rev. Lett.* **113** 034101

[25] Frisquet B, Kibler B, Morin P, Baronio F, Conforti M, Millot B and Wabnitz S 2016 *Sci. Rep.* **6** 20785

[26] Zhao L C and Liu J 2012 *J. Opt. Soc. Am.* B **29** 3119

[27] Zhao L C and Liu J 2013 *Phys. Rev.* E **87** 013201

[28] Guo B L and Ling L M 2011 *Chin. Phys. Lett.* **28** 110202

[29] Baronio F, Degasperis A, Conforti M and Wabnitz S 2012 *Phys. Rev. Lett.* **109** 044102

[30] Ablowitz M J and Horikis T P 2017 *J. Opt.* **19** 065501

[31] Demircan A, Pietrzyk M and Bandelow U 2008 *Opt. Quant. Electron.* **40** 455

[32] Mussot A, Kudlinski A, Louvergneaux E, Kolobov M and Taki M 2010 *Opt. Lett.* **35** 1194

[33] Mussot A, Kudlinski A, Droques M, Szriftgiser P and Akhmediev N 2014 *Phys. Rev.* X **4** 011054

[34] Weerasekara G, Tokunaga A, Terauchi H, Eberhard M and Maruta A 2015 *Opt. Express* **23** 143
[35] Coulibaly S, Liu Z, Taki M and Agrawal G P 2012 *Phys. Rev.* A **86** 033802
[36] Drouzi L, Coulibaly S, Taki M and Laabidi K 2017 *Phys. Rev.* A **95** 063826
[37] Onorato M, Residori S, Bortolozzo U, Montina A and Arecchi F T 2013 *Phys. Rep.* **528** 47
[38] Stacy E 1962 *Ann. Math. Stat.* **33** 1187
[39] Louvergneaux E, Odent V, Kolobov M I and Taki M 2013 *Phys. Rev.* A **87** 063802
[40] Kharif C, Pelinovsky E and Slunyaev A 2009 *Rogue Waves in the Ocean* (Heidelberg: Springer)
[41] Akhmediev N, Soto-Crespo J M and Ankiewicz A 2009 *Phys. Lett.* A **373** 2137
Soto-Crespo J M, Devine N and Akhmediev N 2016 *Phys. Rev. Lett.* **116** 103901
[42] Akhmediev N and Karlsson M 1995 *Phys. Rev.* A **51** 2602

IOP Publishing

Nonlinear Guided Wave Optics
A testbed for extreme waves
Stefan Wabnitz

Chapter 14

Spatiotemporal nonlinear dynamics in multimode fibers

K Krupa, V Couderc, A Tonello, A Picozzi, A Barthélémy, G Millot, S Wabnitz

14.1 Introduction

Multimode optical fibers (MMFs) have been investigated since the early eighties for their capacity to transmit information over multiple spatial channels [1]. The associated technique of spatial division multiplexing in MMFs has recently seen renewed research and technological interest, owing to the capacity limitations of single-mode fibers for the next generation of transparent optical networks. Guided modes of an ideal, longitudinally invariant optical fiber are orthogonal to each other, so that they can provide an alphabet for transmitting information.

However, the refractive index change induced by light intensity itself (the so-called optical Kerr effect) introduces a coupling among the guided modes of an MMF. Several studies have been dedicated to exploring the consequences of this nonlinear mode coupling. For instance, [2] pointed out the existence of asymmetric spectral broadening, induced by cross-phase modulation among the guided modes. The corresponding experimental analysis resolved in time, wavelength, and modal content was carried out in a six-mode optical fiber. More recently, nonlinear mode coupling has been exploited for showing light-induced mode conversion [3]. Whenever MMFs support the propagation of a relatively large number of guided modes, the resulting nonlinear mode-coupling dynamics exhibit a complex behavior. In other words, the dynamics resulting from the collective interplay of many guided modes may reveal novel and unexpected properties that could not be predictable on the basis of the propagation of the individual modes.

An important consequence of complex interactions is the emergence of order in the form of extreme coherent wave structures. One such example is the concept of spatial multimode, temporal soliton pulses. This is a short pulse that propagates with a single group velocity, resulting from nonlinear mode coupling that compensates for modal dispersion. Multimode temporal solitons were predicted long ago [4], and

 © IOP Publishing Ltd 2017

later studied with a variational method in [5], but only recently were experimentally and numerically studied by Renninger and Wise [6] and by Wright *et al* in [7]. Moreover, the dispersive waves generated by these self-sustained nonlinear waveforms were also recently experimentally revealed in [8].

In our chapter, we will overview different manifestations of a recent and spectacular discovery in the field of complex multimode fiber dynamics. Namely, the observation of the so-called spatial beam self-cleaning effect [9], whereby an intense optical beam propagating in a highly multimode graded index (GRIN) optical fiber experiences an irreversible evolution towards a robust bell-shaped transverse profile that remains very close to the fundamental mode of the fiber. Such an effect has been named self-cleaning, since, in the absence of the nonlinearity, the multiple interferences among the many guided modes of the MMF produce instead a highly irregular, noise-like speckled intensity pattern, also known as modal noise, in the context of multimode fiber communications. The first demonstration of Kerr-induced beam self-cleaning was achieved in a virtually lossless propagation regime (a few meters of GRIN MMF). Subsequent experiments have remarkably and surprisingly shown that self-cleaning may be also achieved in a nearly step-index MMF, whenever it is assisted by the presence of strong dissipation (linear loss or gain) [10]. A theoretical description of spatiotemporal wave propagation in GRIN MMF is introduced, which is exploited for full spatiotemporal numerical simulations, as well as for developing a statistical description that links self-cleaning to the phenomenon of wave condensation in the frame of the theory of wave turbulence and rogue waves. Self-cleaning is further investigated in the presence of spatiotemporal instabilities which generate parametric sideband series and supercontinuum [11–14], as well as in the framework of all-optical poling and second-harmonic generation in MMFs [15]. Experiments have shown that self-cleaning can be frequency-converted across the entire spectrum of sidebands and supercontinuum frequencies spanning multiple octaves.

14.2 Spatial beam self-cleaning

It is well known that light experiences an inherent randomization in MMFs, whenever propagation involves a large (i.e. greater than 50) number of guided modes. As a matter of fact, typical fiber perturbations such as stress, bending, or technological irregularities lead to linear mode coupling. As a consequence, an input laser beam of high spatial quality leads to a speckled interference pattern when emerging from an MMF. Our recent experiments have shown that, unexpectedly, when the peak power of input beam pulses of sub-nanosecond duration crosses a certain threshold value, the Kerr-induced complex nonlinear mode mixing in a GRIN MMF leads to a strong reduction of this mode beating (or modal noise). As a result, a bell-shaped beam resembling the fundamental mode of the fiber is formed at the output of the GRIN MMF. It is interesting to note that Kerr self-cleaning disappears by bringing down the power at the input of the fiber below the threshold value, and can be restored as the input power is increased again.

It is also important to point out that the power threshold for Kerr beam self-cleaning is lower than that for stimulated Raman scattering (SRS). It is well known that SRS leads to beam cleaning for the generated Stokes wave, since the fundamental mode has the largest SRS gain. This means that one can observe Kerr-induced cleaning for the pump beam itself well before any spectral broadening or frequency conversion occurs.

Owing to the self-focusing nature of the Kerr effect in fibers, there is a third phenomenon that may lead to a strong spatial beam reshaping, known as catastrophic self-focusing [16, 17]. However, such an effect occurs for powers above a certain critical value, which happens to be orders of magnitude larger than the threshold for Kerr self-cleaning. Kerr beam cleaning was observed so far with relatively long pulses propagating in the normal dispersion regime of few-meter-long GRIN MMFs, so that dispersion does not play a significant role in the pulse dynamics. Kerr beam cleaning has the peculiar property of significantly increasing the beam brightness at the output of a MMF. This situation is in striking contrast to beam clean-up obtained via other nonlinear effects such as SRS or stimulated Brillouin scattering (SBS), whereby the down-converted Stokes waves is preferentially generated in the fundamental mode of the fiber [18]. The phenomenon of Kerr spatial beam self-cleaning turns out to be remarkably robust against the presence of relatively strong external mechanical perturbations, a situation markedly different from what happens in the multimode linear regime.

Kerr-induced beam self-cleaning was first observed in experimental conditions that can be modeled in terms of a conservative or lossless wave system. This is the case of standard GRIN MMFs, where the presence of attenuation can be neglected for wave propagation over a few meters of fiber. More recently, spatial beam self-cleaning was also observed in experimental conditions where nonlinear wave propagation should be modeled as a strongly dissipative system. This is the case of MMFs doped with rare earths, where in the absence of a pump, the attenuation in a few meters of fiber can halve the input power, whereas in the presence of a pump, the input beam experiences exponential amplification, thus reducing the power threshold for Kerr beam self-cleaning.

14.2.1 Experiments in lossless fiber systems

Kerr-induced spatial beam self-cleaning was first theoretically and experimentally investigated in an essentially lossless propagation environment, using standard commercially available GRIN MMFs. Its first demonstration was carried out in the normal dispersion regime with quasi-CW sub-nanosecond pulses [9, 11, 19, 20]. Subsequently, Kerr self-cleaning was investigated in the femtosecond pulse regime, where the power threshold grows larger and becomes comparable with the threshold for catastrophic self-focusing [21].

Figure 14.1 shows the experimental observation of Kerr beam self-cleaning in a 12 m long GRIN MMF with 52 μm of core diameter. Here, the beam is generated by 900 ps pulses at 30 kHz repetition rate from an amplified Nd:YAG microchip laser at 1064 nm [9,19]. Panels (a)–(d) of figure 14.1 show the 2D near-field spatial beam

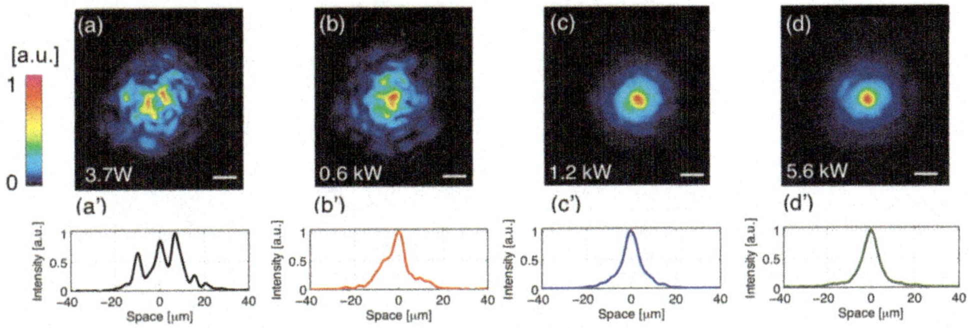

Figure 14.1. Experimental evolution of near-field spatial pattern of GRIN MMF output at 1064 nm as a function of output power Ppp (a)–(d) induced by Kerr beam self-cleaning, and corresponding transverse beam profiles versus x (a')–(d'). Scale bar: 10 μm; Fiber length: 12 m. Reprinted by permission from Macmillan Publishers Ltd: [9] (© 2017).

pattern emerging from the fiber at 1064 nm, as a function of the output peak power (Ppp). Panels (a′)–(d′) of figure 14.1 display the corresponding beam transverse profiles. As we can see, at relatively low power (Ppp = 3.7 W), where propagation is essentially linear, the highly multimode excitation, involving about 200 guided modes when including both polarization components, leads to a randomly speckled beam distribution (figure 14.1, panels (a) and (a′)). However, when increasing the input pulse power, the output beam profile evolves from highly irregular to a well-defined bell-shaped pattern at the center, which is quite close to the fundamental mode of the fiber, sitting on a residual low-power speckled background (figure 14.1, panels (c), (c′) and (d), (d′)). It was experimentally verified that the redistribution of power towards a central quasi-Gaussian profile results neither from loss of spatial and temporal (frequency conversion) coherence, nor from additional transmission losses [9, 19].

Figure 14.2 confirms the significant increase of the spatial quality of the output beam and its brightness: a nearly threefold decrease of the output near-field beam full width at half maximum intensity (FWHMI) diameter was measured (red circles in figure 14.2). The corresponding fraction of output power confined in the central part of the beam was increased by a factor of 2.6. This fraction of power was measured at 1064 nm through the use of a diaphragm with a diameter of 12 μm (blue squares of figure 14.2), as a result of Kerr beam self-cleaning. Numerical simulations predict that, above the threshold power, a continuous energy transfer from high-order modes (HOMs) towards the fundamental or low-order modes occurs in the GRIN MMF. Statistical analyses of a long series of experimental results (involving a large number of input beams, characterized by different orientations of their linear state of polarization) have confirmed that the regime of beam self-cleaning is associated with a significant reduction in the average beam diameter, as well as with a dramatic reduction in the associated standard deviation [9, 19]. It is also important to mention that the Kerr-induced spatial beam self-cleaning was followed by significant temporal reshaping, which results in a shortening of the initial pulse duration.

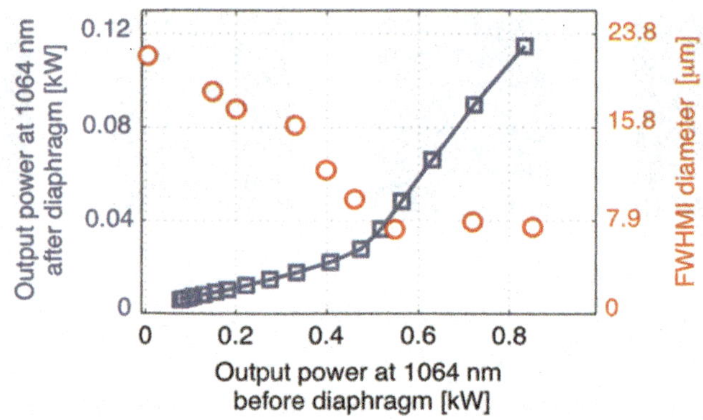

Figure 14.2. Output power at 1064 nm measured in the central part of the fiber as a function of the total output power (blue squares), and corresponding FWHMI diameters of the near-field beam at 1064 nm (red circles). Reprinted by permission from Macmillan Publishers Ltd: [9] (© 2017).

To gain further insight into the mechanism responsible for beam self-cleaning, we proceeded to cut back measurements as shown in figure 14.3. Our results confirm that spatial reshaping occurs along the propagation direction, towards a beam shape that resembles the fundamental mode. These observations also prove that the onset of beam self-cleaning is activated in the presence of a multimode excitation. In fact, beam scrambling always occurs in the initial 20 cm of the MMF, both in the linear as well as in the nonlinear regimes (figure 14.3 panels (a) and (b)) despite the high input beam power (68 kW). These experimental observations are in quite good qualitative agreement with the numerical simulations later discussed in this chapter, based on the solution of the (3 + 1)D spatiotemporal nonlinear Schrödinger equation (14.1).

The Kerr-induced spatial beam self-cleaning may be explained, at least with a very simplified model, in terms of the nonlinear nonreciprocity of the mode-coupling process in a GRIN MMF. The periodic oscillations of the multimode beam intensity that occur along the fiber through mode beating lead to a longitudinal modulation of the fiber refractive index that is induced by the Kerr effect. GRIN MMFs have the property of an equal spacing between the propagation constants of their guided modes, which leads to the longitudinal periodic evolution of the beam profile, or self-imaging effect. The resulting nonlinear (or dynamical) grating will, in turn, provide quasi-phase matching for a variety of nonlinear mode-coupling processes, which results in the energy exchange between the fundamental and the HOMs via nonlinear modal interactions. The situation is analogous to the nonlinear self-switching that is observed in long-period fiber Bragg gratings [22]. The main difference here is that the grating is self-induced by the multimode beam. The effect of nonlinear nonreciprocity on self-switching was first studied in the frame of coherent mismatched directional couplers in [23]. Above a certain threshold power value, the reciprocity of mode coupling is broken. In the highly multimode case, one may extrapolate that a lack of reciprocity leads to the irreversible flow of energy in favor of the fundamental mode. However, it is important to underline that this fact does not directly imply that the entire beam is predominantly carried by the

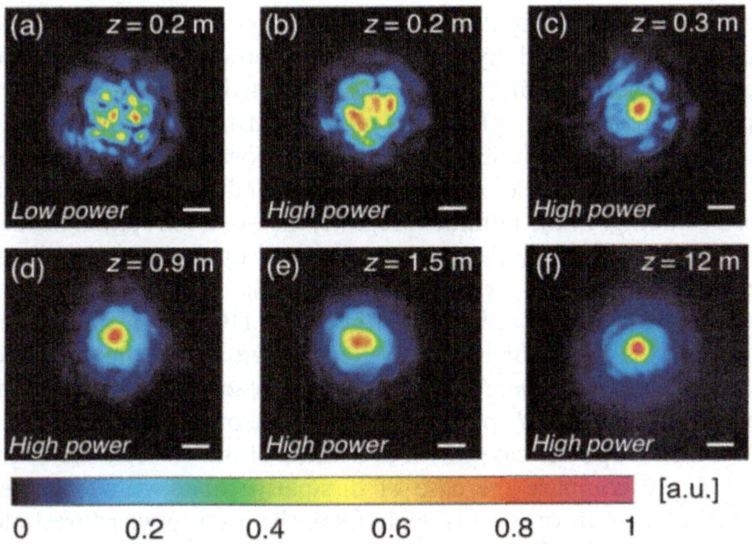

Figure 14.3. Experimental evolution of near-field spatial pattern of GRIN MMF output at 1064 nm as a function of propagation distance z (a)–(f) induced by Kerr beam self-cleaning. High input power = 68 kW; Low input power = 0.007 kW. Scale bar: 10 µm. Reprinted by permission from Macmillan Publishers Ltd: [9] (© 2017).

fundamental mode, since the presence of a dynamic grating always requires the presence of the collective contribution of the other modes. The simplified explanation of the physical mechanism behind the Kerr-induced spatial beam self-cleaning effect is based on a two-mode mean-field model, as described in detail in [9, 19].

14.2.2 Experiments in dissipative fiber systems

Kerr-induced beam self-cleaning was also experimentally studied in a dissipative context, that is, in the presence of strong loss or gain. For this purpose, multimode fibers doped by rare-earth atoms were used. Two different configurations were explored: in the first, passive configuration, beam propagation took place in the presence of strong distributed losses. In the second case, the doped fiber was pumped by a CW laser diode: we will call this the active configuration, owing to the presence of optical gain for the laser at 1064 nm. Experiments used a 3 m long Ytterbium doped, double-clad multimode fiber (Yb-MMF) with a nearly step-index profile, 55 µm core diameter surrounded by a D-shaped inner cladding of 340 µm × 400 µm size, for guiding the pump radiation [10]. To investigate the role played by losses, we simply imaged an amplified microchip Nd:YAG laser beam (pulse duration of 500 ps, repetition rate of 500 Hz, central wavelength 1064 nm) into the Yb-MMF with a spot size of ~22 µm in diameter, in order to excite a large number of modes.

Interestingly, despite the large overall attenuation of about 7.3 dB (corresponding to 2.44 dB m^{-1} at 1064 nm) and the non-parabolic refractive index profile, we observed a beam self-cleaning similar to that obtained in the case of a virtually lossless GRIN MMF. In particular, when increasing the input beam power, the output pattern (which was highly speckled at low powers) evolved into a smooth bell-shaped spot, surrounded by a residual low-power, multimode background. Nevertheless, it is worth noting that in the present dissipative case the nonlinear beam reshaping process is qualitatively different from the one observed in the lossless case, since with the doped fiber the input power not only coupled into the numerous guided modes of the core, but also coupled into the HOMs leaking into the D-shaped cladding. In the nonlinear regime, the observed spatial beam reshaping into the quasi-fundamental mode was accompanied by an increment by a factor of 1.8 of the fractional output power carried by the fiber core and a significant reduction of the degree of coupling into the leaky HOMs. In other words, in the nonlinear regime, the doped fiber was less leaky than in the linear regime [10].

The power-induced improvement of the spatial beam quality was also testified to by the measurements of the M^2 parameter [24] of the output beam, dropping from a value of 16 in the low-power case down to a value of only 2 after the beam cleaning was reached. If we compare the case of dissipative fiber with high losses to the lossless case (i.e. when absorption is negligible), the input power threshold required to obtain Kerr beam self-cleaning increases by 6 times for a fiber of similar core diameter and length. This fact proves that the input peak power should be large enough to reach and preserve spatial beam cleaning up to the output of the fiber, in spite of the relatively large propagation losses. However, it is interesting to notice that if we consider the output peak power instead, the self-cleaning threshold obtained in the doped fiber in the presence of strong linear loss remains close to the one measured in the lossless propagation case.

Next, we investigated the role played by gain in Kerr-induced beam self-cleaning. For this purpose, we fixed the 1064 nm signal of the microchip Nd:YAG laser at the relatively low power level of 0.5 kW, and we pumped the D-shaped inner cladding of the Yb-MMF with a CW laser diode at 940 nm. Figure 14.4 illustrates the evolution of the near-field pattern at the fiber output, upon the optical gain (G), which grows when increasing the pump power. As we can see, the observed spatial modification is very similar to what is obtained in the passive configuration with high losses, and it is also similar to the Kerr beam self-cleaning effect observed in a conservative fiber system such as the standard GRIN MMF. The decrease of the M^2 parameter presented in figure 14.5 further confirms the substantial improvement of the output beam quality thanks to the self-cleaning effect. This remarkable result was reached for a gain $G = 20$, and with a signal pulse peak power of only 0.5 kW (we recall that a peak power as high as 40 kW was required when the Yb-MMF was not pumped, whereas a threshold peak power of 7 kW was observed in the lossless case of a GRIN fiber). The corresponding power threshold, expressed in terms of the output peak power, was once again close to that obtained in the unpumped situation.

However, what is also remarkable is that the power threshold expressed in terms of path-averaged power ($\langle P \rangle_z$) is significantly reduced (~1/6) by the presence of linear gain with respect to the lossy case: this fact shows that the presence of gain can substantially facilitate the Kerr beam self-cleaning. These experiments clearly demonstrate that Kerr-induced beam self-cleaning exhibits a robust wave dynamics that is of a more general nature than what was initially thought. In fact, nonlinear

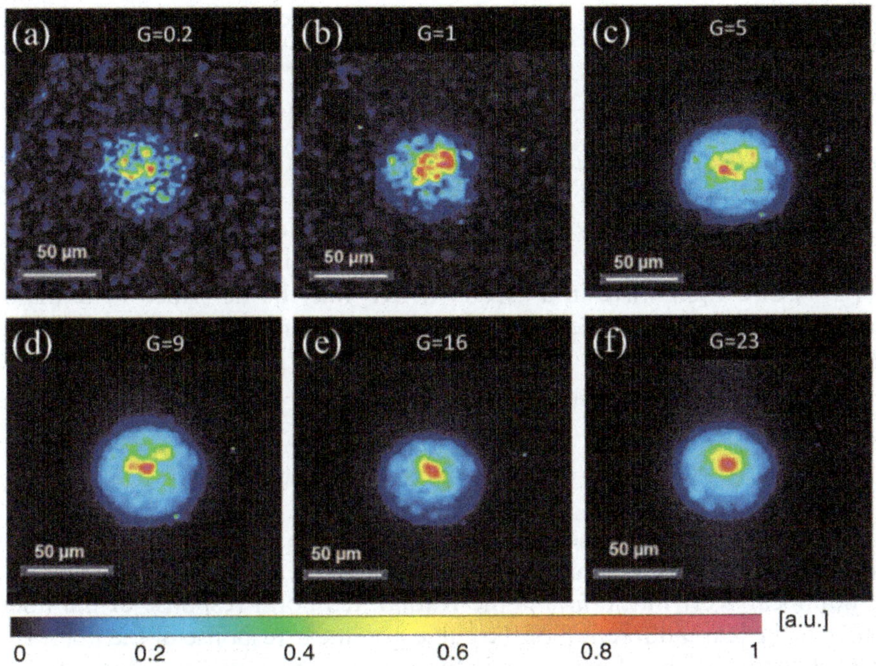

Figure 14.4. Experimental evolution of near-field spatial pattern of Yb-MMF output at 1064 nm as a function of optical gain G (a)–(f) induced by Kerr beam self-cleaning. Input power of signal = 0.5 kW; Fiber length: 3 m. Reprinted figure with permission from [10].

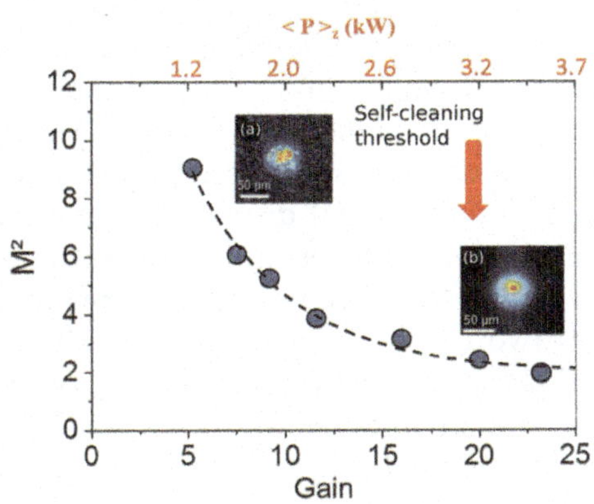

Figure 14.5. M^2 measurement ($1/e^2$ diameter) of the output beam pattern versus gain of the amplified Yb-MMF. Input signal peak power: 0.5 kW. Insets: output beam patterns for low amplification (a) and high amplification (b); $<P>_z$ is a path-averaged power. Reprinted figure with permission from [10].

spatial reshaping is not only limited to strictly conservative systems, since we have experimentally demonstrated that it can occur in dissipative fiber systems and even in lasers [25]. Moreover, our experimental results show also that beam self-cleaning is not limited to the case of GRIN MMFs only, where periodic self-imaging was initially supposed to be the key mechanism for the process of phase-matched nonlinear mode coupling.

14.3 Theoretical models of spatiotemporal dynamics

In this section, we present a theoretical description of pulse propagation in a multimode fiber that permits us to model the beam self-cleaning effect and the generation of new frequency components via spatiotemporal instabilities. The evolution in three space and one temporal coordinates of the complex envelope $\psi(x, y, z, t)$ of a light wave in a GRIN MMF may be described in terms of the generalized 3D lossless nonlinear Schrödinger equation (GNLSE3D)

$$\begin{aligned}\frac{\partial\psi}{\partial z} - i\frac{1}{2k_L}\nabla_\perp^2\psi + i\frac{\kappa''}{2}\frac{\partial^2\psi}{\partial t^2} + i\frac{k_L\Delta}{\rho^2}r^2\psi \\ = i\gamma(1 - f_R)|\psi|^2\psi + i\gamma f_R\psi\int h(\tau)|\psi(t - \tau)|^2 d\tau\end{aligned} \tag{14.1}$$

where ω_0 is the angular frequency of the pump, $k_L = \omega_0 n_{co}/c$, κ'' is the chromatic dispersion, $\gamma = \omega_0 n_2/c$ is the nonlinear coefficient, n_{co} is the core refractive index, and n_2 the nonlinear Kerr index; $h(t)$ is the Raman temporal response of the fiber and $f_R = 0.18$ is the Raman fraction of the cubic nonlinear response of the fiber. Δ is the relative refractive index difference, ρ is the core radius, and r is a transverse radial coordinate, with $r^2 = x^2 + y^2$. Our normalization is chosen so that $|\psi|^2$ has the physical dimension of an optical intensity.

Equation (14.1) contains the principal physical mechanisms at play: diffraction, dispersion, waveguide potential, Kerr effect, and SRS. As the beam power is increased, the first manifestation of nonlinearity in multimode dynamics occurs in the mode-coupling processes, which involve the interplay of transverse beam diffraction, waveguide potential, and the Kerr effect. When the power grows larger, optical pulses will break up in time, thus generating new frequency sidebands. In this regime, eventually chromatic and modal dispersion come into play. Besides parametric mixing effects, in the normal dispersion regime, the pump light experiences frequency conversion to longer (Stokes) wavelengths owing to SRS.

Figure 14.6 illustrates two examples of numerical solutions of equation (14.1). The first example is carried out at low power, while the other example refers to the same input condition, but with an intensity above the self-cleaning threshold, so that the Kerr effect reshapes the beam towards a well-defined and robust, bell-shaped transverse profile after a propagation over just 1 m of fiber. In these simulations, mode coupling was effectively introduced by adding a random input spatial phase distribution. Simulation results confirm the experimental results illustrated in figure 14.1. In both low- and high-intensity cases, the field is spectrally filtered around the pump wavelength, and the resulting intensity is integrated in time.

14.3.1 Wave condensation and turbulence in a multimode fiber

The remarkable phenomenon of beam cleaning in a GRIN fiber discussed above in the lossless configuration may possibly be related to a process of wave condensation.

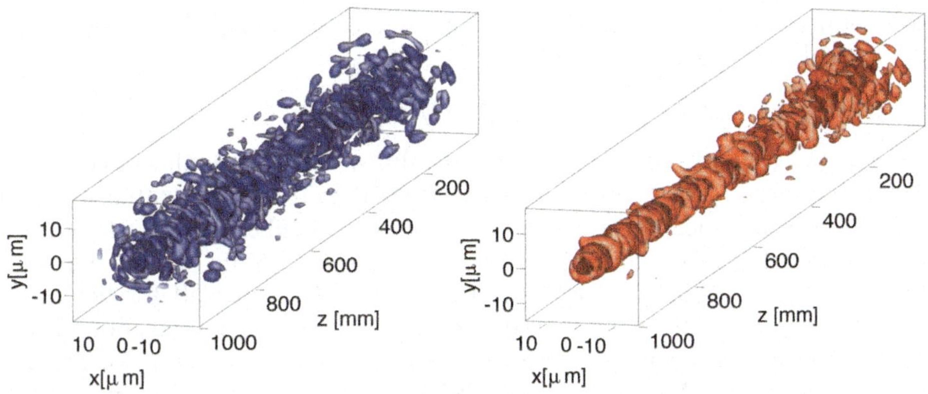

Figure 14.6. Simulations showing beam self-cleaning in lossless GRIN MMF. Left: speckled beam evolution at 0.5 MW cm^{-2} (low intensity); right: self-cleaning at 5 GW cm^{-2} (high intensity).

Indeed, theoretical and experimental studies on wave turbulence revealed that a purely classical system of random waves can exhibit a process of condensation, whose thermodynamic properties are analogous to those of quantum Bose–Einstein condensation, despite the classical nature of the system [26–33]. Classical wave condensation finds its origin in the natural thermalization of the wave system toward the thermodynamic Rayleigh–Jeans equilibrium distribution, whose divergence is responsible for the macroscopic occupation of the fundamental mode of the system. Wave condensation then manifests itself by the spontaneous formation of a large-scale coherent structure (fundamental 'condensate' mode) that remains immersed in a sea of small-scale fluctuations ('uncondensed particles' that populate higher-order modes), which store all the information that is necessary for time reversal.

The wave turbulence description of random waves has been mostly developed in the ideal situation in which the fluctuations of the waves are supposed to be statistically homogeneous in space [34, 35], an assumption that may be considered as justified when the correlation length of the fluctuations is much smaller than the size of the speckled beam. However, the propagation of an incoherent localized beam is eventually affected by incoherent diffraction, which inevitably affects the long-term equilibrium properties of the beam, i.e. it affects the processes of thermalization and condensation. In the following, we derive the wave turbulence kinetic equation by considering the propagation of the incoherent beam in a multimode waveguide configuration. Owing to the guided-wave configuration, incoherent diffraction is compensated by a confining potential, thus allowing us to study the thermalization and condensation of the optical field over large propagation lengths [31]. Accordingly, we consider the NLS equation with a confining potential $V(\boldsymbol{x})$, and we formulate a wave-turbulence description of the random wave into the basis of the eigenmodes of the waveguide, instead of the usual plane-wave Fourier basis relevant to statistically homogeneous random waves [31]. In addition, the recent experiments [9] evidenced that the effect of spatial beam self-cleaning can take place with a negligible temporal spectral broadening, so that in the first approximation we can neglect the temporal dynamics of the random wave. Neglecting furthermore polarization and Raman effects ($f_R = 0$), the NLS equation governing the evolution of the optical wave recovers the form

$$i\partial_z\psi = -\alpha\nabla^2\psi + V(\boldsymbol{x})\psi - \gamma|\psi|^2\psi, \tag{14.2}$$

where $\alpha = 1/(2k_L)$ and γ are, respectively, the diffraction and nonlinear coefficients, k_L being the laser wave number. $V(|\boldsymbol{x}|)$ denotes the effective potential due to the waveguide. For instance, for a graded-index multimode fiber, we have $V(|\boldsymbol{x}|) = qx^2$ if $|\boldsymbol{x}| \leqslant a$ and $V(|\boldsymbol{x}|) = V_0$, if $|\boldsymbol{x}| \geqslant a$, where $q = V_0/a^2$ [31]. Note that the finite depth of the potential $V_0 < \infty$ introduces an effective frequency cut-off for the classical wave. This is due to the fact that the nonlinear coupling among bounded and unbounded modes is negligible because of the poor spatial overlap of the corresponding modes [31]. We recall that the NLS equation conserves the power of the optical field, $N = \int|\psi|^2 d\boldsymbol{x}$, and the total energy (Hamiltonian) $H = E + U$, which has a linear contribution, $E = \int\alpha|\nabla\psi|^2\, d\boldsymbol{x} + \int V(\boldsymbol{x})|\psi|^2\, d\boldsymbol{x}$, and a nonlinear contribution, $U = -\frac{\gamma}{2}\int|\psi|^4\, d\boldsymbol{x}$.

14.3.1.1 Kinetic equation for the modal components

The kinetic equation describing the evolution of the modal components of the random wave can be derived in the weakly nonlinear regime, where $E \gg |U|$. In this strongly incoherent regime, linear diffraction effects dominate over nonlinear effects, which brings the statistics of the random wave close to Gaussianity, a feature that is important in order to achieve a closure of the infinite hierarchy of moment equations [26, 34].

As it propagates through the waveguide, the incoherent field $\psi(\boldsymbol{x},z)$ can be represented as a superposition of modal waves with random coefficients $c_m(z)$, which denote the respective modal occupancies:

$$\begin{aligned}\partial_z n_{\kappa}(z) = \frac{4\pi\gamma^2}{\beta_0^6}\iiint & d\kappa_1 d\kappa_2 d\kappa_3 \delta\left(\beta_{\kappa_1} + \beta_{\kappa_3} - \beta_{\kappa_2} - \beta_{\kappa}\right)|W_{\kappa\kappa_1\kappa_2\kappa_3}|^2 n_\kappa n_{\kappa_1} n_{\kappa_2} n_{\kappa_3} \\ & \times\left(n_\kappa^{-1} + n_{\kappa_2}^{-1} - n_{\kappa_1}^{-1} - n_{\kappa_3}^{-1}\right) + \frac{8\pi\gamma^2}{\beta_0^2}\int d\kappa_1 \delta\left(\beta_{\kappa_1} - \beta_\kappa\right)|U_{\kappa\kappa_1}(\mathbf{n})|^2 (n_{\kappa_1} - n_\kappa),\end{aligned} \tag{14.3}$$

where $U_{\kappa\kappa_1}(\mathbf{n}) = \frac{1}{\beta_0^2}\int d\kappa'\, W_{\kappa\kappa_1\kappa'\kappa'}\, n_{\kappa'}$. We assumed as usual that the modal occupancies $c_m(z)$ are random variables uncorrelated with each other, $\langle c_m(z)c_n^*(z)\rangle = n_m(z)\delta_{n,m}^K$. Note that, although this random phase approximation is usually justified in the presence of a large number of weakly interacting modes [34], here the presence of the parabolic potential has been shown to correlate the modes in the purely linear regime [36].

Also note that the kinetic equation (14.3) differs from the conventional WT kinetic equation in two respects: (i) the presence of the new second term in the right-hand side of equation (14.3), which enforces an isotropization of the mode occupancies amongst degenerate modes, (ii) the presence of the function $W_{\kappa\kappa_1\kappa_2\kappa_3}$ in the collision term that denotes the overlap among the modes.

Equilibrium Rayleigh–Jeans distribution in a MMF

The kinetic equation (14.3) conserves the power $N = \beta_0^{-2} \int d\kappa n_\kappa$ and the energy $E = \beta_0^{-2} \int d\kappa \beta_\kappa n_\kappa$. The irreversible property of equation (14.3) is expressed by a $H-$theorem of entropy growth, $d\mathcal{S}/dz \geqslant 0$, where the nonequilibrium entropy reads $\mathcal{S}(z) = \beta_0^{-2} \int d\kappa \ln(n_\kappa)$.

The Rayleigh–Jeans equilibrium state n_κ^{eq}, corresponding to the condition of maximum entropy subject to the constraints of conservation of E and N, is obtained by introducing the corresponding Lagrange's multipliers,

$$n_\kappa^{eq} = \frac{T}{\beta_\kappa - \mu}. \tag{14.4}$$

The temperature denotes the amount of energy $\mathcal{E}_\kappa$ that is equipartitioned among the modes of the waveguide, since in the tails of the equilibrium distribution (14.4) ($\beta_\kappa \gg |\mu|$), we have $\mathcal{E}_\kappa = \beta_\kappa n_\kappa^{eq} \sim T$. Also note that the equilibrium state (14.4) cancels both collisions terms of the kinetic equation (14.3): the system is statistically stationary once it has reached the most disordered equilibrium state.

The irreversible process of relaxation to the Rayleigh–Jeans thermal equilibrium has been confirmed by the numerical simulations of the NLS equation (14.2) with a truncated parabolic potential, as illustrated in figure 14.7(a) and (b). In the simulations we considered a graded-index multimode optical fiber with a radius of 15 μm, an index difference of $n_1 - n_0 = 10^{-3}$, and a refractive index $n_0 = 1.45$. With these parameters, the number of modes is $N_* = 66$. It is important to note that silica fibers exhibit a focusing nonlinearity, i.e. $\gamma > 0$ in equation (14.2). The incoherent beam may thus exhibit filamentation effects (i.e. speckle beam fragmentation) during its propagation in the fiber. However, as revealed by the numerical

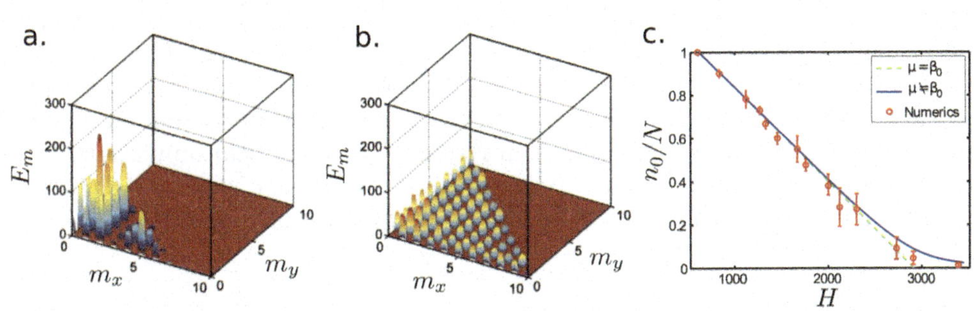

Figure 14.7. Condensation and thermalization in a GRIN fiber: Numerical simulation of the NLS equation (14.2), showing the establishment of energy equipartition among the modes of the waveguide: Energy per mode, $\mathcal{E}_m = \beta_m n_m$ versus the mode $m = (m_x, m_y)$ in the initial condition (a), and averaged over the propagation once the equilibrium state is reached, i.e. $\partial_z \mathcal{S} \simeq 0$ (b). Energy equipartition is almost established, except for the fundamental condensed mode $m_x = m_y = 0$. (c) Wave condensation: Fraction of power condensed in the fundamental mode at equilibrium, n_0/N, versus the energy H. The red points refer to the numerical simulations of the NLS equation (14.2) once the equilibrium state is reached. The continuous blue line refers to the theoretical condensation curve, while the dashed green line refers to the corresponding thermodynamic limit [$\tilde{\mu} = \mu - \beta_0 \to 0$]. For more details, see [31]. Reprinted figure with permission from [31]. Copyright 2011 by the American Physical Society.

simulations, the beam does not exhibit such fragmentation because we consider the weakly nonlinear regime of propagation, in which the linear energy dominates over the nonlinear energy, that is $|U|/E \ll 1$.

We note that the kinetic equation (14.3) is quite general and relevant to different types of waveguide configurations such as the GRIN MMF or a step-index MMF (see [31] for details).

14.3.1.2 Wave condensation

The Rayleigh–Jeans equilibrium distribution (14.4) exhibits a divergence for $\mu = \beta_0$, a property which entails condensation into the fundamental mode of the trapping potential, irrespective of the sign of the nonlinearity.

Starting from the Rayleigh-Jeans distribution, one can easily find a closed relation between the condensate fraction and the energy [31]:

$$\frac{n_0}{N} = 1 - \frac{E - E_0}{NV_0/2}, \tag{14.5}$$

where $E_0 = N\beta_0$ refers to the minimum energy, i.e. the energy of the field when all the power is condensed, $n_0/N = 1$. The condensate amplitude n_0/N increases as the energy E decreases, and condensation arises below the critical energy

$$E_c = E_0 + NV_0/2 = \frac{NV_0}{2}\left(1 + \frac{2\beta_0}{V_0}\right). \tag{14.6}$$

This expression deserves to be commented on in two respects. First, because of the truncation of the waveguide potential ($V_0 < \infty$), the value of E_c does not diverge to infinity. This is in contrast with the homogeneous problem $[V(\boldsymbol{x}) = 0]$, where the critical value of the energy behaves as $E_c \sim Nk_c^2/\ln(k_c)$, where $k_c = \pi/dx$ is the arbitrary frequency cut-off. In the continuous limit in which the spatial discretization of the NLS equation tends to zero, $dx \to 0$, the critical value of the energy E_c diverges to infinity (see, e.g. [27, 30]). A second point that could be remarked in equation (14.6) is that wave condensation is re-established in the thermodynamic limit in 2D [31], in analogy with quantum systems. The numerical simulations of the NLS equation have been found to be in good agreement with the theoretical condensation curve, as illustrated in figure 14.7(c) (for more details we refer the reader to [31]).

14.3.1.3 Perspectives

Wave condensation and beam cleaning: similarities and differences

As discussed in [9], the phenomenon of beam cleaning exhibits some qualitative properties that are apparently analogous to those of wave condensation. In particular, a non-negligible significant fraction of the power appears to concentrate in the low-order modes of the GRIN MMF, and such a cleaned beam remains immersed in a sea of incoherent weakly populated, higher-order modes. However, caution should be exercised when drawing definite conclusions in this respect. For instance, a characteristic feature of wave thermalization is the existence of a random phase dynamics among the modes of the fiber. Such a random phase regime should emerge spontaneously when a large number of modes interact in the weakly nonlinear regime. However, the presence of the parabolic potential is known to deeply affect light propagation, for instance by correlating the modes in the purely linear regime [36]. The parabolic potential may also substantially reduce the effective modal interaction, thus preserving a coherent phase dynamics. This was discussed in the Supplemental of [9] by elaborating a mean-field approach leading to a remarkable nonreciprocal reversible dynamics of the modes. Such a nonreciprocal dynamical behavior of the fundamental mode can explain the phenomenon of beam cleaning from a fundamentally different approach than wave condensation [9]. In addition, the condensation process should be characterized by an incoherent background component that exhibits an energy equipartition at thermal equilibrium. It would be difficult to confirm this property experimentally, since it would require an averaging (during propagation) of the modal occupancies once equilibrium is reached. It is also important to stress that the theory of wave condensation just discussed ignores any detrimental effect due to the spatial disorder during the propagation in the MMF. However, the impact of disorder due to birefringence fluctuations or fiber stretching is known to lead to random polarization coupling and random modal coupling [37], which are supposed to affect light propagation and thus wave condensation in MMFs. Work is in progress in order to elaborate a wave turbulence kinetic theory that accounts for disorder during the propagation.

Rogue waves near the transition to condensation?

A rather commonly accepted opinion is that rogue waves (RWs) can be conveniently interpreted in terms of exact analytical solutions of integrable nonlinear wave equations, the so-called Akhmediev breathers, or more specifically their limiting cases of infinite spatial and temporal periods, the rational soliton solutions such as the Peregrine and its higher-order solutions of the integrable 1D NLSE—see the recent reviews [38, 39]. Rational soliton solutions can be regarded as a coherent and deterministic approach to the understanding of RW phenomena. On the other hand, RWs are known to spontaneously emerge from an incoherent turbulent state [40–45]. This raises a difficult problem, since the description of the turbulent system requires a statistical WT approach, whereas rational soliton solutions are inherently coherent deterministic structures. This problem was addressed in the optical fiber context in [46, 47] by considering a specific NLSE model that exhibits a quasi-soliton turbulence scenario, a feature that can be interpreted in analogy with the previously discussed wave condensation phenomenon.

It was shown that the deterministic description of rogue wave events in terms of rational soliton solutions is not inconsistent with the corresponding statistical WT description of the turbulent system [47]. It is important to stress that the emergence of RW events was shown to solely occur nearby the transition to (quasi-)soliton condensation. From a different perspective, the fluctuations of the condensate fraction in 2D wave condensation have been recently computed theoretically, revealing that large fluctuations solely occur nearby the transition to condensation, whereas fluctuations are significantly quenched in the strongly condensed

Bogoliubov regime (small 'temperature') and almost completely suppressed in the weakly nonlinear turbulent regime (high 'temperature') [48]. This result is consistent with the general idea that, near second-order phase transitions, physical systems are inherently sensitive to perturbations and thus exhibit large fluctuations. One can then address a possible alternative point of view on the question of the spontaneous emergence of rogue waves from a conservative turbulent environment [49]: is it possible to interpret the sporadic emergence of RW events as the natural large fluctuations inherent to the phase transition to soliton condensation? This issue may pave the way for a statistical mechanics approach based on the idea of scaling and universal theory of critical phenomena, applied to the description of RW generation.

14.4 Spatiotemporal instabilities

Parametric wave-mixing effects in multimode fibers have been studied for four decades [50–52]. Nevertheless, not until very recently has it been experimentally observed that, by injecting an intense laser beam in a GRIN MMF, one generates a multi-octave spanning series of unequally spaced sidebands. These sidebands are phase-matched by the previously discussed dynamic long-period grating, induced via the Kerr effect from the periodic oscillations of the beam intensity owing to self-imaging. When using femtosecond multimode solitons as a pump propagating in the anomalous dispersion regime, spatiotemporal soliton oscillations lead to the generation of sideband series via the emission of dispersive waves, as discussed in [8].

Sideband series may also be observed with a quasi-CW pump propagating in the normal dispersion regime, as a result of the self-induced parametric instability associated with the dynamic index grating. Although the beam dynamics can be studied in the frame of guided modes, when the number of interacting modes becomes relatively large (say, greater than ten), it is useful to apply the approach proposed by Longhi in [53]. In particular, it is possible to carry out a linear stability analysis of the pump beam that is periodically evolving along the propagation coordinate. Note that the intensity to observe the effect of sideband generation is well above that of spatial beam self-cleaning. Since the periodic oscillations are due to the geometry of an otherwise longitudinally invariant waveguide, we dubbed this effect geometric parametric instability (GPI). The present type of parametric instability is fundamentally different from all other known types of instabilities. GPI in an MMF requires space–time coupling. As a first approximation, we may imagine that the waveguide acquires a light-induced modulation of the refractive index whose period permits the quasi phase matching between the pump and its primary sidebands. In a GRIN MMF, the self-imaging period reads as $\xi = \pi\rho/\sqrt{2\Delta}$, where ρ is the fiber core radius and Δ is the relative refractive index difference. For standard GRIN fibers, this self-imaging period is sub-millimetric and can provide the momentum for quasi-phase matching at several harmonic orders, as was illustrated by a series of experiments with sub-nanosecond pulses in [11].

The wave-vectors of pump k_P, and of the GPI Stokes k_S and anti-Stokes k_A sidebands are expected to satisfy the QPM condition $2k_P - k_S - k_A = -2\pi h/\xi$, where $h = 1, 2, 3, \ldots$. If we limit the frequency dependence of the refractive index to the contribution of chromatic dispersion at the pump wavelength (κ''), the generated GPI sidebands are detuned from the pump by a discrete set of resonant frequency offsets f_h, which satisfy the condition $(2\pi f_h)^2 = 2\pi h/(\xi\kappa'') - 2n_2\hat{I}\omega_0/(c\kappa'')$, where ω_0 is the angular frequency of the pump, and $\hat{I}$ is the path-averaged intensity of the beam. Following [53], the condition $h = 0$ reduces to the well-known MI of a uniform or non-oscillating CW beam, which requires anomalous group velocity dispersion $\kappa'' < 0$.

In the normal dispersion regime, GPI occurs for $h = 1, 2, 3, \ldots$. In experiments, resonant frequencies depend only weakly upon the quasi-CW beam intensity: this permits us to approximate well their values as $f_h \simeq \pm\sqrt{h}f_m$, where $2\pi f_m = \sqrt{2\pi/(\xi\kappa'')}$. With a standard GRIN MMF with a $\rho = 26$ μm, $n_{co} = 1.470$, $n_{cl} = 1.457$, and $\kappa'' = 16.55 \times 10^{-27}$ s^2 m^{-1} at the pump wavelength of 1064 nm, $\Delta = 8.8 \times 10^{-3}$, so that the self-imaging period is $\xi = 0.615$ mm. These values lead us to analytically predict an extremely large frequency detuning for the first resonant sideband, namely $f_1 = f_m \simeq 125$ THz. Such sideband shift is much larger than the value (8 THz) that was predicted by Longhi [53], owing to the different fiber parameters and pump wavelength. However, as we will see, theoretical predictions agree well with numerical simulations (124.5 THz) and, more importantly, with the experimental value of $f_m \simeq 123.5$ THz.

Note that GPI can be observed in principle both in the normal and in the anomalous dispersion regime. An example of optical spectrum from a 6 m long GRIN fiber is shown in figure 14.8. The inset of the same figure shows the photographic image of the spectral lines in the visible side of the spectrum, as it was obtained after a dispersive grating. To observe the spatial features at different wavelengths, we selected the spectral components by a bandpass filter, and then we analyzed the principal features of the spatial beam shape with a camera.

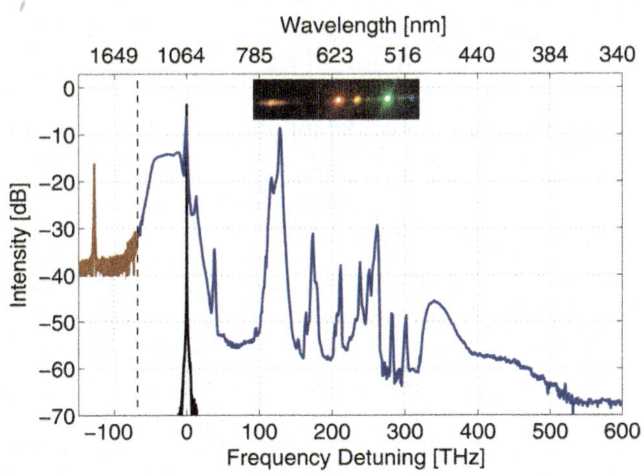

Figure 14.8. Experimental spectra in a 6 m long GRIN MMF with 50 kW of input power, obtained using two different spectrum analyzers. The inset shows an image of the visible part of the spectrum dispersed by a diffraction grating. The input pump spectrum is also reported for ease of comparison (black curve). Reprinted figure with permission from [11]. Copyright 2016 by the American Physical Society

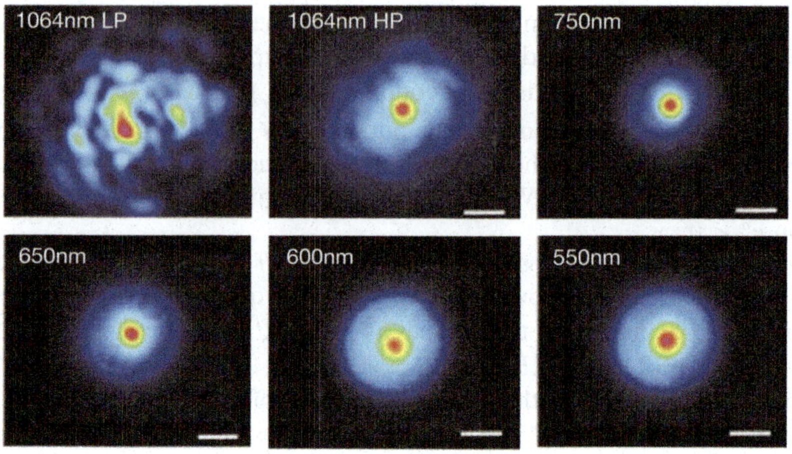

Figure 14.9. Experimental results of output intensity profiles (in linear scale) at the pump wavelength (1064 nm) for a low input power of 0.06 kW (LP) and at high input power (HP) at 50 kW; for the high-power case, the other panels show the beam profiles at 750, 650, 600, and 550 nm. Scale bar: 10 μm. Reprinted figure with permission from [11]. Copyright 2016 by the American Physical Society.

A collection of experimental results is shown in figure 14.9. Remarkably, looking at the first two panels on the top of figure 14.9, we can observe a net change (or self-cleaning) in the spatial beam shape when the input pump power grows from low to high values. In the high-power regime, parametric sidebands were also generated: figure 14.9 shows the spatial beam shape emerging from different sidebands, with the common feature of exhibiting a self-cleaned bell shape. This means that self-cleaning is frequency converted from the pump to the entire series of parametric sidebands. More recently, a similar spectral generation was observed in a GRIN fiber with only a few guided modes [14]. In this case, nonlinear mode interactions led to the generation of an anti-Stokes sideband at 625.2 nm, a wavelength such that the fiber is substantially more multimodal. In turn, the anti-Stokes wave acted as a secondary pump, thus generating its own series of sidebands. In figure 14.10 we show a set of spectra obtained for different values of the input peak power. Recently, this sideband generation has been also discussed as a form of self-organized instability [54].

14.5 Supercontinuum generation

As the peak power of the input laser beam pulses in a GRIN MMF is increased above the threshold for self-cleaning and then for the generation of GPI sidebands, the interplay between GPI and self- and cross-phase modulation acting together with SRS leads to wideband supercontinuum (SC) generation [12, 13].

Figure 14.11 reports the results of the experimental characterization of the optical spectrum at the output of a 14 m long GRIN MMF, when the peak power of the pump pulses is progressively increased. As is shown, by increasing the pump power above 25 kW, a SC develops at visible wavelengths until about 500 nm. In terms of SC spectral flatness, whenever the input beam peak power reaches 35 kW, the SC amplitude remains relatively uniform until about 700 nm, beyond which it rapidly

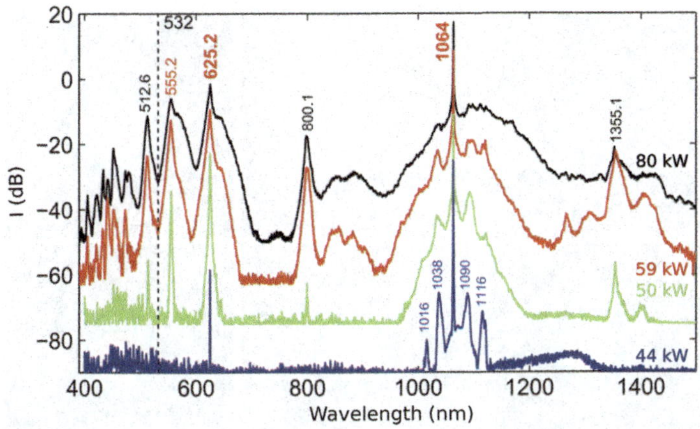

Figure 14.10. Experimental spectra from a few-mode 1 m long GRIN fiber for different pump peak powers. Reprinted figure with permission from [14].

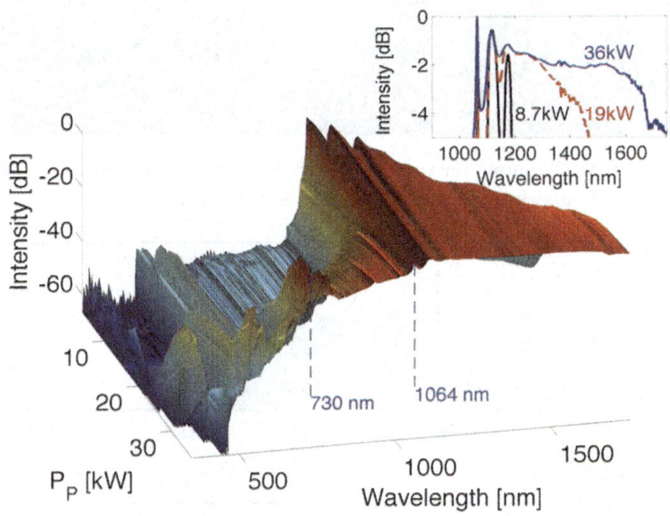

Figure 14.11. Experimental SC spectra versus pump pulse peak power P_p, for a fixed length of 14 m GRIN MMF. Inset shows that the spectral ripple is reduced to 1.1 dB at 1142 nm when increasing the power. Reprinted figure with permission from [13].

drops. Figure 14.11 also clearly shows that spectral broadening on the infrared (IR) side of the pump continuously grows with pump power: above 25 kW of peak power, the SC extends beyond 1750 nm and it exhibits a remarkable spectral flatness.

As shown in figure 14.12, the generated SC is not only spectrally flat, but most interestingly it is essentially carried by the same bell-shaped beam profile, which remains across all wavelengths very close to the beam waist of the fundamental mode.

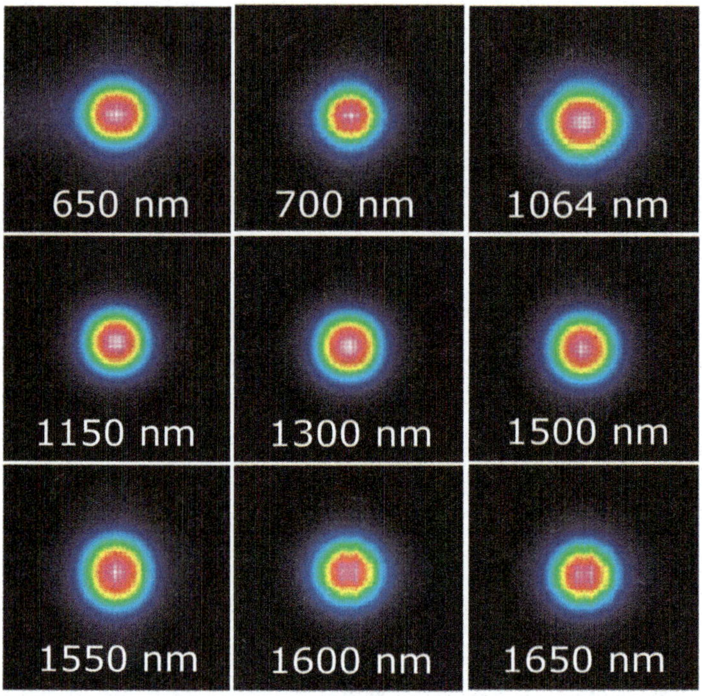

Figure 14.12. Far field at MMF output for L = 30 m, from different bandpass filters (10 nm bandwidth) and 1.4 W input average power. Reprinted figure with permission from [13].

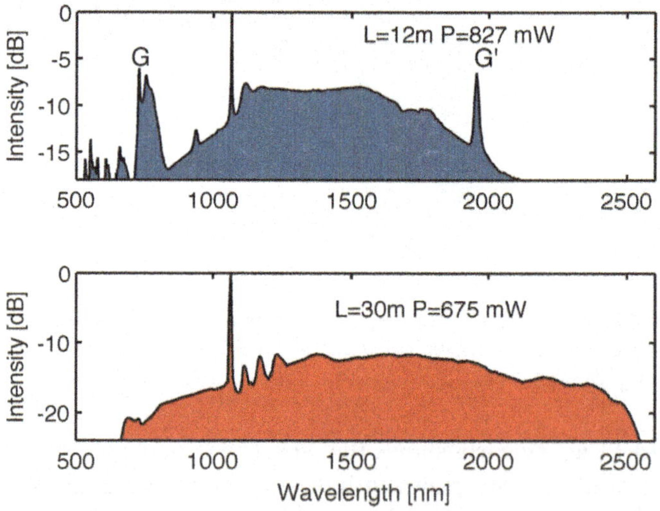

Figure 14.13. Supercontinuum spectral intensity from a 12 m (top) and 30 m long (bottom) GRIN MMF. Reprinted figure with permission from [13].

Figure 14.13 demonstrates that multimode SC may extend over multiple octaves, spanning from the visible till the MIR [13]. Here, we combined the output of the OSA and of the spectrometer, and we observed that as the MMF length increases up to 30 m, SC generation can be extended into the MIR (i.e. up to 2500 nm). The result of figure 14.13 is unexpected and significant, especially when taking into account the large linear absorption of the MMF in the MIR. By comparing the SC intensity profiles obtained for 12 m and 30 m, one can see that the SC bandwidth critically depends upon the MMF length. Moreover, figure 14.13 reveals that the isolated GPI sidebands in the visible domain are better discernible for MMF lengths of about 10 m or shorter (sideband G is at 730 nm and the corresponding Stokes G' is at 1954 nm), whereas SC is dominated by the interplay of Kerr-induced broadening and Raman scattering for MMF lengths above a few tens of meters.

A time-resolved analysis of different spectral regions of the SC, obtained by combining a diffraction grating and ultrafast photodiodes to cover the visible and infrared spectral range, has revealed that a significant (up to tenfold) temporal compression may occur in the GPI sidebands [13].

14.5.1 Second-harmonic generation in multimode fibers

Although silica fibers have mainly a cubic nonlinear susceptibility, the presence of dopants can provide a local microscopic quadratic nonlinear susceptibility that can be strongly exacerbated to produce a strong second-harmonic generation (SHG). One of the possible methods to enhance such quadratic nonlinear response in silica optical fiber is known as optical poling, and it is essentially based on a long exposure to a strong laser beam [55, 56]. The second-harmonic generation from optical poling in multimode GRIN fibers was studied experimentally and numerically in [15], showing new spectral and spatial dynamics. During the process of exposure to the strong laser

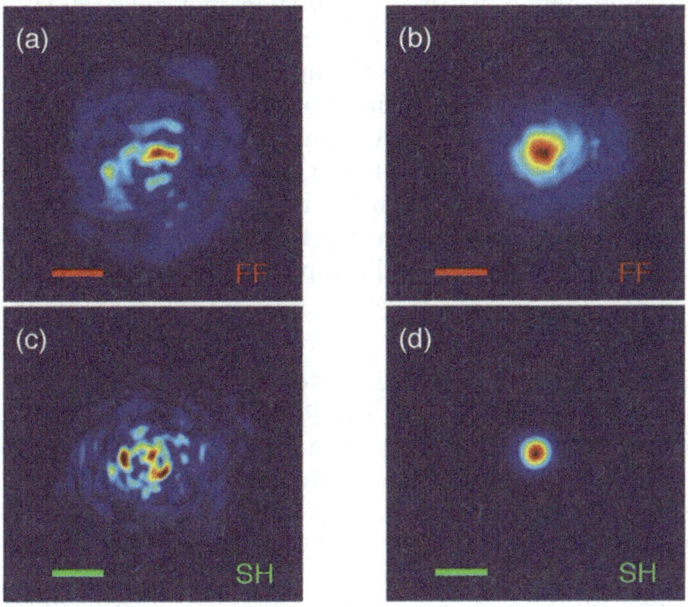

Figure 14.14. Experimental results of fundamental (FF) and second-harmonic (SH) beam shapes at the output of an optically poled MM-GRIN fiber. The pump peak power is 0.94 kW (a) and (c) and 39 kW (b) and (d). Reproduced from reference [15].

beam, which gradually increases the SHG by writing an equivalent appropriate longitudinal grating, the spatial distribution of the quadratic nonlinearity is drastically affected by the spatial Kerr self-cleaning effect of the fundamental beam. Consequently, the SHG reaches its maximum efficiency by assuming a spatial bell-shaped profile. An example of the spatial profiles of SHG at the output of a 2 m long GRIN fiber is shown in figure 14.14: panels (a) and (c) show the fundamental pump and the resulting second-harmonic beam with a relatively low input pump power of 0.94 kW. Panels (b) and (d) show instead the case of an input pump of 39 kW: both fundamental and second harmonic have a bell-shaped beam structure. When the fundamental beam power (the pump) grows larger and spectrally broadens, the SHG takes the form of a series of sharp spectral sidebands, whose presence can be explained by the quasi-phase matching enabled by the sub-millimetric periodic evolution of the intensity at the fundamental frequency. Experimental results and numerical simulations of these effects are provided in [15].

14.5.2 Conclusions

In summary, we have described the effect of spatial beam self-cleaning. In the presence of the Kerr effect, and for intensities above a certain threshold value, an intense optical beam propagating in a highly multimode GRIN MMF evolves into a robust bell-shaped transverse profile, close to the fundamental mode. First, we discussed the experimental characterization of this effect in a conservative propagation context. Next, we have presented experiments showing that self-cleaning may be also achieved in a step-index MMF, in the presence of strong linear loss or gain. A theoretical model was introduced, providing a statistical description that links self-cleaning to wave condensation, wave turbulence, and rogue wave phenomena. In the presence of spatiotemporal instabilities, which lead to multiple sideband series and supercontinuum, self-cleaning is effectively frequency converted across the entire output spectrum. We believe that Kerr beam self-cleaning may be exploited in a variety of technological applications where high power, multimode fiber light sources can be exploited, such as beam delivery for industrial processing and cleaning of cultural heritage, suppressing modal instabilities in large mode area fiber lasers, and the generation of ultrashort pulses via spatiotemporal pulse and beam shaping.

Acknowledgements

This work was supported by iXcore research foundation; ANR Labex Action; Italian Ministry of University and Research (MIUR) (PRIN2015KEZNYM); Ministry of Education and Science of the Russian Federation (14.Y26.31.0017); Horiba Medical and BPI-France; the European Union's Horizon 2020 research and innovation programme under the Marie Sklodowska-Curie grant agreement No. GA-2015-713694 ("BECLEAN" project). All authors contributed equally to this chapter.

References

[1] Berdagué S and Facq P 1982 *Appl. Opt.* **21** 1950–5

[2] Louradour F, Barthelemy A, Shaklan S and Reynaud F 1991 *Opt. Commun.* **82** 245–7

[3] Hellwig T, Schnack M, Walbaum T, Dobner S and Fallnich C 2014 *Opt. Express* **22** 24951–8
[4] Hasegawa A 1980 *Opt. Lett.* **5** 416–7
[5] Yu S S, Chien C H, Lai Y and Wang J 1995 *Opt. Commun.* **119** 167–70
[6] Renninger W H and Wise F W 2012 *Nat. Commun.* **4** 1719
[7] Wright L G, Christodoulides D N and Wise F W 2015 *Nat. Photon.* 1–5
[8] Wright L G, Wabnitz S, Christodoulides D N and Wise F W 2015 *Phys. Rev. Lett.* **115** 223902
[9] Krupa K, Tonello A, Shalaby B M, Fabert M, Barthélémy A, Millot G, Wabnitz S and Couderc V 2017 *Nat. Photon.* **11** 234–1
[10] Guenard R *et al* 2017 *Opt. Express* **25** 4783–92
[11] Krupa K, Tonello A, Barthélémy A, Couderc V, Shalaby B M, Bendahmane A, Millot G and Wabnitz S 2016 *Phys. Rev. Lett.* **116** 183901
[12] Lopez-Galmiche G, Eznaveh Z S, Eftekhar M A, Lopez J A, Wright L G, Wise F, Christodoulides D and Correa R A 2016 *Opt. Lett.* **41** 2553–6
[13] Krupa K *et al* 2016 *Opt. Lett.* **41** 5785–8
[14] Dupiol R *et al* 2017 *Opt. Lett.* **42** 1293–6
[15] Ceoldo D, Krupa K, Tonello A, Couderc V, Modotto D, Minoni U, Millot G and Wabnitz S 2017 *Opt. Lett.* **42** 971–4
[16] Garmire E, Chiao R Y and Townes C H 1966 *Phys. Rev. Lett.* **16** 347–9
[17] Shen Y R 1975 *Prog. Quantum Electron.* **4** 1–34
[18] Terry N B, Alley T G and Russell T H 2007 *Opt. Express* **15** 17509–19
[19] Krupa K, Tonello A, Shalaby B M, Fabert M, Barthélémy A, Millot G, Wabnitz S and Couderc V 2017 *Nat. Photon. (Supp. inf.)* **11** 234–41
[20] Wright L G, Liu Z, Nolan D A, Li M J, Christodoulides D N and Wise F W 2016 *Nat. Photon.* **10** 771–6
[21] Liu Z, Wright L G, Christodoulides D N and Wise F W 2016 *Opt. Lett.* **41** 3675–8
[22] Kutz J N, Eggleton B J, Stark J B and Slusher R E 1997 *IEEE J. Sel. Top. Quantum Electron* **3** 1232–45
[23] Trillo S and Wabnitz S 1986 *Appl. Phys. Lett.* **49** 752–4
[24] Svelto O 2014 *Principles of Lasers* (Springer)
[25] Gurnard R *et al* 2017 *Opt. Express* **25** 22219–27
[26] Newell A C, Nazarenko S and Biven L 2001 *Physica* D **152–153** 520–50
[27] Connaughton C, Josserand C, Picozzi A, Pomeau Y and Rica S 2005 *Phys. Rev. Lett.* **95** 263901
[28] Zakharov V and Nazarenko S 2005 *Physica* D **201** 203
[29] Nazarenko S and Onorato M 2006 *Physica* D **219** 1
[30] During G, Picozzi A and Rica S 2009 *Physica* D **238** 1524
[31] Aschieri P, Garnier J, Michel C, Doya V and Picozzi A 2011 *Phys. Rev.* A **83** 033838
[32] Picozzi A, Garnier J, Hansson T, Suret P, Randoux S, Millot G and Christodoulides D 2014 *Phys. Rep.* **542** 1–132
[33] Nazarenko S, Onorato M and Proment D 2014 *Phys. Rev.* A **90** 013624
[34] Zakharov V E, L'vov V S and Falkovich G 1992 *Kolmogorov Spectra of Turbulence I* (Berlin: Springer)
[35] Nazarenko S 2011 *Wave Trubulence* (Lecture Notes in Physics 825) (Springer)
[36] Fusaro A, Garnier J, Xu G, Conti C, Faccio D, Trillo S and Picozzi A 2017 *Phys. Rev.* A **95** 063818

[37] Sami Mumtaz S, Essiambre R J and Agrawal G P 2013 *Nat. Photon.* **31** 398

[38] Onorato M, Residori S, Bortolozzo U, Montina A and Arecchi F 2013 *Phys. Rep.* **528** 47–90

[39] Dudley J M, Dias F, Erkintalo M and Genty G 2014 *Nat. Photon.* **8** 755–64

[40] Onorato M, Osborne A, Serio M, Cavaleri L, Brandini C and Stansberg C 2006 *Eur. J. Mech.* B **25** 586–601

[41] Akhmediev N, Soto-Crespo J and Ankiewicz A 2009 *Phys. Lett.* A **373** 2137–45

[42] Ankiewicz A, Soto-Crespo J and Akhmediev N 2010 *Phys. Rev.* E **81** 046602

[43] Viotti C, Dutykh D, Dudley J M and Dias F 2013 *Phys. Rev.* E **87** 063001

[44] Walczak P, Randoux S and Suret P 2015 *Phys. Rev. Lett.* **114** 143903

[45] Randoux S, Walczak P, Onorato M and Suret P 2014 *Phys. Rev. Lett.* **113** 113902

[46] Hammani K, Kibler B, Finot C and Picozzi A 2010 *Phys. Lett.* A **374** 3585

[47] Kibler B, Hammani K, Finot C and Picozzi A 2011 *Phys. Lett.* A **375** 3149

[48] Picozzi A and Rica S 2012 *Opt. Commun.* **285** 5440

[49] Akhmediev N *et al* 2016 *J. Opt.* **18** 063001

[50] Stolen R 1975 *IEEE J. Quantum Electron.* **11** 100–3

[51] Hill K O, Johnson D C and Kawasaki B S 1981 *Appl. Opt.* **20** 1075–9

[52] Mafi A 2012 *J. Lightwave Technol.* **30** 2803–11

[53] Longhi S 2003 *Opt. Lett.* **28** 2363–5

[54] Wright L G, Liu Z, Nolan D A, Li M J, Christodoulides D N and Wise F W 2016 *Nat. Photon.* 771–6

[55] Österberg U and Margulis W 1986 *Opt. Lett.* **11** 516–8

[56] Stolen R H and Tom H W K 1987 *Opt. Lett.* **12** 585–7

IOP Publishing

Nonlinear Guided Wave Optics
A testbed for extreme waves
Stefan Wabnitz

Chapter 15

Noise-initiated dynamics in nonlinear fiber optics

Shanti Toenger, Mikko Närhi, Goëry Genty and John M Dudley

15.1 Introduction

There is much continued interest in the connections between nonlinear instabilities in physics, and the emergence of extreme events or rogue-wave-type events. In optics, much of this has been motivated by recent developments of ultrafast measurement techniques which have provided new windows into studying the underlying non-repetitive dynamics of optical signals. A particular class of instability that has attracted significant attention is the Benjamin–Feir or modulation instability (MI) associated with the exponential growth of a noise perturbation on a continuous or plane wave background. The MI results in the emergence of high-contrast random structures and a rich variety of chaotic dynamics. This chapter presents a review of a selection of topics in this field, beginning with a short summary of MI dynamical behaviour in the spectral and temporal domains, with particular emphasis on the interpretation in terms of analytical breather structures. This is followed by a brief review of ultrafast measurement techniques, with special emphasis on techniques that can operate in real-time, including the dispersive Fourier transform and the time lens. We conclude with a discussion of the latest experimental developments that have used the dispersive Fourier transform and time lens to capture real-time dynamics of breather structures generated from noise-seeded MI.

15.2 Modulation instability and breather solutions

We base our discussion of MI considering the cubic nonlinear Schrödinger equation (NLSE), which describes propagation in optics subject to second-order (group velocity) dispersion and third-order (Kerr) nonlinearity. The NLSE can be written in dimensionless form as:

 © IOP Publishing Ltd 2017

$$i\frac{\partial\psi}{\partial\xi} + \frac{1}{2}\frac{\partial^2\psi}{\partial\tau^2} + |\psi|^2\psi = 0, \tag{15.1}$$

with the complex envelope $\psi(\xi,\tau)$ evolving as a function of propagation distance ξ and co-moving time τ. The second and third term of the equation represent the effect of dispersion and nonlinearity, respectively. The NLSE is integrable and can be solved using the inverse scattering transform. Since the pioneering work of Zakharov and Shabat [1], there has been continued interest in the mathematics and physics of the analytical solutions of the NLSE [2–5]. Although the most celebrated NLSE solution is the propagation-invariant bright sech-soliton, there are in fact a much larger number of (families of) solutions. Among these, breathers represent a special class of solutions, in the form of localised nonlinear structures evolving on top of a continuous background, and are also commonly referred to as 'soliton on finite background' (SFB) solutions. The SFB solutions of the *dimensionless* NLSE (equation (15.1)) can be written in the form [2, 6]

$$\psi(\xi, \tau) = \left[1 + \frac{2(1-2a)\cosh(b\xi) + ib\sinh(b\xi)}{\sqrt{2a}\cos(\omega\tau) - \cosh(b\xi)}\right]e^{i\xi}, \tag{15.2}$$

where the second term of the function corresponds to the dynamical structure evolving on top of the background represented by the first term. The properties of the breather solution family are determined by a single positive parameter a ($a \neq 0.5$), through arguments $b = \sqrt{8a(1-2a)}$ and $\omega = 2\sqrt{1-2a}$.

Figure 15.1(a)–(c) shows how the parameter a influences the localisation properties of SFBs. Specifically, for $0 < a < 0.5$, equation (15.2) describes the Akhmediev

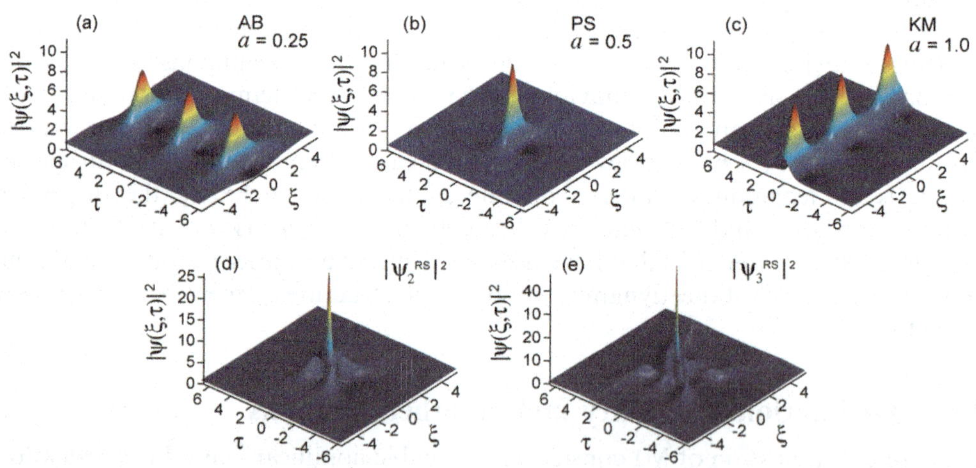

Figure 15.1. SFB solutions for different values of the parameter a as indicated: (a) Akhmediev breather (AB), (b) Peregrine soliton (PS), (c) Kuznetsov–Ma (KM) soliton, (d) and (e) second- and third-order rational solutions corresponding to the nonlinear superposition (or collision) of two and three Peregrine solitons, respectively.

breather (AB), a temporally periodic pulse train that undergoes a single growth–decay cycle along the longitudinal propagation ξ [6, 7]. In the context of nonlinear fiber optics, it is the initial exponential growth phase of the AB solution that describes the evolution of a weak perturbation in the framework of the linear stability analysis of MI, where the real parameters ω and b correspond to the modulation frequency of the perturbed plane wave and instability growth rate, respectively. The MI gain curve is plotted in figure 15.2(a) as a function of modulation frequency ω and parameter a. Maximum growth rate is obtained for $a = 0.25$ ($\omega = \sqrt{2}$).

We now focus on the properties of ABs when $0 < a < 0.5$, since it is these particular solutions that are central to interpreting the evolution of a noisy MI field (as discussed in the next section). As the parameter a increases, the modulation frequency ω decreases and the temporal period of the AB pulse train increases according to

$$\Delta\tau = 2\pi/\omega = \pi/\sqrt{1 - 2a}\,. \tag{15.3}$$

The increase in the temporal period is accompanied by a narrowing of the temporal width $\delta\tau$ and increase in the peak intensity $|\psi|^2_{\max}$, as shown in figure 15.2(b). The temporal width (full width at half maximum (FWHM)) $\delta\tau$ is given by

$$\delta\tau = \frac{2}{\omega}\cos^{-1}\left[\frac{1}{\sqrt{2a}}\left(1 - \frac{2(1 - 2a)}{\sqrt{|\psi|^2_{\rm HM} + 1}}\right)\right], \tag{15.4}$$

where $|\psi|^2_{\rm HM}$ is the half-maximum AB intensity. For $a \leqslant 0.125$,

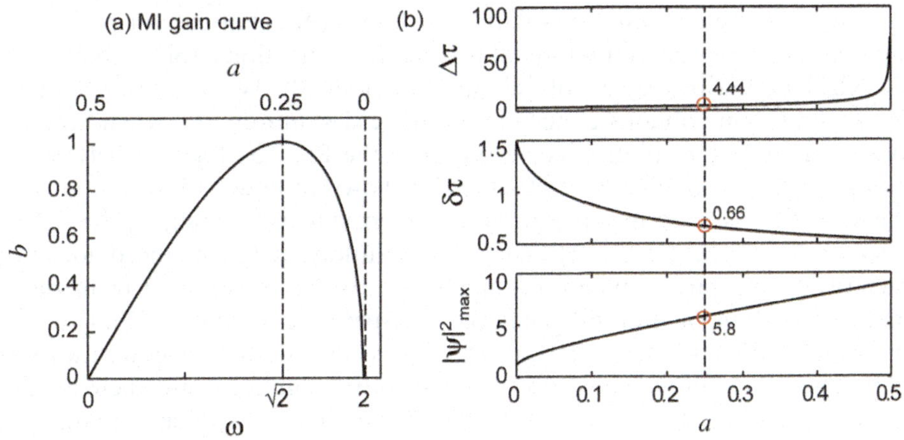

Figure 15.2. (a) Modulation instability gain as a function of modulation frequency ω and parameter a. The dashed lines indicate the modulation frequency at maximum MI gain ($a = 0.25$) and the cut-off frequency. (b) Temporal period $\Delta\tau$, pulse duration $\delta\tau$, and maximum intensity of AB structures as a function of parameter a. The red circles indicate the values of these quantities at the maximum of MI gain (corresponding to $a = 0.25$).

$$|\psi|^2_{\mathrm{HM}} = 8a + 1,$$

while for $a \geqslant 0.125$,

$$|\psi|^2_{\mathrm{HM}} = \frac{1}{2}\left(2\sqrt{2a} + 1\right)^2.$$

The maximum intensity $|\psi|^2_{\max}$ is reached at distance $\xi = 0$ and time $\tau = 0$:

$$|\psi|^2_{\max} = |\,\psi(0{,}0)\,|^2 = \left(2\sqrt{2a} + 1\right)^2. \tag{15.5}$$

In the limit $a \to 0.5$ ($\omega \to 0$), the solution in equation (15.2) asymptotically tends towards the rational Peregrine soliton (PS) [8]

$$\psi_{\mathrm{PS}}(\xi, \tau) = \left[1 - \frac{4(1 + 2i\xi)}{1 + 4\tau^2 + 4\xi^2}\right]e^{i\xi}.$$

The PS solution is a peculiar structure that is fully localised in time τ and space ξ. It possesses the narrowest temporal width $\delta\tau_{\mathrm{PS}} \approx 0.53$ and the highest intensity amongst all the family of elementary AB solutions $|\psi_1^{\mathrm{RS}}|^2_{\max} = |\psi_{\mathrm{PS}}|^2_{\max} = 9$ [7, 9].

For $a > 0.5$, equation (15.2) remains a solution of the NLSE, but the physical nature of the solution changes as circular trigonometric functions become hyperbolic and vice versa (as the parameters b and ω become imaginary). In this case, the solution corresponds to the Kuznetsov–Ma (KM) soliton, which shows localisation in time τ but periodicity along ξ with spatial modulation period given by $\Delta\xi = 2\pi/(ib) = \pi/\sqrt{2a(2a - 1)}$ [10, 11]. As a continues to increase above 0.5, the spatial period $\Delta\xi$ decreases to zero (i.e. no periodicity at all) as $a \to \infty$, and approaches the fundamental sech-soliton solution (a pulse propagating in an invariant way along ξ).

Each of the solutions described by equation (15.2) is a limiting case of a more general two-parameter family of solutions of the NLSE that exhibit periodicity in both transverse time τ and longitudinal propagation direction ξ (bi-periodic solution of the NLSE) described by Jacobi elliptic functions [2]. When one of the periods tends to infinity, one obtains either the AB or KM solution, and when both of the periods are infinite the solution corresponds to the PS. The relation between these solutions can be schematically represented as shown in figure 15.3.

The solutions just described are also often referred to as 'elementary' or first-order SFB solutions of the NLSE. Higher-order solutions with enhanced localisation characteristics and larger maximum intensities can be constructed from the nonlinear superposition of multiple elementary bi-periodic solutions. These solutions can be generated using mathematical techniques such as the Darboux transformation [2]. The exact functional form of higher-order solutions and their maximum intensity depend on the properties of the elementary constituent breathers. For example, the properties of the higher-order solution (or higher-order breather) resulting from the nonlinear superposition or collision of elementary ABs vary as a function of the modulation frequency, 'velocity' (the angle formed by the line of maxima of the AB with respect to the *co-moving* time τ-axis), and the relative phase

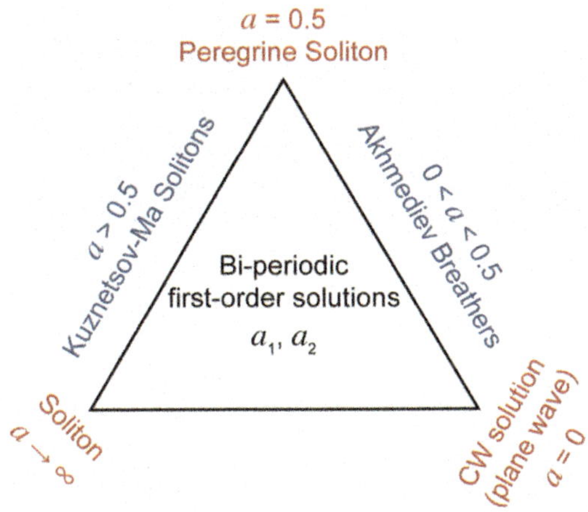

Figure 15.3. Schematic representation of the interrelation between various first-order solutions of the NLSE (adapted from reference [12]).

shift (the difference in their spatiotemporal center position (ξ_0, τ_0)) of the constituent elementary ABs [12]. Following the mathematical formulation of reference [12], each constituent AB can be described by an arbitrary complex number (within certain limits) $l = \mu + i\nu$, whose real part μ corresponds to the 'velocity' and imaginary part ν corresponds to the modulation frequency related to a by $\nu = \sqrt{2a}$). The AB solutions with $0 < a < 0.5$ then correspond to $0 < \nu < 1$.

The maximum intensity of a higher-order breather generally increases with the values of the parameter ν of the colliding ABs. Concretely, this means that a higher-order breather resulting from the collisions of two ABs with $l = 0.8$ will generally exhibit a higher maximum intensity as compared to the collision of two ABs with $l = 0.7$. However (and although this is often the case), one should be careful, as the maximum intensity of a breather collision is also affected by the relative phase (or spatiotemporal centers) of the elementary ABs, which may reduce significantly the maximum intensity at the collision point.

In order to illustrate the subtle role of the constituent AB parameters in the collision dynamics, figure 15.4 shows several examples of second-order breathers ψ_2, all with a maximum intensity $|\psi|^2_{\rm HM}$ around 15 at the point of collision but constructed from two ABs with different parameters. For each case, the spatiotemporal intensity profile is plotted in the top panel and the temporal cross section taken at the collision point is plotted in the bottom panel. The parameters used for each plot, together with the peak intensities $|\psi|^2_{\rm HM}$ and the temporal widths of the corresponding intensity peaks (FWHM) are given in table 15.1. The collision of two ABs with different modulation frequencies, perfectly in phase and with zero velocity, is shown in figures 15.4(a) and (b). One can see how the dynamics are nearly identical but also that, although the maximum intensities are practically the same, there is a small discrepancy in the temporal width at the point of collision. The effects of relative phase and velocity of the constituent ABs can be observed in figures 15.4(c)–(f); (c) and (d) plot two second-order breathers formed by the same

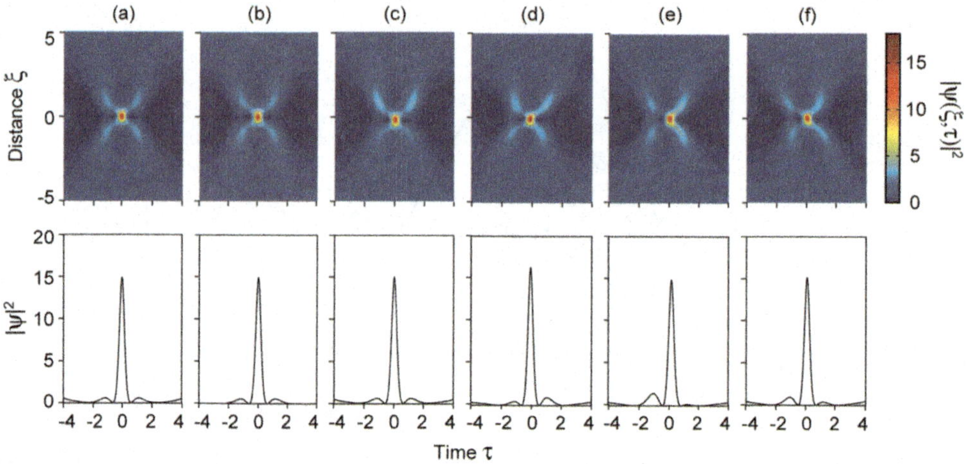

Figure 15.4. Selected examples of second-order breather structures corresponding to different combinations of two AB collisions with parameters as shown in table 15.1. The maximum intensity in all cases is close to 15.

Table 15.1. Different combinations of parameters corresponding to the second-order breathers shown in figure 15.4.

Figure	ν_1	ν_2	μ_1	μ_2	ξ_{01}	ξ_{02}	τ_{01}	τ_{02}	$\lvert\psi\rvert^2_{\mathrm{HM}}$	$\delta\tau$
(a)	0.7	0.73	0	0	0	0	0	0	14.9	0.395
(b)	0.5	0.93	0	0	0	0	0	0	14.9	0.397
(c)	0.7	0.77	0	0	0	$\pi/100$	0	0	15	0.394
(d)	0.7	0.77	0	0.2	0	$\pi/100$	0	0	15.2	0.391
(e)	0.7	0.84	0	0	0	0	0	$\pi/100$	14.9	0.398
(f)	0.7	0.84	0	−0.2	0	0	0	$\pi/100$	16.3	0.377

modulation frequency combination, but one is slightly shifted in space in (c), while in (d) one is slightly shifted in space with a nonzero velocity component. In (c), one can see that a relative phase in space leads to some asymmetry in the longitudinal dynamics, while the presence of a velocity component in (d) also adds some asymmetry in the transverse time dimension. Similarly, (e) and (f) illustrate the effect of a relative phase in time (for an identical modulation frequency combination), where in (e) one AB is simply shifted in time while, in (f), one AB is shifted with a nonzero velocity at the same time. In (e), one can see asymmetry in the time dimension, while in (f) there is some asymmetry both in space and time. As we shall see in the next section, the fact that collision dynamics depend sensitively on the elementary ABs leads to a 'band' formation observed when plotting the temporal width versus maximum intensity for the structures present in a random MI field emerging from noise.

The maximum possible intensity of higher-order solutions corresponds to the nonlinear superposition of multiple Peregrine solitons with zero-phase difference. In this case, the solutions can be written in a rational form

$$\psi_n^{\mathrm{RS}} = \left[(-1)^n + \frac{G_n(\xi, \tau) + i\xi H_n(\xi, \tau)}{D_n(\xi, \tau)}\right]e^{i\xi},$$

where G, H, and D are polynomials [9]. The maximum intensity of the nth order rational solution is given by $|\psi_n^{\mathrm{RS}}|^2_{\max} = (2n + 1)^2$, where the order n signifies the number of elementary constituent solitons, giving $|\psi_1^{\mathrm{RS}}|^2_{\max} = |\psi_{\mathrm{PS}}|^2_{\max} = 9$, $|\psi_2^{\mathrm{RS}}|^2_{\max} = 25$, and $|\psi|^3_{\mathrm{RS}}|^2_{\max} = 49$ for $n = 1$, $n = 2$, and $n = 3$, respectively. Figures 15.1(d) and (e) show the three-dimensional plots of these higher-order solutions.

15.3 Noise-driven modulation instability

Having introduced in the previous section the analytical framework of solitons on finite background solutions of the NLSE, we next study in more detail specific features associated with MI when it is seeded by noise rather than a well-defined periodic modulation. To this end, figure 15.5 shows the evolution of a noise-perturbed plane wave in the NLSE with initial condition $\psi(0, \tau) = 1 + \eta(0, \tau)$ ($\eta \ll 1$). The field is propagated up to $\xi_{\max} = 343\,750$ (using 6 250 000 ξ-points), with a temporal window extending to $\tau_{\max} = \pm 69.42$ (using 2048 τ-points), yielding a total of 2 853 669 spontaneously emergent localised structures with typically 50 temporal points across each peak.

We first analyse the time-domain evolution map in figure 15.5(a). We can see how the spatiotemporal localised structures spontaneously emerge from the noisy input plane wave due to MI. When they first emerge, the localised structures take the appearance of a quasi-periodic temporal pulse train around $\xi \approx 10$, but subsequently evolve into a chaotic field with further propagation, where one observes various forms of interaction (merging, splitting, pseudo-periodic evolution) between the individual structures taking place at random space and time.

Interestingly, although the localised structures emerge from noise, their temporal and spatial characteristics reveal clear signatures of the ideal analytic SFB solutions described previously. For example, at the initial point of emergence around $\xi \approx 10$, the average temporal periodicity from the simulations is $\Delta\tau \approx 4.44$, which corresponds to the calculated AB periodicity at maximum MI gain (at $a = 0.25$). Similarly, the average temporal width and maximum intensity of these AB-like structures are in close correspondence with those calculated before (see figure 15.2(b)). In order to establish a direct comparison between the emergent noisy MI structures and the analytic SFB solutions, figure 15.6 plots the intensity profiles from different regions in the spatiotemporal MI field shown in figure 15.5(a). One can readily identify in the evolution map specific regions where the local properties of the MI field bear close resemblance to the analytic SFB solutions (AB, PS, KM, and AB collisions), as shown in figure 15.6. Of course, exact agreement is not to be expected given the random initial conditions, but these results very clearly show how the analytic solutions map very closely to the noise-generated structures.

More convincing evidence of the correspondence between the spontaneously emerging and seeded MI structures can be seen when comparing the temporal widths (FWHM) and the intensity maxima of the spontaneously generated breathers with

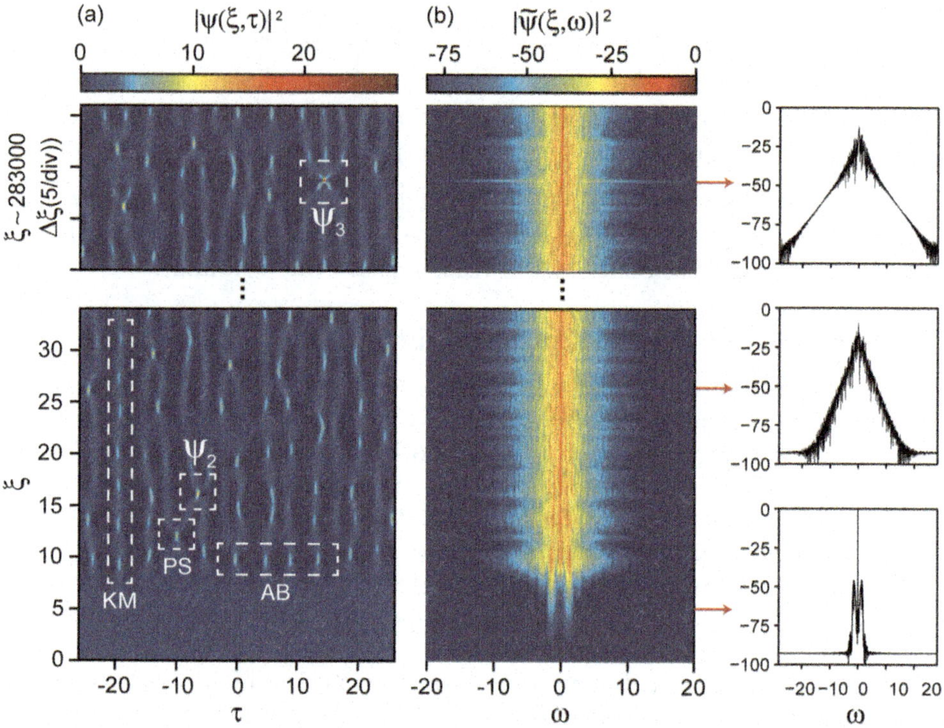

Figure 15.5. (a) Density map showing a selected portion of the long-term temporal evolution of a spatiotemporal MI field triggered by one photon per mode noise superimposed on top of a plane wave background. Local regions highlighted by white dashed lines correspond to the intensity profiles shown in figure 15.6. (b) Corresponding evolution in the spectral domain. Bottom subfigures plot the evolution from $\xi = 0$ to $\xi = 34$; top subfigures plot evolution over a range around $\xi \approx 283\,000$. The right panels show the spectrum at different distances ξ along the propagation in the initial stage at $\xi = 5.5$, when MI is fully developed at $\xi = 26.3$, and at the point of three AB collisions at $\xi = 283211.5$. The noise field $\eta(0, \tau)$ was computed from the inverse Fourier transform of a broadband noise of one photon per mode spectral density, with random phase as is commonly done in fiber-optics [13].

those expected for ideal SFB solutions. The results are shown in figure 15.7 for $\approx 3 \times 10^6$ intensity maxima detected in the random MI field. The scatter plot (grey circles) shows the temporal FWHM plotted against the corresponding intensity maxima. The circles span a continuous range of intensities up to a maximum of ≈ 28 clustering around the theoretical curve for analytical AB solutions (elementary and higher-order). As already mentioned briefly at the end of the preceding section, the scatter about the calculated curve is to be expected since the calculated higher-order solutions assume only one combination of constructing elementary breathers at each intensity maximum (and under the assumption of zero velocity and zero relative phase difference of constituent ABs for the higher-order solutions). For the elementary breather solutions, the observed scatter arises mainly from the fact that the spontaneously emergent breathers are in general not perfectly symmetric, as in the case for the analytical solutions considered here. In the higher-order breather regime, the observed scatter arises from different combinations of elementary ABs,

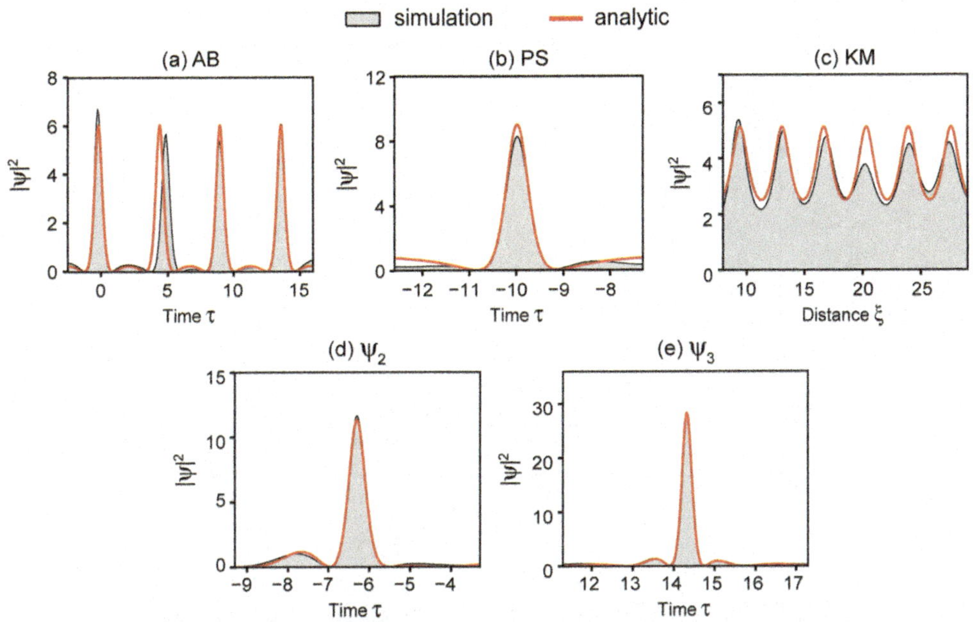

Figure 15.6. The grey shaded plots show the intensity profiles extracted from the regions of the random MI field indicated in figure 15.5 for an AB, PS, KM, a two-AB collision ψ_2, and a three-AB collision ψ_3, respectively. The red solid line shows the corresponding ideal analytical solutions for comparison.

leading to collisions with similar maximum intensity characteristics but different temporal widths (see also figure 15.4 and table 15.1 for some examples). Additionally, one can also compare the scatter plot and theoretical curve with a power function fit $\delta\tau_{\rm fit} = 1.678/(|\psi|^2_{\rm max})^{0.532}$ (blue dotted-dashed line) that describes fairly accurately the relation between the temporal width and the intensity maximum of SFB solutions both in the seeded and spontaneous cases.

15.4 Measuring chaotic dynamics in real time

Characterizing the average spectral and temporal intensity of optical fields with high precision is nowadays relatively straightforward. In the spectral domain, diffraction-grating spectrographs, Fourier transform infrared spectroscopy, or optical spectrum analysers are commonly used, allowing for sub-nm resolution. In the time domain, autocorrelation techniques or electro-optic sampling can characterise temporal intensity profiles on the sub-picosecond time scale. More advanced techniques in the form of frequency-resolved optical gating [14], spectral phase interferometry for direct electric-field reconstruction [15], dispersion scan [16], or their variants [17 can characterise the complex electric field (amplitude and phase) with temporal resolutions that can reach the attosecond level [18], but these techniques also typically require averaging over a large number of temporal periods or shots. Single-shot versions do exist [19–24], but these usually have rather stringent limitations in terms of pulse complexity (time–bandwidth product), or require temporal-to-spatial encoding or multiple beam geometries that make experimental

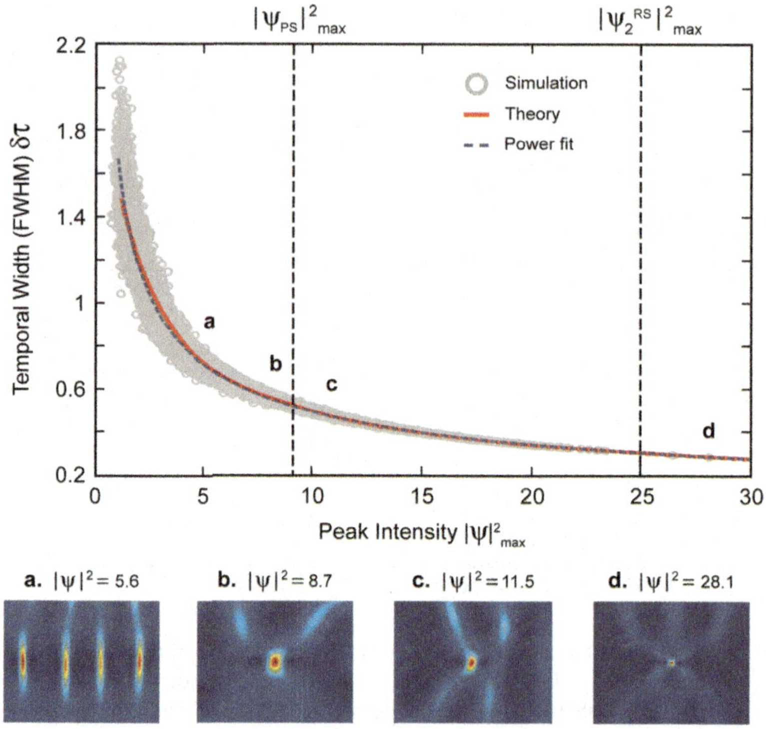

Figure 15.7. Scatter plot of temporal width (FWHM) versus intensity maxima for the 2 853 669 localised structures in the noisy MI field from simulations (grey circles) compared with theoretical predictions for the properties of SFBs (red solid line). A comparison with a power function fit $\delta\tau_{\text{fit}} = 1.678/(|\psi|^2_{\text{max}})^{0.532}$ is also shown (blue dotted-dashed line). The intensity maxima of the spontaneously emergent localised structures are obtained using peak detection over a full two-dimensional spatiotemporal computational window with an 8-connected neighbourhood regional maximum search and a minimum relative threshold of $|\psi|^2 = 1$ (in order to avoid counting subsidiary maxima of primary peaks). For the theoretical prediction, the temporal period $\delta\tau$ and intensity peak are calculated from equations (15.4) and (15.5) for intensities <9 and, for intensities >9, the temporal period and maximum intensity are extracted from numerically constructed higher-order solutions using the Darboux transformation. The bottom panel plots several examples of spatiotemporal structures of the spontaneously emerging breathers to highlight the nature of the breathers typically found around the region. These breathers correspond to the temporal intensity profiles plotted in figure 15.6(a,b,d,e).

systems very complex. They are also generally limited in terms of sensitivity and acquisition speed from hundreds of Hz to a few kHz, which makes them unpractical for measuring the temporal structure of non-repetitive rapidly varying waveforms with features on the femtosecond timescale. A somewhat better approach for temporal (intensity) measurement of non-repetitive ultrafast optical waveforms is a so-called temporal imaging system based on a time lens [25], where the time base of the waveform to be measured is stretched—just as a conventional lens magnifies the spatial dimensions of an object. Magnification of the waveform then allows for the use of ultrafast detectors with tens of GHz bandwidth to obtain sub-picosecond resolution. Measurement of the spectral envelope/spectrum of non-repetitive waveforms, on the other hand, can be performed with sub-nm resolution using spectrum-to-time transformation also commonly referred to as the dispersive Fourier transform (DFT) [26, 27] operating at speeds in excess of GHz. Before we examine in more detail the performance of these systems in the particular context of noise-seeded modulation instability dynamics, we next give a brief overview of the operation principles behind DFT and TL systems.

Both DFT and TL systems operate on the same principle of space–time duality, which reflects the analogy between diffractive propagation of light beams in space and dispersive propagation of short pulses in time. This analogy has been known for many years, and is based on the mathematical equivalence of the propagation equation expressing both phenomena [28–32]. In the paraxial approximation, the electric field of the monochromatic plane wave can be written as $\mathbf{E}(x, y, z, t) = E(x, y, z)\exp[i(\omega_0 t - kz)]$, where $E(x, y, z)$ is a slowly varying envelope function of the transverse beam profile. A direct comparison with dispersion can be easily performed by considering only one-dimensional diffraction, in which case the propagation of the spatial envelope is governed by

$$\frac{\partial E}{\partial z} = \frac{i}{2k}\frac{\partial^2 E}{\partial x^2},$$

where k is the wave number of the propagating wave. Solving this equation in the transverse spatial frequency domain, one obtains

$$\tilde{E}(k_x, z) = \tilde{E}(k_x, 0)\exp\left(\frac{ik_x^2 z}{2k}\right). \tag{15.6}$$

The above equation shows that diffraction can be seen as the result of a quadratic phase-shift imparted onto the transverse k-vector upon propagation.

In the case of narrowband dispersion, the electric field associated with a plane wave can be written as $\mathbf{E}(x, y, z, t) = A(z, t)\exp[i(\omega_0 t - \beta_0 z)]$, where $A(z, t)$ is a slowly varying envelope function of the travelling wave and β_0 is the wave number of the propagation constant at the wave angular frequency ω_0. It can then be shown that the propagation of the wave envelope is governed by

$$\frac{\partial A}{\partial z} = \frac{i\beta_2}{2}\frac{\partial^2 A}{\partial T^2},$$

where β_2 is the group velocity dispersion (GVD) and $T = t - z/v_g$ is the co-moving frame at the velocity of the envelope. Solving this equation in the frequency domain, one obtains

$$\tilde{A}(\Omega, z) = \tilde{A}(\Omega, 0)\exp\left(\frac{-i\Omega^2\beta_2 z}{2}\right), \tag{15.7}$$

where $\Omega = \omega - \omega_0$ is the baseband angular frequency. Similar to the spatial domain, this equation shows that dispersion can be seen as the result of quadratic phase-shift imparted onto the spectral envelope upon propagation.

Comparing equations (15.6) and (15.7), we can see that the scaled propagation distance z/k for diffraction is analogous to the product of the GVD parameter and the propagation distance $\beta_2 z$, corresponding to the total group delay dispersion (GDD) D_2. This equivalence of the mathematical expressions shows that diffractive propagation in spatial imaging can indeed be replaced by dispersive propagation in temporal imaging (see figure 15.8(a)). This also implies that, just as the far-field diffraction of a spatial light distribution yields its Fourier transform, the intensity of a temporal field evolves into its Fourier transform after sufficient propagation in a linear dispersive medium. This analogy allows for mapping the spectrum of a particular waveform into the time domain [33]. More specifically, considering the dispersion of a waveform of temporal width $\Delta\tau$, the condition for the far-field dispersion regime is $|D_2| \gg \Delta\tau$, analogous to the spatial case condition $z \gg k\Delta x^2$, where Δx is the width of the beam profile (figure 15.8(b)). This property can be used for real-time spectral measurement of ultrafast pulses, and can be realised in practice using a highly dispersive single-mode optical fiber [26].

A thin lens is known to impose a quadratic phase-shift directly onto the spatial transverse parameter x, y of the input electric field [34], which, in one-dimension, is described by

$$\phi(x) = -\frac{kx^2}{2f},$$

where f is the focal length of the lens. An analogous time lens can therefore be realised by imparting a quadratic phase-shift to the local time variable T of the input waveform to be magnified:

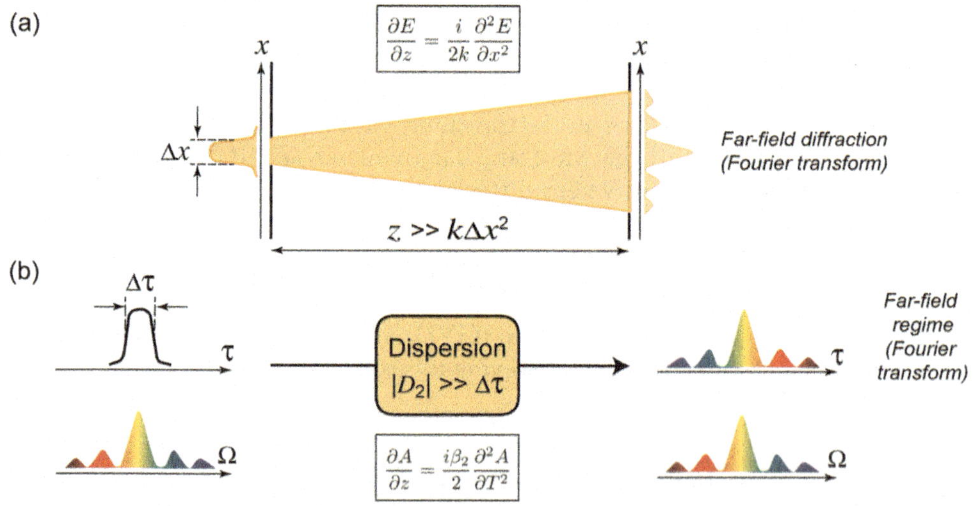

Figure 15.8. Space–time duality of diffraction and dispersion in the far-field regime. (a) Illustrates diffraction of a monochromatic beam, resulting in the broadening of the beam profile. (b) Shows the dispersion experienced by a short pulse, resulting in frequency chirp that broadens the pulse. In both cases, the far-field propagation leads to the Fourier transform of the input waveform (frequency-to-time conversion process). Adapted with permission from reference [25].

$$\phi(T) = \frac{T^2}{2D_f}, \tag{15.8}$$

where D_f is the focal GDD of the time lens.

A temporal imaging system that magnifies a given waveform can then be developed in analogy to a spatial imaging system, using the combination of a time lens with a suitable segment of dispersive propagation preceding and following the time lens [35, 36]. As shown in figure 15.9(a), a temporal image is formed when the input waveform $A_{\rm in}(\tau)$ experiences, consecutively, dispersive propagation GDD $D_{\rm in}$, a time lens with focal GDD D_f, and dispersive propagation GDD $D_{\rm out}$, such that the imaging condition

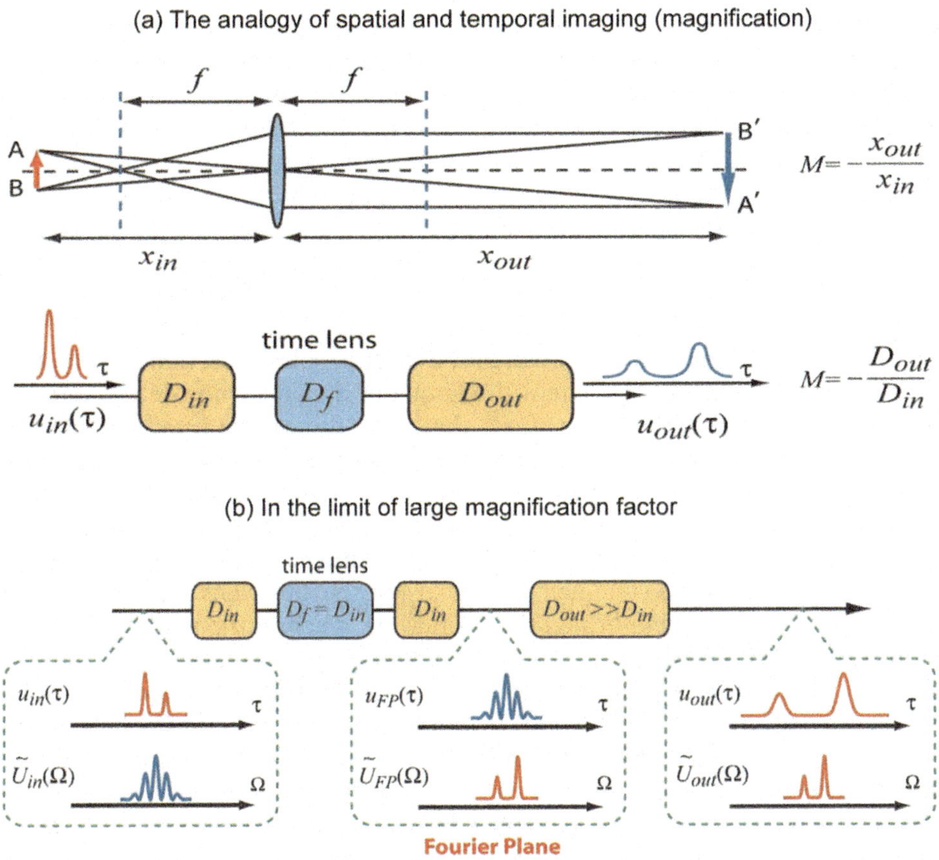

Figure 15.9. (a) Analogy between spatial and temporal imaging. Spatial magnification is realised by cascading input diffraction, lens, and output diffraction. Temporal magnification is realised by cascading input dispersion, time lens, and output dispersion. (b) Spatial and temporal magnification in the limit of large magnification. The system can be seen as a $2f$ Fourier processor time-to-frequency conversion realised by a temporal $2f$ Fourier processor and followed by far-field frequency-to-time conversion. Here u_{in}, u_{FP}, and u_{out} denote the input, Fourier plane, and output electric field, respectively, and $\tilde{U}_{in}$, $\tilde{U}_{FP}$, and $\tilde{U}_{out}$ denote the corresponding Fourier transform. Reprinted with permission from reference [25].

$$\frac{1}{D_{\text{in}}} + \frac{1}{D_{\text{out}}} = \frac{1}{D_f} \tag{15.9}$$

is satisfied. Similar to spatial imaging, the magnification factor of temporal imaging is given by the ratio between the the output and the input GDD [36]

$$M = -\frac{D_{\text{out}}}{D_{\text{in}}}. \tag{15.10}$$

From equations (15.9) and (15.10), we can see that large magnification can be achieved when $D_{\text{out}} \gg D_{\text{in}}$ and $D_{\text{in}} \simeq D_f$, which can be viewed as a $2f$ Fourier analyser followed by a large dispersive segment [37]. The waveform is first transformed from time to frequency by the $2f$ Fourier analyser and subsequently transformed back to the temporal waveform by the large dispersive path, which results in the large temporal magnification of the waveform with factor $M = -D_{\text{out}}/D_{\text{in}}$ (see 15.9(b)).

The first experimental observation of spontaneous modulation instability of continuous waves in the time domain was reported nearly 30 years ago by injecting light from a CW Nd: YAG laser of 1.319 μm into a 5 km depressed-cladding silica fiber [38]. Following this seminal work, other approaches have also been used to study noise-seeded MI, including the use of Fabry–Perot filter in the laser cavity to decrease the MI threshold power [39], passive fiber cavity systems [40], or the use of a partially coherent source injected into a photonic crystal fiber [41]. Although all these observations have improved our understanding of noise-seeded MI dynamics and in particular the (averaged) growth of the spectral sidebands, none of the studies have provided insight into the actual time-domain dynamics nor details about the characteristics of the localised structures that form during the development of the noisy MI field. Certainly, recent studies of the real-time MI spectral characteristics using the DFT technique have proved a significant step forward in developing new insights into these dynamics [26, 27], but the absence of measurements of the unstable time-domain properties of MI has represented a significant bottleneck.

To this end, recent measurement of MI using the time lens has yielded much progress, and has in fact confirmed the framework wherein the analytical breather solutions to the NLSE provide a natural basis with which to interpret the characteristics of a random field generated when modulation instability is seeded by noise. In these experiments, the spontaneous MI field was generated by injecting light from an amplified CW laser at $\lambda_0 = 1550$ nm with 700 mW power into 17 km of standard single-mode optical fiber with nonlinear and GVD coefficients $\gamma = 0.0013$ $(\text{Wm})^{-1}$ and $\beta_2 = -21.4 \times 10^{-27}$ s^2 m^{-1}, respectively. This resulted in the emergence of a chaotic pulse train in the temporal domain with pulse durations in the 2–10 ps range. In order to capture these structures with sufficient resolution, a temporal imaging lens system was used similar to that described above, with a magnification factor of ~76; this, when combined with a 38 GHz photodiode and 30 GHz real-time oscilloscope, yielded an effective temporal resolution of ≈300 fs. The quadratic chirp of the time lens was implemented via four-wave mixing with strong pump pulses in a highly nonlinear silicon waveguide. Because of the pulsed nature of the pump, rather than measuring

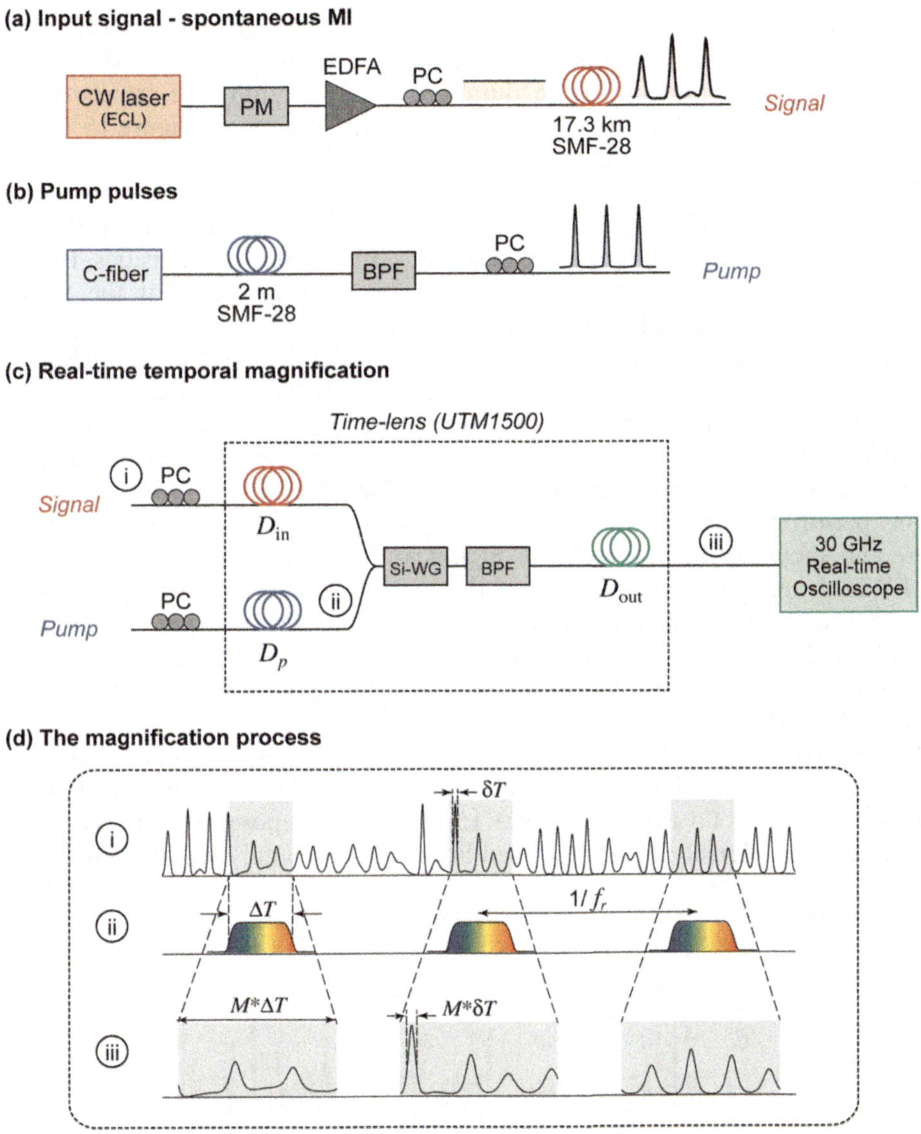

Figure 15.10. Experimental setup of the real-time temporal measurement of spontaneous MI on CW signal using ultrafast temporal magnifier (Picoluz UTM-1500). (a) Generation of the spontaneous MI on CW field. (b) Pump pulses used to implement the time lens. (c) Time lens magnifier system based on four-wave mixing. The signal and idler are passed through dispersive propagation steps (D_{in} and D_{out}) before and after the time-lens system, respectively. The pump pulses are passed through dispersive propagation D_p before being combined with the signal in the silicon waveguide (Si-WG). (d) Schematic illustration of the temporal magnification process: (i) temporal intensity profile of the breather structures with typical width $\delta T \approx 3$ ps, (ii) linearly chirped pump pulses used to implement the time-lens system, and (iii) magnified breather structures (76.4×) measured by the ultrafast oscilloscope. The record length $M\Delta T \approx 5$ ns and refresh rate $f_r = 100$ MHz of the magnified waveforms are determined by the pump pulses.

the characteristics of the continuous field exiting the fiber, the magnification system captures 'snapshots' of the spontaneous MI pulse train in finite windows and at intervals fixed by the pump pulse duration and repetition rate, respectively. A schematic illustration of the experimental setup used is shown in figure 15.10. More details about the parameter values can be found in [42].

A selection of measured temporal profiles after 17 km of propagation is shown in figure 15.11, and one can clearly observe how the initial CW field has developed into a random pulse train, with structures of various durations and intensities. It is these localised structures of durations on the picosecond timescale that the temporal imaging system allows us to resolve in real time.

The large amount of data collected allowed for detailed statistical analysis, and the results of this analysis are shown in figure 15.12. Specifically, the figure plots the temporal structures duration (FWHM) against the corresponding normalised peak intensity $P/\langle P\rangle$, and the result is compared to the theoretical curve calculated from the elementary and higher-order breather solutions. Results from numerical simulations performed using the experimental input parameters are also plotted in the figure and show excellent agreement. Significantly, these plots show that although the FWHM and the peak intensity of the noisy MI breathers are not necessarily measured at the point of maximum localisation, the scatter points clearly cluster around the theoretical curve, confirming that the analytic SFB solutions constitute a natural way to interpret the localised structures spontaneously emerging from noise-seeded MI.

This interpretation is further confirmed when comparing the intensity profiles of the breathers extracted from the experiments with those of the exact analytic breather solutions. To this end, figure 15.12(c) superimposes 10 intensity profiles with maximum intensity $P/\langle P\rangle \approx 9$ (red lines) with the analytical Peregrine soliton

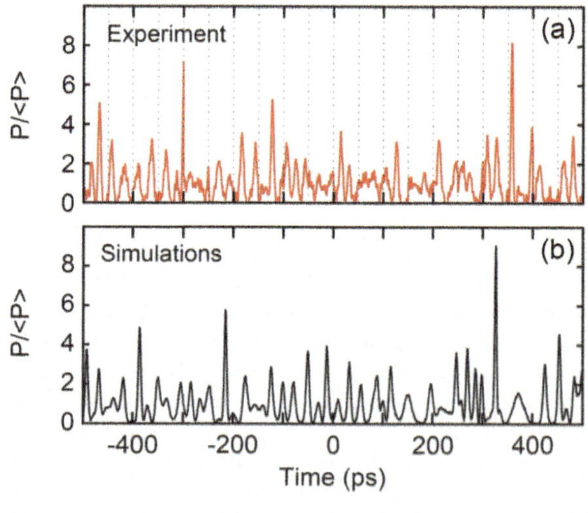

Figure 15.11. Simulated (a) and experimental (b) temporal profiles for spontaneous MI in a single-mode fiber SMF28 after 17 km of propagation. Experimental results are cut in 50 ps windows due to the time lens sampling at 10 ns intervals. Reprinted by permission from [42] (© 2016).

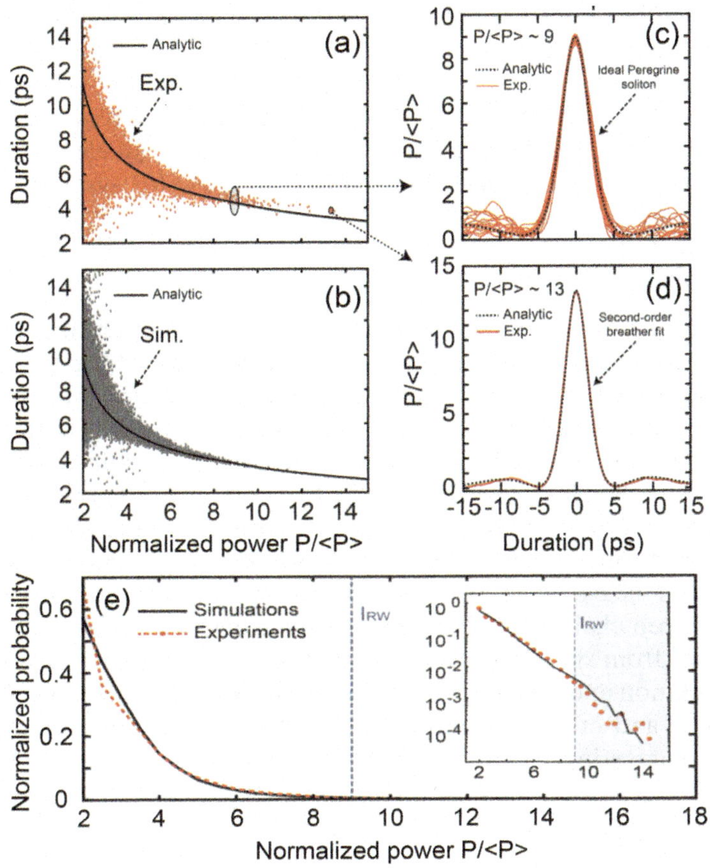

Figure 15.12. (a) and (b) show scatter plots of pulse duration against normalised power from the intensity profiles obtained from the experiments (top, red points) and simulations (bottom, black points). The plots were generated from 30 000 distinct structures for the experiments (only taken from pulses with peak intensity $P/\langle P\rangle > 2$) and 50 000 distinct structures for the numerical simulations. The theoretical curve is calculated at the point of maximum localisation of both the elementary and higher-order (in phase collision) breathers. Right panels (c) and (d) plot the intensity profiles extracted from the experimental results (red lines) compared to the analytical fits (black lines) for (c) $P/\langle P\rangle \approx 9$ and (d) $P/\langle P\rangle \approx 13$. (e) Histogram of the normalised peak intensities with the inset plotted in semilogarithmic scale. Reprinted by permission from [42] (© 2016).

structure (dashed black line) [8], and figure 15.12(d) plots the intensity profile of a higher-order breather with maximum intensity $P\langle P\rangle \approx 13$ (red line) with the analytic second-order breather solution (dashed black line) constructed using the Darboux transformation [2]. Both fits show remarkable agreement between the experimentally measured temporal profile and the analytical solutions. The histogram of the intensity maxima (see figure 15.12(e)) is seen to follow that of a power law, as evidenced by the quasi-linear slope when plotted on a logarithmic scale. Inspired by ocean rogue wave statistical treatment, one can then estimate a rogue wave intensity limit I_{RW} defined as the average of the highest third of the structures intensities, i.e. $I_{\mathrm{RW}} \geqslant 2I_{1/3}$ [43], and corresponding to the intensity limit beyond which the intensity of a given structure can be considered as extreme or rogue. From the experimental data shown in figure 15.12, $I_{RW} \approx 9.2$ with $\approx 0.3\%$ of the peak intensities exceeding this value, which was found to be remarkably close to numerical simulation results [42]. But perhaps even more remarkable is the fact that the rogue waves occur at intensities over 9, which is the highest obtainable value for an elementary breather, confirming that the rogue events observed in a pure NLSE system correspond to collision or nonlinear superposition of elementary breathers [12, 44, 45].

15.5 Conclusions

It is very often the case in physics that theoretical and numerical studies advance years or even decades ahead of experiments. Notwithstanding the fact that many important discoveries in nonlinear optics have resulted from experimental serendipity (e.g. the photonic crystal fiber supercontinuum), it has been far more often the case that theoretical work revealed a rich landscape of dynamical behaviour that was simply not accessible to experiments. As this book goes to press, it is possible to safely conclude that this is no longer the case for the study of modulation instability in nonlinear fiber optics, where recent advances have now permitted direct access to real-time optical fields undergoing dramatic temporal and spectral fluctuations. There remain open challenges of course, and the particular problem of reconstructing the full field from simultaneous time and frequency domain measurements with improved resolution and dynamic range. But we fully expect these to be overcome in the near future, and one can look forward with optimism to a new generation of experiments uncovering behaviour that theoretical work has not previously anticipated.

References

[1] Zakharov V and Shabat 1972 Exact theory of two-dimensional self-focusing and one-dimensional self-modulation of waves in nonlinear media *Sov. Phys. JETP* **34** 62

[2] Akhmediev N and Ankiewicz A 1997 *Solitons: Nonlinear Pulses and Beams* (Chapman & Hall)

[3] Dauxois T and Peyrard M 2006 *Physics of Solitons* (Cambridge University Press)

[4] Wang L H, Porsezian K and He J S 2013 Breather and rogue wave solutions of a generalised nonlinear Schrödinger equation *Phys. Rev.* E **87** 069904

[5] Baronio F, Chen S, Grelu P, Wabnitz S and Conforti M 2015 Baseband modulation instability as the origin of rogue waves *Phys. Rev.* A **91** 033804

[6] Akhmediev N N and Korneev V I 1987 Modulation instability and periodic solutions of the nonlinear Schrödinger equation *Theoret. Math. Phys.* **69** 1089–93

[7] Akhmediev N, Ankiewicz A and Taki M 2009 Waves that appear from nowhere and disappear without a trace *Phys. Lett.* A **373** 675–8

[8] Peregrine D H 1983 Water waves, nonlinear Schrödinger equations and their solutions *J. Austral. Math. Soc.* B **25** 16–43

[9] Akhmediev N, Ankiewicz A and Soto-Crespo J M 2009 Rogue waves and rational solutions of the nonlinear Schrödinger equation *Phys. Rev.* E **80** 026601

[10] Kuznetsov E A 1977 Solitons in a parametrically unstable plasma *Dokl. Akad. Nauk SSSR* **236** 575–7

[11] Kibler B, Fatome J, Finot C, Millot G, Genty G, Wetzel B, Akhmediev N, Dias F and Dudley J M 2012 Observation of Kuznetsov–Ma soliton dynamics in optical fibre *Sci. Rep.* **2** 463
[12] Akhmediev N, Soto-Crespo J M and Ankiewicz A 2009 Extreme waves that appear from nowhere: On the nature of rogue waves *Phys. Lett.* A **373** 2137–45
[13] Frosz M H 2010 Validation of input-noise model for simulations of supercontinuum generation and rogue waves *Opt. Express* **18** 14778–87
[14] Trebino R, DeLong K W, Fittinghoff D N, Sweetser J N, Krumbügel M A, Richman B A and Kane D J 1997 Measuring ultrashort laser pulses in the time–frequency domain using frequency-resolved optical gating *Rev. Sci. Instrum.* **68** 3277–95
[15] Iaconis C and Walmsley I A 1998 Spectral phase interferometry for direct electric-field reconstruction of ultrashort optical pulses *Opt. Lett.* **23** 792–4
[16] Miranda M, Arnold C L, Fordell T, Silva F, Alonso B, Weigand R, L'Huillier A and Crespo A 2012 Characterization of broadband few-cycle laser pulses with the d-scan technique *Opt. Express* **20** 18732–43
[17] Walmsley I A and Dorrer C 2009 Characterization of ultrashort electromagnetic pulses *Adv. Opt. Photon* **1** 308–437
[18] Mairesse Y and Quéré F 2005 Frequency-resolved optical gating for complete reconstruction of attosecond bursts *Phys. Rev.* A **71** 011401
[19] French D, Dorrer C and Jovanovic I 2009 Two-beam spider for dual-pulse single-shot characterization *Opt. Lett.* **34** 3415–17
[20] Fabris D, Holgado W, Silva F, Witting T, Tisch J W G and Crespo H 2015 Single-shot implementation of dispersion-scan for the characterization of ultrashort laser pulses *Opt. Express* **23** 32803–8
[21] Wong T C, Rhodes M and Trebino R 2014 Single-shot measurement of the complete temporal intensity and phase of supercontinuum *Optica* **1** 119–24
[22] De Ninno G *et al* 2015 Single-shot spectro-temporal characterization of xuv pulses from a seeded free-electron laser *Nat. Commun.* **2015** 8075
[23] Anashkina E A, Ginzburg V N, Kochetkov A A, Yakovlev I V, Kim A V and Khazanov E A 2016 Single-shot laser pulse reconstruction based on self-phase modulated spectra measurements *Sci. Rep.* **6** 33749
[24] Bowlan P and Trebino R 2011 Complete single-shot measurement of arbitrary nanosecond laser pulses in time *Opt. Express* **19** 1367–77
[25] Salem R, Foster M A and Gaeta A L 2013 Application of space–time duality to ultrahigh-speed optical signal processing *Adv. Opt. Photon* **5** 274–317
[26] Solli D R, Herink G, Jalali B and Ropers C 2012 Fluctuations and correlations in modulation instability *Nat. Photon.* **6** 463–8
[27] Wetzel B *et al* 2012 Real-time full bandwidth measurement of spectral noise in supercontinuum generation *Sci. Rep.* **2** 882
[28] Tournois P 1964 Analogie optique de la compression d'impulsion *C. R. Acad. Sci., Paris* I **258** 3839–42
[29] Tournois P, Vernet J and Bienvenu G 1968 Sur l'analogie optique de certains montages électroniques: formation d'images temporelles de signaux électriques *C. R. Acad. Sci., Paris* I **267** 375–8
[30] Akhmanov S, Chirkin A, Drabovich K, Kovrigin A, Khokhlov R and Sukhorukov A 1968 Nonstationary nonlinear optical effects and ultrashort light pulse formation *IEEE J. Quantum Electron.* **4** 598–605

[31] Treacy E B 1969 Optical pulse compression with diffraction gratings *IEEE J. Quantum Electron.* **5** 454–8

[32] Caputi W J 1971 Stretch: a time-transformation technique *IEEE Trans. Aerosp. Electron. Syst.* **AES-7** 269–78

[33] Jannson T 1983 Real-time Fourier transformation in dispersive optical fibers *Opt. Lett.* **8** 232–4

[34] Goodman J 1968 *Introduction to Fourier Optics* (McGraw-Hill Physical and Quantum Electronics Series)

[35] Kolner B H and Nazarathy M 1989 Temporal imaging with a time lens *Opt. Lett.* **14** 630–2

[36] Kolner B H 1994 Space–time duality and the theory of temporal imaging *IEEE J. Quantum Electron.* **30** 1951–63

[37] Salem R, Foster M A, Turner-Foster A C, Geraghty D F, Lipson M and Gaeta A L 2009 High-speed optical sampling using a silicon-chip temporal magnifier *Opt. Express* **17** 4324–9

[38] Itoh H, Davis G and Sudo S 1989 Continuous-wave-pumped modulational instablity in an optical fiber *Opt. Lett.* **14** 1368–70

[39] Yoshida E and Nakazawa M 1997 Low-threshold 115-GHz continuous-wave modulational-instability erbium-doped fiber laser *Opt. Lett.* **22** 1409–11

[40] Coen S and Haelterman M 2001 Continuous-wave ultrahigh-repetition-rate pulse-train generation through modulational instability in a passive fiber cavity *Opt. Lett.* **26** 39–41

[41] Mussot A, Lantz E, Maillotte H, Sylvestre T, Finot C and Pitois S 2004 Spectral broadening of a partially coherent CW laser beam in single-mode optical fibers *Opt. Express* **12** 2838–43

[42] Närhi M, Wetzel B, Billet C, Toenger S, Sylvestre T, Merolla J M, Morandotti R, Dias F, Genty G and Dudley J M 2016 Real-time measurements of spontaneous breathers and rogue wave events in optical fibre modulation instability *Nat. Commun.* **7** 13675

[43] Dudley J M, Dias F, Erkintalo M and Genty G 2014 Instabilities, breathers and rogue waves in optics *Nat. Photon.* **8** 755–64

[44] Toenger S, Godin T, Billet C, Dias F, Erkintalo M, Genty G and Dudley J M 2014 Emergent rogue wave structures and statistics in spontaneous modulation instability *Sci. Rep.* **5** 10380

[45] Soto-Crespo J M, Devine N and Akhmediev N 2016 Integrable turbulence and rogue waves: breathers or solitons? *Phys. Rev. Lett.* **116** 103901

IOP Publishing

Nonlinear Guided Wave Optics
A testbed for extreme waves
Stefan Wabnitz

Chapter 16

Cavity soliton dynamics and rogue waves in driven Kerr cavities

Mustapha Tlidi, Krassimir Panajotov and Marcel G Clerc

16.1 Introduction

In their seminal paper, Luigi Lugiato and René Lefever derived a mean field model to describe transverse pattern formation in cavity nonlinear optics [1]. The purpose of their article was to analyze spatiotemporal pattern formation in a dissipative, diffractive, and nonlinear optical cavity submitted to a continuous laser pump. Early works have reported on transverse pattern formation in bistable systems subjected to a Gaussian beam [2, 3]. The originality of the Lugiato–Lefever equation, which is hereafter referred to as the LLE, lies in the fact that it predicted the existence of transverse structures of the electric field whose appearance does not require either a switching process or an inhomogeneous injected beam. More importantly, Lugiato and Lefever have shown that the mechanism of symmetry-breaking leading to the spontaneous formation of transverse patterns is fundamentally analog to the classical Turing instability [4]. Turing's work remained largely unknown over almost two decades, until it received theoretical thermodynamic support by Prigogine and collaborators [5, 6]. The spatial structures discussed hereafter evolve on macroscopic spatial scales and can be only maintained by permanent non-equilibrium constraint. These dissipative structures are characterized by an intrinsic wavelength that is solely determined by the dynamical parameters and not by the system size or the boundary conditions. Localized structures, often called dissipative or cavity solitons, have been experimentally observed in several devices involving nonlinear optical resonators pumped by a coherent laser field. They can be classified into two categories; the first of these is homoclinic, whereby the intensity profile of light evolves towards either a periodic distribution originated from a symmetry-breaking instability or to localized states [7–9]. The latter consists of bright or dark localized spots in one or more dimensions that can be either randomly distributed, self-organized, or independent. The bright dots may be considered as bits or pixels in

doi:10.1088/978-0-7503-1460-2ch16 © IOP Publishing Ltd 2017

a parallel processing of information. The second category of structures can be called heteroclinic, whereby the spatial distribution has localized solutions far from any symmetry-breaking instability. In this case, localized structures result from interactions between switching fronts in bistable systems [10–18]. In this chapter, we focus on the homoclinic localized structures in driven cavities filled with Kerr media.

The mean field approach used to derive the LLE has permitted us to model and analyze intra-cavity structures in various optical systems using liquid crystals, an optical parametric oscillator, second-harmonic generation, left-handed materials, photorefractive materials, fibers, and photonic crystal fiber. The field of cavity nonlinear optics has witnessed a constant increase with the explosive growth of the localized structures theme, as witnessed by the large number of overviews [19–39]. An excellent review on this issue can be found in the book of Lugiato and collaborators [40].

In another line of research, recent investigations of the spatiotemporal dynamics of driven nonlinear cavities indicates that these simple devices can develop a rich spatiotemporal dynamics, such as temporal and spatiotemporal chaos and the formation of rogue waves. Rogue waves (RWs) are pulses whose amplitude or intensity is much higher than that of the surrounding pulses; they belong to an active field of research often called extreme and rare events in physics. The rogue waves have been initially studied in the context of fluid mechanics or an oceanic setting. Later on, rogue waves were generated along an optical fiber by Solli *et al* [41]. They are characterized by a long-tailed probability distribution of pulse amplitude considerably different from the Gaussian distribution. The formation of RWs in optics has been the subject of intense research (see recent reviews on this issue [42–50]).

The chapter is organized as follows. In section 16.2 we discuss a quasi-periodic route to spatiotemporal chaos in the Lugiato–Lefever model. In section 16.3, we first show that optical feedback in LLE leads to the formation of drifting cavity solitons. Next, we demonstrate that delayed feedback may also lead to the appearance of spatiotemporal chaos and to the formation of rogue waves in the paradigmatic Lugiato–Lefever equation. We conclude in section 16.4.

16.2 Spatiotemporal chaos in Lugiato–Lefever model

The LLE describes the space–time behavior of laser fields confined in nonlinear optical resonators that are wider than the spatial Kerr length scale. The LLE reads [1]

$$\frac{\partial E}{\partial t} = i\nabla_{\perp}^{2}E - (1 + i\theta)E + i|E|^{2}E + E_{i}, \tag{16.1}$$

where E is the normalized amplitude of the electric field and θ is the frequency detuning between the injected light E_i and the cavity resonance. E_i is considered as being real without loss of generality.

The LLE applies not only to passive resonators but also to fiber optics [51, 52], whispering gallery mode cavities [53], integrated ring resonators [54, 55], and left-handed materials [56]. In these systems, the time and space are considered to be continuous. A discrete version of the LLE to model coupled-waveguide resonators [57, 58], or extended Josephson junctions [59, 60] has been also derived. The envelope equation (16.1) has been also derived to describe the plasma driven by a radio-frequency field [61, 62] and the condensate in the presence of an applied AC field [63].

Stable two-dimensional patterns in the form of hexagons have been reported in the LLE [64]. Patten selection analysis demonstrates that only hexagons are stable over other two-dimensional patterns such as stripes, rhomboids, or honeycombs [65]. The relative stability analysis has shown that rolls and rhombic structures are unstable with respect to perturbation in the form of hexagonal structures [65]. Secondary bifurcation leading to self-pulsating of hexagonal patterns has been predicted in [66, 67]. The influence of noise in the formation of hexagonal patterns was also studied for that system [68]. Beside periodic patterns, the same mechanism predicts the possible existence of aperiodic, stationary localized structures. They consist of isolated or randomly distributed peaks in the transverse plane surrounded by regions in the uniform state [8, 9]. Such localized structures have been experimentally observed in intracavity liquid crystals [69] and in fiber Kerr type of media [70]. In the conservative limit, where the injection and losses are both small, localized structures have analytical solutions [126]. Self-pulsating localized structures in the one- and two-dimensional setting have been investigated in [71]. The interaction of self-pulsating localized structures that emit weakly decaying dispersive waves can allow for the formation of bound states [72]. The concept of excitability mediated by localized structures has been analyzed [73]. Polarization properties of periodic [74] and localized structures [75] have been also investigated. The effect of periodic modulation either in space or/and in time of the detuning parameter or the injection have been addressed in [76, 77]. In the large-intensity regime, complex spatiotemporal dynamics such as spatiotemporal chaos [78–81] or rogue wave [82] formation have been realized. In a regime devoid of Turing or modulational instability, switching wave fronts connecting the homogeneous steady states of LLE have been studied theoretically and/or experimentally [83–86]. When diffractive and dispersive effects have comparable influence, the spatiotemporal dynamics of the nonlinear Kerr cavity can be described by the three-dimensional LLE [87, 88].

In the case of spatiotemporal chaos, the Lyapunov spectrum has a continuous set of positive values. This matches the definition that has been proposed in [89]. In the case of a low-dimensional chaos, the Lyapunov spectrum possesses a discrete set of positive values. However, the turbulence or weak turbulence are characterized by a power law cascade of a scalar quantity such as energy or norm. On the basis of the Lyapunov spectrum, we cannot conclude that the system develops a turbulence.

In this section, we characterize the complex behavior that has been reported numerically and experimentally in driven cavities such as whispering-gallery-mode microresonators and optical fibers. Studies on optical frequency comb generation in such systems have demonstrated evidence of complex spatiotemporal dynamics [79, 90–92]. However, in most of these studies, complex behaviors are characterized by a power spectrum, filtering spatiotemporal diagrams, embedding dimension, and time series analysis. These tools are necessary but they are not sufficient to distinguish between spatiotemporal chaos, low-dimensional chaos, and turbulence. Classification of complex spatiotemporal phenomena has been reported in the recent book by Pikovsky and Politi [89]. The definition proposed in this reference is that, when spatiotemporal chaos occurs, the Lyapunov spectrum has a continuous set of positive values. Numerical analyzes of the Lyapunov exponents are generally accessible; however, few experimental studies have been conducted on this type of exponent [93]. More recently, it has been shown that the complex behavior has a spatiotemporal chaotic nature. Therefore, we estimate the Lyapunov spectra in our system. The Yorke–Kaplan dimension (D_{YK}) is used as an order parameter to establish the bifurcation diagram of the spatiotemporal chaos. In addition, we show that the spatiotemporal chaos, oscillatory localized state, and homogeneous steady state (HSS) can coexist in a finite range of the pumping intensity. The destabilization of an oscillatory localized state through radiation of counterpropagative fronts between the HSS and spatiotemporal chaotic state is also discussed by estimating the front speed as a function of the pump intensity.

Space–time maps showing the evolution of the intensity as a function of injection field are presented in figure 16.1. This figure is obtained by numerical simulations of the LLE based on the Runge–Kutta method combined with a spectral method or a pseudo-spectral algorithm. The intracavity field intensity $|E|^2(x,t)$ is obtained from

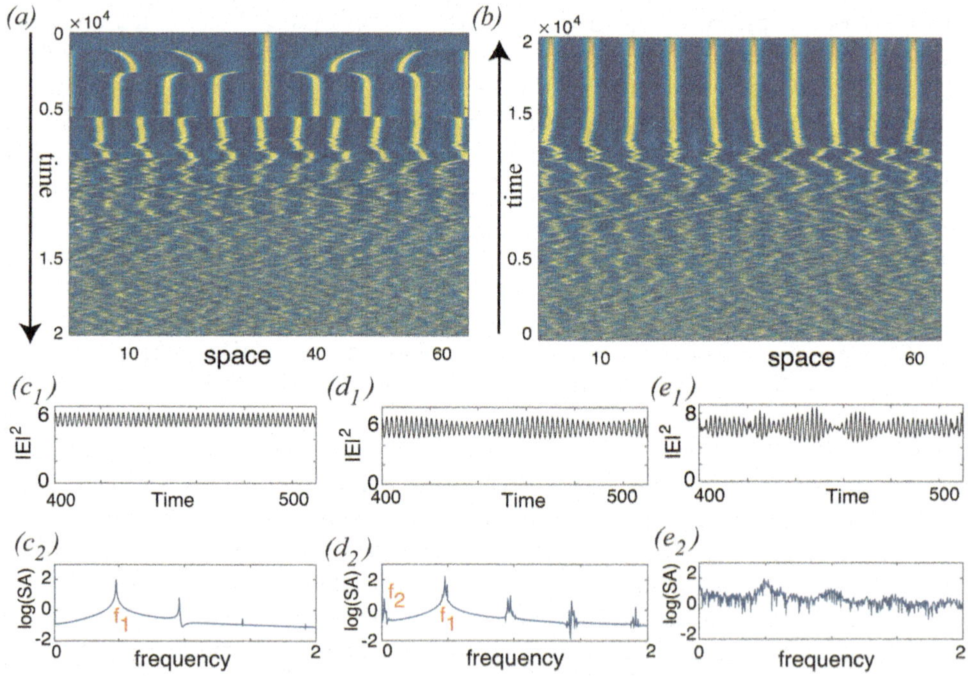

Figure 16.1. Space–time maps showing the evolution of $|E|^2(x, t)$ in the LLE when changing E_i with a step of 0.02 at each 200 time units. Cavity detuning is $\theta = 2$. (a) E_i is increased from 1.3 to 3.3. (b) E_i is decreased from 3.3 to 1.3. (c1), (d1), and (e1) show time traces for an isolated peak from the periodic pattern for $E_i = 1.85$, $E_i = 1.9$, and $E_i = 2.1$, respectively. (c2), (d2) and (e2) show the corresponding power spectra. Reproduced from [101]. With permission of Springer Berlin Heidelberg.

numerical simulations by changing the injected field amplitude E_i at each 200 time units with a step of 0.02. For small values of injection ($E_i = 1.3$), the localized structure is stable; it is similar to the one that has been predicted in [8]. When increasing the injected field amplitude, the transition to a self-organized structure appears at $t \approx 1200$ and $E_i \approx 1.42$, as shown in the space–time map of figure 16.1(a). Switchings between patterns of different wavelength can be seen in figure 16.1(a) for $t \approx 2700$ and 5600. These transitions are associated with an Eckhaus instability reported in the paper by Périnet *et al* [94]. At a time of $t \approx 5700$ the periodic structure becomes unstable and starts to oscillate harmonically in time, as shown in the time trace in figure 16.1(c1). The corresponding power spectrum is shown in figure 16.1(c2). This figure clearly reveals that the pattern maximum experience period one oscillating dynamics, i.e. the first temporal instability encountered, is an Andronov–Hopf bifurcation with a frequency f_1. During these oscillations, the neighboring maxima in the pattern oscillate in antiphase; i.e. when one is increasing, the neighboring ones are decreasing and vice versa. Increasing further E_i, the oscillating pattern undergoes a Torus bifurcation with a new, smaller frequency f_2 appearing as shown in figure 16.1(d1) and (d2) for a time trace and power spectrum of the third peak. Increasing further E_i, the system follows an extended quasi-periodic route to chaos with a similar time dependence for each well-separated peak of the oscillating pattern—see figure 16.1(e1) and (e2) for the time trace and the power spectrum of the fifth peak. For a still stronger injection field, the neighboring peaks in the pattern start to interact strongly, with some of them disappearing and reappearing as shown in the bottom regions of figure 16.1(a). A similar route has been identified from hexagonal patterns in LLE [66], in the non-variational Swift–Hohenberg equation [95, 96, 113], and in semiconductor cavities [97, 98]. Note, however, that in the case of a complex Ginzburg–Landau equation, the route to chaos depends on the way of crossing the boundary of the region of chaos [99]. In the case of LLE, by varying the injected field intensity, the route to chaos depends on the detuning parameter. Two routes have been identified either through period doubling [100] that leads to localized low-dimensional chaos and through extended quasi-periodicity [101] that generates spatiotemporal chaos.

A space–time map for a backward scan of the injection field is presented in figure 16.1(b): the simulation starts from the high-intensity regime ($E_i = 3.3$) and the injected field amplitude E_i is reduced with a step of 0.02 each 200 time units. By reducing the injected field amplitude, a transition from the complex spatiotemporal behavior to a periodic pattern appears. This transition is subcritical and a hysteresis exists, as evidenced by comparing figure 16.1(a) and (b). In addition, the system undergoes a multistable behavior between patterns of different wavelength.

To characterize the complex spatiotemporal behavior just described, we calculate Lyapunov exponents and label them as $\{\lambda_i\}$, with $i = 0, 1, \ldots, N$, N being the grid number and $\lambda_p \geq \lambda_q$ ($p \geqslant q$). The spectrum of Lyapunov exponents, i.e. λ_i as a function of i, is plotted in figure 16.2(a). This spectrum possesses a continuous set of positive values and therefore we can conclude that the complex spatiotemporal behavior described in figure 16.1 has a spatiotemporal chaotic nature. In order to check the convergence of the numerical method used to establish the Lyapunov

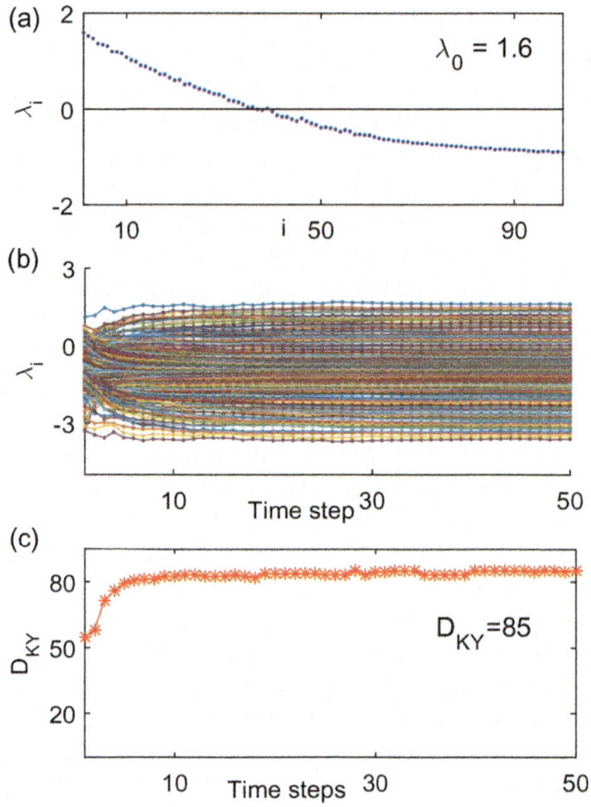

Figure 16.2. LL model with $\theta = 2$, $E_i = 5$. (a) Spectrum of Lyapunov exponents; (b) Lyapunov exponents versus time and (c) Kaplan–Yorke dimension D_{KY} versus time. Reproduced from [101]. With permission of Springer Berlin Heidelberg.

spectra, we plot Lyapunov exponents as a function of the integration time as shown in figure 16.2(b).

A measure of the strange attractor complexity is provided by the Kaplan–Yorke dimension (D_{YK}) defined by [102]

$$D_{KY} \equiv p + \frac{\sum_{i=0}^{p} \lambda_i}{\lambda_{p+1}}, \tag{16.2}$$

where p is the largest integer that satisfies $\sum_{i=0}^{p} \lambda_i > 0$. The Yorke–Kaplan dimension is an extensive quantity which increases with the system size [105]. It is plotted as a function of the integration time in figure 16.2(c) that clearly demonstrates that the Yorke–Kaplan dimension converges towards a value $D_{YK} = 85$.

Finally, figure 16.3 shows the dependence of the largest Lyapunov exponent λ_0 (on the left axis) and of the Kaplan–Yorke dimension D_{KY} (right axis) on the strength of the injection field E_i. While the largest Lyapunov exponent indeed increases almost linearly from 0.27 to 3.19, the tremendous increase of Kaplan–Yorke dimension D_{KY}

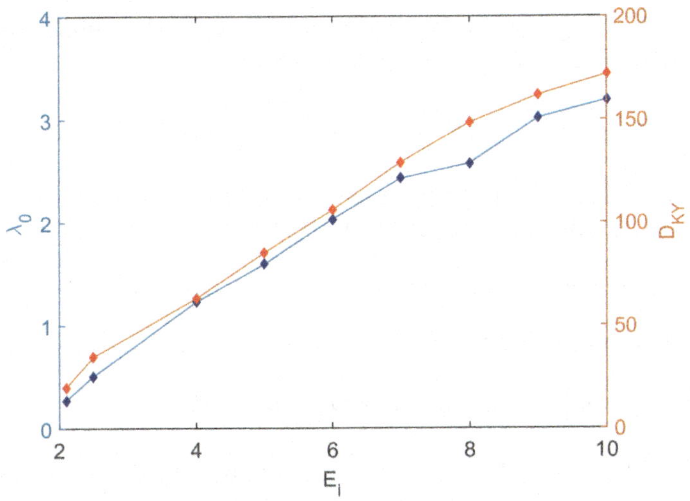

Figure 16.3. LL model with $\theta = 2$. Dependence of the largest Lyapunov exponent λ_0 (left axis) and Kaplan–Yorke dimension D_{KY} (right axis) on the strength of the injection field E_i. Reproduced from [101]. With permission of Springer Berlin Heidelberg.

from about 20 to 170 is mainly due to the development of spatiotemporal chaos, with many more positive exponents appearing in the Lyapunov spectrum as E_i increases.

The preceding analysis allows us to infer that the spatiotemporal complex behavior reported in the LLE and described by figure 16.1 belongs to a spatiotemporal chaos, and it is not a low-dimensional chaos or turbulence. In the case of spatiotemporal chaos, the Lyapunov spectrum has a continuous set of positive values, as shown in figure 16.2(b). This is consistent with the definition proposed in [89]. In the case of a low-dimensional chaos, the Lyapunov spectrum has a discrete set of a few positive values. On the other hand, the turbulence or weak turbulence are characterized by a power law cascade [103]. The power spectrum in the chaotic regime is characterized by displaying an exponential law [104]. On the basis of the Lyapunov and power spectrum, we cannot conclude that the system develops a turbulence. A classification of spatiotemporal complex behaviors can be found in [89, 95, 105, 106].

16.3 Cavity soliton dynamics and rogue waves in the delayed LLE

16.3.1 Impact of time-delayed feedback: a linear stabilty analysis

In this section, we focus our analysis on the formation of cavity solitons in the LLE with delay feedback. The delayed feedback is introduced by an external loop. The delay time $\tau \equiv 2L_{ext}/c$ corresponds to the light travel time in the external loop, with c being the speed of light. A continuous wave of power E_i^2 is launched into the cavity, propagates inside the cavity, and experiences diffraction, Kerr nonlinearity, and dissipation. In what follows, we assume that the diffraction in the external cavity is compensated in a self-imaging configuration. We implement the delayed feedback in the LLE by considering a self-imaged external cavity and a single roundtrip delay term [108]. The LLE with delay feedback reads [109]

$$\frac{\partial E}{\partial t} = i\nabla_\perp^2 E - (1 + i\theta)E + i|E|^2 E + E_i + \eta e^{i\phi} E(t - \tau). \tag{16.3}$$

The feedback is characterized by the time-delay τ, feedback strength η, and phase ϕ.

Homogeneous steady state (HSS) solutions E_S of equation (16.3) are

$$-(1 + i\theta)E_S + i|E_S|^2 E_S + E_i + \eta e^{i\phi} E_S = 0, \tag{16.4}$$

which results in an implicit equation for $I_S = |E_S|^2$:

$$I_i = I_S\left\{[1 - \eta\cos\phi)]^2 + [I_S - \theta + \eta\sin\phi]^2\right\}, \tag{16.5}$$

where $I_i = E_i^2$. The delayed feedback strongly impacts the bistability condition $(\theta - \eta\sin\phi)^2 > 3(1 - \eta\cos\phi)^2$. This condition follows from the expression for the saddle-node (fold) bifurcation on the HSS $dI_i/dI_S = 0$, leading to

$$I_S^{SN_{h1,2}} = \frac{2(\theta - \eta\sin\phi)}{3} \pm \frac{1}{3}\sqrt{(\theta - \eta\sin\phi)^2 - 3(1 - \eta\cos\phi)^2}\,. \tag{16.6}$$

The impact of the optical feedback on the shape of the homogeneous steady-state curve is summarized in figure 16.4.

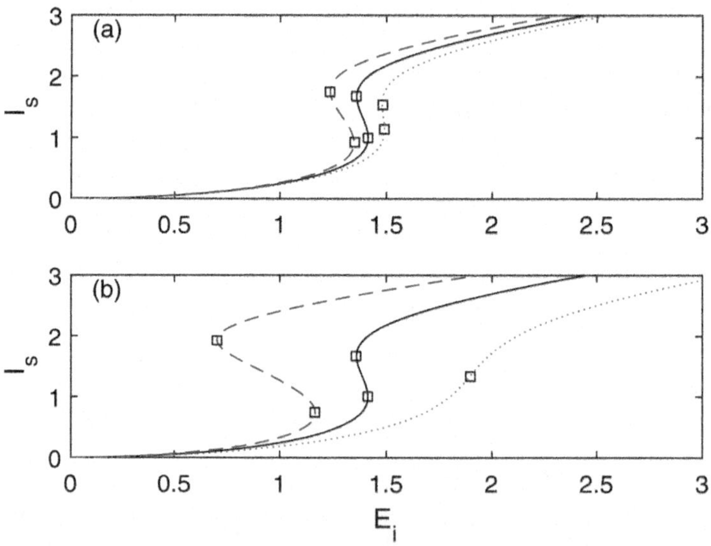

Figure 16.4. Homogeneous steady-state curves of the homogeneous solution of LLE (equation (16.3)) for $\theta = 2$ and different phases of the optical feedback: $\phi = 0$ (dashed blue line) and $\phi = \pi$ (dotted red line). Top (bottom) panel corresponds to the strength of the delay $\eta = 0.1$ ($\eta = 0.5$). Black solid lines in two panels are obtained for the case without optical feedback. Saddle-node bifurcations (equation (16.6)) are denoted by □□□ and coincide with the onset of the Turing instability $I_S^{T(k^2=0)} = I_S^{SN_{h1,2}}$ for the limit $k^2 \to 0$. Reprinted figure with permission from [109]. Copyright 2016 by the American Physical Society.

The linear stability of homogeneous steady states is carried out with respect to small fluctuations; $\delta E(r_\perp, t) = E(r_\perp, t) - E_S$ and $\delta E(r_\perp, t)^* = E(r_\perp, t)^* - E_S^*$; around these flat states of the form

$$\begin{bmatrix} \delta E(r_\perp, t) \\ \delta E^*(r_\perp, t) \end{bmatrix} = e^{(\lambda t + ik\cdot r)} \begin{bmatrix} \delta E \\ \delta E^* \end{bmatrix}. \tag{16.7}$$

This formulation leads to a characteristic polynomial equation for the eigenvalues λ

$$\begin{vmatrix} \lambda + A_r + iA_i + ik^2, & -iE_S^2 \\ iE^{*2}_S, & \lambda + A_r - iA_i - ik^2 \end{vmatrix} = 0, \tag{16.8}$$

where

$$A_r = 1 - \eta e^{-\lambda\tau} \cos\phi,$$
$$A_i = \theta - 2I_S - \eta e^{-\lambda\tau} \sin\phi.$$

Inserting $\lambda = 0$ into equation (16.8) results in a quadratic equation for k^2 with solutions $k_\pm^2$. Above the threshold associated with the symmetry-breaking instability, there exists a finite band of Fourier modes $k_-^2 < k^2 < k_+^2$, with

$$k_\pm^2 = -\theta + 2I_S + \eta \sin\phi \pm \sqrt{I_S^2 - (1 - \eta\cos\phi)^2}, \tag{16.9}$$

which are linearly unstable and trigger the spontaneous evolution of the intra-cavity field towards a pattern that is stationary and spatially periodic, and occupies the whole space available in the cavity. The threshold I_c and the critical frequency ω_c at the Turing instability are obtained when $k_- = k_+$. The expression associated with the critical threshold and the most unstable wavenumber are given by

$$I_c = 1 - \eta\cos\phi, \tag{16.10}$$

$$k_c = \sqrt{-\theta + \eta(\sin\phi - 2\cos\phi)}, \tag{16.11}$$

From the expression of I_c, we see that the threshold associated with the Turing instability is strongly impacted by the phase and the amplitude of the delay feedback. Indeed, for $-\pi/2 + p2\pi < \phi < \pi/2 + p2\pi$ with p is an integer number, the threshold is lowered $I_c < 1$. In terms of the injected intensity, the threshold reads

$$I_{ic} = (1 - \eta\cos\phi)[(1 - \eta\cos\phi)^2 + ((1 - \eta\cos\phi) - \theta + \eta\sin\phi)^2], \tag{16.12}$$

with $I_{ic} = E_{ic}^2$. The dependence of the wavevector square k^2 on the intensity of the steady-state electric field intensity I_S as obtained from equation (16.9) is shown in figure 16.5 for a feedback strength of $\eta = 0.1$ (top panel) and $\eta = 0.5$ (bottom panel) and different feedback phases. As can be seen from figure 16.5, the $k^2 = 0$ points coincide with the turning points of the bistable curve. The upper branch is always Turing unstable for the parameters of figure 16.5.

A map of the saddle-node bifurcation in the θ–E_i plane is shown in figure 16.6. As θ increases, the bistable region widens; however, it is strongly influenced by the

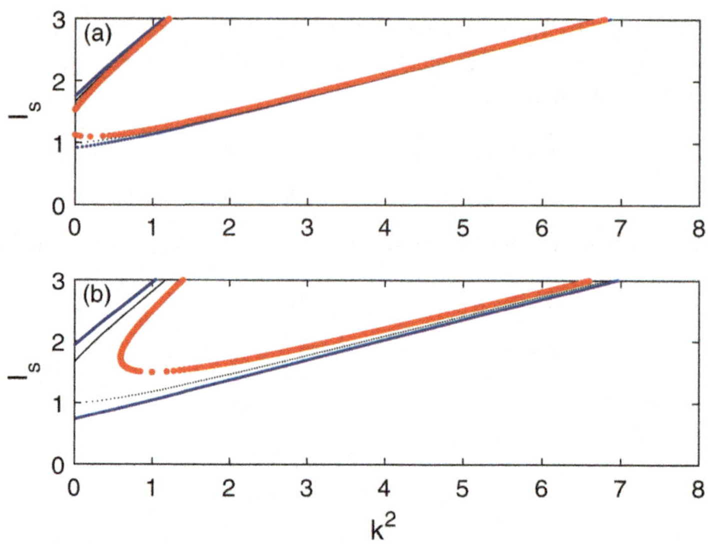

Figure 16.5. Wavevector square k^2 at the onset of modulation instabilities (real eigenvalue $\lambda = 0$) as a function of the steady-state intensity I_S for different phase of the optical feedback: $\phi = 0$ (blue, medium-dot curve) and $\phi = \pi$ (red, large-dot curve). Top panel is for feedback strength of $\eta = 0.1$ and the bottom panel is for $\eta = 0.5$ (see figure 16.4). Black, small-dot curves in the two panels are for the case without optical feedback. Reprinted figure with permission from [109]. Copyright 2016 by the American Physical Society.

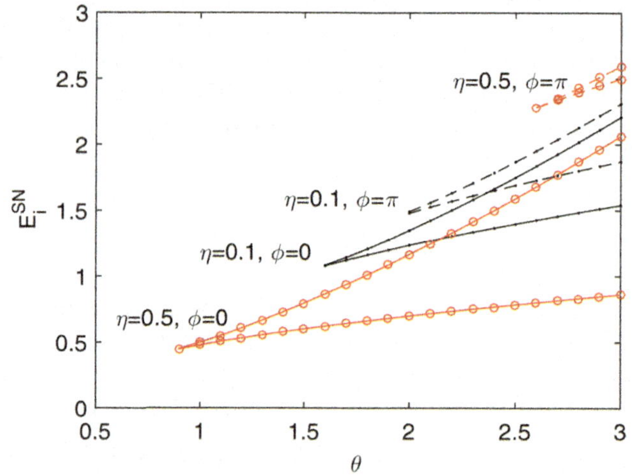

Figure 16.6. Maps of the saddle-node bifurcations in the θ–E_i plane. Black, small dot (red, open circle) curves are for $\eta = 0.1$ ($\eta = 0.5$) and solid (dashed) curves are for feedback phase $\phi = 0$ ($\phi = \pi$). Reprinted figure with permission from [109]. Copyright 2016 by the American Physical Society.

phase of optical feedback being much larger for $\phi = 0$ than for $\phi = \pi$. Such a profound effect of the phase of the optical feedback on the bistable region has been previously observed when studying cavity soliton dynamics in vertical-cavity surface-emitting lasers (VCSELs) [97].

The time-delayed feedback may create new Andronov–Hopf bifurcations in the system due to the exponential term $e^{-\lambda\tau}$ in equation (16.8). Indeed, it has been shown that feedback-induced Andronov–Hopf bifurcation is responsible for time-periodic and eventually more complex dynamics of cavity solitons in VCSELs with saturable absorbers [98]. In order to identify such bifurcations, we solve equation (16.8) for $k^2 = 0$. An example of the appearance of such a feedback-induced Andronov–Hopf bifurcation is presented in figure 16.7 for the case of $\eta = 0.5$, $\tau = 10$, and $\phi = \pi/3$.

A map of the saddle-node and Andronov–Hopf bifurcations in LLE in the parameter space (E_i, η) is shown in figure 16.8 for fixed values of the phase of the delayed feedback $\phi = \pi$ and a fixed delay time $\tau = 10$. When increasing the feedback strength, the bistable region delimited by the two saddle-node (fold) bifurcation lines disappears. However, the region located between the two Andronov–Hopf lines appears and broadens.

There is another bifurcation of the Andronov–Hopf type that takes place at nonzero k^2. In this case, we can express analytically k^2 as a solution of the quadratic equation obtained from the real part of equation (16.8) at a fixed I_S and $\lambda = i\Omega$. Replacing this solution in the equation for the imaginary part of equation (16.8) provides a one-parameter implicit equation for Ω. The obtained expression is cumbersome, and we provide only the plot of the travelling waves' instability threshold in figure 16.9 for $\eta = 0.5$ and $\phi = \pi/3$.

16.3.2 Drift of cavity solitons induced by delayed feedback

Cavity solitons (CSs) typically appear when the Turing instability becomes subcritical, i.e. there exists a hysteresis loop where a homogeneous steady state coexists

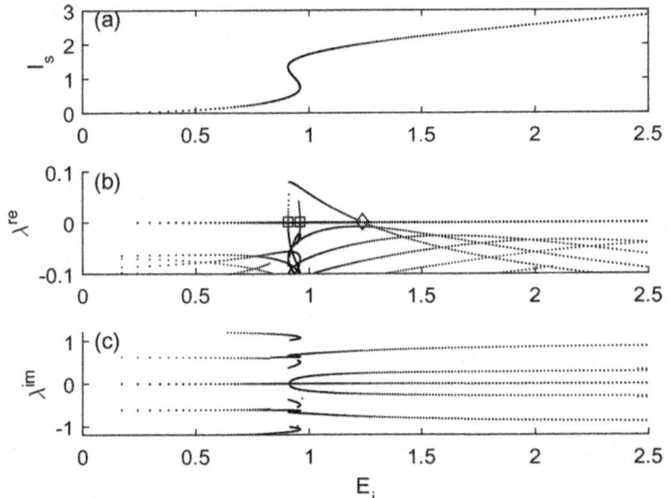

Figure 16.7. Time-delayed feedback-induced Andronov–Hopf bifurcation in Lugiato–Lefever model (16.3) for $\eta = 0.5$, $\tau = 10$, and $\phi = \pi/3$. (a) Homogeneous steady-state solution $I_S(E_i)$; (b) real and (c) imaginary parts of the eigenvalues calculated from equation (16.8) for $k^2 = 0$. Saddle-node and Andronov–Hopf bifurcations are denoted by □□□ and ◇◇◇, respectively. Reprinted figure with permission from [109]. Copyright 2016 by the American Physical Society.

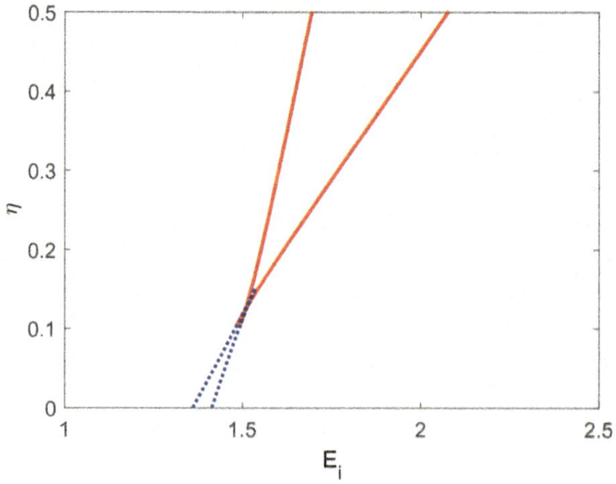

Figure 16.8. Map of the fold (dashed blue lines) and Andronov–Hopf (solid red lines) bifurcations in Lugiato–Lefever model in the plane of (E_i, η) for fixed OF phase of $\phi = \pi$ and $\tau = 10$. Reprinted figure with permission from [109]. Copyright 2016 by the American Physical Society.

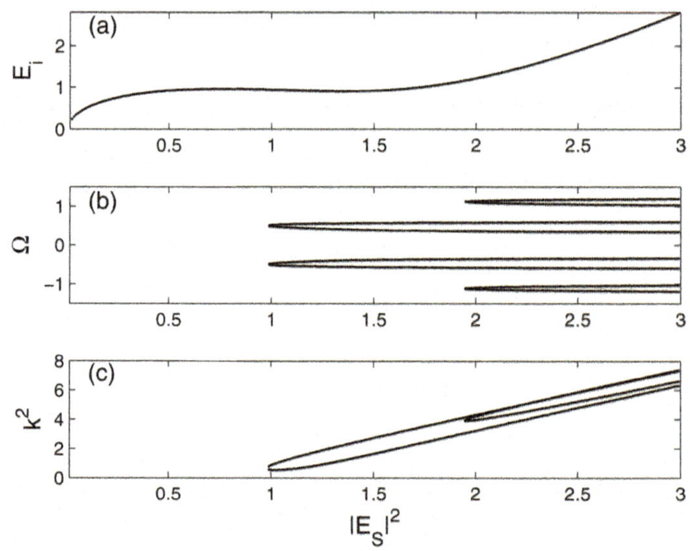

Figure 16.9. Time-delayed feedback-induced traveling-wave (TW) instability in Lugiato–Lefever model for $\eta = 0.5$ and $\phi = \pi/3$. (a) Steady-state homogeneous solution $E_i(I_S)$, $I_S = |E_S|^2$. (b) Imaginary parts of the eigenvalue Ω calculated from equation (16.8) for $k^2 > 0$. (c) Corresponding wavevector square, k^2. Reprinted figure with permission from [109]. Copyright 2016 by the American Physical Society.

with periodic structures such as hexagons and stripes [7, 8]. A CS consists of a region of one state surrounded by another qualitatively different state. When CSs are well separated from each other, they are independent and randomly distributed in space. However, when the distance between peaks decreases they start to interact via their oscillating, exponentially decaying tails [110–114]. This interaction then leads to the formation of clusters through the so-called homoclinic snaking phenomenon [9, 127]. The delayed optical feedback does not affect the clustering behavior; however, it changes the parameter space where this phenomenon appears. Moreover, the snaking homoclinic can now occur as a function of the optical delayed feedback strength. An example of such behavior is illustrated in figure 16.10. Three cavity soliton branches of single, double, and triple CS clusters are shown as a function of the strength of delayed feedback.

It is well known that the delayed optical feedback can induce a drift instability [72, 110–112, 117, 122]. Previous analysis for a driven nonlinear passive cavity close to the nascent bistability described by the delayed Swift–Hohenberg equation supports a drift bifurcation of the CS [108]. Similar investigations have been performed for the vertical-cavity surface-emitting laser model [97, 120], and the LLE describing all fiber resonators. A drift of cavity solitons in the LLE is shown in figure 16.11 for different strengths of the delayed feedback $\eta = 0.11$, 0.12, and 0.15. The time evolution map of figure 16.11 shows how the speed of the cavity soliton is affected by the strength of the delayed feedback. The initial condition consists of a single CS. As can be seen from figure 16.11, after a certain delay, which depends on the feedback strength, the CS starts to move with a constant velocity. The time needed to reach a drift regime decreases as the feedback strength is increased, as shown numerically in figure 16.11, see also [97]. In all our numerical simulations the boundary conditions are periodic. From this figure, we see that the CS moves faster as η is increased.

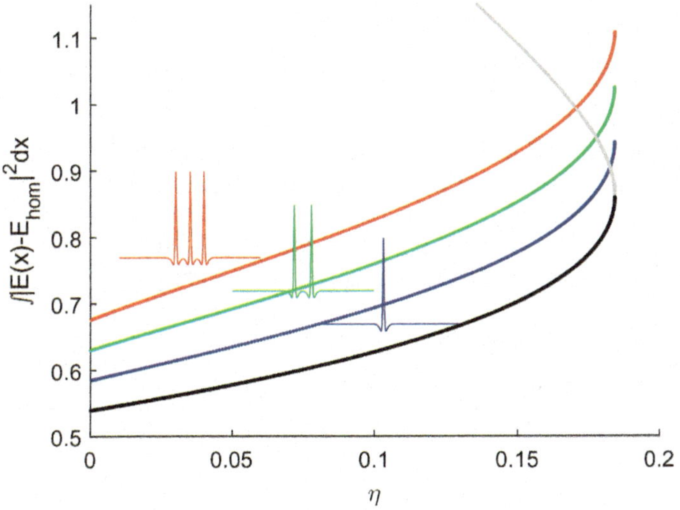

Figure 16.10. Three cavity soliton branches of single CS (blue), double (green), and triple (red) CS clusters as a function of the strength η of the time-delayed feedback. Cavity soliton branches are denoted by the corresponding cavity soliton profiles. Bottom black (grey) solid line is the homogeneous stable (unstable) steady sate. Parameters are $E_i = 1.3$, $\tau = 10$, and $\phi = \pi/3$. Reprinted figure with permission from [109]. Copyright 2016 by the American Physical Society.

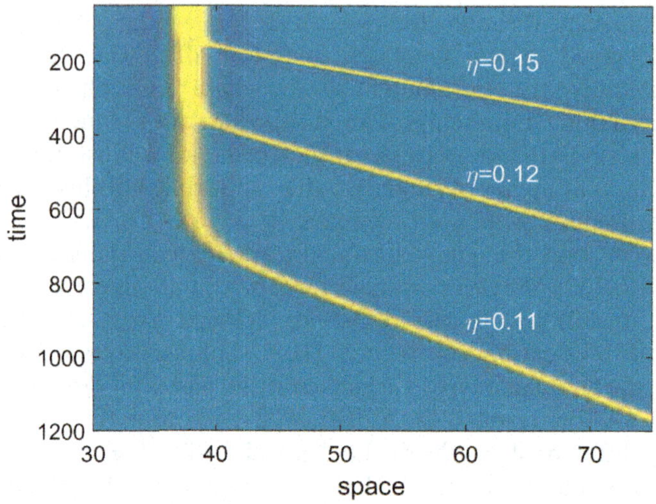

Figure 16.11. CS drift for $\theta = 2$, $E_i = 1.3$ and feedback with different strengths η, fixed delay time of $\tau = 10$, and phase of $\phi = \pi$. The CS is launched at $t = 0$ and starts moving after a certain delay, which decreases as the feedback strength is increased [97]. Notice also the CS moves faster as η is increased. Reprinted figure with permission from [109]. Copyright 2016 by the American Physical Society.

To characterize the drift bifurcation, we perform an analytical analysis that allows us to estimate the threshold of the CS drift bifurcation. For this purpose, we adopt a similar perturbative approach to the one used in [108, 120] to capture the dynamics of CS near the drift instability. Slightly above this threshold, we consider a CS moving uniformly with a small velocity $v = |\mathbf{v}|$, and we perform an expansion the electric field amplitude in power series of v as

$$E = E_0(\boldsymbol{\xi}) + v[E_1(\boldsymbol{\xi}) + vE_2(\boldsymbol{\xi}) + v^2E_3(\boldsymbol{\xi}) + \cdots],$$

where $E = E_0(\mathbf{r})$ is the stationary soliton profile, $\boldsymbol{\xi} = \mathbf{r} - v\,\mathbf{e}\,t$, $\mathbf{r} = (x, y)$, and $\mathbf{e}$ is the unit vector in the direction of the soliton motion. Replacing these expansions in equation (16.3) and collecting the first-order terms in small parameter v, we obtain

$$L\begin{pmatrix} \mathrm{Re}\, E_1 \\ \mathrm{Im}\, E_1 \end{pmatrix} = \begin{pmatrix} \mathrm{Re}[w_{01}(1 + \eta\tau e^{i\phi})] \\ \mathrm{Im}[w_{01}(1 + \eta\tau e^{i\phi})] \end{pmatrix}$$

with $w_{01} = \mathbf{e} \cdot \nabla_{\perp} E_0$. The linear operator L is given by

$$L = \begin{pmatrix} 1 + 2A_0B_0 & \nabla_{eff}^2 + 2B_0^2 \\ -\nabla_{eff}^2 - 2A_0^2 & 1 - 2A_0B_0 \end{pmatrix},$$

where $A_0 = \mathrm{Re}\, E_0$, $B_0 = \mathrm{Im}\, E_0$, and $\nabla_{eff}^2 = \nabla_{\perp}^2 - \theta + |E_0|^2$. By applying the solvability condition to the right-hand side of equation (16.2), we obtain the drift instability threshold

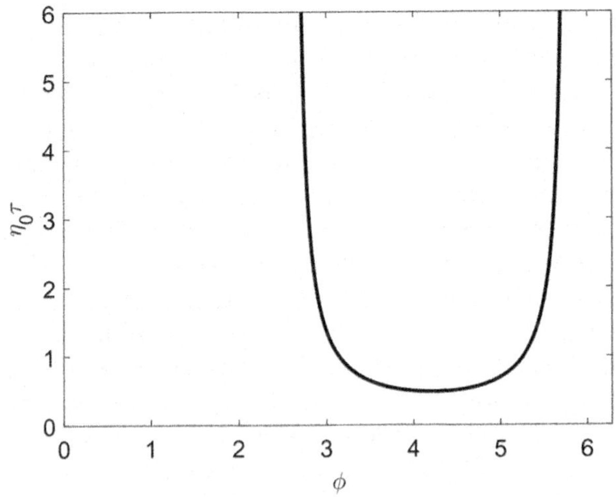

Figure 16.12. Threshold values of the product of the feedback strength and delay time $\eta_0\tau$ at the onset of CS drift instability as a function of feedback phase ϕ. Parameters are $\theta = 2$ and $E_i = 1.3$. Reprinted figure with permission from [109]. Copyright 2016 by the American Physical Society.

$$\eta\tau = -\frac{X_1 + Y_2}{\cos\phi(X_1 + Y_2) + \sin\phi(Y_1 - X_2)} \tag{16.13}$$

with $X_{1,2} = \langle \psi_{1,2}^{\dagger}, \nabla_{\perp} \operatorname{Re} E_0 \rangle$ and $Y_{1,2} = \langle \psi_{1,2}^{\dagger}, \nabla_{\perp} \operatorname{Im} E_0 \rangle$. Here, the eigenfunction $\psi^{\dagger} = \left(\psi_1^{\dagger}, \psi_2^{\dagger}\right)^T$ is the solution of the homogeneous adjoint problem $L^{\dagger}\psi^{\dagger} = 0$, and the inner product $\langle , \rangle$ is defined as $\langle U, V \rangle = \int_{-\infty}^{+\infty} UV \, d\mathbf{r}$.

Note that for a feedback phase of $\phi = \pi$, we recover from (16.13) the threshold condition obtained earlier for the spontaneous motion of cavity solitons in the Swift–Hohenberg equation with delayed feedback, $\eta_0\tau = 1$ [108]. In figure 16.12 we plot the threshold values of the product of the feedback strength and delay time $\eta_0\tau$ at the onset of CS drift instability as a function of feedback phase ϕ. As previously observed for the VCSEL model in [120], the drift instability is only possible in certain intervals of the feedback phase.

16.3.3 Two-dimensional dissipative rogue waves in Lugiato–Lefever model

The 2007 experimental demonstration of rogue waves in fiber optics has motivated several groups to investigate this intriguing phenomenon [41]. Rogue waves correspond to large intensity pulses in the time domain of resonant fiber cavity. RW formation is also a well documented issue in oceans [124]. Among various mechanisms that have been proposed to be responsible for the generation of rogue waves, the modulational instability mechanism together with pulse collisions remains the main ingredient for the creation of rogue waves, as shown by Peregrine [124]. Vector Peregrine solutions in the form of bright and dark rogue waves have been investigated in [125, 126]. The long-tail probability distribution of the wave amplitude is the fundamental characteristic accounting for the generation of rogue waves. In addition, Peregrine solitons are considered as a prototype of rogue waves. Experimental confirmation of Peregrine solitons has been demonstrated in optical fiber [127, 128] and in water wave tank [129, 130] systems. Small-amplitude pulses may grow to large amplitudes if their frequencies belong to a band of unstable modes with a positive gain. Nonlinear interaction between unstable frequencies may lead to a complex wave dynamics. Analytical study of the nonlinear interaction between two frequency solutions of the nonlinear Schrödinger equation in the form of the Akhmediev breathers has been reported in [131].

Most of the above-mentioned studies have been conducted in one-dimensional systems and in the framework of the Schrödinger nonlinear equation. However, when considering the two-dimensional transverse problem, the nonlinear Schrödinger equation is not sufficient to describe rogue waves due to collapse dynamics. Moreover, systems modeled by the LLE are dissipative and operate far from equilibrium regime. In this case, stable two-dimensional patterns are possible in the LLE [64, 65]. Two-dimensional spatial confinement of light in the transverse section of the cavity in the form of localized stationary pulses has been theoretically predicted for LLE [8] and experimentally proved in [69, 70].

We fix the detuning parameter in the LLE and only vary the strength of the injected field. Single or multi-peak stationary localized structures are formed; they can be either isolated or randomly distributed in the transverse plane or peaks appearing in a cluster [8]. When further increasing the injection field E_i, very-high-amplitude pulses appear in the region of well-developed spatial-temporal chaos (see the snapshot of the optical intensity shown in figure 16.13 for $\theta = 2$ and $E_i = 5$). A statistical analysis in figure 16.14 shows that there is a considerable number of events with spatiotemporal maxima of intracavity intensity more than twice the significant wave height (SWH), and even–events with amplitude as high as six times the SWH. This figure shows a non-Gaussian statistics of the wave intensity, with a long tail of the probability distribution typical for rogue-wave formation. We infer that the large-intensity pulses generated in figure 16.13 belong to the class of rogue waves or extreme events. We would like to emphasize that rogue waves are only formed in the LL model when the spatiotemporal chaos is well

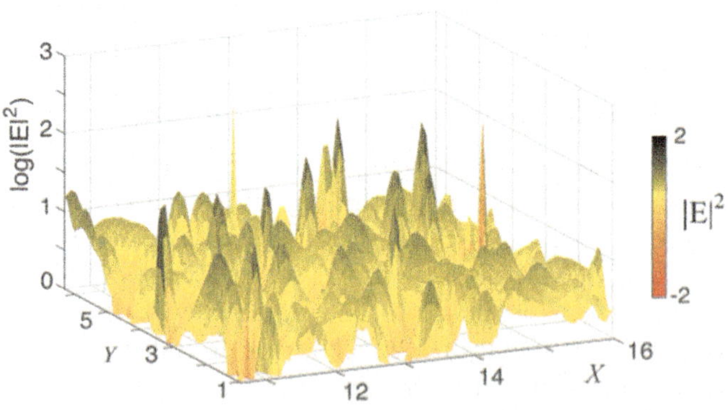

Figure 16.13. A snapshot of the optical intensity in logarithmic scale for the two-dimensional Lugiato–Lefever model with an extreme event captured. The parameters are $\theta = 2$ and $E_i = 5$. Reproduced from [101]. With permission of Springer Berlin Heidelberg.

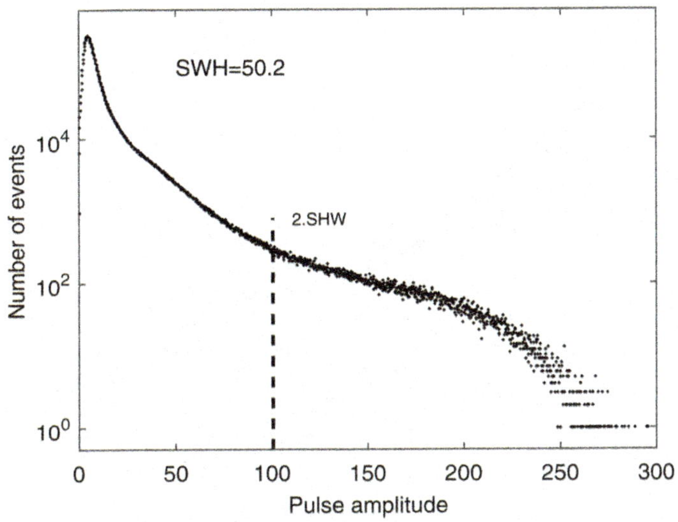

Figure 16.14. Number of events as a function of the intensity of the pulses in semi-logarithmic scale. LL model parameters are $\theta = 2$ and $E_i = 5$. The SWH denotes the significant wave height. The dashed line indicates $2 \times$ SWH. Reproduced from [101]. With permission of Springer Berlin Heidelberg.

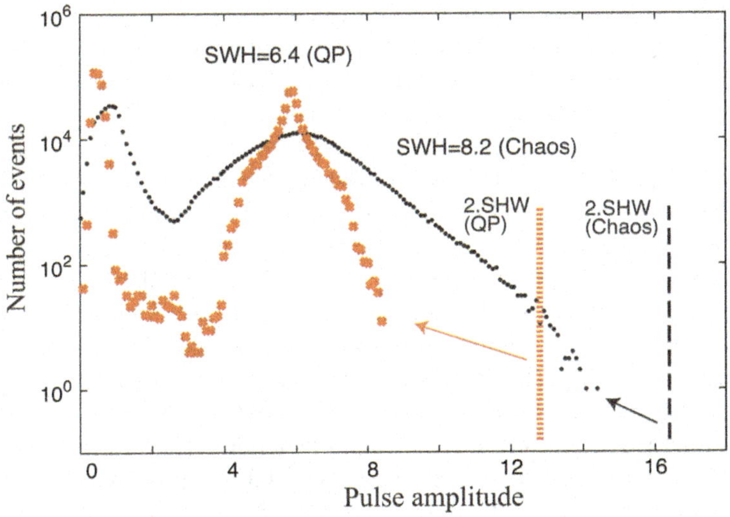

Figure 16.15. Number of events as a function of the intensity of the pulses in semi-logarithmic scale. LL model in 1D settings with $\theta = 2$, $E_i = 1.9$ (crosses, QP regime), and $E_i = 2.1$ (circles, chaotic regime). The SWH denotes the significant wave height and the dashed lines indicate $2 \times$ SWH. Reproduced from [101]. With permission of Springer Berlin Heidelberg.

developed, i.e. when the neighboring pulses in the oscillating pattern are interacting strongly. For example, figure 16.15 displays the statistics of pulse heights in the quasi-periodic and chaotic regimes discussed in the previous section for the 1D LL model. Even when the pulse peaks clearly display chaotic dynamics as in the case of $E_i = 2.1$ (figure 16.1(e)), no rogue waves are formed in the system, that is, the tail of the pulse

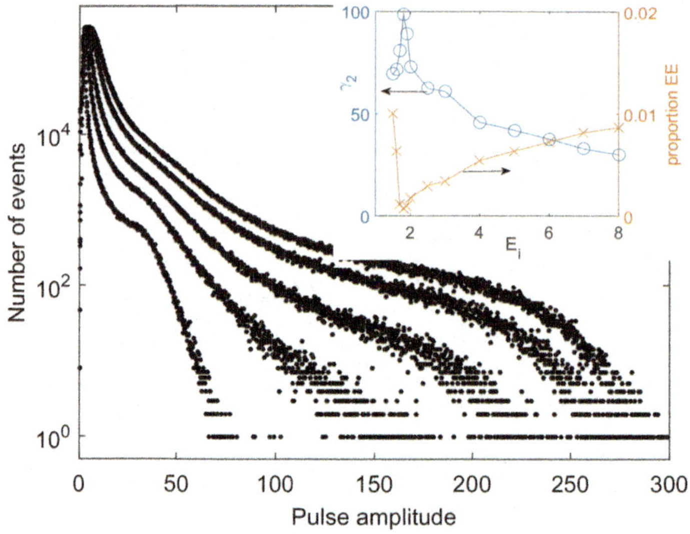

Figure 16.16. Same as figure 16.14 but for different injection field: $E_i = 1.5$, 2, 3.5 and 7 (distribution curves broaden as E_i is increased). The inset shows the dependance of Kurtosis (left axis) and the proportion of EE (right axis) as a function of injection field E_i. Reproduced from [101]. With permission of Springer Berlin Heidelberg.

height distribution stays well below 2 × SWH line. Recently, it has been demonstrated both experimentally and theoretically that intermittency plays a key role for the rogue-wave formation in a spatially extended microcavity laser with saturable absorber [106] and in a liquid crystal light valve with optical feedback [107]. More recently, two-dimensional dissipative rogue waves have been reported in the framework of the LL model and in the laser with saturable absorber with delay feedback [48]. In reference [50], a similar two-dimensional rogue wave for a laser with a saturable absorber without delay feedback is described. For the laser system, it has been shown that after reaching a maximum, both the proportion of extreme events (EEs) and the kurtosis of the pulse height probability distribution start to decrease as a function of laser injection current μ that drives the system into spatiotemporal chaos, i.e. the Kaplan–Yorke dimension increases with μ. It is interesting to check if such a behavior is also observed in the LLE model. We have carried out simulations for increasing injection field for the 2D LL model and we have observed a different tendency: while the kurtosis indeed decreases with E_i, the number of EEs after an initial sharp decrease begins to gently increase, see figure 16.16. Indeed, while the kurtosis is indicative of the deviation of the statistical distribution from Gaussian, the proportion of the EEs is indicative of only the very far tail of the distribution with very-large-amplitude events (if existing). Therefore, kurtosis and EE proportion are different indicators that can, in principle, behave differently. Additional work is in progress to elucidate the differences in this respect between the LLE and the laser with saturable absorber systems.

16.4 Conclusions

In this chapter, we wrapped up the recent results on cavity soliton dynamics and rogue waves in nonlinear optics. We have considered the Lugiato–Lefever equation that constitutes one of the most studied models in the field of nonlinear optics. It is a paradigmatic model equation for the understanding of periodic dissipative structures, stationary or moving localized patterns, and various complex phenomena in nonlinear optical cavities such as spatiotemporal chaos and rogue wave formation. The LLE has become an essential model for investigating the dynamical properties of laser fields confined in nonlinear optical resonators. The applicability of the LLE is wider than the spatial Kerr cavity; it constitutes an essential model to describe Kerr optical frequency comb generation using whispering gallery mode cavities, integrated ring resonators, and all fiber cavities. These Kerr combs can have several possible applications, such as the generation of ultra-stable lightwave and microwave signals that could be useful for aerospace engineering, optical communication networks, and microwave photonic systems.

In the first part of this chapter, we have characterized the formation of spatiotemporal chaos using rigorous tools of dynamical system theory. We have computed Lyapunov spectra and the Yorke–Kaplan dimension, which is considered an excellent order parameter for the characterization of the spatiotemporal chaos bifurcation diagram. The route to spatiotemporal chaos from cavity soliton or periodic structures through an extended quasi-periodicity has also been discussed.

In the second part, we have incorporated a time-delayed feedback loop in the LLE. The linear stability analysis of the LLE with delayed feedback has shown that the modulation instability region depends strongly on both the feedback strength and the feedback phase. The optical delayed feedback induces a traveling wave instability that emerges from the homogeneous steady states. We have shown that there exists a critical injected power over which a single peak cavity soliton exhibits a spontaneous motion with a constant velocity. This behavior occurs in the pinning domain involving a homogeneous steady state and a traveling wave periodic solution. In the absence of delayed feedback control, moving one is excluded. The motion of cavity solitons has been attributed to delay feedback since it breaks the symmetry reflection. Finally, we have shown that the time-delayed feedback can generate dissipative rogue waves in two-dimensional LLEs. These rare events occur for large values of delayed feedback strength. We have characterized the formation of rogue waves by providing a statistical analysis showing a non-Gaussian profile of the probability distribution with a long tail and peak intensity height well above twice the significant wave height, defined as the average of one third of the highest-amplitude pulses.

Acknowledgements

Some of the works presented in this chapter have been undertaken with other past and present colleagues, as acknowledged via the relevant reference citations. We acknowledge financial support of the Interuniversity Attraction Poles program of the Belgian Science Policy Office under grant IAP 7-35 photonics@be. MT is a Senior Research Associate with the Fonds de la Recherche Scientifique F.R.S.-FNRS, Belgium. KP acknowledges the support of Methusalem foundation. MGC acknowledges the financial support of FONDECYT projects 1150507. E.

References

[1] Lugiato L A and Lefever R 1987 *Phys. Rev. Lett.* **58** 2209
[2] Laughlin D W M, Moloney J V and Newell A C 1983 *Phys. Rev. Lett.* **51** 75
[3] Laughlin D, Moloney J V and Newell A C 1985 *Phys. Rev. Lett.* **54** 681
[4] Turing A M 1952 *Philos. Trans. R. Soc. London* **237** 37
[5] Prigogine I and Lefever R 1968 *J. Chem. Phys.* **48** 1695
[6] Glansdorff P and Prigogine I 1971 *Thermodynamics Theory of Structures, Stability and Fluctuations* (New York: Interscience)
[7] Tlidi M, Mandel P and Lefever R 1994 *Phys. Rev. Lett.* **73** 640
[8] Scroggie A J, Firth W J, McDonald G S, Tlidi M, Lefever R and Lugiato L A 1994 *Chaos, Soli. Fract.* **4** 1323
[9] Gomila D, Scroggie A and Firth W 2007 *Physica* D **227** 70
[10] Trillo S, Haelterman M and Sheppard A 1997 *Opt. Lett.* **22** 970
[11] Staliunas K and Sanchez-Morcillo V J 1998 *Physics Lett.* A **241** 28
[12] Tlidi M, Mandel P and Lefever R 1998 *Phys. Rev. Lett.* **81** 979
[13] Tlidi M, Mandel P, Le Berre M, Ressayre E, Tallet A and Di Menza L 2000 *Opt. Lett.* **25** 487
[14] Coullet P 2002 *Int. J. Bifurc. Chaos* **12** 2445
[15] Clerc M G and Falcon C 2005 *Physica* A **356** 48
[16] Fernandez-Oto C, Clerc M G, Escaff D and Tlidi M 2013 *Phys. Rev. Lett.* **110** 174101
[17] Odent V, Tlidi M, Clerc M G, Glorieux P and Louvergneaux E 2014 *Phys. Rev.* A **90** 011806
[18] Tlidi M, Fernandez-Oto C, Clerc M G, Escaff D and Kockaert P 2015 *Phys. Rev.* A **92** 053838
[19] Lugiato L A 1994 *Chaos, Solit. Fract.* **4** 1251
[20] Rosanov N N 1996 *Prog. Opt.* **35** 1
[21] Arecchi L A, Boccaletti S and Ramazza P 1999 *Phys. Rep.* **318** 1
[22] Lugiato L, Brambilla M and Gatti A 1999 *Adv. Atom. Mol. Opt. Phys* **40** 229
[23] Lugiato L 2003 *IEEE J. Quantum Electron.* **39** 193
[24] Kivshar Y S and Agrawal G P 2003 *Optical Solitons: From Fibers to Photonic Crystals* (Elsevier)
[25] Denz C, Schwab M and Weilnau C 2003 *Transverse-Pattern Formation in Photorefractive Optics* (Springer Tracts in Modern Physics vol 188) (Berlin: Springer)
[26] Mandel P and Tlidi M 2004 *Quantum Semiclass. Opt.* **6** R60
[27] Malomed B A, Mihalache D, Wise F and Torner L 2005 *Quantum Semiclass. Opt.* **7** R53
[28] Mandel P 2005 *Theoretical Problems in Cavity Nonlinear Optics* vol 21 (Cambridge University Press)
[29] Tlidi M, Taki M and Kolokolnikov T 2007 *Chaos* **17** 037101
[30] Akhmediev N and Ankiewicz A (ed) 2008 *Dissipative Solitons: From Optics to Biology and Medicine* (Lecture Notes in Physics vol 751) (Springer)
[31] Ackemann T, Firth W and Oppo G L 2009 *Adv. Atom. Mol. Opt. Phys* **57** 323
[32] Purwins H G, Bödeker H U and Amiranashvili S 2010 *Adv. Phys.* **59** 485
[33] Descalzi O, Clerc M and Residori S 2011 *Localized States in Physics: Solitons and Patterns* (Springer)
[34] Barbay S, Kuszelewicz R and Tredicce J R 2011 *Adv. Opt. Technol* **2011** 628761
[35] Rosanov N N 2013 *Spatial Hysteresis and Optical Patterns* (Springer)

[36] Leblond H and Mihalache D 2013 *Phys. Rep.* **523** 61
[37] Tlidi M, Staliunas K, Panajotov K, Vladimirov A G and Clerc M G 2014 *Philos. Trans. R. Soc. Lond.* A **372** 20140101
[38] Tlidi M and Clerc M G 2016 *Nonlinear Dynamics: Materials, Theory and Experiments* vol 173 (Springer)
[39] Mihalache D 2017 *Rom. Rep. Phys.* **69** 403
[40] Lugiato L, Prati F and Brambilla M 2015 *Nonlinear Optical Systems* (Cambridge: Cambridge University Press)
[41] Solli D R, Koonath C and Jalali B 2007 *Nature* **450** 1054
[42] Akhmediev N, Dudley J M, Solli D R and Turitsyn S K 2013 *J. Opt.* **15** 060201
[43] Onorato M, Residori S, Bortolozzo U, Montina A and Arecchi F T 2013 *Phys. Rep.* **528** 47
[44] Dudley M, Dias F, Erkintalo M and Genty G 2014 *Nat. Photon.* **8** 755
[45] Akhmediev N *et al* 2016 *J. Opt.* **18** 063001
[46] Chembo Y K and Yu N 2010 *Phys. Rev.* A **82** 033801
[47] Soto-Crespo J M, Grelu Ph, and Akhmediev N 2011 *Phys. Rev.* E **84** 016604
[48] Tlidi M and Panajotov K 2017 *Chaos* **27** 013119
[49] Tlidi M, Gandica Y, Sonnino G, Averlant E and Panajotov K 2016 *Entropy* **18** 64
[50] Rimoldi C, Barland S, Prati F and Tissoni G 2017 *Phys. Rev.* A **95** 023841
[51] Haelterman M, Trillo S and Wabnitz S 1992 *Opt. Commun.* **91** 401
[52] Haelterman M, Trillo S and Wabnitz S 1992 *Opt. Lett.* **17** 745
[53] Chembo Y K and Menyuk C R 2013 *Phys. Rev.* A **87** 053852
[54] Kippenberg T J, Holzwarth R and Diddams S 2011 *Science* **332** 555
[55] Ferdous F, Miao H, Leaird D E, Srinivasan K, Wang J, Chen L, Varghese L T and Weiner A M 2011 *Nat. Photon.* **5** 770
[56] Kockaert P, Tassin P, Van der Sande G, Veretennicoff I and Tlidi M 2006 *Phys. Rev.* A **74** 033822
[57] Peschel U, Egorov O and Lederer F 2004 *Opt. Lett.* **29** 1909
[58] Clerc M G, Ferré M A, Coulibaly S, Rojas R G and Tlidi M 2017 *Opt. Lett.* **42** 2906
[59] Cuevas-Maraver J, Kevrekidis P G and Williams F 2014 *The Sine-Gordon Model and its Applications: From Pendula and Josephson Junctions to Gravity and High-Energy Physics* (Springer)
[60] Ferré M A, Clerc M G, Coulibally S, Rojas R G and Tlidi M 2017 *Eur. Phys. J.* D **71** 172
[61] Morales G J and Lee Y C 1974 *Phys. Rev. Lett.* **33** 1016
[62] Nozaki K and Bekki N 1984 *Phys. Lett.* A **102** 383
[63] Kaup D J and Newell A C 1978 *Phys. Rev.* **18** 5162
[64] Firth W J, Scroggie A J, McDonald G S and Lugiato L A 1992 *Phys. Rev.* A **46** R3609
[65] Tlidi M, Lefever R and Mandel P 1996 *Quantum Semiclass. Opt.* B **8** 931
[66] Gomila D and Colet P 2003 *Phys. Rev.* A **68** 011801
[67] Gomila D and Colet P 2007 *Phys. Rev.* E **76** 016217
[68] Gomila D and Colet P 2002 *Phys. Rev.* E **66** 046223
[69] Odent V, Taki M and Louvergneaux E 2011 *New J. Phys.* **13** 113026
[70] Leo F, Coen S, Kockaert P, Gorza S P, Emplit P and Haelterman M 2010 *Nat. Photon.* **4** 471
[71] Firth W J, Harkness G K, Lord A, McSloy J M, Gomila D and Colet P 2002 *J. Opt. Soc. Am.* B **19** 747
[72] Turaev D, Vladimirov A G and Zelik S 2012 *Phys. Rev. Lett.* **108** 263906

[73] Gomila D, Matas M A and Colet P 2005 *Phys. Rev. Lett.* **94** 063905
[74] Hoyuelos M, Colet P, San Miguel M and Walgraef D 1998 *Phys. Rev.* E **58** 2992
[75] Averlant E, Tlidi M, Panajotov K and Weicker L 2017 *Opt. Lett.* **42** 2750
[76] Gomila D, Zambrini R and Oppo G L 2004 *Phys. Rev. Lett.* **92** 253904
[77] de Valcárcel G J and Staliunas K 2013 *Phys. Rev.* A **87** 043802
[78] Chembo Y K, Strekalov D V and Yu N 2010 *Phys. Rev. Lett.* **104** 103902
[79] Coillet A and Chembo Y K 2014 *Chaos* **24** 013113
[80] Anderson M, Leo F, Coen S, Erkintalo M and Murdoch S G 2016 *Optica* **3** 1071
[81] Liu Z, Ouali M, Coulibaly S, Clerc M, Taki M and Tlidi M 2017 *Opt. Lett.* **42** 1063
[82] Tlidi M and Panajotov K 2017 *Chaos* **27** 013119
[83] Coen S, Tlidi M, Emplit P and Haelterman M 1999 *Phys. Rev. Lett.* **83** 2328
[84] Coulibaly S, Taki M and Tlidi M 2014 *Opt. Express* **22** 483
[85] Parra-Rivas P, Gomila D, Matias M, Coen S and Gelens L 2014 *Phys. Rev.* A **89** 043813
[86] Parra-Rivas P, Matas M A, Colet P, Gelens L, Walgraef D and Gomila D 2017 *Phys. Rev.* E **95** 020201
[87] Tlidi M, Haelterman M and Mandel P 1998 *Europhys. Lett.* **42** 505
[88] Tassin P, Van der Sande G, Veretenov N, Kockaert P, Veretennicoff I and Tlidi M 2006 *Opt. Express* **14** 9338
[89] Pikovsky A and Politi A 2016 *Lyapunov Exponents: A Tool to Explore Complex Dynamics* (Cambridge: Cambridge University Press)
[90] Mitschke F, Steinmeyer G and Schwache A 1996 *Physica* D **96** 251
[91] Steinmeyer G, Schwache A and Mitschke F 1996 *Phys. Rev.* E **53** 5399
[92] Anderson M, Leo F, Coen S, Erkintalo M and Murdoch S G 2016 *Optica* **3** 1071
[93] Gonzalez-Cortes G, Clerc M G, Odent V and Wilson M 2016 *Opt. Express* **24** 15478
[94] P'erinet N, Coulibaly S and Verschueren N 2017 *Eur. Phys. J.* D in press
[95] Clerc M G and Verschueren N 2013 *Phys. Rev.* E **88** 052916
[96] Verschueren N, Bortolozzo U, Clerc M G and Residori S 2013 *Phys. Rev. Lett.* **110** 104101
[97] Panajotov K and Tlidi M 2010 *Eur. Phys. J.* D **59** 67
[98] Panajotov K and Tlidi M 2014 *Opt. Lett.* **39** 4739
[99] Akhmediev N, Soto-Crespo J M and Town G 2001 *Phys. Rev.* E **63** 056602
[100] Leo F, Gelens L, Emplit P, Haelterman M and Coen S 2013 *Opt. Express* **21** 9180
[101] Panajotov K, Clerc M G and Tlidi M 2017 *Eur. Phys. J.* D **71** 176
[102] Ott E 2002 *Chaos in Dynamical Systems* 2nd edn (Cambridge: Cambridge University Press)
[103] Frisch U 1995 *Turbulence: the Legacy of AN Kolmogorov* (Cambridge: Cambridge University Press)
[104] Coillet A, Dudley J, Genty G, Larger L and Chembo Y K 2014 *Phys. Rev.* A **89** 013835
[105] Ruelle D 1982 *Commun. Math. Phys.* **87** 287
[106] Selmi F, Coulibaly S, Loghmari Z, Sagnes I, Beaudoin G, Clerc M G and Barbay S 2016 *Phys. Rev. Lett.* **116** 013901
[107] Gonzalez-Cortes G, Clerc M G and Wilson M 2016 *Opt. Lett.* **41** 2711
[108] Tlidi M, Vladimirov A G, Pieroux D and Turaev D 2009 *Phys. Rev. Lett.* **103** 103904
[109] Panajotov K, Puzyrev D, Vladimirov A G, Gurevich S V and Tlidi M 2016 *Phys. Rev.* A **93** 043835
[110] Vladimirov A G *et al* 2002 *Phys. Rev.* E **65** 046606
[111] Tlidi M, Vladimirov A G and Mandel P 2003 *IEEE J. Quantum Electron.* **39** 216

[112] Tlidi M, Lefever R and Vladimirov A 2008 On vegetation clustering, localized bare soil spots and fairy circles *Dissipative Solitons: From Optics to Biology and Medicine* (Lecture Notes in Physics vol 751) (Springer) p 381
[113] Clerc M G, Coulibaly S and Laroze D 2010 *Eur. Phys. Lett.* **90** 38005
[114] Vladimirov A G, Lefever R and Tlidi M 2011 *Phys. Rev.* A **84** 043848
[115] Panajotov K and Tlidi M 2010 *Eur. Phys. J.* D **59** 67
[116] Tlidi M, Vladimirov A G, Turaev D, Kozyreff G, Pieroux D and Erneux T 2010 *Eur. Phys. J.* D **59** 59
[117] Tlidi M, Averlant E, Vladimirov A and Panajotov K 2012 *Phys. Rev.* A **86** 033822
[118] Gurevich S V and Friedrich R 2013 *Phys. Rev. Lett.* **110** 014101
[119] Pimenov A, Vladimirov A G, Gurevich S G, Panajotov K, Huyet G and Tlidi M 2013 *Phys. Rev.* A **88** 053830
[120] Vladimirov A G, Pimenov A, Gurevich S G, Panajotov K, Averlant E and Tlidi M 2014 *Phil. Trans. R. Soc. Lond.* A **372** 2027
[121] Tabbert F, Schelte C, Tlidi M and Gurevich S G 2017 *Phys. Rev.* E **95** 032213
[122] Schelte C, Panajotov K, Tlidi M and Gurevich S G 2017 *Phys. Rev.* A **96** 023807
[123] Wabnitz S 1993 *Opt. Lett.* **18** 601
[124] Peregrine D H 1983 *J. Aust. Math. Soc. Ser. B* **25** 16
[125] Baronio F, Degasperis A, Conforti M and Wabnitz S 2012 *Phys. Rev. Lett.* **109** 044102
[126] Baronio F, Chen S, Grelu P, Wabnitz S and Conforti M 2015 *Phys. Rev.* A **91** 033804
[127] Mussot A, Louvergneaux E, Akhmediev N, Reynaud F, Delage L and Taki M 2008 *Phys. Rev. Lett.* **101** 113904
[128] Kibler B, Fatome J, Finot C, Millot G, Dias F, Genty G, Akhmediev N and Dudley J M 2010 *Nat. Phys.* **6** 790
[129] Chabchoub A, Hoffmann N P and Akhmediev N 2011 *Phys. Rev. Lett.* **106** 204502
[130] Chabchoub A, Akhmediev N and Hoffmann N P 2012 *Phys. Rev.* E **86** 016311
[131] Akhmediev N, Soto-Crespo J M and Ankiewicz A 2009 *Phys. Lett.* A **373** 2137

CPSIA information can be obtained
at www.ICGtesting.com
Printed in the USA
BVOW07*1808300118
506320BV00001B/1/P